Thailand's

SUSTAINABLE DEVELOPMENT

Sourcebook

SECOND
EDITION

Thailand's
SUSTAINABLE
DEVELOPMENT
Sourcebook

**SECOND
EDITION**

Issues & Information · Ideas & Inspiration

Editorial Advisory Board

Kasem Watanachai
*Advisor to Thailand Sustainable
Development Foundation and
a member of the Crown Property
Bureau board of directors*

Kitti Wasinondh
*Advisor to the Crown Property
Bureau and former Ambassador
to the United Kingdom*

Priyanut Dharmapiya
*Director of Sufficiency School Center,
Foundation of Virtuous Youth*

Editor-in-Chief
Nicholas Grossman

Project Director
Yvan Van Outrive

Second Edition Editor
Will Baxter

First Edition Editors
Jim Algie
Apiradee Treerutkuarkul
Nina Wegner

Contributing Editors
Alex Mavro
Ingo Puhl

Assistant Editors
Sutawan Chanprasert
Purnama Pawa

Art Directors
Luxana Kiratibhongse
Benjapa Sodsathit
Patinya Rojnukkarin

Designers
Siree Simaraks
Warinthorn Kansupmits
Chanthipapha Sopanaphimon

Studio Manager
Annie Teo

Production Manager
Sin Kam Cheong

*For a list of all writers and contributors,
see pages 424–425.*

First published in 2015 by
Editions Didier Millet (EDM)
Second edition published in 2017

Email: **edm@edmbooks.com.sg**
www.edmbooks.com

Bangkok office
Room 1310, 3rd Floor
8, Sukhumvit 49/9, Klongton Nua
Wattana, Bangkok 10110
Thailand
Tel: +66-2018-7808

Singapore head office
35B Boat Quay
Singapore 049824
Tel: +65-6324-9260

Cover design by
Palotai Design Co., Ltd.
www.palotaidesign.com

FSC
www.fsc.org
MIX
Paper from
responsible sources
FSC™ C103447

Color separation by
United Graphic Pte Ltd.

Printed by
Sirivatana Interprint Public Co., Ltd.

ISBN 978-981-4610-46-9

This project was inspired by the launch of the

and it was made possible thanks to the financial support of the following organizations:

We would also like to thank our project partners:

Page 2: Aerial view of Bangkok, Thonburi and the Chao Phraya River. *Page 5:* Rice fields spread out behind the statue of Buddha at Wat Muang in Ang Thong province. *Page 7:* Coffee beans are dried at Doi Chaang in Chiang Rai province. *Page 12:* Drying fish in the countryside north of Bangkok. *Page 22:* Farmers working in the rice fields between Chiang Mai and Chiang Rai provinces in northern Thailand. Photos on pages 12 and 22 © Yann Arthus-Bertrand

TABLE OF CONTENTS

10   **Foreword** *by*
Dr. Kasem Watanachai
*Chairman of the Editorial Advisory Board, Advisor
to Thailand Sustainable Development Foundation,
and a Crown Property Bureau board member*

11   **Foreword** *by*
Dr. Shamshad Akhtar
*Under-Secretary-General of the
United Nations and Executive Secretary of ESCAP*

13   **Editor's Note**

14   **Maps of Thailand**

16   **Key Performance Indicators**

20   **Thailand's SDG Index Performance**

22   **PART 1**
An Introduction to Sustainable Development

44   **PART 2**
Issues and Information

46   **THE THAI ENVIRONMENT**
48   *Climate Change*
54   *Energy*
62   *Soil*
66   *Water*
70   *Forests*
76   *Oceans and Seas*
82   *Biodiversity*
86   *Urbanization*
90   *Pollution and Waste*

A NOTE ON SOURCES

*The sources for the statistics and information in
this book range from international organizations
such as the World Bank to Thai government
ministries. The editorial team attempted to
identify the most reliable source and check
against other sources whenever necessary. In
many cases, the source is cited. But to cite the
source of every statistic within the text would
encumber the prose and make for very dry reading.*

96 THE THAI ECONOMY
98   Poverty
102   Agriculture
108   Manufacturing
114   State-owned Enterprises
120   SMEs
126   Tourism
132   Finance
136   Trade
140   Transportation
146   Labor
152   Competitiveness
158   Corruption

162 THE THAI SOCIETY
164   Education
170   Health
176   Family
180   Public Participation
186   Inequality
192   Gender Equality
198   Conflict

204 THE THAI CULTURE
206   Monarchy
212   Religion
218   Heritage

224 PART 3
Ideas and Inspiration

**226 THE POWER OF
THE INDIVIDUAL**
228   Responsible Consumption
234   Personal Participation & Awareness
240   Green Homes
246   Commuting

252 COMMUNITY SPIRIT
On the Farm
254   Area-based Rural Development

260   Organic Revolution
268   Integrated Farming/New Theory
By the Forests
276   Reforestation
284   Forest Conservation
292   Wildlife
On the Coasts
300   Saving Marine Habitats
306   Coastal Resource Management
In the City
310   Historic Preservation
314   Urban Development
320   Green Spaces

**326 PRIVATE SECTOR
ENTERPRISE**
328   Sustainable Business
334   Green Buildings
340   Renewable Energy
346   Green Manufacturing
350   Ethical Sourcing
356   Waste Management
362   Sustainable Tourism
370   Social Enterprise
376   Restaurants
382   Green Finance and Banking
388   Indices
392   Countering Corruption

**396 THE ROLE OF
GOVERNMENT**
398   Education for Sustainable Development
404   Energy Conservation
408   Sustainable Transport
414   Sustainable Cities
418   International Partnerships

Appendices
424   Contributors
426   Index
431   Picture Credits

FOREWORDS

It has been more than 70 years since the Kingdom of Thailand was admitted to the United Nations in 1946. During this period the country has followed a development path that mirrors that of many other nations. A once agrarian-based economy and community-based society founded on Buddhist principles, Thailand is now an industrialized country, dotted with densely populated urban areas and Westernized in many respects.

The rewards of this growth, on the one hand, have been clear enough. Millions of Thais have been lifted out of poverty and enjoy a higher standard of living today, including basic security and rights, healthcare and education, than they did previously. However, as experienced by many countries around the world, there have also been costs: environmental degradation, increased income inequality and a questionable and contentious distribution of wealth, opportunities and justice. This growth has also impacted our culture, values and mindset in ways that may be harder to quantify but are impossible to deny. In our hearts, we may feel that this growth is irrevocably changing our very identity.

As a result, here in Thailand as elsewhere, the model of development that emphasizes GDP as the key measure of progress has come under question and inspired calls for a change in direction toward a more balanced and socially inclusive approach, namely growth that measures our success as a society in terms that encompass more than just money. So the international movement distilled into the concept of "sustainable development" resonates well within Thailand.

Because sustainable development requires a people-centered, collaborative and far-reaching strategy on numerous issues, some of the challenges we must overcome to achieve it are substantial. Efforts to strengthen governance and the rule of law and to provide higher quality education have been part of the national agenda for many years. Other challenges are unique to our climate and geography, or to our social and political conditions.

Across Thailand, as this book showcases, there are already thousands of grassroots efforts, entrepreneurs, community and government leaders as well as private sector pioneers, who are sincerely applying the key values and ideas of sustainable development. They are providing us with hope. Many of these actions have been inspired by the decades of leadership, in words and deeds, of the late monarch, His Majesty King Bhumibol Adulyadej. His Majesty's royally-initiated projects and Sufficiency Economy Philosophy have given Thailand a wonderful foundation for its pursuit of sustainable development moving forward.

I hope that this second edition of *Thailand's Sustainable Development Sourcebook* will call attention to His Majesty's ideas, to the solutions we must create, to the opportunities we share and to the sincere efforts we are undertaking. I also hope it will help create further awareness and excitement about the 2030 Agenda for Sustainable Development that has been launched by the UN. Thailand has a firm commitment to promote sustainable development, both at home and abroad. In the spirit of such international cooperation, Thailand is proud to contribute this book to the dialogue on sustainable development in the hope of creating true progress for everyone who shares this planet.

Dr. Kasem Watanachai
Chairman of the Editorial
Advisory Board, Advisor to Thailand
Sustainable Development Foundation,
and a member of the Crown Property
Bureau board of directors

The publication of this book, following the adoption of the 2030 Agenda for Sustainable Development, is highly relevant to countries across the Asia-Pacific and beyond as they work to implement the new agenda. The rich experiences offered by Thailand, documented so well in this sourcebook, will provide guidance and encouragement to other countries tackling sustainable development challenges.

The United Nations Economic and Social Commission for Asia and the Pacific (ESCAP) offers a wide range of platforms to facilitate the sharing of experiences by member States and other development stakeholders. Several countries in the region have adopted unique approaches and have valuable experience in establishing diverse philosophical principles, anchored strongly in Asian culture and values, as the foundation of their development strategies. These nations are leaders in finding ways to implement normative approaches to development that meet the needs of their people.

Thailand provides a good example of this leadership. The Sufficiency Economy Philosophy (SEP), articulated and championed by His Majesty the late King Bhumibol Adulyadej, is one of the most interesting and relevant of these homegrown Asian development approaches, especially now that sustainable development has been adopted as the core of the global agenda. The Thai experience illustrates the power of value-based approaches, with the ability to influence choices and decisions at all levels – from the individual to the family to the national sphere. It is this power to influence that differentiates the philosophy from more top-down approaches, especially since it is so strongly rooted in genuine respect for people and for nature. This power to influence is strengthened by the inclusive way in which SEP is operationalized – it applies equally to the very poor and to the affluent, contributing to livelihoods across Thailand. SEP now represents an enduring legacy of His Majesty King Bhumibol Adulyadej.

Based on these principles, Thailand has been able to establish model development interventions that are influencing the way opportunities are created not only in the country, but also around the world. The focus on schools, through the Sufficiency Economy School Project, could for example be a model for accelerating progress on the Sustainable Development Goals on education and youth. The support provided in expanding the diversity of livelihood opportunities for indigenous peoples – often among the most marginalized groups in many societies – is also impressive.

The development trajectory in Thailand has not been without challenges, but SEP, along with other approaches that seek to integrate the social, economic and environmental dimensions of development, has achieved significant results. The introduction of SEP as part of community and rural development activities in thousands of locations all over Thailand shows the importance of strong governance. The extent to which its programs are administered by various ministries working together is an important demonstration of collaboration based on development consensus.

SEP, as well as the many inspiring examples featured in this book, demonstrates a balanced integration of the three dimensions of sustainable development which ESCAP, and the United Nations in general, has been strongly promoting. Integrated approaches to development that support inclusion and economic progress while strengthening environmental protection, such as those embodied by The Royal Project, are worth emulating. The Royal Project has been internationally recognized and has also provided occupational training, giving people more secure lives. This model, as well as the other models discussed in the sourcebook, hold strong potential for application in other countries and regions as well.

Dr. Shamshad Akhtar
Under-Secretary-General
of the United Nations and
Executive Secretary of ESCAP

EDITOR'S NOTE

The second edition of *Thailand's Sustainable Development Sourcebook* is part of our ongoing effort to support and track the progress of this important movement in the kingdom. Building off the first book, which was released in 2015, we have included new examples of pioneering projects, updated facts and figures wherever possible, and created entirely new content on critical topics such as climate change and inequality. As Thailand continues to gain traction in certain areas and arguably lose its way in others, we hope this second edition continues to achieve our own mission of presenting a balanced but inspirational volume on Thailand's commitment to the movement.

In Part 1, we again offer a broad overview of sustainable development and explain how it has evolved over the decades to become a defining concept of our times. We explain how, in Thailand, sustainable development is often seen through the lens of the late King Bhumibol Adulyadej's Sufficiency Economy Philosophy, with which it shares many principles.

Part II of the book defines the major issues the kingdom is grappling with in its quest for sustainable development. Categorized according to four key areas – the environment, the economy, society and culture – 31 articles that have been written and reviewed by experts present important background information, trends, statistics and challenges. There is much more that could be said about these topics. These articles are merely starting points for any reader who wishes to understand Thailand's unique situation.

Part III of the book showcases many of the programs and groundbreakers who are working to foster sustainable development in Thailand. The articles are organized according to the key audiences and actors, beginning with you, the individual, and followed by the community, private sector and government.

There are many people and projects that have not been featured. The field of sustainable development is vast – it can appear at times to cover every activity on Earth as well as Earth itself – and space is limited. Easily this book could be twice as long. Our goal was to capture an impressive variety of ideas and programs from many different regions and on many different subjects. What this book is not is a technical treatise about how sustainability can be achieved here, nor a speculative account about how it will play out. As a sourcebook, it is intended as an introduction, and as inspiration for further research and action.

Nicholas Grossman
Editor-in-Chief

ACKNOWLEDGEMENTS

The publishing team is grateful to the following individuals for their assistance and advice: Dr. Chirayu Isarangkun Na Ayuthaya whose belief in this project and generous help was essential to its realization. A special thank you to Dr. Priyanut Dharmapiya for her warm and thoughtful guidance, and to Kanchana Patarachoke and Dr. Sooksan Kantabutra for their helpful comments on the text. In addition, we thank Adit Laixuthai, Aj Wisuthithawornwong, Anak Pattanavibool, Anand Panyarachun, Aprurudee Reepol, Ar-tara Satraroj, Ashvin Dayal, Aswin Kongsiri, Atikom Terbsiri, Banthoon Lamsam, Benjamin Schulte, Chaiwat Satyaem, Chanchaya Hadden, Chanin Donavanik, Chartsiri Sophonpanich, Danny Marks, Edward Rubesch, Graham Watts, Grissarin Chungsiriwat, Hasan Basar, Jeff Hodson, Jeff Rutherford, Jittima Srisuknam, Jonathan Grossman, Kampanad Bhaktikul, Kanlayaporn Chonghaisal, Kateprapa Buranakanonda, Katiya Greigarn, Kraiwut Rijaravanich, Krip Rojanastien, Krisana Kraisintu, Lek Sirinya Chaidee, M.L. Radeethep Devakula, Melisa Teo, Nauvarat Suksamran, Nuntakarn Chinprahut, Ongorn Abhakorn Na Ayuthaya , Orawan Yafa, Oraya Sutabutr, Parames Krairiksh, Piti Sithi-Amnuai, Piyaporn Wongruang, Plew Trivisvavet, Ploenpote Atthakor, Prasert Salinla-umpai, Pravit Sukhum , Pruitti Kerdchoochuen, Rapeepat Ingkasit, Richard Mann, Salina Boonkua, Sarah McLean, Sarinee Achavanuntakul, Sawalee Tankulrat, Sek Wannamethee, Sirikul Bunnag, Sirirwat Thiptaradol, Sonchai Nokeplub, Sumneang Raburee, Sunisa Soodruk, Supakorn Vejjajiva, Tanaphol Bangyikhan, Tawatchai La-ongjun, Tevin Vonvanich, Thaninnart Chiewchanpanich, Thapana Sirivadhanabhakdi, Tom Beloe, Varisra Kertsang, Vichit Suraphongchai, Vitool Viraponsavan, Wannapa Bucha, Wayuphong Jitvijak, Werapong Prapha and Yos Euarchukiati.

We also appreciate support offered by the personnel at the following organizations: Anti-Corruption Organization of Thailand, Community Organization Development Institute, Conservation Foundation, Daoreuk Communications Co., Ltd., Earth Net Foundation, Electricity Generation Authority of Thailand, Farmers' Friends Rice, Food and Agriculture Organization of the United Nations, Forest Restoration Research Unit (FORRU), Foundation for Consumers, Green Peace Southeast Asia, Huai Hong Khrai Royal Development Study Centre, Integrated Tribal Development Program, Khao Kwan Foundation, Mab Ueang Agrinature, Organic Agriculture Certification Thailand (ACT), Rambhai Barni Rajabhat University and Marine Science Activity, Rice Department, Roong Aroon School, Royal Discovery Initiative Foundation, Sal Forest, Thai Institute of Director Association. Thailand Environment Institute (TEI), The Association of Siamese Architects Under Royal Patronage, The Thai Silk Company and Wildlife Conservation Society.

MAPS OF THAILAND

NORTH
1   Mae Hong Son
2   Chiang Mai
3   Chiang Rai
4   Phayao
5   Lampang
6   Lamphun
7   Phrae
8   Nan
9   Uttaradit

WEST
10 Tak
11 Kanchanaburi
12 Ratchaburi
13 Phetchaburi
14 Prachuap Khiri Khan

CENTRAL
15 Sukhothai
16 Phitsanulok
17 Phetchabun
18 Kamphaeng Phet
19 Phichit
20 Nakhon Sawan
21 Uthai Thani
22 Chainat
23 Singburi
24 Lopburi
25 Suphanburi
26 Ang Thong
27 Saraburi
28 Ayudhya
29 Nakhon Nayok
30 Nakhon Pathom
31 Pathum Thani
32 Nonthaburi
33 Samut Songkhram
34 Samut Sakhon
35 **Bangkok**
36 Samut Prakan

NORTHEAST (Isan)
37 Loei
38 Nong Khai

39 Nong Bua Lamphu
40 Udon Thani
41 Sakon Nakhon
42 Nakhon Phanom
43 Khon Kaen
44 Kalasin
45 Mukdahan
46 Chaiyaphum
47 Maha Sarakham
48 Roi Et
49 Yasothon
50 Amnat Charoen
51 Nakhon Ratchasima
    (Korat)
52 Buriram
53 Surin
54 Si Saket
55 Ubon Ratchathani
56 Bueng Kan

EAST
57 Prachinburi
58 Chachoengsao
59 Sa Kaeo
60 Chonburi
61 Rayong
62 Chanthaburi
63 Trat

SOUTH
64 Chumphon
65 Ranong
66 Surat Thani
67 Phangnga
68 Krabi
69 Phuket
70 Nakhon Si Thammarat
71 Trang
72 Phatthalung
73 Satun
74 Songkhla
75 Pattani
76 Yala
77 Narathiwat

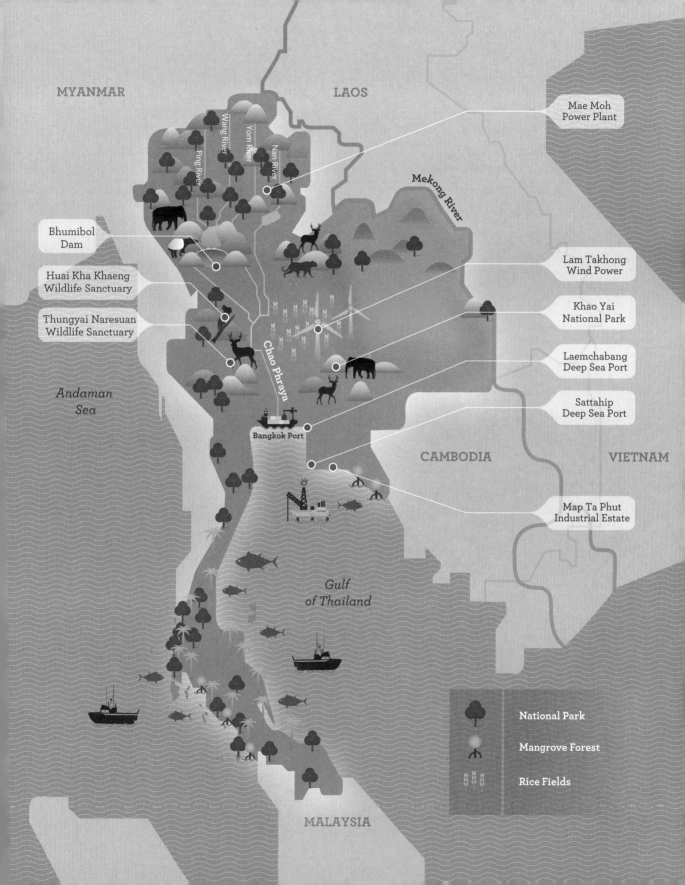

OVERVIEW

PEOPLE

POPULATION (JULY 2016)	68,200,824
URBAN POPULATION (2015)	50.4%
MEDIAN AGE (2016)	37.2
LIFE EXPECTANCY (2016)	74.7

ECONOMY

▶ GROSS DOMESTIC PRODUCT (GDP, US$)
$395 BILLION

▶ GROSS NATIONAL INCOME (US$ PER CAPITA)
$5,620

▶ INCOME LEVEL
UPPER-MIDDLE INCOME

Source: World Bank, 2015

TOTAL AREA

(World Ranking: 51 out of 257 countries)

513,120
Sq Km

LAND
510.890
sq km

WATER
2,230
sq km

LAND USE

Agriculture	46%
Forest	31%
Other	23%

BORDERS

Myanmar	2,416 Km (West)
Cambodia	817 Km (Southeast)
Lao PDR	1,845 Km (Northeast)
Malaysia	595 Km (South)

Source: Office of Agricultural Economics, Thailand

HUMAN DEVELOPMENT RANKING BY COUNTRY
(Out of 188 countries)

Thailand is listed among those countries with high levels of human development. Rankings of other Asian nations include:

Malaysia = 59   China = 90
Sri Lanka = 73   Philippines = 116
Vietnam = 115   Myanmar = 145

 **THAILAND = 87**

Source: United Nations Development Programme Human Development Report 2016

POVERTY

0.6% POPULATION LIVING BELOW **$1.90 A DAY**

This is the international poverty line as determined by the World Bank.

1.0% POPULATION IN MULTIDIMENSIONAL POVERTY

Multidimensional poverty identifies multiple deprivations in the same household in terms of education, health and standard of living.

Source: United Nations Development Programme Human Development Report 2015

ENVIRONMENT & ENERGY

ENVIRONMENTAL PERFORMANCE RANKING (Out of 180 countries)

 78 Thailand **1** Switzerland

Brunei	= 98
Cambodia	= 146
Indonesia	= 107
Lao PDR	= 148
Malaysia	= 63
Myanmar	= 153
Philippines	= 66
Singapore	= 14
Timor-Leste	= 138
Vietnam	= 131

Thailand has improved its track record in areas like Water Resources (66), but still lags behind many nations in terms of Agriculture (106) and Air Quality (167).

Source: The Environmental Performance Index, Yale University, 2016

AMOUNT OF CO₂ EMISSIONS FROM FUEL COMBUSTION IN MT CO₂
(MT CO_2 is metric tons of Carbon Dioxide Equivalent) **243.5**

RANKING IN CO₂ EMISSIONS **25**
(Out of 138 countries)

IN COMPARISON TO OTHER ASEAN NATIONS

Brunei	= 109
Cambodia	= 125
Indonesia	= 16
Lao PDR	= NA
Malaysia	= 29
Myanmar	= 98
Philippines	= 38
Singapore	= 61
Vietnam	= 34

TOP 5 CONTRIBUTORS TO MT CO₂ EMISSIONS WORLDWIDE BY PERCENT OF TOTAL

China **28%**   USA **16%**   India **6%**
Russian Federation **5%**   Japan **4%**

Source: International Energy Agency Energy Atlas of 2014: Database for Global Atmosphere Research, 2014

AMOUNT OF FOSSIL FUELS IN ELECTRICITY PRODUCTION

91% RANK: 40th

Out of 138 economies

1= Most fossil fuel–dependent country

RENEWABLE ENERGY PRODUCTION (MTOE)
(MTOE: Million Tons of Oil Equivalent)

26.24 RANK: 12th

Out of 138 economies

1= Country with the highest production levels of renewable energy

SHARE OF RENEWABLES IN TOTAL ENERGY PRODUCTION

33% RANK: 71st

Out of 138 economies

1= Country with the highest share of renewable energy in total energy production

Source: International Energy Agency Energy Atlas, 2014

CARBON DIOXIDE EMISSIONS
CO₂ emissions per capita (tons) in ASEAN countries

THAILAND **4.6**

Brunei	= 24	Lao PDR	= 0.2	Philippines	= 0.9
Cambodia	= 0.3	Malaysia	= 7.8	Singapore	= 4.3
Indonesia	= 2.3	Myanmar	= 0.2	Vietnam	= 2.0

Source: United Nations Development Programme Human Development Report 2015 for Thailand

TOTAL ELECTRICITY PRODUCTION
(In billion kilowatt-hours)

1990 = 44
2014 = 164

 RANK: 25th out of 220 economies

Source: CIA World Factbook

ECONOMY

GDP GROWTH

Unit %	2013	2014	2015	2016	2017 Forecast
China	7.8	7.3	6.9	6.7	6.5
Indonesia	5.6	5.0	4.8	5.0	5.1
Malaysia	4.7	6.0	5.0	4.2	4.5
Philippines	7.0	6.2	5.9	6.8	6.4
Thailand	2.7	0.8	2.5	3.2	3.5
Vietnam	5.4	6.0	6.7	6.2	6.5
Lao PDR	8.5	7.5	7.4	6.8	6.9
Myanmar	8.4	8.0	7.3	6.4	7.7
East Asia & Pacific	4.7	4.1	4.1	5.1	6.2
World	2.5	2.7	2.7	2.3	2.7

Source: World Bank, Asian Development Bank

THAILAND'S ECONOMIC STRUCTURE BY SECTOR

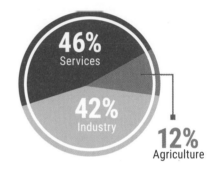

46% Services

42% Industry

12% Agriculture

Source: OECD Thailand Development Indicators from World Bank

ASIA–PACIFIC TOP 10 COUNTRIES IN GLOBAL COMPETITIVENESS

Global rankings out of 138 economies

	Singapore	2
	Japan	8
	Hong Kong SAR	9
	Malaysia	25
	China	28
	Thailand	**34**
	Indonesia	41
	Philippines	57
	Vietnam	60

Source: The Global Competitiveness Index 2016-2017

GLOBAL COMPETITIVENESS RANKING BY SECTOR

The quality of government, financial, legal and administrative institutions.

Macroeconomic environment	13
Infrastructure	49
Institutional environment	84
Trust in politicians	111

Source: The Global Competitiveness Index 2016-2017

EASE OF DOING BUSINESS

Ease of doing business ranks economies from 1 to 189, with first indicating the best and easiest.

2014	18th
2015	26th
2017	46th

Source: World Bank

AIR TRANSPORT

Year	Int'l Carrier Departures and Arrivals	Int'l Freight Departures and Arrivals
2003	135,808	891,414
2010	215,650	1,233,384
2016	402,721	1,282,673

Source: Airports of Thailand

SOCIETY

SOCIAL STATISTICS

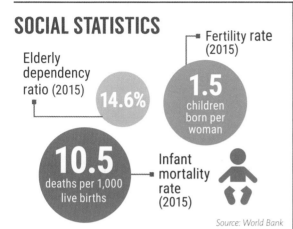

Elderly dependency ratio (2015)
14.6%

Fertility rate (2015)
1.5 children born per woman

10.5 deaths per 1,000 live births

Infant mortality rate (2015)

Source: World Bank

THAILAND IS AN AGING SOCIETY

The year when approximate number of young people and seniors will be the same. **2025**

Source: Thailand Development Research Institute (quarterly review vol.30 No.1 March 2015)

HAPPINESS

East Asian and Southeast Asian countries in Top 50 ranking of happiest nations on earth out of 155 countries.

Source: World Happiness Report 2017

Thailand	**32**
Norway	1
Singapore	26
Japan	51
South Korea	55

EXPENDITURE RANKINGS
(Percentage of GDP)

HEALTH	**4.1%**
EDUCATION	**4.1%**
MILITARY	**1.5%**

Source: World Bank

PEACE RANKING (Out of 163 countries)

125

Ranking is based on three broad themes: the level of safety and security in society; the extent of domestic and international conflict; and the degree of militarization.

Source: Global Peace Index 2016

SOCIAL PROGRESS RANKING
(Out of 133 countries)

61

Social progress includes nutritional and basic medical care, access to basic knowledge, health and wellness, as well as access to information, communications and personal safety.

Source: Social Progress Index 2016

CORRUPTION

Corruption remains a problem in much of Asia Pacific, including Thailand.

OVER 63%

of the region's countries ranked in the bottom half of Transparency International's Corruption Perception Index.

Source: Corruption Perceptions Index 2016

EDUCATION

7.3
Mean Years of Schooling

in comparison to

13.5
Expected Years of Schooling

Source: United Nations Development Programme Human Development Report 2014

ACCESS TO THE INTERNET

67% Internet users (% of population)

Source: We Are Social's Digital in 2017: Southeast Asia Report

THAILAND'S 2017 SDG INDEX PERFORMANCE

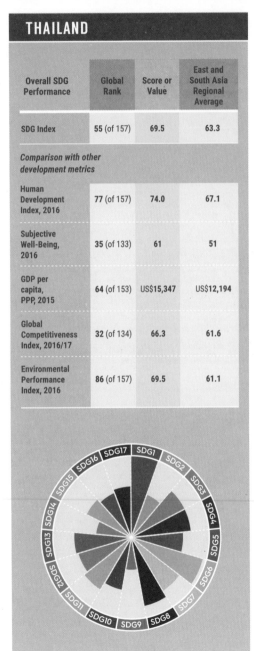

THAILAND

Overall SDG Performance	Global Rank	Score or Value	East and South Asia Regional Average
SDG Index	55 (of 157)	69.5	63.3
Comparison with other development metrics			
Human Development Index, 2016	77 (of 157)	74.0	67.1
Subjective Well-Being, 2016	35 (of 133)	61	51
GDP per capita, PPP, 2015	64 (of 153)	US$15,347	US$12,194
Global Competitiveness Index, 2016/17	32 (of 134)	66.3	61.6
Environmental Performance Index, 2016	86 (of 157)	69.5	61.1

In July 2016, the Sustainable Development Solutions Network (SDSN), headed by expert Jeffery D. Sachs and the Bertelsmann Foundation, launched the SDG Index and Dashboards – an annual global report card to track progress on the Sustainable Development Goals (SDGs).

The SDG Index, released in 2017, ranks countries regarding their status as of 2016 for each of the SDGs. The index is constructed based on a set of indicators for each of the 17 SDGs using the most recent published data. Indicators have been included that offer data for at least 80 percent of all countries with a population greater than one million. Where possible, the SDG Index uses the official indicators. In cases where official indicators have insufficient data available or where indicator gaps remain, the report authors have reviewed official and other metrics published by reputable sources for inclusion in the SDG Index. For example, Goal 12 – Responsible Consumption and Production – takes into account the percentage of wastewater treated and the number of kilograms of solid waste generated per person per year.

Using the gathered data, an adjusted indicator score that lies between 0 and 100 was generated for each country. This adjusted indicator score marks the placement of the country between the worst (0) and best cases (100). Thailand's score of 69.5, for example, signifies that the kingdom is 69.5 percent of the way from the worst score to the best score.

At the same time, the SDG Dashboards highlight each country's SDG progress across the SDGs and specific targets using a color-coded schema.

The color bands are based on absolute thresholds, presented in Table 5 of Part 2 (Page 61 of the report). Green denotes SDG achievement, red highlights major challenges, while yellow and orange indicate that significant challenges remain.

The full report can be downloaded at *www.sdgindex.org*

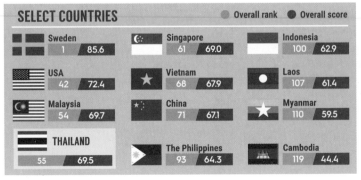

SELECT COUNTRIES   ● Overall rank   ● Overall score

Sweden 1 / 85.6	**Singapore** 61 / 69.0	**Indonesia** 100 / 62.9
USA 42 / 72.4	**Vietnam** 68 / 67.9	**Laos** 107 / 61.4
Malaysia 54 / 69.7	**China** 71 / 67.1	**Myanmar** 110 / 59.5
THAILAND 55 / 69.5	**The Philippines** 93 / 64.3	**Cambodia** 119 / 44.4

THAILAND'S PROGRESS ON THE SDGS

	SDG	Thailand's Score	East & South Asia Regional Score*
Goal 1	No Poverty	100	83.66
Goal 2	Zero Hunger	55.0	47.09
Goal 3	Good Health and Well-being	76.2	61.24
Goal 4	Quality Education	76.2	70.06
Goal 5	Gender Equality	65.7	55.76
Goal 6	Clean Water and Sanitation	95.1	80.97
Goal 7	Affordable and Clean Energy	76.9	58.49
Goal 8	Decent Work and Economic Growth	85.2	59.55
Goal 9	Industry, Innovation and Infrastructure	39.8	24.27
Goal 10	Reduced Inequalities	64.8	68.64
Goal 11	Sustainable Cities and Communities	75.1	56.28
Goal 12	Responsible Consumption and Production	70.4	39.86
Goal 13	Climate Action	73.0	69.84
Goal 14	Life Below Water	45.0	37.29
Goal 15	Life on Land	63.2	46.22
Goal 16	Peace, Justice and Strong Institutions	58.0	56.37
Goal 17	Partnerships for the Goals	62.6	21.18

(Green = achieved / Yellow & Orange = "caution lane" / Red = significant improvement needed)   *2016 Report

THAILAND'S NOTABLE SUCCESSES AND CHALLENGES

	Value	Rating
Poverty headcount ratio at US$1.90 per day (%)	0	
Registered births (%)	99.4	
Freshwater withdrawal (%)	13.1	
Access to electricity (%)	100	
Unemployment rate (%)	0.6	
Terrestrial sites, completely protected (%)	71.7	
Traffic deaths (per 100,000)	66.8	
PM2.5 in urban areas ($\mu m/m^3$)	25.8	
CO_2 emissions from energy (tCO_2/capita)	4.5	
Climate change vulnerability (0-1)	0.2	
R&D expenditures (% of GDP)	0.5	
Women in national parliaments (%)	6.1	
Homicides (per 100,000)	3.9	
Prison population (per 100,000)	398	

PART 1

INTRODUCTION TO SUSTAINABLE DEVELOPMENT

Why has the concept of sustainable development become so popular? What does it mean in the first place? Where does it originate from and how does Thailand fit into this new global development model? From its roots in the environmental activism of the 1960s, sustainable development has grown into a significant international movement, one that is inspiring big businesses, national governments, local communities and individuals to work together and rethink how they approach the pursuit of success.

In Thailand, the urgent need for more balanced development and better resource management holds many commonalities with other recently industrialized countries. Yet the Southeast Asian kingdom also holds certain unique advantages and opportunities. What are they and how can Thailand capitalize on the era of sustainable development?

"**Sustainable development**" is the frequent subject of international conferences everywhere. Corporations and governments alike host them, and in 2015 the United Nations adopted the strategy as the guiding principle of its 15-year global development program. "Sustainability" is also a contemporary catchphrase for groups across many sectors. The term comes up in everything from tourism to manufacturing to architecture, from governmental economic plans to banking seminars to product marketing. It serves as a modus operandi for NGOs and social enterprises. It appears on pamphlets, business cards and restaurant menus. It is taught in classrooms. But what does it mean, really?

One of the most enduring definitions comes from *Our Common Future*, or the Brundtland Report, published in 1987 by the United Nations World Commission on Environment and Development. It arrived at a time when increasing awareness about the limits of natural resources and the impacts of human actions on the environment had begun to inspire calls for a new, more integrated model of development. The report stated, quite neatly, that:

"Sustainable development is development that meets the needs of the present without compromising the ability of future generations to meet their own needs."

Easier said than done. Flash forward three decades and it seems entirely possible that we are leaving our future descendants a negative legacy. Alarm about climate change, health pandemics, economic crises and income inequality is increasing apace with global temperatures. In fact 2014, 2015 and 2016 were each the hottest year on record respectively. Meanwhile, carbon dioxide levels in the Earth's atmosphere have reached a level unknown for 300 million years, fueling more fervent calls for sustainable development.

Big businesses are worried. They are beginning to recognize that they must have dependable human and natural capital to thrive in the future. The fallacy that business is separate from both society and the environment is exposed every time a natural disaster disrupts supply chains, wiping out profits. Governments have taken notice. They see sustainable development

as a way to organize long-term, multifaceted national development agendas for ever-expanding populations. The international community, led by the UN, has been leading the charge for some time. It encourages sustainable development as an encompassing ideology to create more consensus and action on difficult issues such as greenhouse emissions, human rights and poverty. Even in the realm of art, sustainable development has proven to be an inspiration. The world-famous French photographer Yann Arthus-Bertrand flies high over the Earth to capture how unsustainable practices are scarring the planet and leading its steward, humanity, into extreme vulnerability.

Today, the growing popularity of sustainable development suggests a paradigm shift may finally be under way. As one of the world's foremost thought leaders on the subject, Jeffrey D. Sachs, argues in his book, *The Age of Sustainable Development*, that this philosophy and mindset "is a central concept for our age. It is both a way of understanding the world and a method for solving global problems." Underlying all of this is an increasing sense of urgency. Is it an age of sustainable development or an age of anxiety? Are we indeed leaving a negative legacy for future generations? Is the path we are on unsustainable, set only to lead us toward hardship or even calamity? Is it too late to change course? How can we create the change we imagine? How can we unlock the market potential of sustainable development to pay for its costs? We might first ask, though, how did we arrive at these questions in the first place?

SD 1.0: AN EMERGING CONCEPT

As the American economist and Nobel Laureate Joseph Stiglitz has noted, instead of pursuing sustainable development, "We have gone far down an alternative path – creating a society in which materialism dominates moral commitment, in which the rapid growth that we have achieved is not sustainable environmentally or socially, in which we do not act together as a community to address our common needs, partly because rugged individualism and market fundamentalism have eroded any sense of community and have led to rampant exploitation of unwary and unprotected individuals and to an increasing social divide."

This all took place over a relatively short period of time. The Industrial Revolution of the late-18th and 19th centuries, which was heavily reliant on coal as many countries still are, brought about a vast transformation of Western economies, as they shifted from farming to industry. The exploitation of natural resources that fueled the Industrial Revolution and the tremendous economic growth of the West demanded further global exploration for resources. These forays connected peoples, markets and countries like never before, and began an unprecedented wave of global development.

This transition also marked a drastic change in human thought. Over the centuries to come, great ships sailed the seas to open up new trade routes, planes crisscrossed the skies to connect nations, and an American astronaut walked on the moon. All of this seemed to prove that human technology and ingenuity were unstoppable.

At first, that message appeared to be confirmed once again by the "Green Revolution" of the 1950s and 1960s, as the use of chemical fertilizers and pesticides boosted farm yields to feed rapidly growing populations. But this revolution, which was not "green" in spirit at all,

how governments, businesses and consumers had become dependent on cheap oil. (Such crises often spur a rethink, and after 1973, the US increased investment in alternative energy sources, and Japan became famous for producing smaller, more fuel-efficient cars.)

By the 1970s, existing notions of progress, growth and development were being challenged from all directions. Activist groups like Greenpeace and the green parties were convinced that economic growth was not a panacea for all of a country's ills, just as technological progress was not necessarily the solution for environmental problems. Their methods, however, were "anti-cooperation," focused on disrupting mainstream corporate or government agendas.

The disasters at the nuclear power plant of Three Mile Island in the US in 1979 and the release of radioactive contamination at Chernobyl in the Ukraine in 1986, however, put the risks of our search for energy in further relief. Events like these led to a shift toward a new concept of development. In the past, development and conservation had been regarded as conflicting ideas: conservation was understood to be the protection of

By the 1970s, existing notions of progress, growth and development were being challenged from all directions.

eventually catalyzed a backlash in the form of the ecological movement of the 1960s, with telltale texts like Rachel Carson's *Silent Spring* documenting the silencing of songbirds by pesticides, the founding of Greenpeace by a group of counterculture rebels in Vancouver in 1969, and the formation of the world's first green parties.

Another focal point for this nascent movement arrived in the early 1970s, when a group of eminent politicians, economists and scientists who were part of the Club of Rome think tank published *The Limits to Growth*, which warned that the Earth had a limited supply of physical resources and that exceeding those limits could end in disaster. As these limits were considered, events such as the oil crisis of 1973 – when the Organization of Arab Petroleum Exporting Countries (OAPEC) declared an oil embargo that quadrupled the price of a barrel – exposed

resources whereas development was the exploitation of them. Out of this conflict, the concept of sustainable development emerged as a compromise between these two notions, which came to be seen as interdependent.

This idea is reflected in one of the first and most frequently referenced definitions of sustainable development, cited earlier: "Sustainable development is development that meets the needs of the present without compromising the ability of future generations to meet their own needs." This definition, coined by the UN's Brundtland Commission, was published in March 1987. The report sought to investigate the numerous concerns raised in previous decades that human activity was having severely negative impacts on the planet, and if these patterns of growth and development continued unchecked they would be unsustainable in the future.

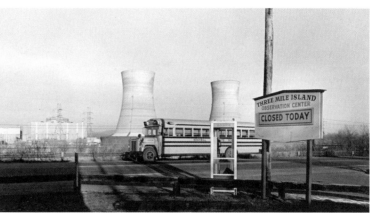

Greenpeace member with a seal pup in 1976.   A school bus passes in front of Three Mile Island in the US after an accident shut it down.

In total, the report took 900 days to compile. It collated contributions both written and oral from experts around the world: scientists, businessmen, activists and bureaucrats. At public hearings the general public also had the chance to make contributions to this dialogue.

Named after the chairperson of the Brundtland Commission, Gro Harlem Brundtland of Norway, the report focused on redefining the relationship between the environment and development: "...the 'environment' is where we live; and 'development' is what we all do in attempting to improve our lot within that abode. The two are inseparable." The report was also concerned with creating a more sustainable business environment and securing global equity for future generations by redistributing resources toward poorer nations to encourage economic growth in a spirit of solidarity that would also bring different peoples and cultures together. Among its major insights, *Our Common Future* noted that it is possible to achieve social equity, economic growth and environmental health at the same time. By doing so, it highlighted the three fundamental components of sustainable development – the environment, the economy and society – which later became known as the "three pillars of SD" or the "triple bottom line."

SD 2.0: CRUCIAL COLLABORATIONS

The Brundtland Commission played a pivotal part in institutionalizing the key concepts of sustainable development and exposing them to a larger audience. This became clear at the 1992 Earth Summit (also known as the UN Conference on Environment and Development) held in Rio de Janeiro. From that historic summit came the action plan dubbed Agenda 21. Finally individual governments and multilateral organizations had a blueprint to follow that was non-binding in legal terms but comprehensive in its far-reaching sweep. What's more, Agenda 21 was intended to be applicable at local, national and global levels. In total, 178 nations pledged their support for the plan.

That conference also resulted in the Rio Declaration on Environment and Development, which set down 27 principles that would serve as guidelines for sustainable development. Right from the start, Principle 1 denotes the importance of humanitarianism and establishing a harmonious existence between humankind and nature: "Human beings are at the center of concern for sustainable development. They are entitled to a healthy and productive life in harmony with nature." It is also easy to glimpse the influence of the Brundtland Commission's report in Principle 3, "The right to development must be fulfilled so as to equitably meet developmental and environmental needs of present and future generations."

World leaders reaffirmed their allegiance to Agenda 21 at the follow up conferences, Rio+10 in 2002 and Rio+20 in 2012. But here's the catch-22 of Agenda 21: top-down development models do not necessarily work from the ground up. While the 1990s saw a lot of discussion about sustainable development, the formation of ambitious agendas and the emergence of ministries of the environment to manage national resources, these

Blueprints for a Greener World

In response to industrialization's impacts on the environment, a new breed of writer laid the groundwork for the ecological movement to come. Rachel Carson's *Silent Spring* (1962) chronicled the disappearance of songbirds in the US due to the toxic effects of different kinds of pesticides like DDT, which was subsequently banned there for use in agriculture, but is still widely used in developing countries like Thailand. Carson's book is also credited with inspiring the founding of the US Environmental Protection Agency. Another influential text was *The Population Bomb* by Paul R. Ehrlich. The doom-monger predicted famines that would cause millions of deaths over the next few decades after the book's 1968 publication. While he was wrong, he did get people talking about overpopulation.

The Limits of Growth, a 1972 book by Donella H. Meadows, Dennis L. Meadows, Jorgen Randers and William W. Behrens III, was one of the first books to consider the fact that the Earth's resources are finite. Still updated every few years, it uses a computer simulation to model five different variables (world population, industrialization, pollution, food production and resource depletion) according to best and worst case scenarios. Controversial for its time, *The Limits of Growth* provided the first warning of what ecologists would later call "planetary boundaries." E.F. Schumacher's *Small Is Beautiful: A Study of Economics As If People Mattered* (1973) informed King Bhumibol Adulyadej's Sufficiency Economy Philosophy, while more recently, *The Ecology of Commerce: A Declaration of Sustainability* by Paul Hawken (1993) emphasized the importance of businesses becoming responsible environmental stewards by slashing their consumption of energy and resources by 80 percent over the next 50 years. Naomi Klein's compelling *This Changes Everything: Capitalism vs. the Climate* (2014) declares "our economic system and our planetary system are now at war."

plans were largely free-standing, and not integrated in a way that would meaningfully impact the non-sustainable activities entrenched in both the private and public sectors. In other words, making pledges was one thing, implementing them proved to be another.

At the grassroots level, where the big development agencies, hobbled by high overheads and bureaucratic procedures, find it difficult to work swiftly or effectively, smaller NGOs stepped in to fill the void and do some of the hands-on work. These were groups often devoted to a single cause, whether reforestation or wildlife conservation, promoting forms of alternative agriculture or helping the poor. Through them the principles underlying sustainable development agendas took root in many different countries. In turn, they spread the word about sustainability to people who had never heard of the Earth Summit or the Brundtland Commission. That's how public participation became another integral aspect of sustainable development. Local communities wanted to arrive at their own outcomes to decide their own fates. To achieve those ends they had to work within the context of their own cultures and communities, while actively participating in the projects.

But where could they find the funding? Who could they turn to for help in such complexities as technology or marketing? As the communities reached out to NGOs and government agencies, or think tanks and other community groups, teamwork came to be an integral part of sustainable development as projects spread to more rural areas, or urban slums, where they were most needed. These kinds of collaborations turned into a defining feature of SD 2.0. UN agencies, NGOs and other public-private partnerships began to own certain topics and causes.

A 2014 World Resources Institute report illustrated a powerful example of such collaborative efforts. It said that deforestation has decreased considerably in places where community forestry rights had been given strong legal recognition as compared to those areas where they had not been granted such rights by the state. In the Bolivian part of the Amazon, deforestation rates were found to be six times lower and in the Brazilian Amazon they were 11 times lower, the report said. Considering that this is the largest forest on the planet, which stores between 80 and 140 billion tons of carbon, and is home to some 30 million people, cutting deforestation rates like this is a significant achievement that shows the potential of such stakeholder engagement.

SD 3.0: INTEGRATION

If SD 2.0 was mostly a period in which sustainable development gained traction within institutions but made little headway into mainstream practice, SD 3.0 is when a truly integrative approach, which embraces the interdependence of economic, social and environmental factors, has begun to take root. This integration is taking place through collaborations between the public, private and community sectors, and features the "greening" of entire supply chains. Economic actors are accounting for their environmental, social and economic footprints in their reporting activities – and the results are part of their key performance indicators.

none at all. The MDGs laid the groundwork for the much more comprehensive and precisely calibrated 2030 Agenda for Sustainable Development, and its "SDGs". The latter places far more emphasis on the involvement of the private sector, which is seen as a key driver for sustainable development. This program consists of 17 goals and 169 targets (see page 31). Many of them, such as, "make cities and human settlements inclusive, safe, resilient and sustainable," as well as "ensure sustainable consumption and production patterns" address urgent problems, while other goals, like eliminating poverty and hunger, carry on from the previous targets agreed on in 2000. As with the MDGs before them, the big question asked by many is who will foot the bill to achieve such ends? Given that government budgets are already strained, that's a good question, which remains a quandary. (By some estimates, half of the money donated by high-income countries to their poorer cousins to help them reach their MDGs went instead to relieving their debts and military spending.) At the same time, transparency and accountability are issues.

Implicit in the green growth model are many of the lessons of the past decades, that consumption has become wasteful and extravagant and that future growth depends on better resource management.

This private sector participation can be traced back to the pledges that companies began to make at the turn of the century, when world leaders gathered in New York to adopt the UN Millennium Declaration. One of the most comprehensive development agendas ever seen, it aimed to reduce extreme poverty and child mortality, to promote environmental sustainability and more equality between the sexes, under the title of the Millennium Development Goals (MDGs). The main eight MDGs also paved the way for a shift in the perception of sustainable development as a movement where social and humanitarian issues are also given their due.

During the Millennium Summit of the UN in 2000 all 189 member states of the UN, along with more than 20 international agencies, pledged to realize these goals by 2015. Some countries like Thailand achieved all of the objectives, while other nations realized only a few or

Sustainable development's proponents see the agenda as having the potential to pay for itself. According to the Business & Sustainable Development Commission, sustainable business models could open the world to economic opportunities worth up to US$12 trillion and increase employment by up to 380 million jobs by 2030. Proponents also argue that, if sustainable development is not widely adopted, the costs of the current path of development and its impacts, such as climate change, will only grow steeper. Moreover, it is argued that those countries which commit to innovation will be rewarded, as the global market is increasingly eager to receive products and services that improve efficiency in the transport, housing, food and energy sectors.

Denmark has seized this opportunity by adopting the concept of "green growth," in which an economy shifts from typical industrial methods of production toward

The 17 goals of the 2030 Agenda for Sustainable Development. The ambitious development agenda was adopted by the UN in September, 2015.

a more sustainable use of its natural capital. Denmark's green growth strategy focuses on "sustainable food production, access to energy and water, and integrated climate efforts," according to Denmark's Ministry of Foreign Affairs. By promoting green growth based on sustainable management and use of natural resources Denmark also supports developing countries in generating sustainable and inclusive economic growth and job creation. Implicit in the green growth model are the many lessons of the past decades: that natural capital is limited, that consumption has become wasteful and extravagant, and that future growth depends on better, more sensitive management of resources.

In the private sector, large firms have embraced the benefits of integrating sustainability into their business models, budgets, practices and value chains through the concept of Creating Shared Value (CSV). This approach is based on increased stakeholder recognition and collaboration between producers, suppliers and consumers. For example, General Electric (GE) has recognized that it can help address the health problems of the world's poor while also taking advantage of previously untapped business opportunities. That's why GE is investing US$6 billion to develop new, inexpensive products and treatments that meet the health needs of low-income populations around the globe, with a goal of reaching

100 million new patients every year. In fact, innovations based on GE's existing technology, like an inexpensive ultrasound scanner that transmits its pictures over the Internet without a computer, are already changing the lives of women in remote villages. GE is also partnering with Grameen Bank, a microfinance institution, to create a sustainable rural health model that can reduce maternal and infant mortality by over 20 percent – not as charitable endeavor, but to make money by finding the profits hidden beneath pressing social issues.

Integrating sustainability directly in operations through sustainable management is also seen as an opportunity for resource efficiency and therefore cost reductions, not to mention a means to create consumer goodwill. Across all business sectors, accountability is on the rise as firms realize the benefits of promoting sustainability. Marketing is certainly a motivator. Through products and practices that are seen as environmentally friendly, companies can build their brands and woo consumers interested in supporting ethical firms. Indeed, entire organizations are founded to certify the sustainability initiatives of the private sector. Organizations like EarthCheck recognize "green hotels," the US Green Building Council developed the Leadership in Energy and Environmental Design (LEED) certification process for "green buildings," and the Dow Jones Sustainability

The 2030 Agenda for Sustainable Development

Without specific targets and goals sustainable development is wishful thinking out of step with reality. The Sustainable Development Goals, or SDGs, adopted by the UN in 2015 are intended to be signposts to show the links between economic, social and environmental issues to better inform international development policies through 2030. They are the follow up to the Millennium Development Goals, or MDGs, introduced in 2001. Thailand achieved all eight MDGs and is confident of achieving all of the SDGs too.

The SDGs are more comprehensive and inclusive than the MDGs, even though some critics have questioned where the money will come from to achieve them and suggested that the 17 goals with 169 targets are too unwieldy.

The Sustainable Development Goals are as follows:

1. End poverty in all its forms everywhere.

2. End hunger, achieve food security and improved nutrition and promote sustainable agriculture.

3. Ensure healthy lives and promote wellbeing for all at all ages.

4. Ensure inclusive and equitable quality education and promote lifelong learning opportunities for all.

5. Achieve gender equality and empower all women and girls.

6. Ensure availability and sustainable management of water and sanitation for all.

7. Ensure access to affordable, reliable, sustainable and modern energy for all.

8. Promote sustained, inclusive and sustainable economic growth, full and productive employment and decent work for all.

9. Build resilient infrastructure, promote inclusive and sustainable industrialization and foster innovation.

10. Reduce inequality within and among countries.

11. Make cities and human settlements inclusive, safe, resilient and sustainable.

12. Ensure sustainable consumption and production patterns.

13. Take urgent action to combat climate change and its impacts.

14. Conserve and sustainably use the oceans, seas and marine resources for sustainable development.

15. Protect, restore and promote sustainable use of terrestrial ecosystems, sustainably manage forests, combat desertification, and halt and reverse land degradation and halt biodiversity loss.

16. Promote peaceful and inclusive societies for sustainable development, provide access to justice for all and build effective, accountable and inclusive institutions at all levels.

17. Strengthen the means of implementation and revitalize the global partnership for sustainable development.

The 169 targets expand on the goals and define them in clearer terms. For instance, under the first goal of ending poverty there are seven targets, including: "1.4 by 2030 ensure that all men and women, particularly the poor and the vulnerable, have equal rights to economic resources, as well as access to basic services, ownership, and control over land and other forms of property, inheritance, natural resources, appropriate new technology, and financial services including microfinance." Besides these targets, the SDGs come with indicators that focus on measurable outcomes. In determining whether or not these goals have been met, this is up to each government in "setting its own national targets guided by the global level of ambition but taking into account national circumstances," the UN said.

Indices began in 1999 as a way of measuring the sustainability performances of around 3,400 companies listed in the Dow Jones Global Total Stock Market Index. New capital is flowing toward the green economy, and private investors and companies see these indicators as key benchmarks for investing in sustainable enterprises. Some, who see themselves as responsible investors, are rejecting investment in fossil fuels altogether.

There is no question that sustainability has become big business, that it is spurring not only multi-million dollar clean energy projects, but also driving innovations like eco-cars, in addition to encouraging the greening of government procurement processes and the building of mass-transit lines to reduce CO_2 emissions. With so much money on the line, it's no wonder that many young people are specializing in sustainable development at universities, and see it as a good career choice.

In the midst of this merger between the profit motive and the ethical mindset lies the next frontier of the movement as individuals and responsible consumers, powered by information technology, rise to the fore to influence the public and private sectors, and show how their voices and purchasing power can help to remake the world into a fairer and greener place.

THE DAWN OF SD IN THAILAND

How does Thailand fit within this global context of sustainable development? In most ways it is no different than any Newly Industrialized Country, with an ever-growing population, and facing the limits of its natural resources, increasing energy demand, climate change, inequality and various other environmental, social and cultural challenges. Like any country though, its situation also has its unique background and attributes, and for Thailand the wake-up call arrived in 1997, when a massive financial crisis saw fiscal institutions go bust, businesses go broke and many people go bankrupt as a bubble economy burst. From its epicenter in Thailand, the ripples spread across the markets of Southeast Asia, causing extensive damage.

Over the decade prior to 1997, Thailand had enjoyed spectacular economic growth, frequently in double digits. Inflation was low and foreign investment poured in for infrastructure projects, automobile and electronics manufacturing, textiles and other light industries, as well as property development and the service industries. Tourism arrivals soared. It was a boom time in the kingdom's history. All over the country, the signs of this newfound prosperity abounded in the form of shopping malls, golf courses, international schools and hospitals. As Thailand enjoyed some of the best credit ratings in its history, the country looked set to join the ranks of the so-called "Asian tiger" economies.

By 1997, however, the baht was overvalued and made an easy target for currency speculators, yields on investments were disappointing, exports were down and foreign debt was sky high. Adding to that volatile mix, the stock market was rife with insider trading and likened to a casino. A panicked withdrawal of credit burst an economic bubble that had been fueled by hot money. When the full-brown crisis hit, 58 local finance companies had their operations suspended by the financial authorities over a two-month period and, after the Bank of Thailand used up 90 percent of the kingdom's foreign reserves defending the currency, the government had no choice but to float the baht. Save for a few areas like tourism, much of Thailand's economy was overwhelmed by the crisis. At its lowest point the baht had plummeted from 25 to the US dollar to 57. In August of 1997 the International Monetary Fund stepped in with a US$17-billion bailout and mandatory economic remedies that were both unpopular and fiercely debated. As the extent of the damage became apparent – millions were forced out of work by 1998 – there was a national consensus that similar catastrophes had to be averted in the future.

In the midst of this watershed moment, King Bhumibol Adulyadej, who had been the country's head of state since 1946, called for a new mindset, a shift in priorities and a return to a more reasonable, prudent and balanced pursuit of growth. Referring to the extravagance and over-indulgence of the 1990s, he said that "to be a tiger is not important." Instead, he extolled the virtues of sufficiency and moderation, an idea he elaborated on in his birthday speech in 1998 as follows:

Manufacturing helped Thailand reach middle-income status and continues to be one of the country's major economic drivers.

"Sufficiency is moderation. If one is moderate in one's desires, one will have less craving. If all nations hold this concept without being extreme or insatiable in one's desires, the world will be a happier place."

It was a philosophy that advocated for a more methodical approach to decision making, following the "middle path" of Buddhism. It recommended that individuals, businesses and state agencies act within their means, making informed and evidence-based decisions that take into account any and all potential repercussions. These same principles would form the basis of his Sufficiency Economy Philosophy, or SEP (see pages 36-37). During the years after the financial crisis, as the country accepted the stark consequences of having recklessly pursued profit and growth, the king's ideas provided Thailand a framework and compelling vehicle for its own form of sustainable development. Adopted by some top firms and influential among policy-makers, SEP would help the country rebound from the debacle of 1997 as well as weather the global financial crises to come.

GREEN BY DESIGN

That the most remembered advice came down from on high was befitting a country that had traditionally looked to its monarchs for leadership and guidance through troubled times. Thailand, or Siam, has always had a monarch since the first seat of power was founded in Sukhothai in AD 1238. To understand Thailand and its potential for sustainable development, it's critical to understand the importance of the country's royal heritage, and the hierarchical system of patronage, which exists to this day. Before the transition to a constitutional monarchy in 1932, commercial enterprises were nominally under sufferance of the monarch, whose officers regulated everyday life. The tradition of strong centralized control has carried over to modern times with the result that the bonds between government and business remain strong. All the earliest Thai-owned commercial initiatives were started or backed by the government or royalty. Many of the descendants of those families and companies continue in business today. Hence, the Thai world view – a world view which encompasses business as well as everything else – revolves around the powerful looking after the weak. Those entrusted with power are expected to wield it with discretion, under obligation to the greater good.

It's a classic patron-client social construct that also means that development has traditionally not been a grassroots-led phenomenon, but largely a top-down one. This model had obvious weaknesses as it typically lacked multi-stakeholder engagement and did not take

At the Heart of It: You, the Consumer

Customers shop at an iStudio store in downtown Bangkok.

We live in a digital age where conspicuous consumption has become a way of life. We are constantly bombarded with advertising for new computers and mobile phones, new cars and clothes. New malls spring up all the time, and the culture of consumerism encourages us to spend and live beyond our means. At the same time, our own behavior, whether through frequent travel or feckless lifestyle habits, also feeds into this cycle of extravagant consumption, bringing us to a precarious place where the planet's population of 7.5 billion people is now using up the equivalent of 1.5 more times the resources than the earth can replenish in a year.

So, not surprisingly, the primary target of one of the UN's 17 SDGs is – quite simply – to "ensure sustainable consumption and production patterns." However, there is only so much that governments, businesses, big international organizations and small NGOs and social enterprises can do. Eventually people of means have to look at themselves and face up to their own habits and consumption patterns. That can be difficult to do. In the face of such Herculean global challenges as lifting the world's billion-odd poor people out of poverty, in the midst of polar ice caps melting, wars and terrorist attacks, corporate expansion and greed, and the daily stresses of one's own life, it's easy to feel powerless – or care less – about making a contribution to solving such colossal problems. But if you do care and adopt a mindset of awareness, your decisions can help decide the future. Most of the time these are not big decisions

and they don't require sacrifice. Indeed, sustainable development as a concept is not about sacrifice but about creating a better society. You can contribute in ways specific to your own occupation or lifestyle. That may be the decision to recycle or compost waste products, taking a trip on a subway as opposed to using a car, refusing to accept that plastic bag in the supermarket, buying a more energy-efficient light bulb, or opting for an eco-tourism excursion rather than a luxurious shopping spree in a foreign country. All these little decisions inform the big picture. When they are multiplied millions of times by other individuals making similar decisions they can be decisive. By recognizing this we can help propel the paradigm shift beyond the boundaries of one's home or office.

In many ways, we see this shift already occurring. Consumer power is pushing the private sector toward practices and products that are environmentally friendly and do not exploit the downtrodden or child labor, or contain harmful chemicals, for example. A wide array of labels, ranging from certifications for organic produce to the fair trade labels on bags of coffee beans, have given consumers unprecedented access to the information we need to make informed choices: where products come from, how they are made and who made them.

When consumers vote with their wallets for this or that product, the big corporations pay attention. In the much-acclaimed 2008 documentary *Food, Inc*, Gary Hirshberg, the hippie environmentalist-turned-millionaire organic yogurt entrepreneur spoke of Wal-Mart's decision to stop selling a brand of milk that contained a synthetic growth hormone because of consumer outrage. "Individual consumers changed the biggest company on earth," he said. Rarely has so much power ever been vested in the pockets of individuals to affect positive change on such an immense number of different levels: fiscal, political, social and ecological. Never has the time been riper – or more urgent – to seize those opportunities and use that power in responsible ways for the betterment of our own lives, our descendants and our planet.

to heart the need for inclusiveness. During his reign, King Bhumibol made an effort to seek the input of farmers and other stakeholders. He also established his own organizations, such as the Chaipattana Foundation (which means "victory of development" in Thai) and the Rajaprajanugroh Foundation, to help the rural poor address perennial challenges such as irrigation, land fertility and crises caused by natural disasters. These organizations stepped in to execute projects the government, whether owing to red tape or a lack of policy initiative, was unable to undertake. The king's Royal Development Study Centers played a similar role by spreading knowledge and development solutions directly to rural areas. In these ways, the monarch was also upholding his own social contract with the people to "rule with righteousness."

as such), a religion whose teachings and concept of deep ecology overlap in many ways with the principles of sustainable development. The religion holds nature sacred. It stresses moderation, immaterialism and the interdependence of all life forms. Buddhist teachings also reveal the importance of cause and effect, and examine the cycles of life, death and rebirth found throughout the world. Not surprisingly, newer sustainability projects often have a distinctly Thai slant that applies homespun wisdom to global concerns, such as the way monks have "ordained" trees to deter timber poachers, wrapping the trees in the same saffron robes the monks wear.

In essence, Thailand's roots and grassroots, its resourcefulness and royally initiated projects as well as King Bhumibol's guidance over several decades could

Still classified as a low-income country through the 1950s, Thailand became one of the "miracle" economies of the 20th century.

These agricultural communities were — and in some areas still are — the pulse of the Thai heartland. Until the manufacturing boom of the 1980s as much as 70 percent of the workforce dedicated themselves to farming and that sector typically accounted for more than 30 percent of the nation's GDP. Farmers' lives revolved around the cycles of the seasons, and the rising and setting of the sun. Most were incredibly resourceful. They used natural fertilizers like manure. They grew their own food. They made their own clothes, tools, baskets and fishing nets. And they also built their own houses - raised up on stilts to prevent them from flooding and to provide a shelter for animals - out of natural materials that were sourced locally. From the forests they picked their own medicinal herbs to use as folk remedies based on local wisdom. Whatever surplus remained from their crops after feeding themselves they sold or traded for other foodstuffs and supplies. In these ways, many farming families, mostly small leaseholders, were following key principles of SEP and sustainable development long before these terms were coined and popularized.

On top of their agrarian origins, Thais were Buddhists (today, some 95 percent of Thais still identify themselves

have provided the proper grounding for sustainable development to flourish in the kingdom, but the reality is that it hasn't, which begs the tricky question: Why not?

As this book discusses, there is no simple answer. But there are some common threads. As former Thai prime minister Anand Panyarachun said in a speech delivered to the Foreign Correspondents Club of Thailand in March 2016, "With the benefit of hindsight, we can see that globalization, consumerism, extravagance, dishonesty and immoderation have led to management failures in both government and business."

"While substantial progress has been achieved in terms of economic development, we have not taken sufficient note of its negative political and social impact," the former prime minister said. In Anand's view, Thailand requires four key improvements: 1) sustainable and widespread economic development, 2) a more open and inclusive society, 3) true respect for the rule of law, 4) and a recalibrated balance of power between the state and people. Such reforms, he said, can engender the type of responsive leadership and sense of collective empowerment essential to sustainability.

THE SUFFICIENCY ECONOMY PHILOSOPHY

King Bhumibol Adulyadej spear-headed numerous development projects.

Thailand has its own framework for sustainable development. It's called the "Sufficiency Economy" (*Setthakit Pho Phiang*), after a phrase coined by King Bhumibol Adulyadej following the Asian Financial Crisis of 1997. *Setthakit* is the Thai word for economic activities; *pho* is the word for "enough"; and *phiang* means "just." So the phrase means a "just-enough economy."

King Bhumibol's strategy was ultimately formalized in the late 1990s into the concept of the Sufficiency Economy Philosophy, or SEP. After a period of reckless speculation and easy credit that led millions into unemployment, the king's idea, which emphasizes balanced development, captured the national mood at the time.

In a way, this approach and mindset represented the culmination of King Bhumibol's development work, linking many of the major facets together in an overarching framework that puts a distinctly Thai spin on sustainable development. Launched over many decades, thousands of development initiatives, commonly referred to as the royally initiated projects, were established to address complex and often interconnected issues such as water and soil management, poverty and education, or opium eradication and livelihood development. The actions taken often combined affordable technology with natural solutions, such as the building of check dams to fend off floods and irrigate fields, or the digging of small reservoirs to secure a water supply for the dry season. Any such strategies also depended on the right technique being deployed with the appropriate technology in the right environment. Long a part of King Bhumibol's projects, this strategy has also become a hallmark of sustainable development all over the world.

It is important to note that although it is often characterized as such, SEP is not a philosophy of austerity, but one that encourages reasonable consumption and expectations. Similar to the sustainable development ethos, SEP does not urge for a return to the past, for Thais to give up all their creature comforts or for development that denies growth or free market mechanisms. It recommends simply that people and likewise businesses live and act within their means. In an article by Professor Harald Bergsteiner and Dr Priyanut Dharmapiya, appearing in the book *Sufficiency Thinking: Thailand's Gift to an Unsustainable World*, the framework of SEP is elaborated as "a state of being that enables individuals, families, organizations and nations to enjoy, at a minimum, a comfortable existence and, if conditions permit, a reasonable degree of luxury that balances economic, social, environmental and cultural conditions."

These days, developing such a "moderation mindset" is of paramount importance, given the world's increasingly stretched global resources. Representing one of the three key decision-making principles of SEP (i.e., moderation, reasonableness and prudence), the kind of moderation that the king called for is also synonymous with the "middle way" of Buddhism. It is applied to avoid extremes by trying to balance necessity and luxury, self-deprivation and over-indulgence, tradition and modernization, as well as self-reliance and dependency. Though there are no hard and fast rules about this, it is usually a question of balancing what we have with what we want. The application of Buddhist principles, and the lessons of karma, which show how positive actions beget positive consequences, also provide a Thailand-specific context widely understood by Thai people from all walks of life.

Students at a provincial school are taught the basics of the Sufficiency Economy Philosophy as part of the national curriculum.

After moderation, the second pillar of SEP is "reasonableness." By this standard we must gauge the impact our actions and decisions have both on others and the world around us. In the context of sustainable development, it's easy to see how everyday decisions — such as opting to put garbage into a proper container rather than littering – do or don't translate to reasonableness, because they either solve or create problems.

Another intrinsic part of the philosophy is prudence, which is all about working carefully, proceeding by stages, growing from an internal dynamic, achieving a level of competence and self-reliance before proceeding further, and taking care not to overreach one's capabilities.

Finally, decisions based on the principles of SEP must be based on virtue and knowledge. Only then can progress with balance, according to SEP's four dimensions – economic, environmental, social and cultural – be truly achieved.

Today, SEP has become Thailand's guiding framework for sustainable development. Books, pamphlets, news articles and websites have all been created to teach the principles of SEP and to publicize the successes of those who follow its tenets. It is also taught in classrooms across the country, and more than 21,100 schools have been certified for their implementation of SEP in their curricula.

In 2007, the United Nations Development Programme (UNDP) devoted its Thailand Human Development Report to explaining the evolution and application of the Sufficiency Economy in the public and private spheres. Perhaps SEP's greatest strength is how it can be applied to so many different areas of life. It's equally valid when it comes to facing such contemporary quandaries as the worrying rise of household debt in Thailand, or when trying to surmount the difficulties faced by big businesses where growth and increasing profits are always priorities.

Many experts also point to Thailand's educational system, which has not widely produced the requisite critical thinking skills and creativity nor highly skilled labor for Thailand to increase its competitiveness in the 21st century. The centralization of wealth and

The owners of capital, who are overwhelmingly based in Bangkok, are monopolizing decision-making and reaping the rewards.

power in Bangkok is also a perennial problem that is further widening the inequality gap. While across the country, the general population finds itself inadequately educated and destined to work in the informal job sector, which offers minimal security, the owners of capital, who are overwhelmingly based in Bangkok, are monopolizing decision making and reaping the rewards. A lack of accountability and transparency combined with a culture of impunity from the rule of law by these "haves" has long been a symptom of Thailand's hierarchical patronage structure, one that has left the "have nots" often excluded from enjoying the rights and opportunities necessary for sustainable development.

THE PURSUIT OF GDP

Since the end of World War II, in particular, Thailand has become a vastly more complex society – and a successful one, at least by contemporary measures. Still classified as a low-income country through the 1950s, Thailand became one of the "miracle" economies of the last half of the 20th century. As American money flowed in to fuel the war effort in Indochina in the 1960s, these funds paved the way for new roads, communications networks, irrigation systems, electricity grids, hotels, airports and other infrastructure projects.

To manage this new era of foreign investment, the Thai government established the Board of Investment (BOI) in 1966. Seven years later, Thailand's first-ever industrial estate was set up just northeast of Bangkok and a concerted drive to promote investment for exports began in 1977. The 1960s and 1970s laid the foundations for the industrial boom to come in the 1980s and for one of the world's most successful tourism industries.

Starting around 1986, the next decade was a golden age of prosperity in Thailand. The weak Thai baht greatly benefited exports, especially in the manufacturing sector, which replaced agriculture as the biggest contributor to GDP by the mid-1980s, while low oil prices spurred on this demand. This decade was also characterized by the relocation of factories and other production facilities from Japan, Taiwan and Hong Kong, to the point where a new Japanese factory was opening every three days in the country by the 1990s, as a strong yen sent Japanese manufacturers in search of cheaper places to make their goods and cars.

Part of this success was fueled by commercial discoveries of gas in the Gulf of Thailand in the 1980s that provided the all-important foundation for the explosive growth in manufacturing that followed. This came in the form of a huge refining and petrochemical industry located on the country's Eastern Seaboard, close to the offshore gas wells and a deep-sea port at Laem Chabang, which became fully operational in 1991, together with the mushrooming of industrial plants churning out electronic components, machinery and cars.

Between 1987 and 1995 the Thai economy grew at an annual compound rate of 9.1 percent with manufacturing contributing some 31 percent to GDP by that last year. This achievement made international headlines with *Newsweek* magazine dedicating a cover story to what it called "Thailand's economic miracle." Even as logging and poaching decimated Thailand's forests and wildlife, social and environmental concerns were rarely raised, which is hardly unusual for developing countries where providing access to the most rudimentary amenities of life, such as electricity, education and proper sanitation, are the initial priorities. The "Green Revolution" that swept Thailand in the 1960s and 1970s, for example, relied heavily on chemicals that boosted yields in the short term but negatively affected soil quality in the long run. This revolution and its side effects were not endemic to Thailand. Many other countries, like India, desperate to feed its poor, pursued similar detrimental agricultural policies that yielded similar results. People were fed, farmers made money, and the environment paid for it.

During the rise of the manufacturing and service sectors in the 1980s and 1990s, the country's mass tourism potential also came on the international radar, thanks in no small part to the "Visit Thailand" campaign organized by the Tourism Authority of Thailand in 1986, when the number of arrivals surged by 24 percent to 3.4 million. Lured by beautiful scenery, friendly people and affordable accommodation, the tourist boom also had its downside as proved by the deterioration of coral reefs, resorts encroaching on national parks and beaches becoming tourist traps abuzz with jet-skis. As tourism became the biggest earner of foreign exchange the question had to be asked: How could the country balance the money-making opportunities of this growth with the maintenance of the natural and cultural attractions which made it possible in the first place? But tourism was far from the only area where this imbalance between fiscal and ecological concerns needed to be addressed.

As the manufacturing sector required more and more laborers, around one million people moved from the countryside to cities each year in search of a better living, according to a report by the United Nations Development Programme. The newfound wealth concentrated in the pockets of once-poor laborers, and often sent back to their families in the provinces, hastened the rise of more materialistic lifestyles in Thailand. So by the time of the economic collapse in 1997, some 90 percent of rural households had a TV set and 60 percent owned a motorcycle. To afford such conveniences they sometimes went into debt. Still, the nation's success in GDP terms was undeniable. Millions were lifted out of poverty and the economy grew 20 times between the early 1960s and 2013, according to the World Bank, but there were costs.

PUSHING THE LIMITS OF GROWTH

As with the 1997 crisis, it took another disaster to bring

Thailand's national symbol remains at risk due to habitat loss.

attention to another timely issue: deforestation. In the southern province of Nakhon Si Thammarat a landslide, exacerbated by illegal logging, tore through a village and claimed some 230 lives in 1988. That same year the government banned logging, though it has continued on a smaller scale. Between 1947 and 2000 more than two-thirds of the country's forests had disappeared. That rate has slowed in recent years. Still, perhaps less than 30 percent of the kingdom's forest cover remains. Among other things, the loss has threatened the integrity of key watersheds not to mention the country's biodiversity. Many keystone species, like the tiger, which have lost much of their old territory, have been hunted to the brink of extinction. Without that forest cover, elephants, the kingdom's most totemic animal, are also running out of places to forage.

The country's rapid industrialization has had other side effects too. The Karen villagers of Lower Klity Creek saw their water supply contaminated by toxic lead from mines in Kanchanaburi province, leading to disabilities. In 1992, thousands of people living near the Mae Moh coal-fired plants suffered from breathing difficulties, nausea, dizziness and inflammation of the eyes and nasal cavities. All in all, half of the surrounding rice fields were damaged by acid rain and some 42,000 people suffered from breathing ailments, after an electrostatic precipitator malfunctioned, causing the release of sulfur dioxide and particulate matter.

Ground zero for such debates about pollution is the

As Thailand's epicenter of trade and business, Bangkok enjoys a disproportionate share of opportunities and wealth.

Map Ta Phut Industrial Estate that opened in 1990. Not only the largest industrial estate in Thailand, it's one of the biggest hubs of petrochemical businesses in the world with a massive port designed to accommodate heavy vessels hauling factory equipment and massive cargoes. In spite of its undeniable success in monetary terms, locals and NGOs claimed that some 2,000 people died of cancer and other diseases in the decade after the complex opened.

Bangkok is yet another flash point for debate about the ravages of unchecked development and unplanned urbanization. From a population of three million in the 1970s the number of inhabitants has swelled to three times that number in just a few decades. The downsides of this expansion are plain to see: chronic traffic jams, air pollution, dirty canals and a dearth of public green spaces.

Until recently, when NGOs and civil society groups rose to the fore, one of the few counterbalances against such unsustainable forms of development has been the afore-mentioned royally initiated projects. These initiatives have consistently addressed many of Thailand's most pressing social and environmental concerns: helping the poor, promoting education, marine conservation, reforestation, developing agricultural cooperatives and

implementing water management systems and other schemes for rural farming and fishing communities who tend to live a subsistence lifestyle.

Often cited as a landmark scheme was the Royal Project, which was initiated in 1969 to help eradicate opium cultivation and slash-and-burn farming in the mountainous area of Doi Angkhang by giving the hill tribes alternative crops to grow like coffee and peaches. The establishment of the Royal Project's first agricultural research station in the highlands also allowed King Bhumibol to test out his ideas about reforestation, such as the building of check dams to help regulate water run-off. Following on from these auspicious beginnings was the Doi Tung Development Project. Established in 1988 on the mountain of the same name in Chiang Rai province, under the royal patronage of Princess Srinagarindra (the reigning monarch's grandmother), the project's goal was to provide better livelihoods to villagers who were living under the cloud of the opium trade. Over time, the Doi Tung brand expanded to encompass four different business units: food, handicrafts, horticulture and tourism. The success of the program has led to it being implemented in other countries afflicted by narcotics such as Afghanistan.

Today, the project is recognized the world over as a paragon of sustainable alternative livelihood development, and all of the project's products sport the seal of the United Nations Office on Drugs and Crime as a hallmark of its success. The project's final phase, set to finish in 2017, is about strengthening the business units so that the brand and the community are sustainable, as well as concentrating on capacity building and education, so that locals can take over the project when it concludes. At its heart the scheme addresses the major concerns of sustainable development: economic, social and environmental.

GREEN LIGHTS AND GRAY AREAS

What prevents Thailand from creating similarly successful initiatives throughout the country and across many sectors? Two of the key issues are good leadership and good governance, both of which were seen as lacking during the 1997 economic meltdown. In taking up the challenge, many Thai institutions rose to the occasion, like the Bank of Thailand. The central bank's emphasis on strong, independent leadership and tough new policy mechanisms shored up the economy against future disasters. By 2003, Thailand had paid back the IMF for the huge loan. Strong governance also came to the rescue again during the economic downturn of 2008 and 2009, triggered by the American subprime mortgage crisis, which did not impact Thailand's economy as it did other countries. In contrast, the last decade of political infighting has created a leadership vacuum in these corridors of state power. As successive governments have come and gone, policies are created, abandoned and revived in turn. Ministries work in silos. The long-term planning and efficient, multiparty collaboration required by sustainable development is lacking.

There have been some examples of strong leadership in the public sector. For decades a key driver of Thailand's manufacturing boom was the building of roads at the expense of railways, in spite of the fact that transporting goods by the former costs more than double the latter. Beginning with the 10th National Economic and Social Development Plan (2007–2011), the road-heavy development model was scaled back in favor of a new

emphasis on mass-transit networks around the capital, and high-speed railways that will connect the country's ports and industrial estates with other countries.

One of the main objectives of the 12th National Economic and Social Development Plan (2017–2021) is to move toward a "low carbon society." But at the same time, state-owned enterprises are still pushing and planning to construct coal-fired power plants both at home and abroad. Such efforts have come under fire from environmentalists, communities, local business operators and civil society groups that see coal projects as both unnecessary and harmful.

In the energy sector, Thailand's current scheme, the Energy Efficiency Development Plan (2015–2036), which is based on tapping a mixed array of sources, calls for more than doubling the country's capacity by 2036. In doing so the government hopes to reduce the use of natural gas to a third of the total while boosting alternative sources by 25 percent over two decades through the Alternative Energy Development Plan (2015–2036). Much of that energy will come from renewable sources like solar, wind and hydro, while the government explores other lesser-known avenues of clean energy such as geothermal and tidal energy.

Reaching that 25 percent goal is not just wishful thinking. Thailand is the solar powerhouse in the ASEAN region, harnessing more energy from the sun than all the other members combined. The kingdom was also one of the first Asian nations to implement a feed-in tariff, or "adder" program, which offers renewable energy producers long-term contracts to sell electricity at attractive rates. Emphasizing the profit motive in tandem with sustainability, the state is serving as a genuine facilitator for such enterprises.

Meanwhile, sustainability has become a benchmark for many corporations that have seen its multifaceted value. The kingdom's largest oil and gas refiner, Thai Oil, topped the 2016 Dow Jones Sustainability Index List (DJSI) as the Energy Industry Group Leader for the third year in a row. All in all, 15 Thailand-based companies were listed on the DJSI as of 2016. For a DJSI-listed conglomerate like SCG, which built a US$200 million headquarters in Bangkok

Good Governance is the Key

Until the 1997 economic crisis one could say that the profit motive had trumped the other democratic pillars of sustainable development in Thailand. But the downturn had an upshot in that it brought the issue of governance into the spotlight as many blamed the economic crisis on careless management at both the public and private levels.

In response, the Thai Institute of Directors (IOD) was established in 1999. Since then this non-profit membership organization has endeavored to teach Thai company directors the imperatives of good governance and best management practices. Through its courses and seminars, like "Ethical Leadership – Creating a Sustainable Culture," awards and publications, the IOD has been responsible for training thousands of company directors, executives and secretaries in the exacting standards of international governance and how to take a firm stand on corruption. It's been particularly successful in working with big listed companies, so much so that Thailand ranked tops in ASEAN for three consecutive years between 2013 and 2015 in corporate governance, according to the IOD's rankings.

Whether or not governance has improved at the state level is questionable. For the most part, political infighting over the last decade has curtailed any positive developments on this front as rival groups fight for power and their own vested interests instead of serving the needs of the people they are supposed to govern. Many business leaders agree that the lack of transparency and accountability, coupled with a lack of public participation in the decision-making process, are still perennial problems, which have weakened the economy, and hurt the kingdom's competitiveness.

that is a towering example of "green architecture," the pursuit of sustainability has inspired innovations like a line of over 80 different "SCG eco value" products, ranging from cement to chemicals and paper.

When it comes to contemporary culture, the popularity of health-conscious lifestyles has inspired a new wave of small-scale entrepreneurs running restaurants, farmers' markets, organic shops and selling a wide range of artisanal products. Among these smaller businesses, one of the newest trends to emerge is the social enterprise (SE). Essentially a cross between an NGO and a small business, the SE tries to strike a balance between pursuing philanthropy and making profits. For example, the Wanita Social Enterprise empowers women in Thailand's Deep South provinces by connecting them to new markets and thus helping them earn more money from products made from locally sourced materials. The milk producer Dairy Home has a fair trade agreement with local dairy farmers to make organic products and sell them throughout the country. At a time when agribusinesses are coming to dominate the agricultural sector these are small but significant moves.

In addition, ground-level work in crucial areas of sustainability consists of community-led projects in places like rural Trat province. There, villagers saw firsthand how their mangrove forests had been decimated by shrimp aquaculture. To right this ecological wrong, they set about implementing a new system of checks and balances with fines for those who disobeyed them. Another project that started from humble beginnings is the Tree Bank in Chumphon province. Variations on the scheme now exist in all of Thailand's 77 provinces. The success of such ventures highlights a key driver in every area of sustainable development: the profit motive. Doing good deeds for the sake of the environment is one thing. Doing good deeds while making money is a much more enticing, and ultimately enduring, proposition.

As such examples indicate, there is a growing commitment to sustainable development in Thailand. Yet clearly there are many challenges. In recent years, for example, big businesses such as seafood companies,

Monks wrap the same saffron cloth as their own robes in a "tree ordination" ceremony to protect the tree.

state-owned enterprises and agribusinesses have been on the receiving end of scathing critiques for supporting a multitude of practices seen as purely profit driven. Examples include the deplorable treatment of fishermen along the supply chain of seafood companies, insider trading by top executives at one of Thailand's largest conglomerates, and the continued facilitation of monocropping at the expense of the forests and soil. Endemic corruption and weak rule of law continue to negatively impact Thailand's image and competitiveness, as well as its people's ability to prosper. The burning of the detritus left over from crops like corn, which are grown for animal feed, has led to major incidents of haze and pollution blanketing the north of the country. Such practices must end if the kingdom is going to become a world-class player in the field of sustainable development. In order to bring about a paradigm shift in consumption patterns and more responsible spending – household debt in Thailand has risen to 82 percent of the country's GDP – the mindset of the public and consumers must also shift.

Thailand clearly recognizes many of these challenges.

The recent "Ya Hai Krai Wa Thai Campaign," for example, which can be translated as "Don't let anyone blame Thais," is using communications and public relations to encourage SEP principles and target four notorious Thai habits: littering, corruption and bribe-taking, careless driving, and overspending. Indeed, pollution, corruption, road safety and debt were also noted as challenges for Thailand according to the SDG Index (see pages 20-21).

How Thailand faces up to these challenges and whether or not it makes a strong and sincere commitment to the 2030 Agenda for Sustainable Development is sure to define the country's progress in the decades to come.

In light of Thailand's achievements in some areas – like leading ASEAN in solar energy, like the many royally initiated projects, like producing companies that are winning international plaudits, and like implementing an internationally renowned Universal Coverage Scheme (UCS) that gives free medical care to almost all Thais – there are grounds for cautious optimism and a viable foundation on which the country can pursue a more enduring form of development.

PART 2

ISSUES AND INFORMATION

Every country faces a distinct set of circumstances and unique challenges. What are Thailand's?

In this section, we present **31 topics** that will necessarily be at the heart of Thailand's sustainable development, such as energy resources and water security, labor and transportation, education and public participation. They are divided into four chapters, following the traditional three pillars of sustainable development – the environment, economy and society – and a fourth, culture, which is particularly important to the kingdom.

Each of these articles is intended as a concise introduction to the subject, providing essential context, statistics and indicators, as well as brief explanations of the **reasons** we should care in the first place.

We also **spotlight** issues, trends and critical challenges, or **"reality checks,"** in the name of giving you a firmer grasp on the key ingredients and obstacles to sustainability in Thailand.

THE THAI
ENVIRONMENT

Climate Change

Energy

Soil

Water

Forests

Oceans and Seas

Biodiversity

Urbanization

Pollution and Waste

The environment is absolutely central to sustainable development because future generations will also depend on having clean air to breathe, clean water to drink, healthy food to eat, and enough energy to power their ambitions. Without these, the world's expanding population will be running on empty.

From the ground up, Thailand, like many other countries, is facing complex environmental challenges and coming to grips with the impacts of its past actions. Put simply, several decades of successful economic growth have come at the expense of the environment. Energy discoveries have powered new industrial development and brought electricity to the entire country as well as pollution, waste and public health scares. Industrial, agricultural and aquacultural output have been boosted but the forests, rivers and soil are less healthy, affecting their ability to host the country's remarkably rich biodiversity. Fish-related exports have soared but stocks in Thailand's beautiful seas are rapidly being depleted. Tourism and urban areas employ millions now but are negatively encroaching on natural ecosystems.

On the horizon is one of the biggest environmental issues humanity has ever faced: climate change. Its more adverse effects, causing extreme weather and natural disasters, are already being seen around the world. Bangkok is in a particularly precarious position. Experts have ranked it as high as number three on a list of the world's cities most vulnerable to rising seas. Climate change could create hardship in a country already prone to droughts and floods, where so much of the agricultural and fishing sectors depend on effective soil, water and marine resource management.

Fortunately, along with this adversity, a far greater awareness of these environmental issues and the need for better balance has evolved. During his reign, King Bhumibol Adulyadej made improving water management and the quality of soil his own life's work. Reforestation projects are beginning to slowly reverse a decades-long loss of forest cover. And in the public sector, much longer-term plans, especially in developing renewable power sources, are being implemented through incentives and private sector collaborations. Thailand is rich in natural resources and lush landscapes. Learning how to preserve these treasures and steward the environment for future generations will be at the heart of the kingdom's sustainable development.

CLIMATE CHANGE

One of the most complex challenges we have ever faced

Decimated rice crops, Bangkok under water and large swaths of coastal tourism wiped out by storm surges: these climate change doomsday scenarios for Thailand may seem like Hollywood blockbuster fodder, if you don't believe in science. Unfortunately, study after study is proving that human activity is altering our planet and environment in unprecedented fashion.

As Earth Institute Director Jeffrey D. Sachs writes in The Age of Sustainable Development: "There has never been a global economic problem as complicated as climate change. It is simply the toughest public policy problem that humanity has ever faced. First, it is an absolutely global crisis. Climate change affects every part of the planet, and there is no escaping from its severity and threat."

A 2014 report from the Intergovernmental Panel on Climate Change warns that if usage of fossils fuels continues to accelerate, the average temperature

of the Earth's surface will spike by 1.4–2.6 degrees Celsius by the middle of the 21st century. Scientists predict this temperature increase will lead to mega-droughts, mega-floods, violent storms, species extinction, crop failures, a massive rise in sea levels and the stunted growth of marine life as the oceans absorb more and more CO2. Higher temperatures may also "adversely affect rice and other crops," the report said.

A regional climate projection conducted by Chulalongkorn University predicted that, if global warming continues, Thailand could have higher rainfall in the order of 10–20 percent in all regions. In addition, all regions will be hotter and the duration of the cold season will be shortened, while extreme events – especially droughts, floods, storms and landslides – will intensify. Unfortunately, the impact of these and other changes will in all likelihood severely disrupt three of Thailand's key economic mainstays: agriculture, tourism and trade.

According to analysis by UNEP, Thailand is one of the countries most at risk when it comes to climate change's impacts on agricultural production, with an estimated loss of between 15 and 55 percent by 2080 possible. As usual in Thailand, floods and droughts of increasing severity will be the likely culprits. Between 1989 and 2012, 227 floods in Thailand claimed more than 4,000 lives and destroyed over a million homes, according to the 2012 Statistics on Disaster by the Department of Disaster Prevention and Mitigation (DDPM) under the Ministry of Interior. Droughts have also plagued the country's agrarian heartland and the arid northeast, costing the country some 15 billion baht and affecting some 10.5 million hectares of farmland. Given that Thailand is the world's largest exporter of rice and the agricultural sector employs nearly 40 percent of the population, drastic disruptions in agricultural production would likely have severe repercussions, hitting the most vulnerable (such as small-scale farmers) the hardest,

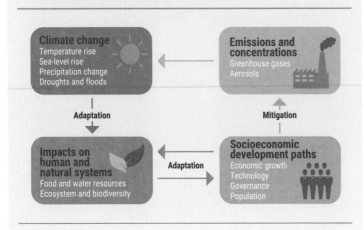

AN INTEGRATED FRAMEWORK MODEL FOR THE RESPONSE TO CLIMATE CHANGE

Climate change
Temperature rise
Sea-level rise
Precipitation change
Droughts and floods

Emissions and concentrations
Greenhouse gases
Aerosols

Adaptation

Mitigation

Impacts on human and natural systems
Food and water resources
Ecosystem and biodiversity

Adaptation

Socioeconomic development paths
Economic growth
Technology
Governance
Population

Source: Intergovernmental Panel on Climate Change

More severe storms and droughts brought on by climate change could imperil Thailand's status as the 'Kitchen of the World.'

and potentially leading to higher food prices and economic hardships that could cause civil unrest.

Thailand's coastal areas also face multiple threats. Through overfishing and overharvesting of marine life, human activity is already putting tremendous pressure on marine ecosystems. Climate change will make matters worse. Violent storms could disrupt Thailand's tourism industry, which is centered around the southern islands and beaches. Tourism as a whole contributed around 11 percent to GDP as of 2016 and employed nearly 11 percent of the total workforce. Even if such severe storms do not occur, the rising concentration of CO2 will be absorbed by the oceans, increasing its acidity, which will further kill off coral reefs and stunt the growth of shellfish crabs and other marine life.

Bangkok, not only a crossroads of trade for the country and region but also one of the world's most low-lying cities, is also vulnerable. In the 2013 Climate Change Vulnerability Index prepared by Maplecroft, Bangkok placed third after Dhaka and Manila in a survey of major cities sure to be affected by climate change. (Thailand ranked 45th most at risk out of all the countries in the world.)

Thailand's policymakers are well aware of the many climate change-related threats it is facing. The Climate Change Master Plan 2012–2050 sets out guidelines on climate change preparedness,

adaptation and other appropriate applications in the name of creating a low-carbon society. Thailand, however, remains heavily dependent on fossil fuels, with 92 percent of electricity production depending on them, and it ranks as the 23rd heaviest emitter. The government also refuses to rule out coal as an energy source despite its deleterious effects on global warming, and has recently gone ahead with plans to construct six coal-powered plants.

At the same time, there are some encouraging trends. The Climate Change Master Plan is committed to the polluter pays principle, and since 2007, the government has been providing **feed-in tariffs** to boost private investment in renewable energy. Today, Thailand ranks 12th in renewable energy production (out of 141 countries), according to the International Energy Agency's Energy Atlas. Thailand's Alternative Energy Development Plan (2015–2036) also calls for boosting use of alternative sources to 30 percent. On the global level, Thailand ratified the Paris Agreement in 2016, pledging to reduce greenhouse gas emissions by 20 to 25 percent by 2030. But the fact that Thailand still relies heavily on fossil fuels and may face opposition from a variety of powerful interests that profit from this industry means that fulfilling the Paris Agreement will be a great political challenge, requiring cooperation from the private sector and the country's state-owned enterprises.

Feed-in tariff (FIT):

The world's most commonly used measure to promote renewable energy. It offers a guaranteed purchasing price for electricity generated from renewable energy sources for a specified period of time so as to ensure cost-effectiveness.

Avoiding the Two-degree Tipping Point

In the field of climate change, two degrees Celsius has become both a warning sign and a magic number, for scientists believe that if global temperatures rise any more than that above preindustrial levels then the planet will be on the verge of a catastrophic tipping point.

As far back as 1975, the Yale professor of economics, William Nordhaus, said that a global mean temperature increase of two or three degrees would be without precedent in the "last several hundred thousand years." In an abstract for his paper, "Can We Control Carbon Dioxide?", he said "It appears that emissions of carbon dioxide particulate matter, and waste heat may, at some time in the future, lead to significant climatic modifications."

In a 1990 report by the Stockholm Environment Institute, the upper limit of two degrees was affirmed, though the report warned that even a rise of one degree "could lead to extensive ecosystem damage." When 115 world leaders and thousands of NGO members, scientists and the media descended on Denmark for the Copenhagen Climate Change Conference in 2009, that figure was the barometer of a debate which disappointed many by not setting any binding targets. In fact, the two-degree limit did not become international climate change policy until the Cancun Agreements of 2010.

For the 2015 UN Climate Change Conference in Paris, as part of the Intended Nationally Determined Contributions (INDCs), the kingdom has declared its commitment to help reduce global warming by cutting greenhouse gas emissions by 20 to 25 percent by 2030.

At the conference, the member states adopted the Paris Agreement, a universal agreement whose aim is to keep a global temperature rise for this century well below two degrees Celsius and to drive efforts to limit the temperature increase even further to 1.5 degrees Celsius above preindustrial levels.

The Agreement, which was ratified by Thailand in 2016, "recognizes that climate change represents an urgent and potentially irreversible threat to human societies and the planet and thus requires the widest possible cooperation by all countries, and their participation in an effective and appropriate international response, with a view to accelerating the reduction of global greenhouse gas emissions."

According to the International Energy Agency (IEA), sticking to two degrees is "technically feasible, but requires a fundamental transformation of the global energy system" and requires that we drastically change the way we use energy by the year 2040. For starters, greater energy efficiency and an end to fossil fuel subsidies would help limit increases in energy use even as the world's population climbs to some nine billion. The world must increase its energy productivity so that each million dollars of wealth requires the equivalent of just 85 tonnes of oil, compared to 184 tonnes in 2014. Additionally, the world must stop burning so much coal and oil, shifting instead to lower carbon alternatives including gas, nuclear and renewables (see graphic). Next, energy emissions must reach net zero by 2070. Transport oil use must also be progressively replaced by biofuels, gas and electric vehicles. Finally hundreds of coal-fired power stations must close while millions of solar panels and hundreds of thousands of wind turbines must be built.

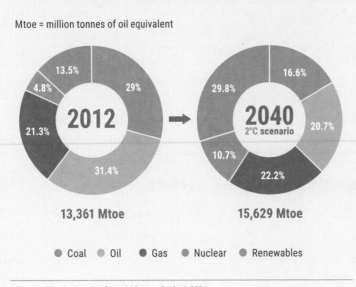

THE GLOBAL ENERGY MIX MUST EVOLVE TO AVOID REACHING 2°C

Mtoe = million tonnes of oil equivalent

2012 — 13.5%, 4.8%, 21.3%, 31.4%, 29%

2040 2°C scenario — 16.6%, 29.8%, 20.7%, 10.7%, 22.2%

13,361 Mtoe

15,629 Mtoe

● Coal ● Oil ● Gas ● Nuclear ● Renewables

Sources: The Carbon Brief, World Energy Outlook 2014

The Challenges of Combatting Climate Change and Mitigating its Impacts

Rising seas With Bangkok and other coastal communities at risk from rising seas caused by climate change, the government needs to reduce greenhouse gas emissions and implement better land use planning and water management.

Nature's breakwaters It is imperative to preserve mangrove forests, which serve as bulwarks against surging waters and coastal erosion.

Shift toward renewables Thailand remains heavily dependant on fossil fuels. To safeguard its energy security and green up its energy consumption, it needs to focus on developing alternative energy sources including solar, wind and bioenergy.

Changing of the guard Political instability makes it difficult for plans to be implemented by successive governments with projects instigated by their predecessors easily scuppered.

Natural drains Enforcing policies to protect forests and swamps would increase their capacity to retain or drain floodwaters.

Prepping the public Educating people about climate change, including its causes and potential perils, would go a long way toward spurring individuals to be more mindful of the impacts their everyday actions have on the world around them and also help mitigate long-term damage from human activities. State media, like radio stations, TV networks and websites, should also provide the public with as much pertinent information as possible.

Managing droughts Government plans to cope with droughts must be moved from the drawing board of hypotheses to workable realities, while farmers should diversify their crops to be less reliant on high levels of water consumption.

Early warning systems Thailand must implement more flood forecasting and early warning systems, and make sure they are regularly tested and maintained.

Climate change deniers Action to combat climate change continues to face opposition from individuals and groups with vested interests in sectors such as fossil fuels which stand to lose significant profits they gain from climate change-causing activities. Such actors can wield significant influence over political leaders and policymakers. Notably, United States President Donald Trump is a vocal climate change denier who has surrounded himself with like-minded advisors who eschew science and make policy decisions favorable to the powerful fossil fuel lobby.

The 2011 floods
impacted communities
across the country.

FURTHER READING

- *The Age of Sustainable Development*, by Jeffrey D. Sachs, 2015
- *This Changes Everything*, by Naomi Klein, 2014
- *Six Degrees: Our Future on a Hotter Planet*, by Mark Lynas, 2007
- "Climate Change Impacts on Water Resources: Key Challenges to Thailand Climate Change Adaptation," by Sudtida Pliankarom Thanasupsin, Environment Project Group, Office of Project Management, RID, 2012
- "The Urban Political Ecology of the 2011 Floods in Bangkok: The Creation of Uneven Vulnerabilities," by Danny Marks, *Pacific Affairs*, 2015
- Thailand Climate Change Master Plan 2012–2050

Climate Change and Rice Production

Leftover rice husks can be used as a source of biofuel.

RICE IS THAILAND'S STAPLE CROP, key to the diets and incomes of tens of millions, and the country has consistently been the world's top exporter. In 2015, the kingdom yielded over 26 million tons and exported 9.8 million tons. Unfortunately, rice is also a contributor to climate change, with the heavy use of chemical fertilizers many farmers employ partly to blame. A move toward organic production of rice would seem to be the ideal solution. However, in an illustration of the complex challenges posed by climate change, even organic rice production would contribute to global warming. This is because, while global warming is commonly associated with carbon dioxide emissions, there are other greenhouse gases that contribute to global warming such as methane, which accounts for 14 percent of emissions worldwide and traps 72 times more heat than carbon dioxide does over a period of 20 years. And half of all methane released in Thailand derives from the cultivation of rice, be it organic or not. That's because when organic matter ferments in flooded paddies, methane is released. So as Corrine Kisner's article "Climate Change in Thailand: Impacts and Adaptation Strategies" points out, "organic fertilizer alone doesn't provide the climate solution for rice (although it greatly improves farmer health and soil fertility)..."

So what to do? As with many climate change issues, the solutions tend to involve the adoption of smarter management and better practices. One option is for farmers to transform the biomass that results from rice cultivation, such as rice husks, as a source of biofuel, or to produce heat or electricity from the biomatter. Indeed, biomass energy is a burgeoning industry in Thailand and the government is supportive. Yet even biomass, if handled carelessly, can have harmful effects that threaten biodiversity and bode ill for public health. Another strategy Kisner points to in her article is to occasionally drain the paddies, thereby eliminating the bacteria that thrive in the oxygen-free setting and that produce methane by decomposing manure or other organic matter. The government has also been supportive of this idea. However, implementing such practices across Thailand's hundreds of thousands of small-scale rice-farming plots is not easy. As the issue of the cultivation of rice in Thailand demonstrates, mitigating climate change has no simple solutions but there are solutions possible. What is often required is awareness, innovation and proper management.

ENERGY

The power behind Thailand's past, present and future

Oil and natural gas platform in the Gulf of Thailand.

Ever since Thailand began industrializing in the 1960s, its steady economic growth has been largely backed by fossil fuels such as coal, oil and natural gas. While the peak levels of economic growth of the late 1980s and 1990s eventually bottomed out, Thailand's energy demand has consistently risen. From 2004 to 2016, the Thai economy grew at an average rate of 3.6 percent a year while energy consumption rose by an average rate of 2.3 percent. On the plus side, proactive initiatives launched in the 1980s to aggressively expand the nation's energy infrastructure mean that nearly 100 percent of the population has access to electricity today. However, on the negative side, the resulting spike in demand has increased the kingdom's reliance on fossil fuels, which still account for 85 percent of domestic energy consumption.

Though it is a net energy importer, Thailand has significant native energy resources. The kingdom contains substantial deposits of coal, and lignite has also long been a part of Thailand's energy mix. The former was first discovered in Thailand around the turn of the 20th century in the southern province of Krabi, though the largest deposits were later unearthed in the Lampang basin of the north. Because Thai coal is poor in quality, it is used mainly for power generation. While the country is not awash in oil and natural gas, both of these commodities can be found underground or buried deep under the Gulf of Thailand's seabed. Oil was first discovered near the Fang district of Chiang Mai province in 1954, but it was not until the 1980s, during the so-called "era of luminosity" (*choat chuang chatchawan* in Thai) that the kingdom began to shine as a source of oil and gas. These discoveries helped power Thailand's impressive economic boom that resounded in the late 1980s and 1990s.

While developing domestic fossil fuel sources remains a key priority for Thailand, the government has created attractive incentives to entice forward-thinking businesses to invest in producing clean energy. Among Asian nations, Thailand was one of the first to implement a feed-in tarrif, or "adder" program, incentivizing the development of renewable energy to encourage public participation and boost private sector investment, especially in solar. Since 2007, Thailand's adder program has offered renewable energy producers long-term contracts to sell electricity at attractive rates. For example, solar producers are eligible for subsidies of up to 6.85 baht per kilowatt-hour paid out over 25 years. Companies that generate power through biomass, biogas, hydro, wind, waste energy and solar are all eligible for the adder program.

Additionally, in 2015 Thailand's Board of Investment designated the renewable energy sector a priority industry for development and investment. As such, the government has introduced grants and tax exemptions toward the purchase and import of renewable energy equipment, and it offers foreign investors a number of other incentives to finance renewable energy production. Thailand is also tapping into the full potential of bio-economy-based energy sources. The use of biogas and biomass has already transformed the nature of competitiveness in a number of agro-industries. For example, no starch producer can afford to not produce and use biogas from its wastewater because of its impact on energy-related production costs.

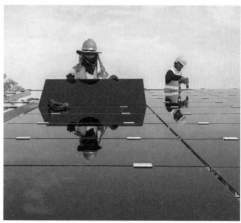

While developing domestic fossil fuel sources remains a key priority for Thailand, the government has created attractive incentives to entice forward-thinking businesses to invest in producing clean energy.

All in all, Thailand is at a pivotal – and exciting – point in its development as a clean energy producer. That said, its reliance on fossil fuels remains firmly entrenched, and with the nation's Power Development Plan (2015) calling for roughly doubling its installed energy capacity by 2036, the ever-growing need for energy has left policymakers with some difficult choices ahead.

For starters, Thailand has a stake in various hydropower projects throughout the region, including the controversial Xayaburi Dam on the Mekong River in Laos, as well as the Hat Gyi Dam and Mong Ton Dam on the Salween River in Myanmar. While on the surface these hold appeal as cheap sources of energy, critics warn that they conceal devastating regional impacts that have yet to be taken into account.

Then there is the issue of coal power plants. In 2016, the state-owned enterprise, the Electricity Generation Authority of Thailand (EGAT), confirmed it would construct six new coal-fired power plants by 2025. But scientists, activists and UN leaders see coal power as antithetical to a commitment to sustainable development, given coal's environmental and human impacts and its contribution to climate change. Indeed, plans for two coal-power

plants in the southern Thai provinces of Krabi and Songkhla have been met with stiff resistance by the local communities and NGOs concerned about their impact on the region's major revenue earners: marine resources and tourism.

Given its significant native energy resources, it is debatable whether the country needs to develop these coal-power plants at all. Some argue that EGAT and other state-owned enterprises shaping these choices are motivated by profit, not the public interest, and wonder how such plans to build coal-power plants are compatible with Thailand's pledge through the Paris Agreement to cut greenhouse gas emissions by 20 percent by 2030.

One reason Thailand may need to hedge its bets is that its most abundant domestic energy resource, natural gas, which accounts for as much as 45 percent of the country's total energy consumption mix, is projected to dry up in the next decade. Already heavily dependent on gas imports from neighboring Myanmar, Thailand has tasked PTT, the country's state-owned, SET-listed conglomerate to help shepherd the country out of this energy hole. As a part of the solution, PTT and its subsidiaries have aggressively acquired fossil fuel sources abroad such as coal mines and natural gas fields.

A Brief History of
Key Discoveries and Issues

1921–1954

The search for petro-
leum begins in 1921
after local residents
report oil seeps in
the Fang basin of the
north. But it's not
until 1954, with private
sector involvement,
that the first oil dis-
covery is made in the
Chai Prakan district of
Chiang Mai.

1959

A small, privately
run refinery starts to
produce 1,000 barrels
of oil a day, though the
exploration rights are
given to the Defense
Energy Department.

1960s

Union Oil (which later
becomes Unocal) is
granted exploration
rights to the Korat
Plateau in 1962. Two
years later, several
foreign companies
apply for offshore
exploration rights, but
it's not until 1968 that
foreign companies are
invited to bid for them
in the Gulf of Thailand
and the Andaman Sea
under the Minerals
Act.

1979–1981

Unocal signs a sales
contract to supply gas
to the domestic mar-
ket. Shell and Esso are
granted large blocks in
the Phitsanulok basin
and Korat Plateau,
respectively, in 1979.
Both companies
strike it rich quick as
Esso discovers the
Nam Phong gas field
in 1981, and Shell
discovers the Sirikit, or
Lan Krabue, oil and gas
field in Kamphaeng
Phet province.

1971–1974

The Petroleum Act is
promulgated. Known
as Thailand I, the law
governs concessions
granted before 1989. In
1973 Unocal makes the
first discovery in the
offshore area called
the "Erawan Field" in
the Gulf of Thailand,
which is followed by
a number of other
discoveries.

1982

A new fiscal regime,
known as Thailand II,
is introduced, limiting
the cost recovery of
annual gross revenues
while increasing
royalties as global oil
prices surge. Thailand
II is short-lived as the
cost of petroleum dips
in 1985.

1985–1986

Several oil field dis-
coveries result in an
upturn in petroleum
production from 4.97
million barrels of oil
equivalent (BOE) per
day to 108.5 million.
Thailand and Malaysia
reach an agreement to
jointly explore petro-
leum resources along
the Thai-Malay border
in late 1986.

1997–2001

The Asian Financial
Crisis in 1997 hits the
petroleum industry
hard. As a result, PTT
buys the Thai Oil refin-
ery plant. In 2001 PTT
is partially privatized
to become a publicly
listed company.

1989

A new fiscal regime,
Thailand III, comes
into effect, revising
the royalty rate to
a sliding scale that
enables commercial
production for all sizes
of fields.

2006–2007

Consumer activists
file a petition with the
Supreme Administra-
tive Court to delist PTT
on the stock market.
The move follows their
successful court action
to halt the listing of the
Electricity Generating
Authority of Thailand
(EGAT). A year later,
the court rejects the
suit but demands that
PTT transfer part of its
assets, including the
3,000 kilometers of gas
transmission pipelines,
to the state.

2014–2017

A military coup sets
the stage for major na-
tional reforms of many
sectors including ener-
gy. A new Alternative
Energy Development
Plan and Energy Effi-
ciency Development
Plan are drafted with
new targets for the
years 2015–2036.

*Prime Minister Prem Tinsulanonda (right) starts a new
era of energy production in Thailand.*

ENERGY CONSUMPTION IN THAILAND

Final Energy Consumption by Fuels in 2015

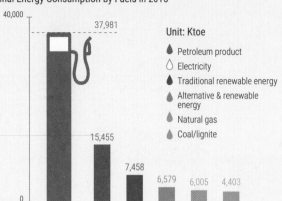

40,000

37,981

15,455

7,458

6,579   6,005   4,403

0

Unit: Ktoe

- Petroleum product
- Electricity
- Traditional renewable energy
- Alternative & renewable energy
- Natural gas
- Coal/lignite

Final Energy Consumption by Economic Sector in 2015

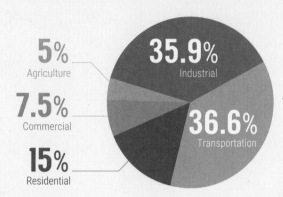

5% Agriculture

7.5% Commercial

15% Residential

35.9% Industrial

36.6% Transportation

Source: Energy Balance of Thailand 2015 by Department of Alternative Energy Development and Efficiency

ALTERNATIVE ENERGY CONSUMPTION

SOLAR POWER

STATUS AS OF 2016 **2,149** MW

TARGET IN 2036 **6,000** MW

WIND POWER

STATUS AS OF 2016 **507** MW

TARGET IN 2036 **3,002** MW

HYDRO POWER

STATUS AS OF 2016 **3,088** MW

TARGET IN 2036 **3,282** MW

BIOMASS

STATUS AS OF 2016 **2,815** MW

TARGET IN 2036 **5,570** MW

BIO GAS

STATUS AS OF 2016 **435** MW

TARGET IN 2036 **1,283** MW

WASTE TO POWER

STATUS AS OF 2016 **145** MW

TARGET IN 2036 **550** MW

NEW ENERGY

STATUS AS OF 2016 **0** KTOE

TARGET IN 2036 **10** KTOE

TOTAL

STATUS AS OF 2016 **9,139** MW

TARGET 2036 **19,687** MW

Source: Department of Alternative Energy Development and Efficiency

The Five Major Players in Domestic Energy

PTT PLC is synonymous with Thai oil and gas. It operates in upstream and downstream petroleum sectors and coal businesses in Thailand and abroad. Domestically, the company is the dominant power player, with oil refineries, factories for petrochemical production, and retail petroleum products and services. It also owns extensive submarine pipelines for gas in the Gulf of Thailand. The company has investments in subsidiaries, associates and joint ventures that operate in petroleum exploration and production activities, as well as natural gas operations, oil marketing and international trading.

Under its subsidiary, PTT Exploration and Production, the conglomerate has invested in Myanmar, Vietnam, Cambodia, Malaysia, Indonesia, Algeria, Australia, Canada and Mozambique. PTT also owns 95.3 percent of Sakari Resources, a Singaporean coal company that operates two coal mines in Indonesia and holds an 80 percent share of a coal mining company in Madagascar. In 2012 PTT became the only Thai firm to ever be listed in the top 100 of the Fortune Global 500.

THE ELECTRICITY GENERATING AUTHORITY OF THAILAND (EGAT) is a state enterprise that once had a monopoly on Thailand's electricity generation capacity and its transmission network. In 1992, independent power producers (IPPs) and small power producers (SPPs) were allowed into the Thai market, ending the monopoly. In response, EGAT created the Electricity Generating Public Company Ltd (EGCO) as the first IPP. Today, EGAT operates three thermal power plants, six combined cycle power plants, 24 hydropower plants, eight renewable energy power plants, and four diesel power plants with a total installed capacity of 16,385MW as of January 2017.

THE BANGCHAK PETROLEUM PUBLIC COMPANY LIMITED (BANGCHAK) refines crude oil and markets the finished products. It then sells these products directly to consumers through its nationwide network of service stations.

Created as a state enterprise in 1967 but partially privatized many years ago, the company has also branched out into the fields of solar energy and oil and gas exploration. Its subsidiary, Bangchak Solar Energy, operates several solar power plants with a combined capacity of 118MW. In 2014, it acquired an 81.41 percent interest in Nido Petroleum Ltd, an independent oil and gas exploration and production company listed on the Australian Securities Exchange. In 2016, it acquired SunEdison Japan, which has a total capacity of 13MW.

BANPU PCL is Thailand's biggest coal producer and has since expanded into Australia, China, Laos, Indonesia, the United States and Mongolia. The company provides exploration, drilling, mining, and trading-related services for coal.

Banpu owns a 50 percent stake in a 1,434MW coal-fired power plant in the Map Ta Phut Industrial Estate in Rayong province. More recently, in 2016 the company acquired a 29.4 percent share in a low-cost, unconventional shale gas operation in Pennsylvania.

SOLAR POWER COMPANY GROUP (SPCG) is a relatively new player in the field of clean energy, but its future looks bright. Since 2010 the company has constructed a network of more than 30 utility-scale solar farms across Thailand's northeast with a combined capacity of 260MW and connected them to the grid.

Power Struggles: A Demanding Balancing Act

Thailand's Power Development Plan 2015–2036 (PDP) – which is a blueprint of the country's efforts to generate enough electricity to keep pace with economic growth – calls for more than doubling its capacity by 2036. But planning such a dramatic increase is not a matter of merely putting the right figures in the right columns. Even if the officials work in climate-controlled offices, they would still have felt the heat that their actions have generated on the ground so far.

The current PDP will almost certainly run into some ardent opposition from a variety of groups. Indeed, protests have already been sparked, with a government plan to build a coal-fired power plant in Krabi province meeting stiff resistance from local citizens and NGOs concerned about the impact it will have on the region's lifeblood: marine resources and tourism. The controversial plant, if it does indeed go ahead, is to be built by EGAT not far from the Mu Koh Phi Phi National Park.

Thailand's PDP addresses concerns that over the past two decades Thailand's rapidly expanding economy has sparked the need for more power sources. Yet, the overall capacity is greater than demand, even

as power consumption has grown by an average of five percent per annum – to the point where the kingdom now has one of the highest electrification rates in Southeast Asia with nearly 100 percent of the population possessing access to electricity.

The problem lies in the less developed parts of the country, such as the south, where there is a shortage. This area needs 2,500MW per day while local sources can supply only 2,000MW. To make up the shortfall, 500MW are supplied via high-voltage cables from the central region. Hence, the state's initiative to build two new power stations, one in Krabi and one in Surat Thani.

The PDP, in order to ensure power supply security, also promotes careful planning in expanding the capacity, the diversification of fuel sources and a boost in alternative fuels, demand-side management, and staving off dependence on electricity imports. Major hydropower projects in Myanmar and Laos are also under construction.

To deal with possible disruptions, the scheme provides for a minimum reserve margin of 15 percent over peak power demand. Thanks to the authorities' vigilance in increasing

capacity, the reserve margin has generally stayed well above the target. In 2013, for example, the margin stood at 26 percent over the peak demand of 27 gigawatts (GW).

Natural gas makes up around 45 percent of the fuel mix in power generation. Coal, hydro and renewable energy make up most of the rest. But if the new coal-fired plants down south go ahead, it will set the stage for a leap in greenhouse gas emissions.

By 2036, coal is expected to contribute 7,390MW of capacity. Originally, nuclear power was to supply 5GW, but the Fukushima Daiichi nuclear disaster in 2011 forced the Energy Ministry to reconsider its plan of building five nuclear power plants over the next two decades.

On the bright side, the Energy Efficiency Development Plan (2015–2036) aims at reducing energy intensity by 30 percent over 20 years, while the Alternative Energy Development Plan (2015–2036) calls for boosting the use of renewable sources to meet 30 percent of the country's needs by the year 2036.

In 2016, renewable energy contributed 9,139MW of electricity to the grid. Biomass and solar power were the largest contributors (if large hydro projects are excluded), providing 2,815MW and 2,149MW, respectively. But it's solar, both on an industrial scale or through rooftop panels on houses and other buildings, that is being promoted by the current military-installed government as a potential game-changer.

Thailand's power development plan calls for solar power to be increased to 6,000MW and biomass to 5,570MW in the next two decades. Scaling up these renewable energy sources is essential if the kingdom is to reach its goal of reducing its reliance on natural gas from 45 percent to around 30–40 percent by 2036.

Anti-coal banners in Krabi's Nua Khlong district.

FURTHER READING

- "Thailand 4.0: Powered by Renewable Energy?" in the March 2017 edition of *Thai-American Business Magazine,* published by the American Chamber of Commerce in Thailand
- *The Great Transition: Shifting from Fossil Fuels to Solar and Wind Energy,* by Lester R. Brown, 2015
- *Between Populism and Price Increases: Who Will Pay for the Cost of Renewable Energy?* by Davida Wood et al., 2011

REALITY CHECKS

Clean Energy: The Pathway to a Sustainable Future

Biomass and Biofuels

• Biogas, made from the decay of organic matter in the absence of oxygen, is a gaseous fuel high in methane and carbon dioxide. It can be produced from substances such as agricultural waste, manure, plant matter, sewage, food waste, etc.

• Biomass is made of organic matter such as animal waste, rice husks, cassava, palm oil and algae, which can be converted to energy. It can be converted indirectly into energy by distillation into biofuels, such as ethanol and biodiesel. It can also be turned directly into energy by incineration or combustion to produce heat or electricity. Considered a key renewable resource of the future, it already supplies 10 percent of the world's primary energy supply.

• Biomass energy can be made in ways that help reduce or increase global warming. It can help clean the air, water and soil, or have harmful effects that also threaten biodiversity and bode poorly for public health.

Solar and Wind Power

• Power from the sun is the most abundant energy resource. If only 0.1 percent of the energy reaching the Earth could be converted at an efficiency of ten percent, it would be four times more than the world's electricity generating capacity of 5,000GW.

• The Renewables 2015 Global Status Report found that 22.8 percent of the world's electricity was generated from renewable sources in 2014. Researchers

Odorless charcoal is a biomass innovation.

claim that harnessing the power of the wind and sun could be enough to meet the world's energy demand.

• 2011 marked the first time global investments in renewable energy surpassed investments in fossil fuels. As of June 2016, clean energy investment had increased to US$286 billion, with solar accounting for 56 percent of the total, and wind accounting for 38 percent. Overall, more than twice as much was spent on renewables than on coal- and gas-fired power generation (US$130 billion in 2015).

• Solar and wind power are not without their critics. Both sources of power can require large tracts of land that could be farmed or reforested.

Fuel Cells

• Fuel cells, once the stuff of science fiction, now generate pollution-free power from hydrogen and oxygen, leaving only water and waste heat as byproducts. Their use is on the rise, with worldwide sales of fuel

cells exceeding US$2 billion in 2015. According to a report by the Fuel Cell & Hydrogen Energy Association, in 2015 nine percent of Fortune 500 companies and 23 percent of Fortune 100 companies were using fuel cells in some aspect of their operations.

• Fuel cell efficiencies range between 40 and 60 percent, depending on the type. When waste heat is captured the overall efficiency can be 85 percent.

• Hyundai, Honda and Toyota have all announced their own brands of hydrogen-heavy Fuel Cell Vehicles.

"Clean" Coal

• Coal provides 40 percent of the world's electricity while producing almost the same amount of global CO_2 emissions. In Thailand, coal supplied 20 percent of the fuel for power generation in 2016. Currently, three coal-fired plants generate 4,186MW. To reduce dependence on natural gas, new coal-fired power plants will be built to provide up to 4,400MW more capacity by 2030. Facing opposition to this "dirty" energy, the government is considering constructing more plants using a new "clean coal" technology.

Hydropower

• Hydropower provided Thailand with 3,088MW of power in 2016. But as a result of opposition to new domestic dams, Thailand has sought to secure additional power from projects in neighboring Laos and Myanmar. However, these projects have also stirred a significant level of controversy in their respective countries.

SOIL
The lifeblood of agriculture

Farmers in the northwest seed soy beans in burnt soil after a rice harvest

Soil is much more than dirt. It's a living entity: a combination of organisms, minerals, liquids and gases that nurses plants and nourishes crops. Vegetation dies off without such life-sustaining nutrients.

In short, soil is as complicated as the wildly different terrains that it supports. In the north of Thailand and to the west bordering Myanmar are highlands of sandy soil. The central plains consist of low-lying farmland characterized by alluvial soils fed by five rivers from the north, with mudflats at the mouth of the Gulf of Thailand. The gulf-facing parts of the Eastern Seaboard and southern provinces are filled with saline soil, or alkaline soil, unsuited to crop cultivation. And in the famously arid northeast, prone to **desertification**, salt is also an issue, as swathes of the region sit atop enormous salt deposits buried under the Korat Basin and the smaller Sakhon Nakhon Basin. Both were seabeds in prehistoric times.

Given Thailand's reputation as one of the world's top producers of agricultural products, one would think that the country's bedrock must be its nutrient-rich soil. Not so, says Chalermpol Kirdmanee, principal researcher and head of Plant Physiology & Biochemistry Laboratory in the Agricultural Biotechnology Research Unit at the National Center for Genetic Engineering and Biotechnology (BIOTEC), the country's research arm on biological science. "Soil conditions in Thailand are like famished, mal-nourished kids who eat too much junk food," he said.

It's a provocative way of saying that Thai soil has five times less the vital nutrients like nitrogen than the accepted global standard, which makes much of the land unsuitable for farming. Chalermpol, who has spent more than 20 years working with royally initiated projects on soil salinization in the northeast, explains that many of the problems are man-made. Ignorance and greed top the list of culprits. Both are to blame for the monocropping that became a staple of local farms in the 1970s. This style of cultivation uses the same plot of land to plant the same type of crop over and over again, until the nitrogen and

other essential elements are leached from the soil. The intensive use of chemical fertilizers, which began in the 1960s as part of the "Green Revolution" and reached Thailand a decade later, has also had a harmful effect on soil, while modern farming equipment like tractors has compacted the earth, preventing moisture and organic matter from penetrating it.

The damage has been devastating to say the least. According to the latest survey by the Land Development Department (LDD), under the Ministry of Agriculture and Cooperatives, more than 54 percent of the total land in the country is low grade, while the amount of **soil organic matter (SOM)**, a key indicator of the soil's health and fertility, is too low nationwide. Among these substandard varieties is acidic soil – meaning soil with a pH level lower than 5.5. This is mostly caused by the misuse of chemical fertilizers and by the air pollution created by burning coal and bunker oil (a sludgy extract).

Globally, Thailand keeps good company in facing these issues. According to figures released by the World Economic Forum in 2016, some 52 percent of the planet's soil is degraded. The main causes have been identified as high-yield agricultural practices requiring the intensive use of chemicals and

Desertification:

A type of land degradation in which a relatively dry region becomes increasingly arid, typically losing its bodies of water, vegetation and wildlife. It is caused by a variety of factors, such as climate change and manmade activities.

Soil organic matter (SOM):

The organic matter component of soils, consisting of plant and animal residue, cell and cell tissues of decomposed soil organisms. SOM is a proven indicator of soil quality.

over-plowing, which strips away topsoil, in addition to removing crop stubble by burning or animal grazing. At this rate of loss, the world's topsoil, which is vital for sustaining plant life, will be gone within 60 years.

However, over generations, Thailand has been able to overcome some soil issues and create a dynamic agricultural sector due to the skill and knowledge of farmers and farmhands, rather than the richness of the earth. And thanks to scientific innovations and new projects aimed at encouraging organic crops, as well as King Bhumibol Adulyadej's royally initiated projects, the country does not lack in good, grassroots examples for farmers and officials to follow down a more sustainable path.

"Soil conditions in Thailand are like famished, malnourished kids who eat too much junk food."

Chalermpol Kirdmanee, a lead figure at the National Center for Genetic Engineering and Biotechnology (BIOTEC)

Types of Soil in Thailand

ACIDIC SOIL, or soil with a pH level lower than 5.5, is the most widespread issue in Thailand, accounting for some 29 percent of the land. More than half of that total has severe quality issues. The problem is caused by the misuse of chemical fertilizers and air pollution caused by burning coal and bunker oil.

SHALLOW SOIL, often located on hill slopes, suffers from a proliferation of gravel and pebbles, which make up more than a third of its content. Located close to the surface, hence its name, shallow soil is infertile because it cannot retain water. Around 14 percent of the land in Thailand is classified as shallow soil.

ALKALINE SOIL contains too much salt. This makes it bad for cultivation purposes as the salt content remains in the plant and prevents the earth from absorbing water. It is found in coastal areas and in the northeast basins where large deposits of salt remain from prehistoric seas.

SANDY SOIL is loose since sand makes up 85 percent of its structure. As the dregs of rivers and seas, sandy soil cannot hold water but certain plants like sugarcane, cassava and pineapple thrive in it. Mostly found in coastal areas, riverbanks and around sandstone mountains in the northeast, almost four percent of the soil in Thailand can be classified as sandy.

ACID SULFATE SOIL boasts high amounts of these substances and a pH level lower than 4, as a result of being deluged by salty seawater. Naturally, this strain of soil, which makes up less than 2 percent of the total, proliferates around the Southern and Eastern Seaboards.

PEAT is made up of decomposed vegetation and organic matter that has accumulated over thousands of years. Also known as mires or bogs, peat is good for growing grass and is one of the most efficient carbon sinks on the planet. These bogs are found around the sea-hugging parts of the south and east coasts.

HARDPAN accounts for around 8.5 percent of the soil in Thailand. Beneath the topsoil is buried a thick layer of earth from whence its name is derived and which yields little in the way of agricultural products. Hardpan often results from farming and tilling the same plot of earth over and over again.

Solving Nature's Dilemmas from the Ground Up

While most monarchs are associated with opulently appointed palaces, those of King Bhumibol Adulyadej in Bangkok and Hua Hin long served as labs and launching pads for environmental projects. These were not mere hobbies. King Bhumibol spent so many hours working on issues of soil, for example, that his birthday, December 5, is now celebrated as World Soil Day by the UN.

One of the most groundbreaking initiatives was the Land Development Project in the Cha-am district of Phetchaburi province. Created in the 1960s, the project is renowned as the first to deal with land management through the introduction of proper irrigation systems into this drought-prone area as well as the application of organic fertilizers and soil treatments. Establishing this agricultural cooperative in the village of Hoob Krapong was only the start.

From there, the king and his crew set up soil treatment projects across the country. In the Phanom Sarakham district of Chachoengsao province, for example, the arid land with salty soil became ground zero for a pilot program designed to show

King Bhumibol with officials at the Pikun Thong Royal Development Study Center in Narathiwat.

how micro-reservoirs can be used as a "water buffer" to slow down the spread of alluvial deposits brought about by rivers and streams in floodplains.

Over the years to come, more pilot projects took off. Up north, in the Huai Hong Krai village of Chiang Mai province, the program introduced check dams and vetiver grass to solve the perennial trouble of soil erosion. "When we talk about soil and His Majesty the King, people usually think of vetiver grass," said Dr Chalermpol Kirdmanee of BIOTEC. At the top of the mountain, King Bhumibol advised the project coordinators to reforest the area at the watershed level to create more moisture. On the slope, they planted vetiver grass, famous for its deep and strong roots, to stop soil erosion. Then the monarch suggested that they build check dams in the small ravines to slow down the water run-off and keep the good alluvial soil from being washed away.

Down south in Narathiwat province, the Pikun Thong Royal Development Learning Center broke ground by using different techniques to "trick the soil" into becoming more fertile and less acidic. Across

the northeast, which is notorious for its salinized soil, dozens of projects have been implemented using King Bhumibol's belief that solving nature's dilemmas requires natural solutions. That was the modus operandi of the collaborative efforts between SCG and the Crown Property Bureau Foundation to launch more soil treatment projects that used saline-resistant plants and trees such as ironwoods to siphon off saltwater from the ground.

For the late king, soil conservation was a philosophy in which he used nature to control nature. He would "look at the big picture and then use his understanding of science and local culture and traditional wisdom," said Dr Chalermpol.

To pay tribute to his many contributions to environmental science and soil management, in 2002 the International Union of Soil Sciences (IUSS) made a resolution to celebrate World Soil Day on His Majesty's birthday. That motion was unanimously adopted by the UN's Food and Agriculture Organization (FAO) in 2013, only a year after the IUSS gave its first "Humanitarian Soil Scientist" award to King Bhumibol.

A man plants vetiver grass.

FURTHER READING

- "What if the World's Soil Runs Out?" by Professor John Crawford, interview on *Time* magazine website, December 14, 2012
- *The Ideal Soil 2014: A Handbook for the New Agriculture*, by Michael Astera, 2014
- *Soil Erosion Problems in Northeast Thailand: A Case Study from the View of Agricultural Development in a Rural Community Near Khon Kaen*, by Kosit Lorsirirat and Hideji Maita, 2006

REALITY CHECKS

Actions Needed to Enrich the Earth

Seedlings grown at Thung Song cement plant in Nakhon Si Thammarat province.

Nipping monocropping in the bud
The overreliance on monoculture, which is characterized by an overuse of pesticides, is not a sustainable form of farming in the long term.

More carbon Getting more carbon back into the soil through the elimination of bad farming practices like overgrazing, over-plowing and burning off crop stubble is crucial for nourishing the earth's nutrients.

Soil erosion As a result of land mismanagement, the Land Development Department (LDD) estimates that one-third of Thailand suffers from soil erosion, leading to the loss of organic matter and nutrients.

Cultivating agriculture The LDD's latest survey from 2009 showed that the amount of agricultural land increased by almost six percent between the years 2004 and 2009. According to the Office of Agricultural Economics (OAE), some 149,225,195 rai of land was used for agriculture in Thailand as of 2014.

Lack of knowledge LDD surveys consistently find that many farmers misuse the land by farming crops not suited to their particular type of soil. Introducing land zoning could help solve this problem. Proper land zoning and education could also encourage farmers to improve soil nutrients and the conditions of the land before planting.

High water As seas rise, the amount of saltwater inundating coastal areas and farmlands has also risen to the point where it has penetrated some 70 to 80 kilometers inland.

This worrisome trend looks set to continue thanks to global warming.

Soil conservation Launching more pilot projects on soil conservation around the country that are adapted to regional needs and include local participation is a necessity.

Low SOM SOM (soil organic matter) is one of the most important factors in soil quality. The more organic matter that is found in soil, the more fertile the soil is. However, low SOM, which leads to the deterioration of the physical and chemical properties that make up soil, has been found to be a widespread problem in Thailand. Sustained efforts in techniques such as crop rotation, using compost and growing cover crops can help raise SOM levels.

Mountainous soils According to a report by the UN's Food and Agriculture Organization (FAO), mountainous soils, which means soil on slopes that are steeper than 35 degrees, have been classified as problematic due to a variety of environmental woes such as erosion.

Farmers learn about fertilizer technology

WATER

Managing a fundamental element

For Thais, water is much more than just a natural resource. It is integral to the lifeblood of their culture, coursing through the New Year celebrations in April when Buddha images are washed and revelers doused, at festivals such as Loy Krathong, where offerings are made to the Goddess of Water, in wedding rituals like the anointing of the couples' hands by guests, or through the bathing ceremony performed on bodies before cremations.

As a country once dominated by agriculture and still dependent on it for food and income, water makes all the difference between bumper crops and broken dreams. The country's factories and industrial complexes would lose their liquidity without it and tourism would also dry up if it were not for this most fundamental of elements. In fact, without water all of us would go hungry – UN estimates state that it takes 3,500 liters to produce one kilo of rice and 15,000 liters to produce the same amount of beef. Such figures make up part of our **water footprint**.

Thailand is fortunate to have been endowed with an abundance of water resources due to its geographical position in the tropics, which feature monsoons that result in a six-month-long wet season. According to "Droughts in Thailand," a 2012 report prepared by leading Thai experts, the average precipitation in Thailand is 1,374 millimeters per year, well above the global average of 990 millimeters. Thailand also scores highly on the SDG Index in areas like access to improved water and sanitation, and freshwater withdrawal. But supply is not the issue. Instead, distribution and resource management are the key challenges.

The kind of adequate irrigation infrastructure – such as dams, dykes, moats, canals and floodgates – to help manage the ebb and flow of water unfortunately covers only a small fraction of Thailand's land area, and the distribution of these facilities is uneven. Overall, only one-sixth of the country's farmland is irrigated, according to the Hydro and Agro Informatics Institute (HAII) under the Ministry of Science and Technology. The vast majority of that lush farmland is in the central region, famous as the country's "rice bowl." Meanwhile, farmers in other parts of the country – where rainfall can be unreliable, waterways non-existent and irrigation systems rare – face the regular threat of drought and must rely on whatever strategies they can devise to eke out a living. For decades, successive governments have advised farmers in these drought prone areas to diversify away from water-intensive crops, the most demanding being rice. In 2016, these efforts were ramped up and state-funded training sessions carried out to educate farmers on alternative crops like fruit trees and peas.

Drought, however, is only one part of a triple threat that beleaguers the country's authorities on water management. Dealing with floods and treating wastewater are the two other main issues. The lower part of Thailand's central region, irrigated by a low-lying river basin and floodplain, is especially susceptible to rising waters and is often inundated by run-off from northern waterways. During the rainy season, these rivers tend to swell, bursting their banks and overflowing into urban areas. The last big deluge was in 2011 when the unusually

WATER DEMAND/USE IN THAILAND

(69,811 Mm³ per annum)

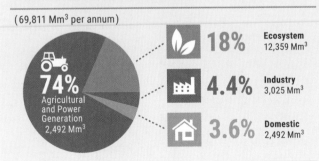

74% Agricultural and Power Generation 2,492 Mm³

18% Ecosystem 12,359 Mm³

4.4% Industry 3,025 Mm³

3.6% Domestic 2,492 Mm³

Sources: Department of Water Resources, Ministry of Natural Resources and Environment

Completed in 2005, the Lat Pho canal (left) in Samut Prakan has helped to mitigate the impacts of seasonal flooding.

heavy rainfall – 20 percent more than the average – combined with mismanagement and perennial problems, such as flagrant land misuse and outdated drainage systems, to set the stage for what the World Bank classified as the fourth costliest disaster in the history of the planet.

While floods represent a sporadic threat, far more chronic is the hazard of wastewater. As Thailand continues to industrialize and urbanize, this toxic tide is on the rise. It is especially problematic in and around industrial estates. By law, factories are required to install wastewater treatment facilities; however, in reality, the laws are not enforced and some factories discharge wastewater directly into rivers and canals. The problem is compounded by a lack of wastewater treatment plants, and the fact that half of Thailand's treatment facilities are offline or experiencing technical difficulties. As a result, in Bangkok only about half of wastewater is treated. And outside the capital, the total is much lower. This statistic is notable as it also highlights the issue of unequal access to clean water in the kingdom.

But even these concerns are mild compared to the storm clouds massing on the horizon in the form of climate change. Among climatologists, the consensus is that water will be impacted more than any other element by global warming. Dr Royol Chitradon, director of the HAII, which monitors data on water and advises other agencies on management issues, said rising sea levels have already begun inundating coastal areas with saltwater, contaminating both farmland and freshwater sources.

Yet, the most worrisome aspects of climate change are its volatility and unpredictability. Entire weather patterns are changing. Storms may come out of the blue to blacken the skies and dampen the earth. The cycle of floods and drought may become even more severe in the future. Such changes, said Dr Royol, will make it harder for the powers-that-be to predict and map out accurate water management plans. "The problem is that rain will fall in unusual places at unusual times. We might not be able to use old models and analytical methods to predict these kinds of patterns in the future," he said.

As successive Thai governments struggle to come up with a feasible, comprehensive water management plan to forecast such dilemmas and ward off future floods and droughts, the subject of water management has taken on a grave and tremendous significance: part science, part prophecy, and all important.

Water footprint:

The amount of water you use in the course of a day, not just directly from drinking or flushing the toilet but indirectly. According to National Geographic magazine, 95 percent of our water consumption comes from the products we purchase and the energy required to produce them and the food that we eat.

The King and the Water Paradox

"From that point on," King Bhumibol Adulyadej wrote in his journal, "I have thought about this seemingly insoluble and paradoxical problem: when there is water, there is too much, it floods the area; when the water recedes, it is drought."

"That point" which King Bhumibol referred to was a 1955 tour of Thailand's northeastern region, the country's poorest, where farmers were barely able to produce one successful crop. Traveling over 680 kilometers by train and 1,592 kilometers by road, inundated by stories and images of hardship, the king was moved. The effective management of the country's water resources became an abiding preoccupation for him throughout his reign.

Over the ensuing decades, a wide variety of resources, ideas, projects, programs and technology were directed toward improving this fundamental underpinning of rural existence. "If we take proper care of the environment, there will be water for many hundreds of years," he once observed, echoing the long-term thinking that is at the heart of sustainable development. "By that time our descendants might be thinking of some new ways to solve the problems." As much as any story of his reign – the rise and fall of Communism or the rising skyline of Bangkok – the late king's attempts to improve the security of the farmer was the one he personally led.

In May 1960, the king revived the Royal Ploughing Ceremony, an ancient fertility ritual that gave farmers hope. In 1962, near his summer residence of Hua Hin, he invested his own money to build an earth-fill dam near the coastal village of Khao Tao. The poor living conditions of the villagers were due, in part, to soil erosion and fresh water scarcity. The king's dam blocked seawater

"If we take proper care of the environment, there will be water for hundreds of years."

King Bhumibol Adulyadej

from flowing into a natural channel and helped trap rainwater flowing in. The 600,000 cubic-meter-capacity reservoir known as Khao Tao reservoir would be the first of many localized projects the king helped create.

It was also typical of his development model: establish direct contact with the villagers and learn about their problems, offer practical solutions the villagers themselves could adopt, cut through bureaucratic red tape to implement a solution, and then monitor the progress.

In order to help villagers better manage their natural water supplies, throughout the country he advised them to protect watersheds and build weirs as miniature dams to help regulate a river's flow. He also advocated the construction of reservoirs and ponds dubbed "monkey cheeks" to store floodwaters during the rainy season so that it could be used later when other sources had dried up. The idea, which came to him in the wake of the 1995 floods, was based on the way that monkeys keep special

reserves of food and other nutrients in their cheeks to use in times of emergency. In essence, the ponds that the villagers dug functioned in the same way. To treat wastewater in the canals of Bangkok, the king told the authorities to divert freshwater containing high amounts of oxygen into them to dilute the polluted water.

Known as *kaem ling* in Thai, the concept has gained notoriety nationwide as a cheap and environmentally friendly way to manage water and mitigate the effects of floods and drought. Today there are some 190 monkey cheek water retention areas across Thailand and in the wake of the 2016 drought the government has called for the construction of more *kaem ling* projects.

When coming up with these plans his modus operandi was to deploy "appropriate technology" that harnessed the raw power of nature. As Pramote Maiklad, the Royal Irrigation Department's former secretary-general explained, the king often used "nature to solve natural problems", or at least technology he had invented along with some natural elements. Whatever the initiatives were, he always made sure that his so-called appropriate technology was simple and doable so locals could use it.

King Bhumibol was also an innovator. Of his inventions, few are more renowned than the artificial rain-making technique he patented some 30 years ago, and the Chaipattana aerator, patented in 1993, to recycle wastewater by adding oxygen to it. In 2009 the World Intellectual Property Office (WIPO), a United Nations organization tasked with advocating intellectual property rights, bestowed the WIPO Global Leader Award on King Bhumibol as a champion of intellectual property rights and a prolific inventor, with some 20 patents and 19 registered trademarks to his name.

FURTHER READING

- "History of Water Resource and Flood Management: A Policy Brief," by Deunden Kilomborirak and Kittipong Ruenthip, 2013
- *National Stakeholder Consultations on Water: Supporting the Post-2015 Development Agenda*, by Global Water Partnership Southeast Asia, 2013
- *New Studies on Age-Old Water Management Issues*, by Derek Sanderson, 2012

REALITY CHECKS

Measures to Improve Water Management

A check dam at Huai Kong Khrai Royal Development Study Center in Chiang Mai.

Go to the source To preserve the sources of this precious resource, sensitive areas like river basins and watersheds must be protected and kept off-limits from potentially harmful developments. Villagers should be encouraged to build small weirs along local rivers to preserve upstream watersheds and replenish the underground water supply.

Flood prevention measures To preserve natural drainage systems, proper land zoning measures need to be rigorously enforced. For example, low-lying areas on Bangkok's eastern edge, which are earmarked for agriculture and flood drainage, have instead been converted into real estate developments. The flood plain in the area of Ayudhya province is filled with industrial estates. Retention areas like natural ponds and swamps must be protected. Canals and rivers need to be dredged regularly

and the building of infrastructure that encroaches upon these natural resources must be barred.

Leadership glut To cut through all the red tape that ties up the decision-making process, the government needs to streamline bureaucratic processes and legal apparatuses. According to a report by TDRI, unlike most countries, Thailand does not have a law governing the management of water resources and thus the management of water resources and floods is highly fragmented between many bodies. Currently, 30 different state organizations under seven different ministries apply more than 50 different laws to the multiple issues of water management.

Preparing for climate change To deal with the shifting weather patterns associated with climate change requires firm policies and

well-formed plans to be put into place while there is still time.

Public participation To win the trust and cooperation of the public, the private and public sectors must allow locals to have a say in issues that impact their water security. Locals must also be able to trust that firms and authorities are conducting and taking into account honest environmental impact assessments.

Hazardous wastewater To keep industrial-strength wastewater from poisoning rivers and canals, authorities must strictly enforce the law by imposing fines on companies that flout the regulations.

Crop selection Thai farmers need to break away from water-intensive crops and try growing produce that isn't quite so draining on water supplies.

Water footprint Start campaigns to raise awareness about everyone's "water footprint," 95 percent of which comes from the products and energy we consume and the food we eat.

Tree planting Planting trees to reduce deforestation and ease the effects of floods can only be undertaken if the local communities involved stand to earn some money.

Cleaning up the dregs The sediments clogging reservoirs, canals and irrigation moats must be dredged to improve distribution, and drainage systems maintained or updated.

FORESTS

We depend on them for food, water and oxygen

FOREST COVER

THAILAND
1945

THAILAND
2014

- 🌳 **70%**
- 🌲 **50-70%**
- 🌿 **25-50%**
- 🍂 **LESS THAN 25%** | **NONE**

Source: Royal Forest Department

Carbon sinks:

Rivers, forests, soil and the atmosphere are carbon sinks, areas that absorb and store carbon for long periods of time, thereby taking it out of the atmosphere.

The world's most productive ecosystems are the tropical forests that stretch across South America's Amazon and through parts of Africa, India and Southeast Asia. Not only do these greenbelts provide shelter and sustenance for around half of the world's species, they also serve as vital watersheds and **carbon sinks**. Thailand has a wealth of such forests. But those in the northern provinces of Chiang Mai, Chiang Rai and Nan, which are key watershed forests for Thailand, are under threat.

In general, Thailand's forests can be divided into two main categories: evergreen and deciduous. In the north, the latter species cling to the highlands and are made up of such subspecies as cloud forests. In the northeast, dry dipterocarp and dry deciduous forests make up much of the tree line, whereas evergreen forests enrich the provinces of the south and the east. On the isolated fringes of the Eastern Seaboard and the southern flanks of the Andaman Sea are the vital yet much-decimated mangroves.

All of these are threatened by deforestation. Until the late 19th century this was not an issue in Thailand. That changed with the arrival of the British who were already carving up the teak forests of Burma as the fodder for beautiful hardwood furniture and the raw material for the ships that kept their empire afloat. These days, logging is well down the list of deforestation drivers in Thailand. Instead, commercial agriculture and subsistence farming vie for the top spots, followed by mining, major infrastructure projects like dams and roads, and urban encroachment. Drivers like these also help explain why deforestation accounts for some 10 percent of greenhouse gas emissions around the globe.

In developing countries the root causes behind deforestation are not difficult to nail down: short-term profits trump the long-term benefits of human health and ecological welfare. It is not only large landowners who clear forests. In Thailand, unable to secure title deeds, for decades small-scale

Huai Kha Khaeng Wildlife Sanctuary is an important nature reserve near the border with Myanmar.

farmers have headed into the forests to clear plots for cultivation. These days, agribusinesses, for instance, encourage northern farmers to clear trees to grow corn and other monocrops. The firms offer a good price for the crops, which are used as animal feed. After the harvest, the farmers torch the field and refuse, creating a toxic haze that has periodically plagued the northern region.

As David W. Pearce summed up in his paper, "The Economic Value of Forest Ecosystems," "rapid population change and economic incentives...make forest conversion appear more profitable than forest conservation." This philosophy is endemic to most, if not all, of Southeast Asia. In Thailand's case it's had some disastrous repercussions. According to the latest government statistics, Thailand's forest cover is roughly around 31 percent, though some experts insist it may be only 27 or 28 percent. The only way to get these statistics is through satellite images, which take about two to three years to compile. Whatever the figure, it still represents a steep decline from 1945 when around 60 percent of the forests were still standing. Though the rate of deforestation has slowed to around 0.2 percent per year on the low end, the implications are still dire, with floods, landslides and soil erosion taking their toll on communities and crops, while the decline

of keystone species such as the tiger and hornbill threatens the country's biodiversity.

Among the bright spots on the horizon is the spread of **community forestry**, which allows villagers who live in and depend on these ecosystems to use their homegrown wisdom to manage them in more sustainable ways. This strategy has paid off in South America, where countries like Bolivia and Brazil have drastically reduced deforestation rates in the Amazon. Already the Thai government has granted legal status, if not stewardship, to some 9,000 villages situated in or near forests. At least a thousand other communities exist in a no man's land that has more grey areas than green spots. With reforestation a priority under the current government, which hopes to increase forest cover to 40 percent over the next 10 years, the state has begun reclaiming land listed as forest reserves from villagers, resort moguls and plantation owners.

"An intact forest cycles nutrients, regulates climate, stabilizes soil, treats waste, provides habitats, and offers opportunities for recreation."
Janet Larsen, the Worldwatch Institute

Community forestry:

Coined back in the 1970s and popular in places like North America, Brazil and India, community forestry describes how local villagers work with NGOs and the state to manage the forests near where they live. It's still a relatively recent trend in Southeast Asia.

Why Forests Matter to Sustainable Development

Watersheds Forests act as watersheds, meaning they soak up and discharge rain into the tributaries that make up the veins of bigger water bodies like rivers. Protecting these watersheds helps to strengthen the natural ecosystems on which we rely and also helps mitigate disasters such as mudslides and floods.

Medicine chest The active ingredient in Aspirin, discovered in ancient Greece by the famous doctor Hippocrates, came from the bark of a willow tree, as do many other time-tested medicines.

Deep breaths Put simply, trees help us to breathe. Through photosynthesis, trees suck up carbon dioxide, serving as natural air purifiers in the process of producing

oxygen, the reverse of human breathing in which oxygen is inhaled and carbon dioxide is exhaled.

Darwin's theater The "food web," a term that may have originated from Charles Darwin, describes the feeding connections between all of Earth's creatures, whether gentle grazers or carnivores, ghastly scavengers or parasites. Think of the forests as a stage where these battles for survival are enacted on a daily basis.

Raw materials Forests provide us with the raw materials with which we build and furnish our homes, and so much more.

Great outdoors The phrase "human nature" denotes an alliance between people and the natural world that has its roots in our evolution from cave dwellers to urbanites that still seek refuge from our frenetic existences by retreating to the great outdoors.

Karen villager with cumin grown in the forest.

Forest Facts

OUR GREAT ANCESTORS began farming some 11,000 years ago. Since then half of the globe's forest cover has been lost, the majority of it in the last 50 years. On a global level, agriculture drives around 80 percent of deforestation.

THE WORLD'S MOST productive ecosystems are tropical forests, which stretch across South America, especially the Amazon, and through parts of Africa, India and Southeast Asia. These greenbelts provide shelter and sustenance for around half of the world's species.

OF ALL THE DIFFERENT types of forests, primary forests occupy about one-third of the total. Sometimes referred to as old growth forests or virgin or primeval forests, they are renowned for their age, beauty and biodiversity, whereas woodlands recovering from a forest fire are referred to as second-growth forests.

Vital Mangroves Cleared for Commercial Use

A 2014 report from the United Nations Environment Program (UNEP) called "The Importance of Mangroves: A Call to Action" states that mangrove forests are being cut down at a rate three to five times faster than any other forests. The emissions resulting from the burning and cutting of them now accounts for one fifth of the total gases produced by deforestation in the world today.

This is an ecological disaster in the making. Forming a wooded wetland habitat along coastal communities and waterways, mangroves are vital ingredients in a healthy marine ecosystem. They serve as breakwaters against floods, nurseries for small fish, crabs and shrimp, nests for birds, and hotspots for eco-tourists.

Expounding on the ramifications of the report, UN Under-Secretary-General and UNEP Executive Director Achim Steiner said, "The escalating destruction and degradation of mangroves, driven by land conversion for aquaculture and agriculture, coastal development and pollution is occurring at an alarming rate, with over a quarter of the Earth's original mangrove cover now lost. This has potentially devastating effects on biodiversity, food security and the livelihoods of some of the most marginalized coastal communities in developing countries where more than 90 percent of the world's mangroves are found."

Thailand is very much a part of this alarming story. In isolated areas of the country's Eastern and Southern seaboards, mangroves are hanging on for dear life. Their importance to the local ecology only became known to many locals after the 2004 tsunami, when fishing villages along the Andaman coast that were protected by mangroves suffered far less damage and fewer fatalities than the towns where the forests had been chopped down and carved up to make charcoal.

Although there are many different causes of mangrove destruction here – such as timber production, the encroachment of tourism, construction of industrial parks and the intensive operation of fishing boats – aquaculture is clearly a key culprit.

One of the world's top five shrimp exporters, Thailand ramped up shrimp aquaculture in coastal areas when a 1981 fishing policy limited shrimp capture at sea. According to the Office of Agricultural Economics, there were 33,000 hectares of land under shrimp cultivation in 2016.

While this meant profits for the local shrimp farmers, it has spelled disaster for mangrove forest conservation. Hundreds of thousands of acres of mangroves were clear-cut to make room for short-term aquaculture ponds, while chemical-intensive farming practices contaminated waters.

Since the 1960s, Thailand has lost more than half of its mangrove forests; the southern province of Phatthalung has lost all of them.

Recently, modern farming techniques have helped push aquaculture away from the coastal areas of mangroves, and the government has established two mandatory certification schemes for Good Aquaculture Practice and the Code of Conduct for Responsible Shrimp Farming.

In addition, community initiatives in southern provinces such as Trat and Surat Thani are helping to protect and reclaim mangroves not only for the forests themselves, but also for the health of the community at large.

Shrimp farms (left) have had a disastrous impact on coastal mangroves (right) in Thailand.

The Many Pressures Facing the Forests

Agricultural fodder The cutting down of forests to make rice fields, orchards, plantations and other plots of arable land is the single largest driver of deforestation in the world. In Thailand, the northern and north-eastern regions were denuded due to post–WWII government initiatives to populate the highlands and grow agribusiness.

Cash crops Oil palms and the eucalyptus plantations used for pulp paper are cash crops that are environmentally unfriendly. These plantations destroy native forests, ecosystems and biodiversity, yet oil palm plantations have been expanding in Thailand at an average rate of roughly nine percent per year.

Northern hillsides bear the scars of deforestation.

Better enforcement Thailand has an impressive list of laws on the books to protect forests but enforcing them or prosecuting those who break them remains a challenge.

Habitats under threat The loss of natural habitat is the biggest threat to wildlife in the world today. Areas like the Western Forest Complex are breeding grounds and lairs for some of Asia's most endangered species.

Vested interests The state has frequently been lax about conducting transparent Environmental Impact Assessments (EIAs). In many cases, the companies sponsoring the EIAs are the same ones that stand to benefit from this or that particular project.

Degraded forests Part of Thailand's reforestation plan revolves around reviving "degraded forests" – areas where there have been fires or periods of drought. The theory is sound, but the reality is that "degraded forests" are often sold to the highest bidder for plantations and resorts.

Community forestry More community forestry projects need to be cultivated. When discussing the government's reforestation policies in the wake of the 2011 floods, officials said the plans could not succeed without increasing local input.

Integrated approach As experts like Louis Verchot have said, forestry needs to be integrated into other research fields like agriculture to bring about a more holistic approach that involves all stakeholders.

On the front lines According to statistics from the Department of National Parks, Wildlife and Plant Conservation, as many as 50 forest rangers have been murdered in the line of duty and 49 others injured, 23 of them seriously, since 2009. The main culprits are poachers hunting for rare species or valuable trees like Siamese rosewood. Rangers deserve higher pay, better equipment and less hostile working conditions.

Blood timber Illegal logging along the Thai-Cambodian border is bringing Thailand's rare Siamese rosewood forests near extinction. China's demand for luxury furniture is driving deforestation, increasing logging by 850 percent in the past several years despite the trees being on a protected species list.

Confiscated illegal rosewood.

FURTHER READING

- *Forest Guardians, Forest Destroyers*, by Paul Fanlol, 2013
- *An Overview of Thai Forestry*, by Ken Black, 2011
- "Sustainable Developments in Thai Forests," United Nations, 2014
- "Thailand's Deforestation Solution," by Evan Gershkovich, World Policy Blog, www.worldpolicy.org/blog/

TIMELINE

Groundbreaking Developments in Thai Forestry

1000 AD

Long before the kingdoms of Sukhothai and Ayudhya are enthroned, farmers work together to manage water and other resources in a pioneering display of what comes to be dubbed community forestry.

1956

The Thai Forestry Organization is formed to let the government exert more commercial control over Thai forests.

1970–76

Political instability in the ranks of the militaristic governments allows timber poachers free reign to decimate the countryside and mountains. Some estimates put the percentage of illegally felled timber coming out of Thailand at 60 to 70 percent of the total during this time.

2005

Thailand's first national park of Khao Yai is given UNESCO World Heritage Site status as the Dong Phaya Yen–Khao Yai Forest Complex for its phenomenal range of flora and fauna.

2011

Massive floods kill more than 800 people and inundate the country for months on end. The surging waters could have been alleviated to some extent by forest watersheds. As a result, the government announces that it would spend some three billion baht on reforestation projects.

1890

By this time British companies hold the majority of teak leases in Thailand, which also act as a bargaining chip that King Chulalongkorn uses to keep the French, beckoning from the nearby borders of Indochina, at bay.

1896

The British bring with them modern forms of forestry management. From 1896 to 1925, a Brit served as the chief conservator for the newly established Royal Forestry Department (RFD).

1976–80

After the massacre of pro-democracy protestors at Thammasat University on October 6, 1976, many flee to the jungles to join the Communist Party of Thailand, which wages guerilla warfare from rural bases, where the authorities set fire to a number of forests to smoke them out.

1990

To slow the rate of deforestation, the government introduces new policies like the Forest Plantation Act and the Enhancement and Conservation of National Environmental Quality Act.

2013

Environmental activists march some 250 kilometers from the proposed site of the Mae Wong Dam in Nakhon Sawan province to Bangkok. Their complaints that the project would harm the Huai Kha Khaeng Wildlife Sanctuary and the Klong Lan National Park in Kamphaeng Phet province are heeded by the authorities who put the project on hold.

"I think the days of forests for the sake of forests is over, and now we have to have forests integrated into other research agendas. For example, we have groups that deal with forests and groups that deal with agriculture and that's not productive. We have to bring them all together in landscapes in an integrated way to make them useful to each other."

Louis Verchot, principal scientist for the Center for International Forestry Research

OCEANS AND SEAS

New waves of development bring prosperity, tourists and challenges

The story of Thailand's oceans, seas and coasts captures the tensions of sustainable development in a clamshell. On the one hand, the rich natural resources of Thailand's coastal areas have been instrumental in driving positive growth and development, especially since the 1970s. Oil and natural gas reserves, the enormous tourism industry and one of the world's biggest seafood industries are now the basis of millions of jobs and are all major contributors to the modern-day Thai economy.

In their wake, however, these developments have left a significant level of environmental degradation, which Thailand now finds itself struggling to address adequately. According to the SDG Index, Thailand's "Ocean Health Index" scores, which cover the health of its waters, biodiversity and fisheries, are all firmly middle-of-the-pack when compared to the rest of the world. In addition, 43 percent of fish stocks have collapsed or are over-exploited. Overall, Thailand does score higher than regional averages. But the issues it faces illustrate one of the common global scourges in the rush for growth: environmental exploitation. In short, wherever there are natural and financial riches at stake, expect unsustainable practices fueled by short-term opportunism.

Thailand's 22 National Marine Parks, encompassing more than 6,000 square kilometers, are among the most beautiful tropical sites in the world and studies in biodiversity, with coral reefs, mangrove forests and seagrasses inhabited by endangered species like the dugong (or sea cow), marine turtles, manta rays and whale sharks. But the authorities have rarely strictly regulated the tide of tourists or the construction of tourism infrastructure. As a result, biomass in and around many parks has declined due to this encroachment as well as illegal fishing.

Islands like Koh Chang, Koh Samet, Koh Lanta, Koh Phi Phi and the archipelagos of the Surin and Similan island chains, which all reside within the confines of supposedly protected parks, are inundated with tourists. Only the latter two, because they are closed during the six months of the monsoon season from May to November, have truly retained their pristine nature. But tourism is far from the only culprit in what is a multi-pronged issue, not just for Thailand but for many other countries surrounded by water.

Around the world, **ocean acidification** is contaminating marine ecosystems, just as overfishing has decimated fish stocks. (Marine biologists believe that 90 percent of the bigger species like tuna have been fished out already.) Destructive fishing methods such as bottom trawling have destroyed many coral reefs, the world's second most productive ecosystems after rainforests, or substantially reduced the **coral cover**, and the purse seine nets that some industrial trawlers use scoop up loads of other marine creatures known as by-catch. Coral bleaching and the dip in bigger marine creatures like mantas and whale sharks have threatened the kingdom's reputation as one of the world's best dive spots. Coastal erosion has also been taking a serious toll. According to Thailand's Marine Department, of the country's more than 3,200 kilometers of coastline, around 670 kilometers

Ocean acidification:

At least one-third of the carbon dioxide emitted from cars, factories and buildings is absorbed by the oceans, which raises their acidity while lowering their ability to safely absorb more carbon. This process of acidification has had a profoundly negative effect on marine life like corals.

Coral cover:

The amount of stony coral that exists on a reef. It's both the reef's main building block and an important habitat for many marine creatures.

Koh Phi Phi Leh is part of a National Marine Park near Phuket in southern Thailand.

are experiencing severe erosion with land being lost at a rate of more than five meters per year.

Compounding these challenges is the high level of industrial activity along the coasts. In previous decades, discoveries of natural gas and oil in the Gulf of Thailand fueled the industrialization of the country's Eastern Seaboard and the growth of the nation as a whole, with the state-run Map Ta Phut Industrial Estate and deep-water port in Rayong province built to host petrochemical companies and other heavy industries. Unfortunately, since 2004, there have been more than a dozen oil spills and leakages that have caused significant damage. As with so many other environmental issues either on land or at sea, the long-term effects of these pollutants on their respective environs and the creatures that live there are as yet unknown.

With the fate of the the contentious Southern Seaboard Development Plan still in limbo, concerns abound that the resource-rich waters off the coasts of Songkhla and Satun provinces could be damaged by a new wave of development. As alternatives to Map Ta Phut, the new deep-sea ports could be much-needed sources of employment and revenue. But locals worry that the mega-projects may threaten two of the region's biggest earners – fishing and tourism – displacing many community members who rely on these sectors for their livelihoods.

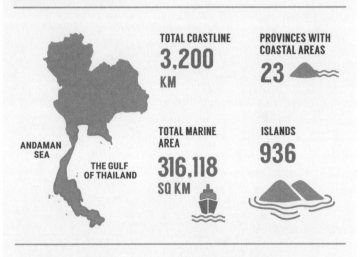

MARINE RESOURCES

ANDAMAN SEA

THE GULF OF THAILAND

TOTAL COASTLINE
3,200 KM

PROVINCES WITH COASTAL AREAS
23

TOTAL MARINE AREA
316,118 SQ KM

ISLANDS
936

Andaman Sea Coast is characterized by deep oceanic waters and a narrow, rocky and coral-reef-associated continental shelf, with a thick mangrove belt protecting the coastline.

The Gulf of Thailand has a shallower profile with a combination of mangrove forests, mudflats and sandy beaches. It is highly productive for fishing due to its shallow depth and high influx of nutrients and freshwater from regional rivers.

Source: Department of Marine and Coast Resources; Greenpeace

Why Oceans and Seas Matter to Sustainable Development

Soaking up carbon Oceans and seas are the biggest "carbon sinks" on the planet, meaning they soak up a quarter of all the carbon dioxide emitted by cars, factories and cities.

Net profits Fisheries and aquaculture are two of the country's economic lifelines with Thailand ranking number three out of the world's top seafood producers. Among the country's food exports, prawns and canned tuna are both in the top five.

Food security More than three billion people globally depend on the oceans for their primary source of protein.

Shelter from storms Seagrass beds, coral reefs and mangroves provide sanctuaries for marine animals and are some of the most biodiverse and productive ecosystems on Earth. They also enhance nutrient circulation in the ecosystem, filter wastes and protect shorelines from the intensity of water currents.

Tourism hotspots Many tourists visiting Thailand make a beeline for the country's fabled islands, caves and beaches. According to government statistics, in 2016 Thailand's international tourism revenue reached 1.6 trillion baht, while tourism arrivals reached 32.6 million.

HABITATS UNDER THREAT

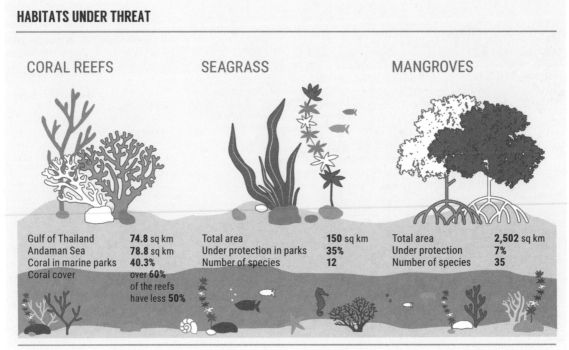

CORAL REEFS

Gulf of Thailand	**74.8** sq km
Andaman Sea	**78.8** sq km
Coral in marine parks	**40.3%**
Coral cover	over **60%** of the reefs have less **50%**

SEAGRASS

Total area	**150** sq km
Under protection in parks	**35%**
Number of species	**12**

MANGROVES

Total area	**2,502** sq km
Under protection	**7%**
Number of species	**35**

Source: Greenpeace

Development Wave for the South?

For more than two decades, the government has been trying to develop an alternative to the Map Ta Phut Industrial Estate in Rayong province to generate income and jobs for the south. For the residents of Satun and Songkhla provinces waiting for the controversy-plagued Southern Seaboard Development Plan to go ahead, Map Ta Phut's own record of creating hazards to human health and the environment has created unease. Yet such development schemes are

deep-sea ports on either coast in the provinces of Songkhla and Satun.

Locals fear that if the project goes ahead it will open the floodgates to more such schemes that will decimate fisheries and tourism – the area's biggest revenue spinners and sources of employment.

The Pakbara area is both a spawning ground for marine creatures and a launching pad for embarking on odysseys to the Tarutao National Marine Park, which encompasses

The Pakbara area is both a spawning ground for marine creatures and a launching pad for odysseys to the spectacular Tarutao National Marine Park. Locals fear that if the development project goes ahead it may cause damage to the area's biggest revenue spinners and sources of employment: fisheries and tourism.

Greenpeace Mini-Marathon to oppose the coal-fired power plant and seaport in Krabi.

nothing new in the south. Since the late 1990s, locals have protested these energy projects (mostly oil, gas and coal) largely to no avail. The net result tends to be that the government pushes ahead with such endeavors.

The proposed Southern Seaboard Development Plan is one of the most ambitious schemes to date. Part of it entails building a "land bridge" that will link the Andaman Sea with the Gulf of Thailand to facilitate trade with neighboring countries and the rest of Asia. That part of the project will necessitate constructing two

51 islands in the Andaman Sea. The vast marine park is the first such park in Thailand. Experts have estimated the park to be home to roughly 25 percent of the world's fish species. It is also where the ever-popular Koh Lipe is located, serving as a boon in tourism revenue for the local community.

According to Sakanan Plathong, a researcher at Prince of Songkla University, the new ports have engendered widespread skepticism on the part of locals. "From what has happened at Map Ta Phut, people here do

not trust the state's ability to prevent environmental problems," he said.

Fishing is another quandary. Up and down the coasts, development projects have threatened this traditional occupation, thanks to waste water discharges that pollute the seas. The small-scale fishing communities have suffered the most because their low-tech vessels cannot compete with the bigger trawlers, according to Greenpeace.

Following the construction of the two ports, industrial estates will be built to house petrochemical factories and other such heavy industries reliant on energy from newly constructed power plants. Local residents, who have spoken at public forums, are worried about the potential impacts on the environment from all of these mega-projects.

Even though the economic incentives may be tempting enough to lure some of the better-educated fishermen or young factory workers away from their usual occupations, Sakanan pointed out that most of the locals, who are undereducated and lack the skills to capitalize on these opportunities, are unlikely to benefit much from any of the new projects.

Forced Labor on the High Seas

"The use of trafficked labor is systematic in the Thai fishing industry. The industry would have a hard time operating in its current form without it."

Phil Robertson, deputy director of Human Rights Watch's Asia division

The Thai fishing boats plying local, regional and international waters contain riches in their cargo holds that have made the country into a captain of the seafood industry but, in a grave irony, are staffed by some of the poorest of the poor. The vast majority of the fishermen on Thai trawlers are actually migrant laborers from Myanmar and Cambodia caught in a net of exploitation.

Recruiting men like these to work in the fishing industry began in 1989 after Typhoon Gay ravaged parts of Southeast Asia and India, racking up a body count of some 800 people (many of them fishermen) working in the Gulf of Thailand. In the typhoon's wake, Thai brokers began smuggling foreign men into the country to take up jobs on fishing trawlers that many Thais would no longer do.

A 2011 report from the International Organization for Migration (IOM) revealed that many migrants arrive in the country to seek work through middlemen.

While some got work in food processing factories or in other menial positions, many ended up on long-haul fishing trawlers. Having no contracts and owing huge debts to the brokers, the men were sold to the captains or boat owners and forced to work on their vessels for months or years at a time, trawling all the way from Indonesia to India and as far away as Somalia. At sea, many were resold to other captains to prevent the men from escaping back to the mainland and reporting crew leaders to the police or other authorities.

The report estimated that some 300,000 migrants are employed in the fishing industry, accounting for 90 percent of the sector's workforce. Some 17 percent can be classified as forced labor. Even though about 50,000 fishing trawlers are registered, there are half that many unlicensed vessels or those lacking proper registration (known as "ghost boats") at sea. For the migrants staffing them, the potential occupational hazards are 20-hour shifts, beatings, malnutrition and, in some cases, execution-style killings.

Thanks to their efforts and sacrifice, the seafood business is an important economic driver for Thailand, pulling in some US$7 billion per year. According to the Food and Agriculture Organization, Thailand is one of the world's top five exporters of fisheries products with the US, UK and EU as the main markets.

In theory, migrants are protected by labor laws and regulations that prohibit human trafficking. But in reality existing legislation offers migrants scant protection and no recompense. State agencies overseeing this sector often lack the resources and manpower to stop exploitation and trafficking, or they collude with the culprits.

In 2014, *The Guardian* reported that some Thai authorities were not just lenient but also complicit in such abuses. The negative publicity from that exposé, and other damning reports by news agencies such as the Associated Press, put a dent in bilateral relations with the US, which downgraded Thailand to Tier 3 (the lowest ranking) on the State Department's "Trafficking in Persons" index in 2014. In the middle of 2015, the EU followed suit, giving Thailand six months to address the problem of illegal fishing or face a potentially devastating seafood embargo.

To its credit, Thailand has shown some improvement in addressing the issue, announcing the implementation of a new system for registering previously undocumented migrant workers that in theory would also protect them from gross injustices and prosecute human traffickers. The kingdom also introduced a system to track fishing vessels. As a result, the US bumped Thailand back up to Tier 2 in 2016, but not without a significant degree of grumbling from rights monitors, many of whom feel that the kingdom could do far more to protect migrant workers.

Some point out that doing so would also be in Thailand's best interests. According to IOM, the sustainability and profitability of Thailand's fishing sector depends on it making systemic changes that will protect nomadic fishermen and promote a fairer, more carefully regulated industry.

FURTHER READING

- *State of World Fisheries and Aquaculture: Contributing to Food Security and Nutrition For All*, by the Food and Agriculture Organization of the United Nations, 2016
- *Oceans in the Balance, Thailand in Focus*, Greenpeace, 2013
- *Coral Planting in Thailand*, by Chayanun Kongfahpakdipong, 2016
- "Trafficked into Slavery on Thai Trawlers to Catch Food for Prawns," by Kate Hodal and Chris Kelly, *The Guardian*, June 10, 2014

REALITY CHECKS

Undercurrents of Concern in Local Waters

Better coordination Laws and policies on marine and coastal resource management and planning are not in sync, resulting in ineffective and unsustainable development that fails to address such major issues as coastal erosion.

Out of their depth Limited research on oceans and seas means that bureaucrats possess insufficient knowledge to draw up and implement proper legislation.

Industrial weakness Despite the existence of many laws and regulations, the wastewater released by many industrial plants often exceeds the permissible limit.

Guardians of the seas Some 12,000 sea gypsies, or *Chao Lay*, who for

Rayong's Koh Samet after the oil spill in August 2013.

A sea gypsy catches fish with a spear.

generations lived a nomadic life on the Andaman Sea and now often stay on islands, face pressure to settle elsewhere. Like many traditional communities, their livelihoods have been disrupted by industry.

Finite fish Illegal fishing and overfishing have caused a dip in fish stocks. In 1961, fishing capacity was 298 kilograms per hour compared to only 17.8 kilograms per hour in 2011.

Coral collapse An estimated 96 percent of coral reefs are threatened by destructive fishing methods. Rising sea temperatures are leading to coral bleaching.

Unregulated labor The labor shortage in the fishing sector has been addressed through exploitation of migrant labor instead of better remuneration and stricter laws to attract more locals who currently see the profession as unsafe and ill-paid.

Teeming with tourists The presence of so many tourists and tour operators has taken its toll on coral reefs, and other sensitive marine environs, across the Gulf of Thailand and the Andaman Sea.

Marine stewardship To manage national marine parks, the authorities must shift from an exploitative culture to a more stewardship-like model of sustainable development.

Tide of litter Garbage, such as plastic water bottles and Styrofoam boxes, is increasingly finding its way into Thailand's coastal waters.

BIODIVERSITY

The biggest threat to flora and fauna is humankind

Wherever you go in Thailand, it is virtually impossible to ignore the country's incredible biodiversity: the rich variety of animals, trees and plants combined with the wild range of ecosystems, landscapes and habitats.

Scientists estimate that between five to eight percent of all known animal and plant species on Earth are found here. The list includes 10,250 kinds of vascular plants, 1,010 species of birds and 336 species of mammals, according to the Convention on Biological Diversity Fifth National Report – Thailand 2014. Amongst these strange and rare creatures, Thailand boasts the world's smallest bat and possibly the tiniest mammal, known as the Kitti's hog-nosed bat (*Craseonycteris thonglongyai*) or bumblebee bat, which weighs a mere two grams.

This diversity is due, in part, to the country's location at a bio-geographic crossroads between the Indochinese region in the north and the Sundaic region to the south. The problem is that this biodiversity is under constant threat from deforestation, urban encroachment, pollution, the poaching of exotic species and climate change.

In some ways, you could argue that the loss of flora and fauna is inevitable given Thailand's rapid transformation from a largely agrarian nation to a Newly Industrialized Country. Until the mid-20th century, the majority of Thai people lived in small communities in harmony with nature. The forests provided them with a constant source of food, clean water, firewood and herbal medicine, instilling a profound sense of wonder and respect for their surroundings. Villagers worshipped Mae Phosop, the goddess of rice, as well as the spirits who inhabited the trees and the mountains.

But reverence for those natural riches has not been enough to prevent the loss of wildlife or the felling of forests. Starting in the 1960s, vast tracts of jungle were cleared to make way for commercial crops, such as rice, cassava and sugar cane. Between 1961

Clockwise from top left: Forest frog, Kitti's hog-nosed bat, banteng, pitcher plant, and a lotus flower with bees.

Endangered pangolins are hunted for their meat and scales; the latter is used in Chinese medicine.

and 2014, forest cover shrank from 53 percent of the country to 31.5 percent. The impact of deforestation on wildlife populations has been startling. Elephants, which numbered in the range of 100,000 individuals at the beginning of the 20th century, have fallen to around 3,000–3,500 in the wild today, while the population of tigers is estimated at less than 250.

Because forests act as carbon sinks, this rapid deforestation has also contributed to Thailand's rising greenhouse gas emissions. Meanwhile, factories illegally dump waste into the country's once pristine rivers whilst mangroves have been cleared to make way for shrimp farms or luxury hotels, posing threats to marine diversity. Without more concrete actions, this cradle of biodiversity could very well reach a **tipping point.**

The real question is: Can Thailand change its mindset to put sustainable development ahead of short-term profit before it is too late?

The country's National Biodiversity Action Plan, covering the period from 2015–2021, sets lofty goals. Among the most important are a reduction in the rate of habitat loss by 50 percent, the improved protection of threatened species, and a concerted campaign to better educate the public about the importance of preserving biodiversity throughout the nation. There are also plans to increase forest

coverage to 40 percent of the total land area within 10 years. Collaborative efforts involving wildlife officials, NGOs and local communities have also demonstrated that multi-stakeholder initiatives can successfully curb poaching and deforestation.

Many inside and outside government know these measures are essential, but there is often a gap between good intentions and implementation. Will these initiatives be enough to protect Thailand's creatures and ecosystems from the onslaught of commercial development and climate change?

Tipping point:

The term refers to the critical juncture at which the collapse of biodiversity could irreversibly change life on Earth and threaten the survival of many species, including humans.

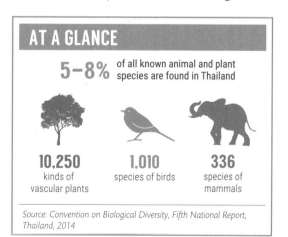

AT A GLANCE

5–8% of all known animal and plant species are found in Thailand

10,250 kinds of vascular plants

1,010 species of birds

336 species of mammals

Source: Convention on Biological Diversity, Fifth National Report, Thailand, 2014

Why Biodiversity Matters to Sustainable Development

Sustaining life Healthy ecosystems where the food web of plants and animals interact are essential for the world's future. While industrial-scale development brings short-term profits, biodiversity ensures the long-lasting health of our planet by providing us with clean air and water, fertile soil, sources of sustenance and a stable climate.

Source of livelihoods Many rural communities depend on forests and waterways – and the fish and animals that live in them – for their work, diets and livelihoods.

Combating climate change Scientists may disagree on the extent to which biodiversity loss causes climate change, but the overwhelming majority still see it as a key weapon in the fight against global warming.

Healing power More than 1,157 plant species in Thailand are reputed to have medicinal properties. Many are still used in traditional medicine. Now pharmaceutical companies are carrying out research to determine if some of these could serve as key ingredients for the wonder drugs of the future.

Deep gene pool Biological diversity is the key to life. A strong and diverse gene pool ensures a strong and diverse range of species that dwell in healthy ecosystems.

Fascinating Facts and Stats About Biodiversity

- The World Wildlife Fund's Living Planet Index, which tracks more than 10,000 representative populations of mammals, birds, reptiles, amphibians and fish, has revealed a 58 percent decline in world animal populations from 1970 to 2012. WWF warns that we could see a two-thirds decline in the half-century from 1970 to 2020 unless we act now to "reform our food and energy systems and meet global commitments on addressing climate change, protecting biodiversity and supporting sustainable development".

- A 2013 study on global biodiversity by German biologist Peter Uetz estimates that man has discovered 99.8 percent of all mammal species together with 99.9 percent of all bird species. By contrast, he postulates that we have identified a mere 20 percent of all insects and 31 percent of crustaceans. But how many species are there on Earth? Using new analytical techniques, a groundbreaking study published in the journal, *PLOS Biology*, puts the total number of species on Earth at 8.7 million, of which close to 90 percent have yet to be discovered. Such is the richness of global biodiversity.

Great hornbill at Khao Yai National Park.

- Thailand boasts an array of natural attractions and refuges that nurture biodiversity. In total, the country has 128 national parks, 60 wildlife sanctuaries and 60 non-hunting areas.

- Hotspots refer to areas of outstanding yet threatened biodiversity. Conservation International (CI) has identified 35 hotspots around the world, which comprise just 2.3 percent of the Earth's land surface, but support more than half of the world's plant species together with an estimated 43 percent of bird, mammal, reptile and amphibian species. Scientists working with CI argue that by conserving these hotspots, we can protect the Earth's biodiversity with only limited financial resources. Critics of the plan say it is man attempting to play god.

FURTHER READING

- *World Wide Fund for Nature Thailand Annual Report, 2016*
- *Cost of Implementing National Biodiversity Strategies and Actions in Thailand: A Conceptual Approach,* UNDP, 2016
- ASEAN Centre for Biodiversity, www.aseanbiodiversity.org

REALITY CHECKS

Challenges Facing Our Natural Ecosystems

The Malayan sun bear is the smallest bear species.

Habitat loss Thailand has suffered severe habitat loss as a result of urban encroachment. Such shrinking habitats are widely recognized as one of the primary forces driving the extinction of many species. A logging ban has been in place since 1989, but government laws to safeguard the nation's forests and biodiversity have not yet been sufficiently enforced.

Poaching High prices for rare animals and plants drive a thriving, fast-growing black market for endangered species. Sadly, Thailand remains one of the major wildlife trafficking centers in Asia.

Toxic agriculture The widespread use of pesticides, herbicides and other agricultural chemicals can wreak havoc on plant and animal populations. The good news is that there are plenty of sustainable alternatives to these chemicals. These can protect crops against known diseases whilst providing a livelihood for farmers and their communities.

End of the line Overfishing has drastically reduced fish stocks, especially in the Gulf of Thailand, with some scientists warning that the country's marine ecosystem is on the verge of collapse. While a new Fisheries Act passed in 2015 promises to reform fisheries management and control illegal fishing, without proper enforcement it is likely to prove as ineffective as the previous act.

Invasive species Thailand is threatened by so-called "invasive species" which are not native to a specific location, but can take over the habitat of indigenous species. The Global Invasive Species Database lists 138 invasive species in Thailand ranging from the Caiman crocodile to the hispid palm leaf beetle. Article 8(h) of the Convention on Biological Diversity calls on member governments to "prevent the introduction of, control or eradicate those alien species which threaten ecosystems, habitats or species."

Energy demand Thailand's search for new sources of energy to offset the declining production of oil and gas could unfortunately put vital ecosystems at risk.

Raising awareness Better education and more public awareness are vital if Thailand is to create a greater understanding about the necessity of biodiversity. Students must be taught to appreciate the immense value of nature rather than to exploit it.

In 2017, Thailand seized over three tons of pangolin scales with an estimated value of US$800,000.

"We should judge every scrap of biodiversity as priceless while we learn to use it and come to understand what it means to humanity."

Edward O. Wilson, biologist, professor and author

URBANIZATION

Because cities consume a large amount of resources, they must be managed efficiently

A popular Thai saying claims that Bangkok is Thailand. Indeed, as the hub of the kingdom's politics, commerce, media and much more, the capital is what is known as a "primate city," dominating over the rest of the country. With no other city boasting more than 500,000 residents, Bangkok accounts for 80 percent of Thailand's total urban area, according to the World Bank, and is the fifth-largest urban area in East Asia. The capital's glitz and prosperity may be envied upcountry, but the drawbacks of this urbanization – the crowds, the pollution and the traffic – are just as notorious.

As the city expands – it's now home, temporary or permanent, to around 9.6 million people – Bangkok has also become the country's capital of consumption. A third of the electricity generated in Thailand goes to feed the power-hungry needs of urbanites, with shopping malls alone devouring more electricity than all of Cambodia, and Siam Paragon complex using twice as much electricity

as Mae Hong Son province. Moreover, Bangkok's carbon dioxide emissions per person are more than ten times those of a Northeasterner and the city consumes more water than the rest of the country combined. According to the Asian Green City Index of 2011, Bangkokians use 340 liters of water per person per day, higher than the 276-liter average of residents in the 22 other Asian cities surveyed. That means much of the city's water has to be pumped in from upcountry reservoirs, shortchanging farmers who need it for their crops. Because of a lack of wastewater treatment plants, the discharge (mostly untreated) goes straight back into the Chao Phraya River and various canals.

Air pollution is another perpetual hazard. While Bangkok's air quality has improved and ranks better than many other major cities, the country needs to reduce the concentration of its fine particulate matter (or "PM2.5"), according to the SDG Index. In a 2013 study, the Pollution Control Department (PCD) claimed that Bangkok has the third worst air quality in Thailand, ranking behind the district of Na Pralarn in Saraburi province, home to the country's biggest gypsum factories, and the Map Ta Phut Industrial Estate on the Eastern Seaboard.

The city's sanitation capacity has also been pushed to its maximum capacity. The excess refuse must be dumped at landfill sites in nearby provinces, such as Samut Prakan, where locals have protested after big fires at garbage dumps left an acrid, eye-watering stench in the air for weeks.

All these symptoms are suggestive of poor urban planning. Instead of expanding northwards, as city planners have recommended, Bangkok's mid-section continues to widen, spilling over into its eastern fringes as urbanites follow the expansion of the Skytrain. Condos, housing estates and malls spring up in their wake. This haphazard pattern of urban development is being replicated across the country. Urban encroachment in floodplains and agricultural areas in Nakhon Ratchasima, for example, was

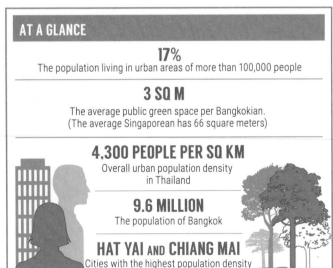

AT A GLANCE

17%
The population living in urban areas of more than 100,000 people

3 SQ M
The average public green space per Bangkokian.
(The average Singaporean has 66 square meters)

4,300 PEOPLE PER SQ KM
Overall urban population density in Thailand

9.6 MILLION
The population of Bangkok

HAT YAI AND CHIANG MAI
Cities with the highest population density

Source: World Bank Report, 2015

People walk across Chong Nonsi pedestrian bridge in Bangkok, which is home to at least 9.6 million people.

blamed for exacerbating the epic floods of 2011. Up in Chiang Mai, still without a popular public transport system, traffic jams are a growing problem and air pollution is rising to the forefront of ecological issues.

All these cities have essentially emulated the **ribbon development** style. This model, pioneered in the UK in the 1920s and '30s, spawned both urban sprawl and suburbs as people acquired cars that enabled them to live farther away from the town center. As a result, land prices in the middle of the city shoot up beyond the means of the majority, so the middle classes and the poor have to move out to the city's outer limits and suffer through long commutes in heavy traffic. This is also why Bangkok has such an excess of motorcycles and cars. Of the 34.5 million private automobiles registered in Thailand almost a quarter of them are in Bangkok. But the city has far too little road area – only 10 percent compared to 23 percent in Tokyo or 38 percent in New York City – to handle the number of vehicles.

That said, there are some bright spots on the horizon. In 1999, the Skytrain, or BTS system, got online after more than 20 years of plans derailed by successive governments and corruption. The BTS was followed by the MRT, or subway, five years later. Both of those mass-transit systems are being expanded now to cover the city's outer reaches.

The BTS and MRT may allow commuters the chance to rise above or duck under the city's gridlock for short intervals, but it's impossible to ignore Bangkok's other gritty realities and pressing problems for too long. In many areas, the city's sidewalks are in need of an upgrade and more shade, while green spaces are also sorely lacking.

As long as the capital remains the biggest city of opportunity, political power is centralized and material benefits and decent jobs are kept within city limits, it is likely that this slapdash style of urbanization will prevail. Vested interests and political interference will continue to upend proper planning, suspend projects and, worst of all, prevent other municipalities from following a better model of more sustainable urban growth.

Given all these challenges, what's the best solution? According to Associate Professor Ariya Arunin, who works in the Department of Architecture at Chulalongkorn University, one possibility is to upgrade the Division of Town and Country Planning, currently under the Department of Public Works and Town and Country Planning, into a full-fledged national think tank devoted to devising policies, programs and laying down plans for the whole country, dispelling the notion once and for all that Bangkok is Thailand.

Ribbon development: This term, which came into vogue in the 1920s, describes the pattern of developing houses, buildings and infrastructure projects along transportation routes. It is most often cited as the cause of unregulated urban sprawl.

According to Plan? Proper Blueprints Needed

In looking for the underlying causes to the unique issues facing Bangkok, such as too few parks and too many cars, policymakers and activists have typically fingered the same culprit: town planning, or a lack thereof.

Basically a blueprint for how a city should be developed, these master plans are drafted in bigger Western countries or Asian nations like Japan and China years before projects are constructed. But in Thailand the town plan is not completed until after the projects are up and running, if there is any such plan at all.

That makes it easy for corrupt businessmen and politicians to get in on the ground floor of such developments, said Srisuwan Janya, who works as a pro-bono lawyer for communities affected by environmental woes. In 2014, he helped residents in Soi Ruamrudee in Bangkok to win a case against a high-rise building, because the developers had not constructed the building according to the town plan.

This is commonplace, he said. "Development policies in Thailand are carried out in piecemeal efforts. Many development plans are there not because they should be, but because landowners and investors want them to be there. Roads are built to follow in the path of real estate speculators and those influential people who will get benefits from the construction. Then we have many buildings and infrastructure projects built in the wrong environment which leads to various problems such as traffic congestion and floods."

A lack of planning and overdevelopment are the causes of many environmental woes in Bangkok, where roads block drains, swamps and canals have been filled in for commercial properties, and zoning laws are often ignored. It is also a key reason Bangkok has such notorious traffic. In 2016, Bangkok drivers spent an average of 64.1 hours stuck in traffic jams making Thailand's capital the 12th most congested city

A resilient Bangkokian uses a damaged phone booth.

of 1,064 surveyed, according to the INRIX Global Traffic Scorecard.

It's not that the authorities have never paid any attention to urban planning (as far back as 1935 the government first attempted to draft a town plan), it's just that any such attempts to draw them up rarely go according to the plan. In fact, the first town plan for Bangkok was not drawn up until 1960 by an American firm.

The few town plans in existence today expire after just five years. Meanwhile, if a new plan has not been put in place then developers can apply to construct new projects with little in the way of scrutiny.

Across the country, only nine provinces have drawn up master plans to allocate land and aid development, wrote Ratchatin Sayamanond, former director-general of the Department of Public Works and Town and Country Planning, in an article. The rest of the kingdom's 77 provinces have no such plans in place and little interest in drawing any up in the near future.

The Bang Bua Canal in Bangkok.

FURTHER READING

- *Urbanization in Thailand*, by Jeff Romm, Ford Foundation, 2009
- "Regional Rapid Growth in Cities and Urbanization in Thailand," by Thanadorn Phuttharak, *Journal of Arts and Humanities*, Vol 3, No 1, 2014
- *Suburban Nation: The Rise of Sprawl and the Decline of the American Dream*, by Andres Duany, Elizabeth Plater-Zyberk, Jeff Speck, 2001
- *Urbanization Trends in Asia and the Pacific*, by UNESCAP, 2013

REALITY CHECKS

The Road to a Cleaner, Happier Bangkok

Capping car emissions Promoting cleaner fuels in addition to raising the taxes on older vehicles responsible for a higher level of emissions could help to clear the air.

Better zoning Zoning laws designed to keep factories and other commercial developments from encroaching on communities need to be enforced.

Rusty pipes Many water pipes in Bangkok need to be fixed or replaced because they have a leakage rate of around 35 percent, well above the average 22 percent rate of the 22 Asian cities included on the Asian Green Index 2011.

Treating wastewater The Bangkok Metropolitan Authority (BMA) needs to build more wastewater treatment plants and enforce existing laws on the discharges from factories.

Cutting energy consumption Energy conservation should be encouraged through the use of clean energy sources like solar panels, while people should be discouraged from using too much air-conditioning.

Green lungs Provide tax incentives for building owners who preserve big trees and create new green spaces when they develop their projects.

Urban planning Extending the expiry date of town plans from five years to 10 years or more would allow for more strategic long-term planning and better ensure the plan's sustainability. It is also of crucial importance to prevent any further revision or amendments to the plans that allow for questionable projects to be built during their implementation period.

"Improving the quality of data to understand trends in urban expansion is important, so that policy makers can make better-informed decisions to support sustainable communities in a rapidly changing environment, with access to services, jobs and housing."
Marisela Montoliu Munoz, director of the World Bank Group Social, Urban, Rural and Resilience Global Practice

POLLUTION AND WASTE

Few issues have such direct bearing on
human health and the environment as these two

At the heart of all definitions of sustainable development is the idea of leaving the world a better place for future generations. Few issues generate alarm and galvanize people into action more than man-made disasters because they suggest our legacy may be the opposite. Incidents such as Three Mile Island in the United States in 1979 or Chernobyl in the former Soviet Union in 1986 revealed the possible prospect of leaving our children with a negative burden. Oil spills, poisoned rivers and toxic leaks leave in their wake indelible images of human tragedy and tarnished ecosystems.

In Thailand, air, water and noise pollution as well as the disposal of waste have been major concerns for decades. A growing population means higher consumption and increased waste to manage. In addition, the economy's increasing reliance on manufacturing and industry means more people are living in proximity to factories and dump sites.

Illnesses and disabilities suffered by ethnic Karen living by the lead-contaminated Klity Creek in Kanchanaburi province, exposed since the late 1990s, put a spotlight on the devastating consequences of pollution and led to a landmark ruling in 2013 instructing the authorities to act.

In the northern provinces the burning issue is the smoky haze that descends during the hot season every year, which blurs vision, clogs up lungs and causes fits of coughing that leads, in extreme cases, to terminal diseases like lung cancer. From January to July 2016, the five cities with the highest annual average concentrations of PM2.5 in Thailand were Chiang Mai, Lampang, Khon Kaen, Bangkok and Ratchaburi. Seven of the eleven cities measured did not reach the National Ambient Air Quality Standard annual limit of 25 µg/m3 for PM2.5 and all 11 cities measured failed to meet the World Health Organization's (WHO) guideline of an annual limit of 25 µg/m3, according to Greenpeace Southeast Asia.

Too often the farmers burning rice straw and husks become the scapegoats, when in fact forest fires sparked by the hot-and-dry season are partly to blame, as are the shifting weather patterns that make the air stagnant at that time of year, and the big agricultural conglomerates planting monoculture farms of maize and other products. By the Asia Foundation's 2014 estimate, in the district of Mae Chaem in Chiang Mai province alone, some 37,000 tons of corncob waste are burned each annum.

Down south, it's a different strain of air pollution that has locals in a chokehold. The smoke blowing over from Indonesia, especially northern Sumatra, comes from the clearing of forests for oil palm plantations and the pulp paper industry. These black clouds have cast a pall over diplomatic relations between half a dozen different ASEAN members, showing how pollution and waste issues, like climate change, know no borders and can become serious regional issues.

A 2014 report released by WHO revealed that air pollution could be linked to some seven million deaths around the world in 2012 as an indirect cause of strokes, heart attacks and cancer. Of the victims, some six million resided in the poorer and middle-income countries of South and Southeast Asia, as well as the Western Pacific regions. Many of the deaths were related to indoor air pollution such as soot and smoke from coal or wood-coal stoves. But there are other factors to consider. Dr Carlos Dora, WHO Coordinator for Public Health, Environmental and Social Determinants of Health, said of the report's findings, "Excessive air pollution is often a by-product of unsustainable policies in sectors such as transport, energy, waste management and industry. In most cases, healthier strategies will also be more economical in the long term due to healthcare cost savings as well as climate gains."

To be fair, the Thai government has notched up some notable successes in putting such policies into action. In 2005, the Pollution Control Department (PCD) recommended a drastic lowering of the

Volunteer prison workers remove sewage and plastic waste from drains outside Bangkok.

permitted emission rates for two-stroke motorcycles in order to eliminate them from the streets. The plan worked. Since it would be too costly for manufacturers to rebuild the motorcycles to comply with the new standards, they stopped making them. To fill the market gap, manufacturers then made four-stroke motorcycles more affordable.

At the beginning of 2012 another PCD recommendation came into law as Thailand adopted Euro IV standards for fuel quality that cut levels of sulfur dioxide and vehicle emissions. Just as the law allowed commuters to breathe a little easier, the government rolled out a huge rebate program for first-time car buyers in 2012 that flooded the streets with over one million new vehicles, basically counteracting the previous effort. Unfortunately, this type of inconsistency in policy is common around the world.

Capitalism and consumerism have increased the use of plastic goods, which, after being discarded, are increasingly finding their way into the oceans. Thailand ranks sixth globally (of 192 countries) in the mismanagement of plastic waste, which has resulted in 1.03 million tons of plastic debris sinking to the bottom of the ocean along the country's coastlines, according to a study published in *Science Magazine* in 2015. Another study, conducted by the Ocean Conservancy and McKinsey Center for Business and Environment, puts Thailand in the top five of worst offenders, along with China, Indonesia, the Philippines and Vietnam, who, together, account for 60 percent of plastic pollution in the oceans.

In terms of fresh water, the quality in many of Thailand's main rivers and canals has dipped, with contamination caused by the discharge of untreated wastewater. One of the key challenges is that the kingdom has only 101 wastewater treatment plants to accommodate around 2,500 municipalities. The former director of the PCD, Supat Wongwangwattana, whose recommendations to other ministries paved the way for new laws cutting down on motor vehicle emissions, said that money for these wastewater treatment plants must come from government coffers. But maintaining them should be the responsibility of the municipalities who must collect fees paid by local communities or businesses, under the **polluter pays principle** of international environmental law.

This approach, he said, of providing economic incentives or disincentives, as the case may be, is the only way to get the public interested in recycling and reducing their own household wastes. Appealing to the bottom line is a stratagem that can also pay dividends when dealing with companies and plants that illegally dump their hazardous waste products. However, until these types of programs and policies are implemented, waste will continue to pile up.

Polluter pays principle:

One of the most basic and globally accepted tenets of environmentalism is that whomever causes the mess should pay to have it cleaned up.

The Secret Costs and Visible Impacts

Red tape The Pollution Control Department (PCD), under the Ministry of the Environment and Natural Resources, has little power to implement its recommendations. Instead it must coordinate with other ministries, like Transport and Industry, that may face competing interests from the private sector.

Illegal dumping The illegal dumping of toxic waste is a major issue in and around industrial areas. According to the Department of Industrial Works, plants produce some 30 million tons of waste per year, out of which three million tons are classified as hazardous. According to ThaiPublica, an investigative news site, of the total hazardous waste, only 900,000 tons were properly managed and trackable, suggesting much of the rest may have been dumped illegally, likely in the northeast and east.

Providing incentives Public campaigns for the 3R's (Reduce, Reuse and Recycle) are not enough. Incentives, such as buy-back programs for electronics goods, fees levied on individual households for waste, or charging extra for plastic bags in convenience stores, could make individuals more responsible.

Northern haze The smog floating over the northern provinces every February through April is a hazy issue. Technology exists to help farmers use agricultural refuse as fertilizer in their fields, but without government subsidies for programs or equipment, the issue will not be cleared up.

Promoting SEAs Strategic Environmental Assessments (SEA) are helpful in weighing possible impacts before a plan or program reaches the policy level. The SEA is different from an Environmental Assessment Impact (EIA) in several important ways, such as assessing alternatives and getting environmental authorities involved at the screening stage. Signed into law by the European Commission in 2001, the SEA has yet to gain traction in Thailand.

Polluters must pay Thailand's 101 wastewater treatment plants have to pay their own operating and maintenance costs by collecting fees from users. Yet very few do, leaving plants in a bad state of disrepair.

Public skepticism A lack of outstanding governance of industrial estates and other big plants has eroded the public's confidence in both bureaucrats and corporations. It has also made people skeptical about future mega-projects.

Intimidation In Chachoengsao province, which has 13 toxic dumpsites, the Nong Nae community has fought for years against the impacts of toxic waste. But two community members have been killed since 2010.

Help the garbage men Most of the separation of trash for recycling is done by garbage men as they make their rounds. If people pitched in before they took out their trash, this process could be more efficient.

E-waste More than 590,000 tons of e-waste were discarded in Thailand in 2015. These surplus electronics end up buried in e-wastelands in the northeastern provinces of Buriram and Kalasin, waiting to be scavenged for precious metals, while contaminating the environs with toxic substances such as cadmium and lead.

The power of trash The Ministry of Natural Resources and Environment is dealing with growing landfills by building waste-to-energy plants. So far, there are 23 such facilities and the ministry plans to build nine more.

Law enforcement Strict laws regarding industrial estates, mining concerns and other pollution-heavy enterprises are in place, but proper enforcement remains a challenge.

In northern Thailand, the burning of farmland has led to an annual haze and pollution crisis.

TOXIC HOTSPOTS

As a result of increasing industrialization and mineral extraction, Thailand faces rising concerns about health impacts from pollution in numerous sites around the country.

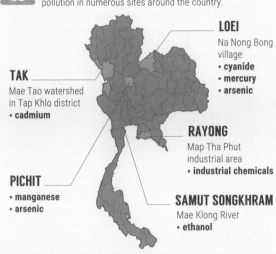

LOEI
Na Nong Bong village
• cyanide
• mercury
• arsenic

TAK
Mae Tao watershed in Tap Khlo district
• cadmium

RAYONG
Map Tha Phut industrial area
• industrial chemicals

PICHIT
• manganese
• arsenic

SAMUT SONGKHRAM
Mae Klong River
• ethanol

Source: Human Rights Watch

THAI WASTELAND

1.15 kg per day
The average waste produced by each Thai

3.9 billion per year
The number of plastic bottles of water consumed by Thais

81%
of the total 2,490 waste disposal sites nationwide are considered substandard

61 million per day
Styrofoam boxes used by Thais

Up to **75%**
Amount of household waste that can be recycled

Of the almost
27 million tons
of household waste produced in Thailand

64% of that rubbish was food

Source: Pollution Control Department, Food and Agriculture Organization of the United Nations (FAO)

Improving waste management in China, Indonesia, the Philippines, Thailand and Vietnam can **reduce global plastic pollution in oceans by**

45% by 2025

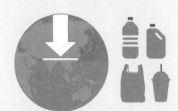

Raising collection rates across China, Indonesia, the Philippines, Thailand and Vietnam to 78% would lead to a

23% reduction in plastic leakage into the oceans.

Improved collection infrastructure and plugging collection gaps can **reduce annual leakage by nearly**

50% by 2020

Source: Ocean Conservancy

SPOTLIGHT

Ground Zero for Industrial-Strength Waste

The discovery of natural gas in the Gulf of Thailand in the early 1970s fueled the plan to build a massive industrial complex and port on the Eastern Seaboard, though it did not get off the drawing board until the fifth National Economic and Social Development Plan (1982–1986).

Map Ta Phut opened in 1990. Not only the largest industrial estate in Thailand, it's one of the biggest petrochemical hubs in the world. The industrial port was designed to accommodate heavy vessels hauling factory equipment and massive cargoes to hone the country's competitive edge, to bulk out production capacities, and to facilitate exports.

It worked. The estate has been extremely successful in monetary terms and created tens of thou-

The controversial Map Tha Phut Industrial Estate in Rayong.

In the long term, the fallout from these scandals may have more beneficial outcomes, as locals have awakened to the dangers posed by these estates and are now banding together to make their voices heard, just as the companies are realizing the significance of their corporate social responsibility to local communities.

sands of jobs. Locals and NGOs, however, have raised concerns about pollution, claiming that some 2,000 people died of cancer and other diseases in the decade after Map Ta Phut opened. The corporations contested those claims. But a lawsuit, launched by 27 villagers in 2009, after years of protests achieved nothing, confirmed some of their worst suspicions. In a move that startled the business sector and galvanized the country's eco-advocates, the Supreme Court upheld the suspension of 65 of 76 projects under scrutiny due to inadequate Health Impact Assessments (HIA).

The ruling was a landmark decision. In a *New York Times* story about the case. Anthony Zola, an American environmental consultant, said of Thailand, "In rural areas, there is almost no enforcement at all. Water pollution, air pollution, noise pollution – you can make all the complaints you want, and no one pays any attention to you."

The victory was short lived. Within a year the court ruled that 74 of the 76 projects could proceed. Three years later, a fire broke out at one of the petrochemical plants, killing 12 and injuring 129. The factory was closed down and residents of many nearby communities evacuated. But business at the other factories carried on as usual.

These kinds of calamities are not unique to Map Ta Phut and occur around the world. But they have led industrial parks all over the country to come under criticism and scrutiny. In 2013, a government study undertaken in the Ta Thoom community of Prachinburi province revealed high levels of mercury – a dangerous heavy metal – in villagers' hair and the fish in the river that runs through the 304 Industrial Park. The park's spokesperson denied any wrongdoing, saying that all their efforts to manage and treat the waste products were carried out in accordance with the law. Government officials sent out to investigate the claims of malfeasance sided with the business owners.

In the long term, the fallout from these scandals may have more beneficial outcomes, as locals have awakened to the dangers posed by these estates and are now banding together to make their voices heard, just as the companies are realizing the significance of their corporate social responsibility to local communities.

In what is perhaps a positive sign of the changing times and recognition of the need for action, the PTT Group, SCG, Dow Thailand and the Glow Group formed the Community Partnership in 2014 with the goal of turning Map Ta Phut into a greener, eco-industrial town by 2018.

FURTHER READING

- "Transboundary Pollution in Northern Thailand Causes Dangerous Levels of Smog" by Gennie Gebhart, The Asia Foundation.org, 2014
- *Thailand State of Pollution Report 2015*, by the Pollution Control Department
- "The Sun Is Still Shining at Map Tha Phut," National Health Foundation, 2010
- *Domestic Wastewater Treatment in Developing Countries* by Duncan Mara, 2009

REASONS

Why Reducing Pollution and Waste Must Be a Priority

Future burdens Polluting the soil, air and water, or illegally dumping toxic wastes, are crimes that claim many potential victims in the long run and also take their toll on future generations through the damage done to the environment and public health.

Global warming After ratifying the Kyoto Protocol in 1999 and pledging in 2015 to uphold the Paris Agreement, Thailand has an international obligation to do its part to cap emissions and keep the world's temperature from rising by more than two degrees Celsius.

Regional flashpoints Pollution and waste issues can transform from local to regional issues. The smoke from burning crops, for example, can travel across borders and create political tensions.

Health hazard Air pollution is now the biggest environmental risk to human health, claiming 5.5 million lives a year according to 2016 data from the Global Burden of Disease Study. Air quality has a direct impact on our health and can lead to strokes, coronary diseases, lung cancer and acute respiratory infections.

Quality of life Few issues have such a direct effect on the public's quality of life, whether rich or poor, urban or rural, and whatever the ethnicity. A government or corporation's decision-making upstream can have a direct impact on the livelihood of the farmer, worker, student, or person from any other walk of life downstream.

Cutting-edge technology The very nature of sustainable development depends on controlling pollution and containing waste by employing advanced technologies and cutting-edge concepts, such as zero-waste management. Under the zero-waste management model, production and distribution systems are drastically restructured to reduce waste from the start, rather than implementing management practices at the end. Ideally, the system, like in nature, finds a useful purpose for all products.

Thick smoke billows from a fire blazing through a rubbish dump on Phraeksa Road in Samut Prakan province.

"*In rural areas, there is almost no enforcement at all. Water pollution, air pollution, noise pollution – you can make all the complaints you want, and no one pays any attention to you.*"

Anthony Zola, environmental consultant

THE THAI
ECONOMY

Poverty

Agriculture

Manufacturing

State-owned Enterprises

SMEs

Tourism

Finance

Trade

Transportation

Labor

Competitiveness

Corruption

Thailand's economy may be at a turning point. A "business as usual" approach is likely not the best strategy moving forward. Farmers and many others are saddled with one of the highest rates of debt in the region. The kingdom's manufacturing sector, which transformed the country from a developing nation to an upper-middle-income one, faces difficult challenges. With a lack of skilled workers necessary for more high-tech industries and the emergence of cheaper labor in neighboring countries, Thai manufacturers are at risk of losing their competitive niche. The country's many state-owned enterprises are enormous contributors to GDP but not as efficient as they might be, and its SMEs are always clamoring for more government support.

Across the board, labor is in need of an upgrade: improved training in vocational occupations, English and so-called 21st-century skills, such as technology literacy, are necessary for Thailand to maximize the great potential of its people. Although the kingdom has the second lowest unemployment rate in ASEAN, this is largely due to the staggering number of workers in the "informal sector."

Tourism, one of the Thai economy's mainstays, has proven remarkably resilient. But as the country nears the 35-million mark in arrivals, its attractions – the coral reefs, mountains, forests and traditional culture – that inspired this boom must be protected from tourism's harmful footprint.

In Thailand's financial sector, bankers and economists learned a great deal from the 1997 financial crisis. Since then they have implemented a system of checks and balances that has created strong fundamentals and more immunity to risk. This is part of the reason why Thailand remains an attractive place for foreign investors, even as corruption and the lack of political stability conspire against investment.

Overall, sustainable development opportunities abound in Thailand and the values that drive them could provide a rich way forward for the country. The resourceful and entrepreneurial spirit of the kingdom's population as well as its modern infrastructure, business-friendly environment, and geographic position at the crossroads of trade routes, make it an excellent launch pad for sustainable development initiatives, and for commerce and innovation in general.

POVERTY

Leaving no one behind is Thailand's next big challenge

Poverty eradication has always been a key pillar of the sustainable development agenda, which is an inclusive model of progress that seeks to improve the lives of everyone. Mired in the daily struggle of making ends meet, the poor, in general, have less access to healthcare and educational opportunities. They are the most vulnerable to natural disasters and their voices and needs are often ignored in the name of larger economic growth.

Fortunately, in Thailand, such extreme poverty is no longer a pressing issue. According to the SDG Index, Goal 1 ("No Poverty") is the lone Sustainable Development Goal that Thailand has already achieved. One key reason for this is that the immense economic development the kingdom has enjoyed over recent decades has been generally inclusive, providing jobs and improving standards of living.

Initiatives old and new such as The Doi Tung Development Project and Wanita Social Enterprise show how Thailand is committed to **Leave No One Behind** and has even tackled poverty in remote areas, targeting particularly vulnerable populations and providing them with the kind of knowledge and skills that allow them to thrive in their unique local environments without disrupting their traditional culture. This philosophy of helping people to become self-reliant has informed Thailand's poverty eradication efforts for decades.

Leave No One Behind:

A key rallying cry and feature of discussions of the Sustainable Development Goals (SDGs). The idea is that "no goal should be met unless it is met for everyone."

Thailand is actually one of the world's great success stories in terms of poverty eradication, with only six out of 1,000 people (or 0.6 percent of the population) living below the poverty line in 2012. As defined by the World Bank, living below the poverty line means subsisting on less than US$1.90 per day (about 67 baht). Meanwhile, Thailand's Office of the National Economic and Social Development Board (NESDB) defines poverty as living off less than 2,572 baht per head per month (approximately US$2.50 per day). By this measurement, as of 2015, an estimated 7.3 million live in poverty or about 11 percent of the population, a figure that is still relatively low compared to other upper-middle income economies around the globe.

By materialistic standards, Thailand also appears to be doing well. In urban areas, the signs of consumer culture are as bright as neon: shopping malls are filled to the brim with luxury brands, omnipresent billboards advertise glitzy condos and car showrooms gleam with the latest deluxe models. Newfound disposable income in the pockets of once-poor laborers, which is often sent back to their families in the provinces, has hastened the rise of more materialistic lifestyles in rural areas as well. By the time of the economic collapse in 1997, some 90 percent of rural households had a TV set and 60 percent owned a motorcycle.

While this rise in the country's consumerist fortunes has been accompanied by a drop in the overall poverty rate, in reality, these aggregate figures conceal some hard truths. In particular, three groups of people are at the most risk of being left behind by this prosperity: single women, the disabled and the elderly. In rural areas especially, these groups are increasingly vulnerable and isolated. For the elderly in particular, the traditional safety net of being supported by relatives or taken in by the local temple no longer proves adequate in the absence of proper government programs.

There are also considerable and important differences in the incidence of poverty across subnational regions and demographic groups. For example, some 80 percent of the country's downtrodden reside in rural areas, a recent NESDB report noted. About 44.8 percent of the country's destitute live in the northeast, 26.5 percent live in the north, and around 13.6 percent reside in the south. Almost half of these impoverished households are engaged in the agricultural sector. The rest of the poor are part of Thailand's so-called "informal workforce," comprised of part-time employees, self-employed individuals, informal

Youth from economically disadvantaged households grow up having far fewer opportunities than those from well-off families.

small-and medium-sized enterprises (SMEs), retirees and landless laborers.

In rural areas, it has been found that residents can suffer a "cycle of deprivation." This theory suggests that the key ingredients of poverty, such as bad housing and a lack of access to quality education and employment opportunities are transmitted through families over generations, ensuring that future generations remain in poverty.

Despite Thailand's remarkable progress in joining the ranks of the upper-middle income economies in a brief space of time, the income and opportunity gaps between the haves and the have-nots persist. The question of how to bridge the urban-rural gulf through decentralization or other means and divvy up the spoils of Thailand's larger economic success more fairly and equitably remains one of the country's chronic challenges as it pursues a path toward further development. Many argue that addressing this conundrum is also essential if the country is to achieve stronger national unity.

"The educational and opportunity gap must be reduced in order for Thais to build trust among each other," Bank of Thailand Governor Veerathai Santiprabhob said in a 2016 speech.

POVERTY IN THAILAND IN PERSPECTIVE

12.7% OF THE WORLD'S POPULATION
0.6% OF THAIS

ARE LIVING ON
US$ 1.90 PER DAY (67 BAHT)

7.3 MILLION THAIS
11% OF THAIS (APPROXIMATELY)

ARE LIVING ON
US$ 2.50 PER DAY (87 BAHT)

Source: The World Bank, 2015; NESDB, 2015

MAPPING OUT THE KINGDOM'S POOR

NORTHEAST 21,094 BAHT PER MONTH

NORTH 18,952 BAHT PER MONTH

41,022 BAHT PER MONTH **BANGKOK** & 3 Neighboring Industrialized Provinces

MAE HONG SON has been Thailand's poorest province for several years.

Source: NESDB, NSO, 2015

Empowering an Aging Population

These days, declining fertility rates aren't just a phenomenon in wealthy nations. Thailand's birthrate has dipped dramatically from 7 children per woman in the 1970s to an average of just 1.5 children today, according to the World Bank. It's also the third-most-rapidly aging society in the world. That makes it one of the less wealthy countries facing the daunting challenge of a shrinking labor pool coupled with a greying population.

In Thailand, the official retirement age for government employees and people employed by state-owned enterprises is 60, while some private sector firms also enforce a compulsory retirement age. Due to a range of factors such as a dearth of surplus income or a lack of planning, most Thais do not set aside savings for retirement. Instead, they rely heavily on their children or local temples, which offer them refuge.

For decades, a hefty portion of the kingdom's rural working-age population has flocked to Bangkok to seek higher-paying work. Though jobs are abundant, younger Thais are still struggling economically. A consequence of this is that the long-held tradition of Thais caring for their elderly parents has declined. Indeed, the share of older persons relying on children as their primary source of income fell from 52 percent in 2007 to 35.7 percent in 2014.

While it is still common practice for children to send a percentage of earnings home to their parents, the amount often falls short of covering basic needs. Today about half of Thailand's elderly do not have a child living in the same village or municipality, and 16 percent have no living children to offer support. To make ends meet, a 2014 national survey found that 38 percent of older people still work regularly. Some 90 percent of these individuals work in the informal and self-employed sector.

In rural areas, the strain is particularly acute, especially among elderly farmers. Community leaders are trying to offset these shifts by devising new ways to make the elderly more self-sufficient. One such forward thinker, Mechai Viravaidya, has pioneered a project in Buriram province to help elderly individuals cultivate new skills, increase their productivity, find markets for their products and, as a result, earn more income. And the lessons have certainly taken root.

In local villages and on small informal cooperatives, enterprising seniors have adopted innovative techniques such as constraining the roots of lime trees in cement containers in order to make it possible for the trees to yield fruit year round. They've also built special shelters, where they grow cash crops like mushrooms and orchids.

The government is well aware of its population's aging issue and has made steady progress on developing policies to support older people. The Old Age Allowance (OAA) was revamped in 2009 to provide a modest, but nonetheless universal pension to Thais age 60 and over. The government has also supported the establishment of senior citizen clubs that act as self-help groups. By 2012, there were over 23,000 registered elderly groups with a total membership of 1.6 million. While it's a good start, experts say the government needs to increase spending on existing elderly support and pension systems. And time is running out. By 2040, Thailand's aging population is set to increase to 17 million, accounting for 25 percent of the population. That means that one out of every four Thais will be a senior citizen.

NGOs working on aging issues recommend that the government scale up programs that facilitate income generating opportunities for older people, modify the labor protection law to include older workers, and promote the practice of saving for retirement. Introducing a more flexible retirement age, creating incentives to hire individuals over age 60, and offering flexible hours and part-time arrangements for the elderly would also go a long way toward improving their station. Last but not least, the OAA benefit level should be increased and standardized by linking it to the nationally defined poverty line, and it should be regularly adjusted to reflect inflation.

Temples in Thailand have often acted as an unofficial safety net for the poor and elderly.

FURTHER READING

- *Human Development Report 2016: Human Development for Everyone,* by UNDP
- *Statistical Yearbook for Asia and the Pacific 2015,* by the United Nations Economic and Social Commission for Asia and the Pacific
- "Poverty, Income Inequality, and Microfinance in Thailand," by ADB Southeast Asia Working Paper Series, Asian Development Bank, November 2011
- *Ageing Population in Thailand,* by HelpAge International, 2017
- *Inequality and Injustice in Access to Resources and Fundamental Public Services in Thailand,* edited by Apiwat Ratanawaraha, 2013 (in Thai only)
- "Analysis Report on Poverty and Inequality in Thailand 2012," by Office of the National Economic and Social Development Board, Bangkok, August 2014 (in Thai only)

REALITY CHECKS

Challenges to Leveling the Playing Field

Primate city Being the center of Thailand's political and business culture as well as the largest market for jobs, Bangkok has long held inordinate sway over policy making and wealth creation. Decentralizing power and redistributing opportunities and development to rural areas has been much discussed though little headway has been made.

Wage security Farmers may be called the "backbone of the country" but they are not always highly rewarded for their efforts. While 40 percent of Thailand's people work in agriculture-related business and their contribution to overall GDP is 13 percent, the wages received are lower than other sectors and they remain vulnerable every year to climate change and uncertain weather patterns resulting in poor output.

Educational opportunities Only Thailand's rich can afford private education at the nation's best schools or have the means to send their children overseas for highly valued international university degrees. More equal access to better education would not only increase the skills of workers but also prompt an intergenerational transmission of educational achievement among households, creating wider access to better opportunities and helping to shrink the ominous wage gap.

Nepotism and patronage In many businesses and institutions, value is placed on connections,

social standing, and family name and background over quality and merit. So workers from poorer backgrounds face many entrenched obstacles to moving up the social hierarchy.

Social insecurity Poor Thais receive few social benefits from the state and many resort to the informal economy for their livelihoods, which generally offers little to no job security.

Spend first, save later Thailand's household debt is nearing 90 percent of total GDP. Attracted to easy credit and consumer goods and with less access to formal financial services, Thai families generally lack savings and strong financial planning, and are vulnerable to falling into poverty.

> *"Almost half of the population of the world lives in rural regions and mostly in a state of poverty. Such inequalities in human development have been one of the primary reasons for unrest and, in some parts of the world, even violence."*
>
> **A.P.J. Abdul Kalam, former president of India**

At the Thai-Cambodian border, itinerant traders pass by an ad for easy credit.

AGRICULTURE
The changing face of farming

Thailand's importance regionally and globally as an agricultural producer is significant. At the heart of that story is rice. One of the world's top producers and a leading exporter, the kingdom has long held the role as the rice basket of Southeast Asia. In 2015, for example, it sold an estimated US$4.7 billion worth of rice. Other important crops for the economy include natural rubber, sugar cane, cassava and tropical fruits.

Concerns abound, however, about the plight of the Thai farmer. Burdened with mounting debt and losing ground to faceless conglomerates, will the so-called "backbone" of the country benefit from Thailand's status as a Newly Industrialized Country? It is a question with not only economic implications – the agriculture sector contributes between 11 to 13 percent of GDP – but also political, social, environmental and cultural ones. While the number is down markedly from 76 percent of the workforce in 1966 to around 40 percent today, millions of Thais outside Bangkok still have livelihoods, incomes and ideas shaped by what happens on the farm. And some of the trends impacting these individuals are alarming.

All over Thailand, and other parts of the world, the challenges are similar. The first chapter of the contemporary tale was authored during the so-called "Green Revolution" of the 1960s and 1970s. Through the use of herbicides and pesticides, combined with more mechanization and better land management, scientists figured out how to increase crop yields. This farming revolution reaped plenty of rewards: many of the world's hungry were fed; and many of the world's destitute found employment. However, there have also been damaging side effects. The reliance on chemicals has been linked to health afflictions and ecological degradation. Many chemicals like DDT, banned in Western countries, are still used here. In 2005, Thailand imported 78,000 tons of farming chemicals. By 2015 that total had shot up to 173,000 tons. Lacking the proper schooling on application, Thai farmers have often over-used pesticides and over-fertilized their crops.

Bio-economy:

The bio-economy encompasses the production of renewable biological resources and their conversion into food, feed, bio-based products and bioenergy via innovative and efficient technologies. It is already a reality and one that offers great opportunities and solutions to a growing number of major societal, environmental and economic challenges, including climate change mitigation, energy and food security, and resource efficiency.

A Thai farmer uses a modern combine harvestor.

Moreover, the chemicals are not cheap. In particular, the rising price of fertilizers has driven Thai farmers deeper into debt. According to government statistics for 2015, the average debt carried by farmers was a staggering 237,000 baht. Many are also increasingly turning to unscrupulous lenders. In 2015, nearly 150,000 farmers borrowed 21.59 billion baht from loan sharks who are known to charge as much as 20 percent interest, according to the Provincial Administration Department. And clawing their way out of this debt is difficult. Lacking organizational clout, farmers have almost no negotiating power over the trade of their own goods. Terms are instead dictated by the international market price, controlling middlemen and conglomerates.

Attempts by various Thai governments to shore up the sector have occasionally resulted in blunders with some experts arguing that less government intervention, not more, is the solution. Nipon Poapongsakorn, a senior fellow at the Thailand Development Research Institute, who has also served as a consultant to the government on rice and other commodities, said, "The future of Thai agriculture is in big family farms, not government policies. If the government gets their way they will kill agriculture." Yet those family farms are threatened by the disinterest of the younger generation and the rise of agribusinesses, which are vertically integrated

and able to control the entire supply chain, providing farmers with seeds, chemicals, livestock and feed. These conglomerates also practice **contract farming** with farmers, who are often landless (only 29 percent own land titles). By renting land to them, companies cut their costs. They have what amounts to free laborers to work the land, and it is the farmers who become the most vulnerable to risk.

Is there a sustainable solution on the horizon? That's not entirely clear. In the mid-1980s, a loose amalgamation of NGOs joined to form the Alternative Agriculture Network, bringing together small-scale and organic farmers with cooperatives and like-minded groups. At their first forum in 1992 they defined "sustainable agriculture" as a more holistic approach toward "agricultural production and farmer livelihood that contributes to the rehabilitation and maintenance of ecological balance and the environment, with just economic returns, promoting a better quality of life for farmers and consumers and fostering the development of local institutions for the benefit and the survival of all human kind."

Such lofty ambitions may run counter to some ground-level realities. Nipon argues that farmers must take a more Darwinian approach to farming to adapt to both market and weather conditions in an age of turmoil and uncertainty. "Flexibility is the

> *"Someday we shall look back on this dark era of agriculture and shake our heads. How could we have ever believed that it was a good idea to grow our food with poisons?"*
>
> **Jane Goodall,** *conservationist and author*

key to sustainability," said Nipon. As an example, Thailand was once the biggest cassava supplier to the EU, but when that tax loophole closed, Thai farmers turned to China where they have come to dominate the market. Subscribing to that logic, if the price of rice is low, then farmers must turn to other crops, just as many rubber plantation owners are now switching to palm oil.

Another potential solution is innovation. Some experts believe Thailand can follow in the footsteps of European farmers and become a regional leader in the emerging "**bio-economy**," which is seeing rising demand for products such as biomass – organic matter that can be used as an alternative energy source. The government has also thrown its support behind the Food Innopolis Project, which aims to position Thailand as a global food innovation hub. How well such initiatives fare is certain to have repercussions all the way from the farm to Bangkok.

Contract farming:

A type of agricultural production in which the buyer and farmer reach an agreement stipulating the price and/ or quality and timeframe for a specific amount of an agricultural product to be delivered. Normally the buyer is an agribusiness that provides such things as seeds, fertilizers and technical know-how on credit to the small-scale farmer.

RICE FARMING AT A GLANCE

4.8 MILLION
RICE FARMERS IN THAILAND

 29% own land titles

 71% have a mortgage or rent

 4.8 rai average plot for rice farming

 25 rai average plot for double crop farming

 28,036 **baht per year:** average profit from rice farming, which is 51% less than Vietnamese farmers

Why Agriculture Matters to Sustainable Development

Cash harvest Between 2009 and 2016, the agricultural sector contributed between 11 and 13 percent of Thailand's GDP and employed roughly 40 percent of the country's population.

Seed hub Goals for Thailand's agricultural sector include pushing for the implementation of more biotechnology and building up Thailand's role as "the seed hub of Asia."

International reputation One of Thailand's most famous exports is its piquant cuisine, which depends on fresh ingredients, herbs and spices that are anything but garden variety. The kingdom's reputation as the "Kitchen of the World" depends on the stability and prosperity of Thailand's robust agriculture sector.

Food industry Around 80 percent of the ingredients used in the Thai food processing industry come from natural ingredients. With around 10,000 food-processing companies in Thailand employing some 800,000 people, this revenue stream is increasingly significant.

Climate change Climate change may negatively impact crop yields. Sustainable agricultural practices must be considered hand-in-hand with environmental threats.

Roots of Thainess So much traditional Thai culture, from festivals and performing arts, to textile designs, animism and music, revolves around farming. Like the Buddhist temple, the farm is also a pillar of village life and communities upcountry.

THAILAND'S TOP COMMODITIES BY VALUE

(in US Dollars)

RICE PADDY
9.43 billion

NATURAL RUBBER
4 billion

SUGAR CANE
3.17 billion

CASSAVA
2.21 billion

CHICKEN
1.8 billion

PORK
1.64 billion

EGGS
1.14 billion

TROPICAL FRESH FRUIT
0.91 billion

PINEAPPLES
0.76 billion

Source: Food and Agriculture Organization of the United Nations, data for 2012

"The future of Thai agriculture is in big family farms, not government policies. If the government gets their way they will kill agriculture."

Nipon Poapongsakorn, a senior fellow at the Thailand Development Research Institute

Growing the Seeds of an Agrarian Revolution

1910

King Rama V organizes first trade exhibition to promote agricultural goods and sends Thai students to study modern farming techniques in Europe and the US.

1917

Phraya Phojakara returns from Cornell University with a PhD to become the chief breeder at the Rangsit Rice Experiment Station, charged with carrying out the first agricultural research in the country.

1933

Thanks in part to Phraya's groundbreaking work, Thailand takes the top honors and 10 other awards at the World Grain Exhibition conference held in Canada.

1961

King Bhumibol names the first Thai-style tractor, designed by Debriddhi Devakul, the "iron buffalo" after it underwent extensive testing on the experimental farm at Chitralada Palace to seek sustainable solutions for farming dilemmas.

1962–1968

As the so-called "Green Revolution" sweeps the world, increasing crop yields through the use of pesticides and herbicides, more mechanization, better management and superior irrigation techniques, Thailand develops a new hybrid of rice that becomes a global success.

1969

The Royal Project, designed to replace opium with other crops suitable for the mountainous climate of northern Thailand, bears fruit in the form of strawberries, peaches, apples, tea, coffee, lettuce, cabbage and flowers. This acclaimed program has been emulated in parts of South America and more recently in Afghanistan.

1975

In the mid-1970s, growing cassava for animal feed helps shore up the struggling economy of Thailand's northeast where half of the country's farmers eke out a living from soil that is either too sandy or salty to be really profitable. Cassava, known as the "poor man's crop," requires little water and few fertilizers.

1980s

The agrarian sector is transformed by land shortages, urban migration, the loss of competitiveness in the global market, and unsuccessful government schemes, like importing cows from Australia which became known as "plastic cows" because they failed to breed under tropical conditions.

1991

Thailand takes over from Malaysia as the world's largest rubber producer and exporter. The year-round rainfall down south is ideal for rubber trees. Almost 90 percent of the country's rubber is exported.

1994

King Bhumibol unveils what is dubbed the "New Theory," an integrated approach to farming that leads to self-reliance.

2001–2006

As the first elected prime minister to serve out a full four-year term, Thaksin Shinawatra's success is largely dependent on mobilizing poor farmers from the northeast with such populist schemes as village loan funds. In spite of the advent of microfinance most farmers in this region are still small landholders with poor access to credit.

2007

Originally founded in 1995, the Organic Agriculture Certification Thailand (ACT) is the first and the only such body in the kingdom that offers certification recognized the world over. In 2007 ACT branches out to form a regional body to certify organic farmers in Southeast Asia and transfer knowledge.

2013–2014

The government's rice-pledging scheme, which offered farmers fixed prices for different kinds of paddy, crashes amid political turbulence and financial unsustainability. This leads to rice rotting in warehouses and many farmers going into further debt.

REALITY CHECKS

Challenges Facing Thai Agriculture

Stable prices The state must implement market-friendly policies to keep the price of agricultural goods stable.

Mounting competition As the ASEAN Economic Community (AEC) achieves a greater degree of integration than ever before, Thailand will have to stave off challenges from competitors like Myanmar, as well as remove trade barriers that will threaten its competitiveness.

Rural development Many farmers working the fields today can only afford to do it part-time and must migrate to Bangkok when the harvests are complete to make ends meet. Accelerating the pace of rural development and providing new drivers for the agriculture sector will provide an economic incentive for rural people to continue working on their countryside farms.

Aging farmers Only 12 percent of farmers are younger than 25 while their average age has risen from 31 in 1985 to 51 in 2010, according to the Thailand Research Fund. What is clear is that Thailand's younger generation has less interest in such fieldwork.

Climate change Agriculture, whether on a small or large scale, is vulnerable to the most pernicious aspects of climate change: rising seas, wild alternations in weather patterns, and increases in carbon dioxide can all have a significant effect on crop yields. Global warming also threatens livestock populations.

Growing demand Fair trade products are diversifying from coffee into other agricultural commodities such as sugar, fruit, tea and flowers. With better certification for farmers and more awareness among consumers, fair trade goods could morph into a market force to be reckoned with and make sizable contributions to sustainability on a ground level.

Quality control For quality control and competitiveness – both necessary for a healthy agriculture sector – a system of standards for rice paddies needs to be introduced and followed to the letter.

Farmer spraying insecticide.

Healthy dirt Soil quality is a serious concern. Agriculture, grazing, monocropping, plowing and deforestation are some of the main culprits to blame for the degradation of soil as it is leached dry of essential attributes. Crop rotations, managed grazing and other solutions are needed.

Low pay and landlessness A huge percentage of Thai farmers are landless or small landholders working part-time because they have to do other jobs to make ends meet.

Chemical overload Heavy use of agro-chemicals has increased dramatically in recent years, as imports rose from 102,000 tons in 2006 to 199,000 tons in 2013, dipping to some 173,000 tons in 2015. Discouraging their use through taxation – at present they are not taxed at all – while holding them to strict regulations about advertising their wares – at present they can make any claims they want – would be beneficial to the environment and human health.

Better R&D The amount of government spending for agricultural research amounted to 0.9 percent of the GDP in 2004. Now that figure has dropped to 0.2 percent.

Fall of rubber prices As the world's largest producer and exporter of rubber, Thailand's rubber farmers have been hit hard by falling prices. The price of the commodity dropped 27 percent in 2014 alone, only bouncing back in late 2016 due to a shortage caused by floods. The government is now encouraging them to cut down those trees and replace them with oil palms, which can be harvested in only three or four years, though they come with their own set of environmental hazards.

FURTHER READING

- *Thai Agriculture, Golden Cradle of Millennia,* by Lindsay Falvey, Kasertsart University Press, 2001
- *Thailand's Progress in Agriculture: Transition and Sustained Productivity Growth,* by Henri Leturque and Steve Wiggins, Overseas Development Institute (ODI), 2011
- "The Future of Farms," by Thana Poopat in *The Scientist* magazine, 2010

SPOTLIGHT

The Original Sustainable Development Project

Thailand's first real foray into "sustainable development" actually took place long before that term was even coined. Known as the Royal Project, it was created to help solve a complex and unique set of problems in a remote, opium-producing region of northern Thailand. These issues included border disputes, drugs, poverty among the hill-tribe people who lived in infertile areas, and deforestation due to the practice of slash-and-burn farming, which was negatively impacting the region's natural resources.

Thanks to the Royal Project, many of these complex problems have been alleviated. It all started formally in 1969, when King Bhumibol Adulyadej established and funded the Royal Project to help increase the feasibility of highland agriculture. The king's intention was to help Thai people living in mountain areas through job creation, to eliminate opium production through crop substitution, and to improve water resource restoration, which incidentally also impacted people living in other parts of the country.

In 1992, the Royal Project Foundation was set up as a legal entity and led by HSH Prince Bhisatej Rajani. Describing the project, the prince says: "Following His Majesty's advice, the Royal Project was to find cold weather plants to grow on the mountains. Back then, besides opium, no one knew what to plant. So we started a number of research projects, under which we conducted many experiments. Experiments entail manpower and money. As for manpower, there were many scientists specializing in agriculture who were ready to work for him. His Majesty himself

King Bhumibol Adulyadej in the northern highlands in the 1960s.

made working under the Royal Project less stressful by curtailing hierarchical procedures and eliminating unnecessary red tape. Research findings from the Royal Project can now be used as findings for teachings. We no longer have to rely on textbooks written by people from other countries."

Currently, the Royal Project Foundation has four research centers and 38 Royal Project Development Centres in Chiang Mai, Chiang Rai, Mae Hong Son, Phayao and Lamphun, working to develop appropriate plants and animal breeds for each specific area, to transfer knowledge to local farmers, and to help restore natural resources.

The numbers speak for themselves. More than 350 kinds of plants, vegetables and fruits have been developed. More than 30,000 families

or 150,000 people, including farmers from 13 hill tribes and urban areas, have joined the Royal Project, covering an area of 24,000 hectares. With a better management system and a more systematic allocation, more than 450 million baht returns to these farmers each year. On average each house is making 70,000 baht annually, which is more than ten times that earned from growing and selling opium.

In 2011, there were Royal Initiatives Project Centres covering an area of 272,000 hectares. Some 37,561 families, or 172,309 people, were able to sell agricultural products including vegetables, fruits, flowers, tea, coffee, processed goods and handicrafts of more than 1,702 types to markets in and outside the country, earning a total of 629 million baht.

MANUFACTURING

To continue the success story, the country must innovate

When Thailand suffered its worst floods in half a century in 2011, the world took notice, not just because widespread damage impacted millions of people, but also because they caused a global shortage of computer hard disk drives and a slowdown in production at Japanese car plants around the world. As *The New York Times* reported at the time, "The image of Thailand as a land of temples, beaches and smiles has over the years been reinforced by the country's tourism advertising campaigns. But the flooding has revealed to the world the scale of Thailand's industrialization and the extent to which two global industries, computers and cars, rely on components made here."

That growth seems all the more remarkable when you consider that the first real chapter in Thailand's industrial development opened with the establishment of the Board of Investment (BOI) in 1966. Seven years later, Thailand's first-ever industrial estate was set up just northeast of Bangkok and a concerted drive to promote investment for exports began in 1977, with the BOI as the cheerleader beckoning to foreign investors.

The late 1980s marked a significant turning point when a strong yen sent Japanese manufacturers in search of cheaper places to make their goods and cars. Thailand welcomed them with open arms. Commercial discoveries of gas in the Gulf of Thailand in the same decade provided the all-important foundation for the explosive growth in manufacturing that followed. This came in the form of a huge refining and petrochemical industry on the country's Eastern Seaboard, close to the offshore gas wells and a deep-sea port at Laem Chabang, together with the mushrooming of industrial plants churning out electronic components, machinery and cars.

It was textbook development. A country with a prodigious agriculture sector to feed itself – while exporting healthy surpluses – shifted much of its resources and people into strategically located manufacturing plants serviced by good roads and ports as well as pro-business policies. Thailand was a champion among what were called the NICs (Newly Industrialized Countries). Its pre-eminence in vehicle manufacturing among such countries even earned it the epithet "the Detroit of Asia." But in other crucial ways it has failed to move up the **value chain**.

Between 1987 and 1995 the Thai economy grew at an annual compound rate of 9.1 percent. Manufacturing's contribution to this outstanding achievement rose to 31 percent of total GDP by 1995, shortly before the Asian financial crisis of 1997. That crisis laid bare some of the less visible, noxious aspects of Thailand's manufacturing boom. Most notable was the fact that despite its impressive export performance, it was unable to generate net foreign exchange earnings. Throughout the 16-year boom from 1980 to 1996, the country suffered from a growing trade deficit – not least in manufactured goods – that contributed to a grotesque current account deficit of more than eight percent of GDP in each of the two years before the crisis. One of the main reasons was a persistent reliance on imported capital goods, intermediate inputs and technology.

Since the crisis, aided by a more flexible exchange rate, Thailand's manufacturing sector has continued

THAILAND'S MANUFACTURING SECTOR AT A GLANCE			
(out of 138 countries)			
16th	largest manufacturer of goods	**60**th	for R&D investment at 0.2% of GDP
24th	largest in export volume	**54**th	for innovation capability
14th	most competitive manufacturing nation	**63**rd	in technological readiness

Source: Thailand Board of Investment (BOI); World Bank Ease of Doing Business; Global Manufacturing Competitiveness Index; IMD World Competitiveness Yearbook 2013; World Economic Forum Global Competitiveness Report, 2016-2017; The Economist Pocket World in Figures 2016

Thailand must upgrade its key industries to move up the value chain.

to thrive. Today it accounts for about 30 percent of the country's US$395 billion GDP – by far its biggest contributor – and the country is now number 16 on the list of the world's top manufacturing countries.

But today, "the Detroit of Asia" is in danger of losing its foothold as a growing list of cheaper competitors with similar skills rise among the ASEAN nations. Notably, Thailand lags behind in research and development (R&D) spending. According to the World Economic Forum's Global Competitiveness Report 2016–2017, Thailand innovates less than countries with comparable education ratings – investment in R&D has been stuck for decades at just 0.2–0.3 percent of GDP. This low percentage is one reason Thailand's score in the SDG Index for Goal 9 was its second lowest for any of the Goals.

Recognizing the need to do better is why the kingdom is now pushing full-steam-ahead on Thailand 4.0, an economic model based on creativity, innovation and high-level services. Thailand 4.0 is designed to transform the kingdom into a value-based economy by reforming its existing major industries (i.e., automotive, electronics, medical and wellness tourism, food, agriculture and biotechnology); scaling up the development of new sectors such as robotics, digital, aviation, logistics, biofuels and biochemicals; and solidifying Thailand as a medical hub.

The blueprint for Thailand 4.0 singles out innovation, entrepreneurship, sustainability, community-led development and inclusive growth as all being essential to its success, which is all fine in theory. However, it remains to be seen just how fast Thailand's industrial model can undergo a makeover and whether the nation can do so while maintaining an adequate degree of policy continuity.

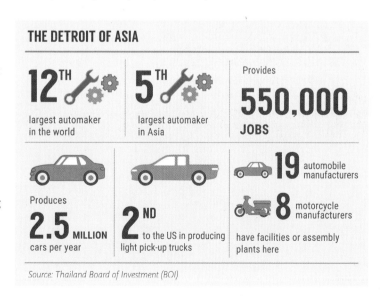

THE DETROIT OF ASIA

12TH largest automaker in the world

5TH largest automaker in Asia

Provides **550,000** JOBS

Produces **2.5 MILLION** cars per year

2ND to the US in producing light pick-up trucks

19 automobile manufacturers

8 motorcycle manufacturers

have facilities or assembly plants here

Source: Thailand Board of Investment (BOI)

Why Manufacturing Matters to Sustainable Development

Boosts development "Developing" used to be synonymous with "industrialization" because, as *The Economist* put it, "For most of recent economic history, 'industrialized' meant rich. And indeed most countries that were highly industrialized were rich, and were rich because they were industrialized." In the more globalized world we live in today, this equation is less precise and economists point to services and innovation as a new and necessary route to an improved standard of living. But there is no doubt about the vital part the manufacturing industry plays in lifting a country from subsistence and low incomes to greater prosperity.

Multiplier effect Manufacturing has what economists call a "multiplier" effect on overall economic activity. For every baht spent on making something, further spending goes on down the line – logistics, marketing, retailing, finance and other activities that generate jobs.

Positive ramifications Foreign investment in manufacturing brings with it skills and technology that spill over into the rest of the economy, helping to underpin and nourish an economy's know-how and proficiency across many other sectors, from finance to medical services. The extent to which this is achieved depends a lot on whether the "host" economy's policymakers use these opportunities to full effect or become complacently dependent on foreign firms to take the lead.

Global trade agreements Manufactured goods have been at the frontier of global trade agreements. Ever since the General Agreement on Tariffs and Trade (GATT) set about bringing down tariffs on industrial goods in 1949, world policymakers have viewed access to each other's markets for industrial goods as an indispensable boost to all countries' development and modernization prospects. Where manufacturing goods lead, other sectors will follow.

Thung Song cement plant of Siam Cement Group in southern Thailand.

"For the past 50 years, Thai manufacturing has developed under the circumstances of passive and slow technological learning of firms, ineffective and incoherent government policies, isolated education and training institutes, technologically unsupportive and risk-averse financial institutions, incompetent and politicized trade/industry associations, and an unfavorable institutional context."

Peera Charoenporn, "Thai Manufacturing at the Crossroads," professor of economics, Thammasat University

Promoting Public-Private Partnerships and R&D

The collective belief among foreign investors and firms operating in Thailand has long been that to actually promote substantial R&D activity, Thailand needed to do more than just offer tax breaks and duty exemptions on imported equipment. For starters, there needs to be dedicated R&D funding, and an environment where public-private partnerships can flourish.

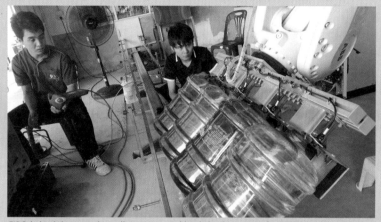

FIBO helped design and install an automated bottle loading system for a client in Buriram.

Typically only the biggest players in Thai industry (e.g., SCG, PTT) have been willing to assume the financial risk to adequately invest in R&D and innovation. Thailand's dearth of qualified researchers has not helped matters. For years the country has struggled to pair its limited number of researchers with companies that can fund their work and help them hone in on areas that benefit Thai industry. There's no question about the potential financial benefits such pairings can engender. World Bank data from a sampling of 66 countries shows that the average return on an R&D investment is 168 percent. But when you look at countries with below US$12,000 GDP per capita (such as Thailand, where GDP per capita was $5,775 in 2015), that return on investment skyrockets to 333 percent.

The good news is that companies interested in R&D, including Small and Medium-sized Enterprises (SMEs), are starting to get the support they need. In February of 2017, the government passed the National Competitive Enhancement Act for Targeted Industries (2017), which aims to enhance Thailand's competitiveness within the Thailand 4.0 core industry clusters by waiving a company's corporate income taxes for up to 15 years and offering a subsidy of 10 billion baht to cover investment expenses for R&D and innovation.

Meanwhile, the Ministry of Science and Technology recently set up Talent Mobility, a program that is designed to foster cooperation between the private sector and top university research programs and public research institutes in Thailand. The aim of the program is to encourage the development of applications for new technology in order to improve efficiency, productivity, and R&D capabilities. Talent Mobility has already proven beneficial at helping to match research projects to the needs of major private sector corporations and SMEs. In 2016, around 50 companies lent support to some 70 projects.

Thailand's National Innovation Agency (NIA) has also launched Innovation Coupon, which focuses on stimulating the development of innovative projects at SMEs. The initiative provides SMEs with a 100,000-baht coupon to cover the cost of a feasibility study for their intended project, which, if all goes well, qualifies them for an additional 400,000-baht coupon to support implementation of the project. Coupons can be used for the assessment of new technology; early stage R&D and prototyping; adoption and development of new products, processing technology, applications, practices, and/or operations; upgrading of existing products, processing technology, applications, practices, and/or operations. NIA's initial goal is to support more than 200 projects per year.

In addition to facilitating R&D, public-private partnerships can also prove invaluable in terms of preparing future workers for high-tech industries. One program that is taking advantage of such partnerships is the Institute of Field Robotics (FIBO) at King Mongkut's University of Technology Thonburi (KMUTT). Dr. Supachai Vongbunyong, assistant director of research and international affairs at FIBO, said in 2017 that he and his students were collaborating with companies big and small on projects such as robotics-related product R&D, industrial automation, the application of industrial robots in production lines and logistics.

"Project-based learning is one of the key mechanisms to prepare our students to be able to handle real world problems," said Dr. Supachai, adding that hands-on experience is key to producing graduates who are able to continuously adapt, develop new skills and think systematically.

Challenges Facing the Manufacturing Sector

Made in Thailand Thailand's first stage of industrialization in the 1980s and the first part of the 1990s succeeded because it was founded on low-cost, semi-skilled labor and a spirited welcome to foreign investors. A vigorous subcontracting sector grew up to serve the big names of the global garments, auto and electronics sector and make the components they specified. But the next stage has proved much more difficult. Thailand is trailing in research and development (R&D) spending, the foundation of the value-creating innovation needed to drive productivity, new technology, designs and create a competitive "home" brand. According to the World Economic Forum, Thailand innovates less than countries with comparable education ratings; investment in R&D has been stuck for decades at just 0.2–0.3 percent of GDP.

The "smile curse" The implementation of Thailand's industrial policies has been sporadic at best and counterproductive at worst. The reasons are those that plague other elements of the country's economic life – political and governance failures, lack of continuity, lack of coordination and what might be called the "smile curse." Like the "resource curse" (the fate suffered by countries that are rich in a natural resource, such as oil, and consequently neglect other, more sustainable aspects of their economy), the "smile curse" is Thailand's welcoming attitude to foreign capital. While it has attracted massive Foreign Direct Investment and brought rapid growth, at the same time it has led policymakers to neglect more self-reliant economic options, in particular the development of its own original design and brand manufacturing. One example is the electronics industry where Thailand has been a global leader in the manufacture of hard disk drives. But the world has moved on over the past few years: mobile and other consumer markets are far outpacing the market for such computer hardware, meaning the annual global demand for them has fallen by more than five percent on average since 2011. By comparison, annual global demand for flash memory solid state drives (SSD), in which Thailand has no comparable share, has risen by over 100 percent on average.

It must be said that the decision about whether to make these technologies in Thailand can't be left to foreign tech companies alone.

No longer cheap When Thailand's minimum wage was increased in 2013 to 300 baht (US$8.50) the country seemed to be officially conceding that it would no longer try to compete on costs in the global marketplace for manufactured goods. That raise lifted the country's average manufacturing wages past the Philippines, to around 50 percent more than those of Indonesia and Vietnam, and many times more than those of Cambodia and Myanmar. By the end of 2014, the average wages in Thai manufacturing were more than double what they were at the turn of the century, and way ahead of inflation during the same period. The minimum wage increase was very much a catch-up because Thai manufacturing labor productivity has outstripped real wages (adjusted for inflation) over the past decade and a half. But the broader question is whether the improved labor productivity has kept pace with other manufacturing economies.

Rubber sheet maker.

Quality control check.

Inspector at CP Foods.

FURTHER READING

- *Industrial Development Report 2016: The Role of Technology and Innovation in Inclusive and Sustainable Industrial Development,* by the United Nations Industrial Development Organization
- *Thailand: Industrialization and Economic Catch-Up,* by the Asian Development Bank, 2015
- The "Industry 4.0" issue, *Update Magazine,* by the German-Thai Chamber of Commerce, Q4/2016
- "Thailand in the AEC: Myths, Realities, Opportunities, and Challenges," by Somkiat Tangkitvanich et al, *Quarterly Review of the Thailand Development Research Institute,* 2013
- *Asian Development Outlook 2016: Asia's Potential Growth* by the Asian Development Bank

Governance matters No one could accuse Thailand's policymakers and academic advisers of failing to see what needs to be done. Governments over the years have set up ministries and institutions, passed laws and appointed experts on just about every conceivable issue related to the manufacturing industry and economic development. Among the many official bodies, there are the National Research Council of Thailand, the Ministry of Science and Technology, the National Science Technology and Innovation Policy Office and the National Science and Technology Development Agency. The problem, as with much else in public life in Thailand, is implementation. In addition to a bureaucratic labyrinth of corruption and vested interests, the lack of coordination and continuity that arises from the frequent shuffles and reshuffles that go on at the top of government – whether within one party, between coalition partners or changes of regime – tends to result in erratic and sometimes dysfunctional policy and governance decisions. This also leaves much of the direction of crucial parts of the economy at the discretion of the corporate strategy of foreign firms.

"The transition from resource-driven, export-led economies to more sustainable growth models based on human capital development, new technology and innovation will be a key challenge for many Asian countries over the next decade."

"Creative Productivity Index: Analysing Creativity and Innovation in Asia," a report by The Economist Intelligence Unit for the Asian Development Bank, 2014

ASEAN Economic Community (AEC) Some commentators have warned of grave threats to the Thai economy, but a more measured assessment by four leading scholars from the Thailand Development Research Institute (TDRI) set about to dispel what it said were four myths about the AEC. Based on their research, they concluded that there would be no significant changes in the way Thailand conducts business because it has already made the changes required. "Despite many rounds of negotiation, ASEAN has remained mostly just a free-trade area," they wrote, and de facto economic integration has already taken place. By contrast, the TDRI authors say that Thailand will become one of the main beneficiaries of the regional integration because it will be able to use neighboring countries as "extended production bases," as well as to leverage the AEC to attract FDI and to build regional brands for Thai products. But there are still hurdles to leap in the long run, as the "benefits from integration will not come automatically. Thailand must formulate and implement proper policies and take appropriate action to enjoy the benefits. Structural reform, especially the liberalization of the service sector, is a top priority. It is also important to promote responsible investment and sustainable development, and to nurture an international mindset among the Thai business community and the general public."

STATE-OWNED ENTERPRISES

The clash of public and private interests

The role that state-owned enterprises (SOEs) play in the course of a country's economic and social development is pivotal. In most countries, essential services such as electricity, water, telecommunications and transport are provided by state-owned companies during the early stages of economic development. The role of SOEs often becomes less prominent in more advanced economies where the private sector is sufficiently sophisticated to undertake large-scale and complex investment projects in lieu of their state counterparts.

Struggling to break free from the middle-income trap to become a high-income country, Thailand has to redefine the role of SOEs in keeping with the changing economic environment. Even today, the definition of an SOE most often referred to is in the Budget Act of 1959. The law stipulates that a state-owned enterprise is an enterprise in which the majority of the capital is held by the state, including state departments and government pension or insurance funds.

Currently, there are 55 Thai SOEs, not including their subsidiaries. According to the Budget Act of 1959 definition, these subsidiaries are also SOEs, but the State Enterprise Policy Office (SEPO), the official regulator of SOEs, does not regulate these.

Interestingly, the number of state-owned enterprises as declared by the SEPO has not changed much during the last two decades. That's because as the Thai private sector became capable of attracting investment without state support or partnerships, there really wasn't a need to create more SOEs. At the same time, only a handful of SOEs have been dissolved and not a single entity has been privatized. Major SOEs (namely, the Airports Authority of Thailand, the Mass Communication Organization of Thailand and the Petroleum Authority of Thailand) that became listed companies between 2001 and 2004 remain state-owned as only a minority equity stake is floated.

This almost unwavering SOE count conceals the real truth: that state enterprises are rapidly expanding through more than 300 subsidiaries. According to statistics, the total revenue of Thailand's SOEs increased from 1.8 trillion baht in 2004 to five trillion baht in 2015. This translates into an average revenue annual growth rate of around nine percent per annum – an enviable number given that the average GDP growth for this period was only around 3.65 percent. As a result, SOEs' revenue as a percentage share of GDP increased from 28 to 36 percent in just over a decade and they now employ almost 270,000 people.

In terms of profits, the top five for 2015 were: the Electricity Generating Authority of Thailand (EGAT), PTT, Krung Thai Bank, Government Savings Bank, and the Provincial Electricity Authority respectively.

So why should policymakers be concerned about revenue-generating, state-owned enterprises that provide so many jobs? Unfortunately, behind the glossy figures of profits lurk grave problems of mismanagement, inefficiency and even corruption in certain SOEs. The stellar financial performance is often a product of mere monopolistic power rather than superior management. Those that face direct competition from their private competitors often fare much worse than their counterparts because of the intrinsic problems mentioned.

AT A GLANCE

NUMBER OF STATE-OWNED ENTERPRISES	**55**
TOTAL REVENUE	**5** trillion baht
EQUIVALENT TO	**36.2%** of GDP
EXPENSES	**4.6** trillion baht
EMPLOYEES/STAFF	**268,162**

Source: State Enterprise Policy Office, 2015; Labour Relations Bureau

Suvarnabhumi International Airport in Bangkok is run by the Airports of Thailand (AOT), which listed on the stock market in 2002.

Reforms have been on the government's agenda since 1998 when the cabinet endorsed the first and, thus far, only comprehensive State-owned Enterprise Reform Master Plan. The plan envisioned the separation of regulatory functions from policy and operational ones, as well as separating monopolistic activities from competitive ones. To date, things have not gone according to the plan. There has been very little progress except for the establishment of two independent regulatory bodies for the telecommunications and energy sectors.

If reforms are still a pipe dream, there have been notable successes in the sector over the years. An SOE whose role is still as relevant to the development of the Thai economy as at its founding 50 years ago is the Bank of Agriculture and Agricultural Cooperatives (BAAC). The bank, established in 1966, provides loans to more than 5.6 million farm households today. With over 1,300 branches nationwide, it is one of the most ubiquitous central government agencies and certainly one of most knowledgeable when it comes to Thai farmers, who are its clients. The chief strength of the organization comes from its proximity to and familiarity with its clientele.

Even more prominently, state enterprises such as EGAT and the Provincial Water Authority have played important roles in providing citizens with essential services. Today, 99 percent of Thai households have access to electricity and 98 percent have access to improved water. The success of these enterprises may be linked to capable management. Most of the managers come from engineering backgrounds. They are attracted by the professional challenges and higher compensation compared with other civil servants, plus the prestige of public service. But in general, the role of SOEs has fallen prey to increasing skepticism, as the private sector can easily replace them. With almost daily reports of alleged corruption in various SOEs, the question of whether these enterprises still matter needs to be answered.

"I believe that eventually, one of the next big issues that will have to be addressed globally is the role of state-owned enterprises. If you can't compete fairly, honestly, effectively, no government should intervene."

Hillary Clinton, former US Secretary of State

115

STATE-OWNED ENTERPRISE REFORM

WHY REFORM IS NEEDED FOR STATE-OWNED ENTERPRISES

Expansion of SOEs over the last 10 years

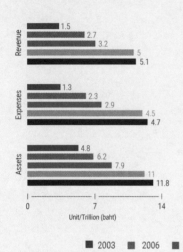

Asset volume among State Financial Institutes except Krung Thai Bank

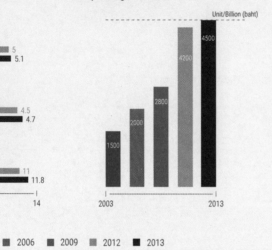

Comparison of Return on Assets: SOEs and Private Companies vs. SOEs Overseas During 2009–2013

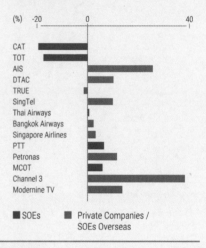

STATE-OWNED ENTERPRISES IN THAILAND

OFFICE OF THE PRIME MINISTER
- The Mass Communication Organization of Thailand
- The Thailand Research Fund

MINISTRY OF DEFENCE
- The Battery Organization
- The Tanning Organization

MINISTRY OF FINANCE
- The Government Lottery Office
- Thailand Tobacco Monopoly
- The Government Savings Bank
- The Government Housing Bank
- Krung Thai Bank Public Company Limited
- Bank for Agriculture and Agricultural Cooperatives
- Export-Import Bank of Thailand
- The Thai Liquor Distillery Organization
- The Playing Card Manufacturing Factory

MINISTRY OF TOURISM AND SPORTS
- Sports Authority of Thailand
- Tourism Authority of Thailand

MINISTRY OF SOCIAL DEVELOPMENT AND HUMAN SECURITY
- National Housing Authority

MINISTRY OF AGRICULTURE AND COOPERATIVES
- The Forest Industry Organization
- Rubber Estate Organization
- Fish Marketing Organization

- Dairy Farming Promotion Organization of Thailand
- Office of the Rubber Replanting Aid Fund
- The Thai Plywood Company Limited
- The Marketing Organization for Farmers

MINISTRY OF TRANSPORT
- Port Authority of Thailand
- The State Railway of Thailand
- Express Transportation Organization of Thailand
- The Bangkok Mass Transit Authority
- Thai Airways International Public Company Limited
- Airports of Thailand Public Company Limited
- New Bangkok International Airport Co., Ltd.
- Aeronautical Radio of Thailand Limited
- The Transport Company Limited
- Thai Maritime Navigation Co., Ltd.
- Civil Aviation Training Center
- Expressway and Rapid Transit Authority of Thailand
- Mass Rapid Transit Authority of Thailand

MINISTRY OF NATURAL RESOURCE AND ENVIRONMENT
- Zoological Park Organization Under the Royal Patronage of H.M. the King
- The Botanical Garden Organization
- Wastewater Management Authority

MINISTRY OF INFORMATION AND COMMUNICATION TECHNOLOGY
- TOT Corporation Public Company Limited
- The Communications Authority of Thailand

MINISTRY OF ENERGY
- The Electricity Generating Authority of Thailand
- PTT Public Company Limited
- The Bangchak Petroleum Public Company Limited

MINISTRY OF COMMERCE
- Public Warehouse Organization

MINISTRY OF INTERIOR
- The Metropolitan Electricity Authority
- The Provincial Electricity Authority
- The Metropolitan Waterworks Authority
- The Provincial Waterworks Authority
- The Marketing Organization

MINISTRY OF SCIENCE AND TECHNOLOGY
- Thailand Institute of Scientific and Technological Research

MINISTRY OF EDUCATION
- Secretariat Office of the Teacher's Council of Thailand
- Kurusapa Business Organization

MINISTRY OF PUBLIC HEALTH
- The Government Pharmaceutical Organization

MINISTRY OF INDUSTRY
- The Industrial Estate Authority of Thailand
- Narai Phand Co., Ltd.

FURTHER READING

- *Corporate Governance in Thailand*, edited by Sakulrat Montreevat, 2006
- "State-owned Enterprises: Challenge to Regional Integration," by Deunden Nikomborirak in *Benefiting from Globalization: Transport Sector Contribution and Policy Challenges*, 2008
- "Raising Thailand's Competitiveness," a paper presented at TDRI's Annual Academic Conference, by Deunden Nikomborirak and Sumet Ongkittikul, 2014

TIMELINE

A Shifting State of Affairs

1932

Thailand changes from an absolute monarchy to a constitutional one after a coup d'état brought about by a small group of military and civilians, who form Siam's first political party called Khana Rasadorn. In order to control the country's economy, the new government sets up several state enterprises including banks, insurance companies, rice and fishery trading firms and a shipping line. Before the outbreak of World War II state enterprises number 30 in total.

1946–1952

After the war, 19 additional state enterprises are founded in order to revive the economy, including those devoted to energy.

1953

Under the government of Field Marshal Pibulsongkram, the State Enterprise Establishment Act is promulgated to facilitate the creation of state enterprises in multiple industries, including textiles, glass, sugar and paper. From 1953 to 1956, a total of 37 new SOEs are created.

1993

Private sector operators are allowed to provide telecom services but only under concessions handed out by the SOEs that hold the statutory right to operate at the time. These concessions allow SOEs in the telecom sector to collect handsome "revenue shares" from the private concessionaires. Although the scheme leads to the rapid development of fixed line and mobile voice services like never before, over time it serves to weaken the state operators that grow more complacent because of the easy money.

1958

Under the government of General Sarit Thanarat, the role of state enterprises is confined to the provision of basic services such as transport, telecommunications, water and electricity only. The private sector is encouraged to invest in manufacturing industries, a move that was also a reaction to the inherent inefficiency and corruption among SOEs at the time.

1998

The cabinet, under the Chuan Leekpai administration, approves the "Master Plan for State Enterprise Reform," which is the only document to date that lays out a comprehensive strategy and timeline for the restructuring and privatization of state enterprises in all sectors.

1999

The Corporatization Law is passed to allow the mass corporatization of state enterprises in preparation for eventual privatization. The law provides a regulatory framework for the conversion of state enterprises to private enterprises or publicly listed companies without having to amend the *sui generis* laws.

2001

The government of Thaksin Shinawatra launches the first partial privatization of an SOE, the Petroleum Authority of Thailand (PTT). To accomplish this, the Thai government sells 30 percent of its stake on the stock market. Two more SOEs are listed in the next three years. As the government retains a majority equity share in all three SOEs, the publicly listed companies continue to be plagued by political meddling.

2006

The Supreme Administrative Court revokes the two royal decrees that led to the corporatization of the Electricity Generating Authority of Thailand (EGAT) in 2005. The reason for the verdict is the conflict of interest inherent in the corporatization process. This case marks the end of the "privatization episode" not only for the Thaksin administration, but also for subsequent governments, because poorly implemented privatization is perceived by the public as serving the interests of politicians rather than those of the country.

Before a 1953 Thai Airways flight. The airline remains a state-owned enterprise.

Challenges Facing State-Owned Enterprises

Increased competition As Thailand joins the globalization bandwagon, it is under constant pressure to open up many of the key service sectors in which state-owned enterprises (SOEs) operate, be they telecommunications, transport or energy. But even when foreign competitors are still kept at bay by stringent foreign investment regulations certain Thai SOEs have been facing fierce competition from local private competitors. For example, TOT PCL and CAT Telecom, the two state enterprises in telecommunications, have had to compete with private players such as DTAC, TRUE and AIS. Hobbled by political agendas and bureaucratic rigmarole, they cannot keep up with their rivals. In the past, when private operators were allowed to compete only under concessions handed out by SOEs, these two enterprises survived on "concession income." But as the concession era comes to an end, replaced by a licensing system, their financial survival is threatened.

Monopolistic practices Several SOEs have not yet had to stave off such fierce competition, thanks to state laws and regulations that either prohibit or do not accommodate effective private competition. These enterprises include PTT PCL and EGAT, which remain the sole buyers and sellers of natural gas and electricity, respectively. But it's only a matter of time before they are pushed out of their comfort zones and have to do battle with private firms.

Boards of directors Making an SOE as efficient as a private enterprise is hampered by the composition of the board of directors. If you compare the members (and their qualifications) of Singapore Airlines and Thai Airways, you will find some glaring

differences although both are state-owned. Singapore Airlines mobilized businessmen of the highest caliber to be its directors, such as executives, or even CEOs, of well-known

multinational firms like BMW, Li & Fung or the Blackstone Group. The board of Thai Airways, however, is staffed mostly with high-ranking civil servants from various offices such as the Air Force, the Ministry of Finance, the National Economic and Social Development Board, and the Office of the Public Prosecutor, who do not have any experience in managing large businesses.

Flat pay State enterprises pay junior staff much higher than do private companies, while they underpay senior staff, in particular those at the managerial level. This is because SOEs' pay scale is linked to that of civil servants, which is relatively "flat." So it's no surprise that many SOEs have highly paid clerks while experiencing a "brain drain" of senior staff, leaving the enterprise "bottom-heavy."

Political scrutiny Besides problems with personnel, SOEs also face rigid

procurement and hiring regulations and come under intense political scrutiny, which in turn leads to political interference. Without strong political will, it is hard to imagine how Thai SOEs can be effectively reformed.

Inspiring corruption SOEs are ripe for exploitation by politicians bent on corruption or playing populist politics and investing in huge projects. Unlike the usual state budget, an SOE budget does not have to be debated and approved by the lower and upper houses. The fact that the enterprises fall under different ministries also means that they are administered by different politicians in vastly different ways.

Mismanaged mega-projects When dealing with such huge budgets there is a great chance of mismanaging mega-projects, which adds up to mounting debts and could lead to economic crises.

Loss of opportunities Because SOEs are frequently built on such monopolistic business structures, players in the private sector lose out on potential opportunities, impacting competition and consumer options.

Why Competitiveness and Innovation are Key

The creation of an SOE "Holding Company" The "State-owned Enterprise Superboard," a committee set up by the National Council for Peace and Order (NCPO) to supervise SOEs, has proposed the creation of an "SOE Holding Company" that will oversee the reform of state enterprises in order to boost their efficiency and effectiveness. The proposal was based on models in neighboring countries such as Singapore's Temasek Holding and Malaysia's Khazanah Nasional. A similar idea was proposed by former prime minister Thaksin Shinawatra back in 2001, but the general public strongly opposed the proposal for fear that the government would use the company as a vehicle to privatize SOEs en masse. Regardless, at the time of writing this much-touted SOE holding company had yet to materialize.

Banning public prosecutors from the directorship of state-owned enterprises The fact that public prosecutors and even an attorney general have sat, or continue to sit, on state-owned enterprises' boards of directors has always been a contentious issue. Several years ago, the National Anti-Corruption Commission (NACC) proposed to elected governments that such practices should be disallowed due to the inherent conflicts of interest as the Office of the Attorney General has to scrutinize SOEs' contracts in order to ensure that they are consistent with public interest.

Reviving Thai Airways In the past decade, the country's national airline has suffered from several severe financial crunches. Back in 2009, Piyasvasti Amaranand was appointed its CEO in order to save the company from the looming operational losses resulting from questionable aircraft purchase deals, which were influenced by political meddling and had resulted in grounded planes. During his time at the top, Piyasvasti cut costs sharply and helped put Thai Airways back on world-class standards as the fifth-best airline ranked by Skytrax. He and a reform-minded board also put in place clear procurement and hiring rules and regulations that would prevent corruption and promote efficiency and transparency. However, as the political tailwinds changed, Piyasvasti was dismissed in 2012 for his supposed "inability to effectively communicate with the (new) Board of Directors." After that, the airline descended into the doldrums again with investment analysts predicting darker days ahead. Given that Thai Airways has struggled to stay in the green in recent years, it remains to be seen just how long the airline will be able to remain aloft in the intensely competitive global aviation market.

Creating a level playing field between state and private enterprises Many SOEs compete neck and neck with private enterprises in fields like telecommunications, broadcasting and transport. Yet, the Trade Competition Act of 1999, a law that prescribes rules to ensure fair competition in the market, provides a blanket exemption for state-owned enterprises. The implication is that public enterprises, in particular those with monopolistic power, are able to carry out trade practices that may unfairly harm competitors and consumers. Attempts to abolish the exemption in the past proved futile due to resistance from both the SOEs and those that benefit from them.

Thailand Post pushes the envelope By 2003, when the lackluster postal service was spun off from the Communications Authority of Thailand (CAT) – the state enterprise engaged in postal and telecommunications services – the deal looked like a write off. The postal service was mired in red ink and required cross-subsidies from more profitable telecommunications services. Stripped of those handouts, Thailand Post had to seek new revenue-generating services besides delivering letters and postcards, such as the delivery of parcels. But with stiff competition from multinationals such as DHL, Fedex or Courier, the SOE had to boost its game by introducing new services such as online tracking and trade facilities for customers. It also exploited its expansive post office network by introducing peripheral services such as bill payments, the sale of pre-paid phone credits and money orders. When those services ran up against competition from convenience stores, the company turned to other niche markets, such as stamp collectors. Thailand Post even partnered with insurance companies to sell their products. The key drivers in this state-owned enterprise's success were a capable CEO and the fact that it was spared from the usual political meddling as politicians had not perceived it to be a potential money-maker.

SMEs
Smaller businesses are big deals

In Thailand's folklore, few economic endeavors receive more adoration and respect than rice farming. In Thailand's business news pages, it is the big corporations (usually listed on the stock exchange) that get the most attention. In the global popular imagination, tourism is tops. But in the real world, the true economic heroes are the almost countless businesses that go by the prosaic designation of "small and medium-sized enterprises," or SMEs.

In fact, there are around 2.9 million of them across Thailand. That figure is based on a dual classification of "small" being not over 50 employees and "medium" from 51 to 200. Altogether these outfits make up 99 percent of all enterprises in Thailand and account for nearly four out of five of the nation's jobs. They also contribute around 40 percent to the country's annual economic output and 30 percent of export earnings. These numbers are very similar to those of other countries, even in the wealthier nations whose policy research unit, the Organization for Economic Co-operation and Development

One Tambon, One Product (OTOP):

Inspired by Japan's successful One Village One Product program, OTOP is a local entrepreneurship stimulus program aimed at encouraging tambon (sub-district) communities to improve the quality and marketing of local products, like foodstuffs and handicrafts.

A guitar shop in Hua Hin.

(OECD), has surveyed SMEs and entrepreneurship in many countries, including Thailand in 2011. In fact, Thailand's so-called SME density (number of SMEs per 100 population) of 4.2 is only slightly less than the typical density of 5 in OECD countries.

Thai SMEs are involved in the full gamut of business activities. Most are in services (about a third, mainly hotels and restaurants), manufacturing (30 percent), and trade and maintenance (retail and wholesale, and vehicle maintenance, 28 percent), according to the Office of Small and Medium Enterprises Promotions (OSMEP). Many operate in the booming tourism industry. This sector's vital importance to sustainable development is recognized in Target 8.9 of the UN's Sustainable Development Goals, which calls on nations to "devise and implement policies to promote sustainable tourism that creates jobs and promotes local culture and products".

Closely related to the presence of SMEs in a country is the level of entrepreneurship, and Thailand scores highly on this front. The Global Entrepreneurship Monitor found that Thailand exhibits one of the highest entrepreneurship activity rates of the 70 economies it surveyed around the world in 2013. The study found that 46.3 percent of Thailand's adult population is involved in such activities (18.3 percent started or ran "new businesses" and 28 percent identified as "established business owners" – the second highest in the world after Uganda).

AT A GLANCE

SMEs IN THAILAND

Criteria: "Small" means less than 50 employees while "medium" means 51 to 200

2.9 MILLION	**NUMBER OF SMEs**
99%	**PERCENTAGE OF TOTAL ENTERPRISES**
4 out of **5** JOBS	**PROVIDES EMPLOYMENT**
40%	**CONTRIBUTION TO ECONOMIC OUTPUT**
30%	**CONTRIBUTION TO EXPORTS**

KEY SME SECTORS

Services make up
33% of all Thai SMEs

Manufacturing accounts for
30% of SMEs

Trade and Maintenance (retail and wholesale, and vehicle maintenance) account for **28%**

Source: Office of Small and Medium Enterprises Promotion (OSMEP)

Over the years, various Thai governments have recognized the value of SMEs, but it wasn't until the 1997 Asian Financial Crisis knocked a hole in Thailand's economic model that anything substantial was done about it on the policy front. Before the boom went bust, SMEs fell partly under "co-oper-atives" policy and partly under industries, buried in the National Economic and Social Development Plan. But then the **Small and Medium Enterprises Promotion Act B.E. 2543** was passed in 2000 and OSMEP was set up in 2001, with the prime minister or deputy as chair. A year later, the SME Development Bank of Thailand was established. In another crucial development, the Market for Alternative Investments (MAI) – intended to provide a simpler, lower-cost alternative for smaller firms to list on the main board of the Stock Exchange of Thailand (SET) – had its first stock listed in 2001.

Another important initiative has focused on upgrading producers of the **One Tambon, One Product (OTOP)** community enterprise program to become SME operators. More recently, the National Innovation Agency (NIA) and the Federation of Thai Industries have collaborated to create "Innovation Coupon," an initiative that focuses on stimulating SMEs to pursue R&D and the development of innovative projects. The initiative provides SMEs with a 100,000-baht coupon to cover the cost of a feasibility study for their intended project, which, if all goes well, qualifies them for an additional

400,000-baht coupon to support implementation of the project. According to the NIA, the current goal is to support more than 200 projects per year.

Digital startups are the latest incarnation of SMEs in Thailand. Recognizing the potential of startups, the government has announced that building digital communities, establishing digital parks for SMEs and creating digital innovative startup networks are its three major digital initiatives for 2017. Hoping to attract heavyweights like Google, Facebook and Amazon, the government is also planning to eventually open Digital Park Thailand and offer tech companies incentives such as tax breaks, unlimited bandwidth, and submarine cables to Europe and China's "One Belt, One Road" project. In 2016, the government also established a US$570-million venture fund for the development of a startup ecosystem. The goal is to spur IoT research, aviation collaboration, e-commerce, e-payments, as well as the development of encryption technologies and hardware and software solutions.

In addition, a Thai branch of Silicon Valley's 500 Startups – 500 TukTuks – is also now managing a multi-million dollar fund to groom Thailand-based startups. Supporting SMEs in areas such as this is vital as the OECD study points out: "Productive entrepre-neurship and innovative and internationalizing SMEs will be key drivers of future economic growth and must be given due attention in policy reform."

The Small and Medium Enterprises Promotion Act B.E. 2543:

Established in 2000, this act provided clear definitions and classifications of SMEs and outlined programs for their promotion, including the establishment of the Office of Small and Medium Enterprises Promotion.

Why SMEs Matter to Sustainable Development

Creating jobs and wealth In economies all over the world, as in Thailand, firms with fewer than 200 employees, in particular those with less than 50, account for the greatest number of jobs and account for a third or more of total economic output and a similar proportion of exports.

International standard Thailand's SME sector is in line with the average of countries in the Organisation for Economic Co-operation and Development (OECD), which consists of 35 of the world's richest economies. Thailand has a "density" of 4.2 SMEs per 100 of the total compared with the OECD average of five. Entrepreneurship is a vital and driving force of SME activity and Thailand has one of the highest entrepreneurial rates in the world.

Knowledge and technology transfers SMEs make it possible for less technologically advanced firms to link up with established companies, particularly multinationals,

and to provide a channel for the absorption of new technologies and know-how for smaller Thai companies.

Alternative to agriculture SMEs provide a safety net and source of employment for Thais migrating from the volatile and seasonal agriculture sector. When the harvest is done, farmers are able to shift to SMEs, which also help absorb first-time workers in rural areas, providing them with a closer-to-home employment alternative to commercially dominant Bangkok and its urban environs.

Fallback plan In the absence of a comprehensive social security or welfare system, unemployed and underemployed low-skilled workers can seek to make a living on the margins of the economy, especially in services industries such as restaurants, hotels and retail. Well-considered government policies can help these workers hone their skills and boost their productivity levels to benefit both themselves and the economy as a whole.

Facts You Need to Know About SMEs

- By global standards, Thailand is a very entrepreneurial nation. Close to half the population is engaged in new business ventures or the expansion of an existing business. Almost one in five of the adult population (18–64 years) is involved in early-stage entrepreneurship and 28 percent are classified as an "established business owner."

- About a third of entrepreneurs say they are motivated more by necessity than opportunity.

- There are proportionately fewer enterprises (0.4 percent) in Thailand employing between 51 and 200 employees ("medium-sized") than in OECD countries (around two percent) and other non-OECD Asian countries (Korea about six percent; Japan around 10).

- Bangkok has an SME density of about ten per 100 people, whereas the north and northeast regions have a density of less than three. Two-fifths of SMEs are located in greater Bangkok where only a 10th of the population is located.

- Thailand has one of the highest female entrepreneurship rates in the world, though this has been declining over the past few years, according to the Global Entrepreneurship Monitor's (GEM) Global Report 2015/2016. Total early-stage entrepreneurial activity (TEA) by females was 14.8 percent in 2015, down from a high of 20 percent in 2011, while the rate for men rose slightly over the same period.

- The vast majority (93 percent) of Thai entrepreneurs are doing business domestically. That percentage is relatively high compared with other Asian nations. Only Indonesia boasts a higher rate at 98.8 percent.

Backing the Underdog: Bringing SMEs Out of Hiding

Loud footsteps in the corridors of power do not always resonate through other walks of life. Even less so when the subject in question generates only small or medium-scale recognition in political circles. This is the struggle with which the SME sector has long had to contend in Thailand.

That is not to say that there isn't an agency accorded responsibility for nurturing it. Indeed, the Office of SME Promotion (OSMEP), which was set up in 2001 under the SME Promotion Act of 2000, is chaired by the prime minister, though in truth the deputy prime minister in charge of economics is assigned to perform on his or her behalf. While this appears to accord the OSMEP some stature, the agency lacks the authority of a ministry. The 2011 OECD study of Thailand's SME sector, while recognizing that OSMEP has "played an important role in SME and entrepreneurship policy coordination in Thailand and made an important contribution to national development," concluded, "There is a limit to the ability of OSMEP to achieve policy coherence across government ministries and agencies... [and it] lacks the operational clout and authority to achieve this objective fully."

Other pundits have weighed in on the subject. One academic described OSMEP's status as "a small agency that piggybacks on the Ministry of Industry." Meanwhile, Chayut Setboonsarng, writing in a review published by the Elliott School of International Affairs at George Washington University, said OSMEP "lacks operational space" and remains limited in its ability to support SMEs.

After General Prayuth Chan-ocha became prime minister, he then assumed the chairmanship of the Board of Directors of OSMEP. His government has promised to provide greater support to the SME sector and in August 2014 allocated a budget

A watchmaker uses a steady hand to make repairs.

of 726.7 million baht to help SMEs improve their products, services and competitiveness. However, this figure – even assuming it is deployed effectively – reflects the lowly status of the SME sector compared with other priorities of the state. It is about one-tenth of that allocated in the 2015 national budget for the Ministry of Culture and about a hundredth of the agriculture budget.

Policy coordination and continuity are made more difficult by lack of information and communication between the sector, other businesses and government agencies. Most SMEs are institutionally invisible. Only a fifth are registered and of those only half pay any taxes. As the chairman of the Federation of Thai Industries, Supant Mongkolsuthree, said, "If the government wants to change this, it needs to offer special tax rates to lure these businesses to register in the formal system. As a former SME [operator] myself, I propose that SMEs must come up above the ground. Now they are hiding from everything, including taxes."

Indeed, bringing them out of hiding and integrating them more

into the visible economy is crucial to formulating good policies and providing guidance. As an OECD study observed: "Timely and reliable information is critical to OSMEP's performance in improving SME and entrepreneurship policies and programs." In particular, the study said that the definition of SMEs needed to include a micro-level and a non-employing level and that policymakers should "collect, harmonize and report data on this group."

Those who know the real economic needs of SMEs, such as the Asian Development Bank, point in particular to the lack of finance and the high cost of funds. However, the kingdom is finally taking some measures to address this. In 2015, the Business Collateral Act (BCA) was passed into law and the majority of its provisions came into effect on July 1, 2016. The BCA aims to establish a legal system that allows borrowers in Thailand to use their assets as collateral to secure loans. If effective, the system will provide Thai borrowers with greater access to credit, which should promote investment in domestic ventures.

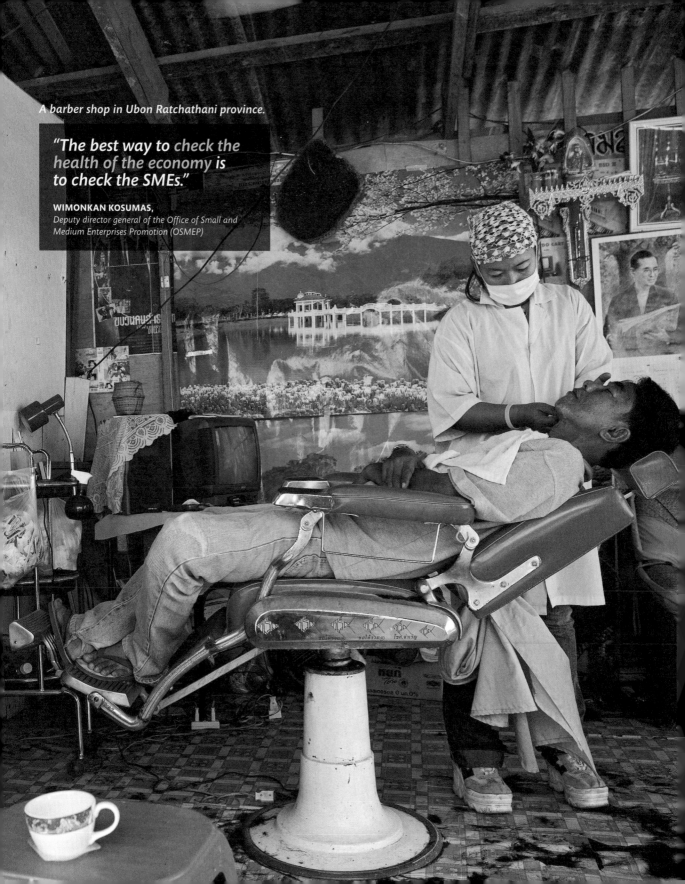

A barber shop in Ubon Ratchathani province.

"**The best way to** check the
health of the economy is
to check the SMEs."

WIMONKAN KOSUMAS,
Deputy director general of the Office of Small and
Medium Enterprises Promotion (OSMEP)

FURTHER READING

- *Thailand: Key Issues and Policies, OECD Studies on SMEs and Entrepreneurship,* by OECD Publishing, 2011
- *The Impact of Finance on the Performance of Thai Manufacturing Small and Medium-Sized Enterprises* by the Asian Development Bank, 2016
- *Strategies for Growth in SMEs: The Role of Information and Information Systems,* by Margi Levy and Philip Powell, 2004

REALITY CHECKS

Challenges Facing the SME Sector

Missing middle Thailand lags behind in the proportion of SMEs that can be classified as medium sized (51 to 200 employees). When only a small proportion of the most innovative, competitive and globally connected SMEs is in the medium category, this has been termed the

A store in Bangkok's Chinatown.

"missing middle." This points to one of the most important features of Thailand's SME and entrepreneurship landscape – a comparatively high percentage of almost one-third of those engaged in this sector are motivated by "necessity" rather than "opportunity." The OECD study points out that these kinds of entrepreneurs "frequently have low productivity, lack growth potential and offer poor income and employment conditions."

Access to capital In spite of the existence of the SME Development Bank and an increasing interest in SME loans among commercial banks, small firms struggle to find access to capital and other funding support because of the lack of collateral laws. However, the kingdom is finally taking some measures to address this. In 2015, the Business Collateral Act (BCA) was passed into law and the majority of its provisions came into effect on July 1, 2016. The BCA aims to establish a legal system that allows borrowers in Thailand to use their assets as collateral to secure loans. If effective, the system will provide Thai borrowers with greater access to credit, which should promote investment in domestic ventures.

Bangkok glut SME densities are much greater in Bangkok than in the north, northeast and southern parts of the country, where entrepreneurial activity is only about half that in the area around the capital. Targeted efforts are needed to encourage SMEs on the peripheries.

Horizontal networking While there has been an increase in "linkages" between multinational firms and Thai SMEs in the country's automobile and parts industry, this important benefit of Foreign Direct Investment is limited. In fact, in many instances this so-called "horizontal networking" and subcontracting is weak, and local SMEs, because of their lack of technological and managerial competitiveness, find themselves losing out to foreign firms. This is made worse by Thailand's

low level of investment in R&D. The OECD study concludes: "A narrow telecommunications infrastructure and limited collaborative links between SME agencies, universities, research institutions and the business community have compounded the weakness of Thailand's innovative capacity. The development of innovative SMEs is equally hindered by educational and training systems, which though endowed with appropriate resources, have failed to produce qualified labor commensurate with market demands."

Productivity gap A large productivity gap exists between SMEs and larger companies. In fact, output per employee in large companies is more than four times that of small ones and nearly twice that of medium-sized enterprises.

Stiffening AEC Competition The ASEAN Economic Community (AEC), established at the end of 2015, places Thai SMEs under even more competitive strain. Very few Thai SMEs have any presence beyond national borders and do not have the capacity to research how they might take advantage of the opportunities provided by the AEC. As Chayut Setboonsarng, writing in *The International Affairs Review* of the Elliot School of International Affairs at George Washington University wrote in 2012, "High overhead costs, complicated rules of origin and an absence of ASEAN and trade specialists contribute to the overall malaise."

TOURISM

The travel business is an economic mainstay

Foreign tourists take part in Songkran festival in Bangkok.

Thailand's rise from an isolated outpost for backpackers and hedonists to one of the world's top 10 travel destinations is one of the country's greatest success stories. Tourism is now the country's biggest earner of foreign exchange and a major source of employment. In 2016, tourism generated 1.64 trillion baht in business and directly employed 4.2 million people, or 11 percent of the workforce, with many more Thais and other industries benefiting indirectly.

In 2013, Thailand became the 10th-most-visited country in the world; the year after that Bangkok was the globe's second-most visited city, trailing London and just ahead of Paris. Some of that success came hot on the heels of a massive surge in Chinese tourists who accounted for almost 27 percent of the total number of arrivals in 2016 – an increase of 64 percent versus 2013. These numbers promise to only mushroom in the coming years. With ASEAN countries cooperating on visa policies and the

growth of middle classes throughout Asia, Thailand is likely to continue to be seen as a welcoming destination that excels at providing value for money.

Meanwhile, the demographic shift from backpackers and Europeans to visitors from Russia, India, South Korea and China, who bring new cultural differences and expectations with them, are posing their own challenges, from issues of language and etiquette to the necessity of developing new tourism products, services and promotions that appeal to them.

In a way, it's nothing new for Thailand and a test it should pass easily. For hundreds of years, Thais have been famously hospitable and outwardly tolerant of the many foreigners, whether merchants, diplomats, missionaries or miscreants, who passed through their borders. When those winning traits were combined with the country's geographic position in the middle of Southeast Asia, endowed with an abundance of natural marvels, from green mountains to white sand beaches and with wild jungles in between, all the elements were in place for a tourism revolution.

During the upheavals of the Vietnam War-era and left-wing uprisings in Southeast Asia, Thailand was a safe haven of relative stability. The influx of US

THAILAND'S WORLD RANKINGS
(OUT OF 141 COUNTRIES)

#35	OVERALL COMPETIVENESS RANKING	🙂
#16	ABUNDANCE OF NATURAL RESOURCES	😁
#21	TOURISM INFRASTRUCTURE	😄
#36	PRICE COMPETITIVENESS	🙂
#38	BUSINESS ENVIRONMENT	🙂
#116	ENVIRONMENTAL SUSTAINABILITY	🙁
#132	SAFETY AND SECURITY	😞

Source: Travel & Tourism Competitiveness Report 2015 by the World Bank

> ***"Research shows that for every 30 new tourists to a destination one new job is created."***
>
> **"Travel & Tourism Competitiveness Report 2015" by the World Bank**

money aimed at maintaining Thailand's position as a bulwark against the red tide of communists, as well as a base for launching American attacks on Vietcong troops, triggered a transformation of the country's underdeveloped infrastructure. Roads were tarmacked and railways laid, American GIs on rest and recuperation tours demanded more Western comforts, while locals profited from providing these services. Meanwhile Bangkok grew as a hub of both airlines and hotels.

Such rapid progress comes with a price tag that is often paid at the expense of the environment. In an ironic twist, the very same scenery that attracts all these tourists, both domestic and international, has been despoiled by some of them, or by unscrupulous tour operators and authorities who have turned a blind eye to violations of environmental protection laws. On Koh Tao, for example, the number one destination in Southeast Asia for scuba divers to get certified, water pollution and the destruction of coral reefs have put the island's premier lure in jeopardy.

The World Bank's Travel & Tourism Competitiveness Index 2015 ranked Thailand 16 out of 141 countries in terms of "Natural Resources" but 116 out of 141 in terms of "Environmental Sustainability." In the same report, Thailand ranked in the lower tiers in "Coastal Shelf Fishing Pressure" (93), "Stringency of Environmental Regulations" (103), "Threatened Species" (109), and "Particulate Matter Concentration" (123).

The spike in new arrivals has also put pressure on the infrastructure, which now lags behind the expansion of the industry. The first airport on Koh Phangan is under construction. A second airport on Koh Samui is on the drawing board. And many of the country's biggest attractions, such as the Grand Palace in Bangkok and Doi Suthep in Chiang Mai, are under siege from a growing number of tourists.

By continuing to promote more untrammeled destinations and niche markets like **sustainable tourism**, the Tourism Authority of Thailand (TAT) can play a dramatic role in relieving pressure and spreading visitors across the country, bringing jobs and development to parts of the country that really need them.

As Thailand's tourism brand continues to grow and global travel remains on the upswing, this quandary looms the largest: How do we balance the economic opportunities of this growth with the maintenance of the natural and cultural attractions that made growth possible in the first place?

Sustainable tourism:

A growing niche in travel often based on excursions into nature that revolve around low-impact trips that benefit local communities and promise authentic experiences.

Why Tourism Matters to Sustainable Development

Foreign exchange bonanza Tourism generates more foreign exchange than any other industry in Thailand. In 2016, it earned around US$45.7 billion, and this figure is projected to rise as the industry expands and arrivals continue to increase.

Promoting the country While Thais may be sensitive to negative news reports about the kingdom, the fact is 55 percent of tourists are return visitors, meaning most people truly enjoy the country and their experience here. Greater awareness of the kingdom has positive repercussions that reverberate through many other sectors such as trade and finance.

Pillar of growth Tourism has always been one of the kingdom's most resilient industries. Whether it was after the tsunami of 2004, the worldwide financial crisis of 2008, or the political squabbles of the last few years, tourism has bounced back quickly, providing a stabilizing influence on the economy.

Preserving Thainess Much of the interest in traditional Thai culture, from the performing arts to handicrafts to food, is inspired by tourism. Visitors find the country's exoticism and native creations appealing and different, leading Thais themselves to appreciate the value of protecting and promoting their heritage.

Spurs development Tourism can be a catalyst for new infrastructure projects from airports to roads, and often boosts the development of rural areas, providing much-needed jobs in the far-flung parts of the country. SMEs, in particular, tend to spring up in tourist areas.

Job creation The substantial contribution that tourism makes to the Thai economy, largely through the influx of tourist dollars, is well known, but it also serves as a major catalyst for job creation. As of 2016, more than 4.2 million people in Thailand, or about 11 percent of the total workforce, were directly employed in the tourism industry. Tourism tends to employ more women and young people, and is increasingly offering jobs in high-skilled areas like ICT, management and marketing.

TOURISM BY THE NUMBERS

32.6 MILLION TOTAL ARRIVALS IN 2016

BY REGION

EAST ASIA	SOUTH ASIA
21.6 million	1.5 million
EUROPE	THE AMERICAS
6.1 million	1.4 million

BY COUNTRY

 CHINA **8.7** million

 MALAYSIA **3.5** million

JAPAN **1.4** million

INDIA **1.2** million

Source: Ministry of Tourism and Sports

Medical Tourism Market Healthier than Ever

Almost three million foreign patients arrive in Thailand each year to seek quality medical care for a fraction of the price it would cost in other parts of Asia, the West or the Middle East. Many used the surplus to relax and recover in the kingdom. Whether it's for cosmetic surgery, cutting-edge operations, dental care or gender-reassignment surgery, the kingdom is arguably the world's leading destination for medical tourism, boasting a 40 percent share of the market, with Singapore, India and Malaysia making up the rest of the top four.

The facility at the forefront of this burgeoning niche market is Bumrungrad International Hospital (BIH), which is more akin to a five-star hotel than your typically sterile medical center. BIH is the epitome of Thailand's competitive advantage in this field of tourism: a one-stop medical center that boasts more than 45 specialty centers and clinics, a kitchen for preparing halal meals, a Japanese restaurant and a staff of translators for overseas patients.

All in all, Thailand has 53 hospitals and clinics affiliated with the Joint Commission International (JCI), the gold standard in worldwide healthcare. Its doctors and nurses are renowned for their competence and compassion. And one-month visas are available on arrival. In turn, these levels of quality and service have given medical tourism a shot in the arm that has caused a massive spike in arrivals over the last decade.

Nation-wide the healthcare system in Thailand treated 2.81 million foreign patients in 2015, an increase of 10.2 percent from the previous year. Of the total number of patients, Japan is the number one source market, followed by the US, the UK, the Middle East and Australia. Together they pumped more than US$4.7 billion into the country

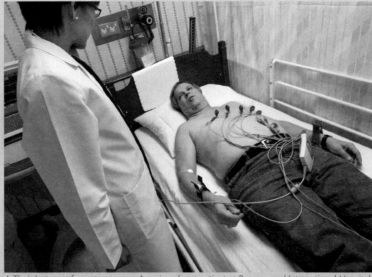

A Thai doctor performs tests on an American heart patient at Bumrungrad International Hospital.

in 2013, the government claims. According to Kasikorn Research Center, in the coming years medical tourism will have a major role in generating revenue for private hospitals, given that their total earnings from international patients rose from 25 percent in 2011 to more than 30 percent in 2015.

Thailand's rise in this field came from an unlikely source. Many people see it as an outgrowth of the Asian Financial Crisis of 1997. After the baht was devalued, hospitals tried to tap into different revenue streams. Businessman Bernard Chan, whose family owns a part of Bumrungrad, said of Thailand's success to the *Financial Times* in 2014, "I'm not sure it can be replicated in other markets."

There are some bones of contention though. The growing market for foreign patients has driven up prices for middle-class Thais who prefer private healthcare to the country's less advanced public hospitals. The World Health Organization also

warns that, perhaps unsurprisingly, more and more doctors at public hospitals are moving into private healthcare where the benefits are better and the salaries are higher.

The country's wellness industry is another attraction for medical tourists and also those looking for training. As the fountainhead of the region's spa and wellness industries, offering traditional Thai treatments that date back to the time of the Buddha, as well as everything from colonic detox programs to new-age treatments, Ayurveda programs, yoga vacations and Chinese acupuncture, Thailand has largely cornered the market for healthy holidays in Southeast Asia.

In 2016, the Ministry of Public Health introduced new health packages to promote medical tourism under the slogan, "Visit Thailand Enhance Your Healthy Life." The scheme is expected to gain the support of some 70 hospitals and clinics certified by JCI international standards.

Challenges Facing the Tourism Industry

Shifting demographics With a rising population of elderly travelers and a huge upsurge in tourists from China, Russia, India and South Korea, new tourism products, services and promotions must be developed that appeal to these diverse consumers.

Marine parks Managing water supplies and the marine environment remains a perennial challenge on many islands. As the number of marine creatures dips and water pollution rises, key island-based industries may be impacted.

Building regulations The enforcement of regulations for property development in Thailand can be haphazard, which means that national park areas like Koh Samet and Koh Lipe – supposedly protected from developers – suffer from the encroachment of resorts.

Balanced development Critics say that major development projects, such as coal-fired power plants in Krabi province and a "land bridge" linking the provinces of Satun and Songkhla, could threaten some of the country's most pristine national marine parks and islands.

New promotional campaigns The TAT has yet to recapture the magic of the "Visit Thailand" and "Amazing Thailand" campaigns. And while the idea behind the 2017 campaign is well-meaning, the "Unique Thai Local Experiences" slogan doesn't exactly inspire awe.

Political stability Bangkok, in particular, has been interrupted by street protests and coups that lead to travel warnings and advisories that thin the tourist crowds. After the political squabbles and coup of 2014, the number of arrivals dropped by 6.6 percent that year.

Sex tourism The stigma of Thailand's reputation as the sex capital of the world still persists. Cracking down on high-profile red-light districts, such as Patpong, Soi Cowboy and Nana Plaza in Bangkok, as well as the bars along Bangla Road in Phuket and Walking Street in Pattaya, could potentially change this damaging perception.

Safety concerns Thailand's lowest ranking in any category in a major World Bank report on global tourism was in "Safety and Security" in which it placed 132 out of 141 nations. The World Bank report also pointed to the low reliability of police services and incidents of terrorism. In addition, other negative events, such as the murder of two British backpackers on Koh Tao in 2014, as well as the high number of road deaths and other accidents, have put Thailand's reputation as a safe destination in jeopardy. Some of the media reports suggesting Thai tourism is unsafe may be unfair and deceptive as Thailand hosts over 32 million tourists per year with a high rate of return visitors.

Surging arrivals With tourist arrivals in Thailand set to reach 35 million in the coming years, there is a real need to develop infrastructure projects to cope with surging numbers by expanding the capacities of airports and improving bus, ferry and rail systems.

FURTHER READING

- *The Travel & Tourism Competitiveness Report 2015*, by the World Bank (Available online)
- "Medical Tourists Flock to Thailand Spurring Post-Coup Economy," by William Mellor, *Bloomberg*, November 19, 2014
- *Travel & Tourism: Economic Impact 2014*, by the World Travel and Tourism Council, 2014
- *Ecotourism and Sustainable Development*, by Dr. Martha Honey, 2008
- *Tourism Policy and Planning: Yesterday, Today and Tomorrow*, by David L. Edgell Sr and Jason Swanson, 2013

TIMELINE

The Tourism Boom in Thailand

1959

Around 70,000 international air travelers arrive in Bangkok

1960

The government establishes the Tourism Organisation of Thailand (later to become the Tourism Authority of Thailand or TAT). At that time, Bangkok had less than 1,000 rooms fit for tourists and little in the way of restaurants, bars and travel agencies to cater to them.

1960s– 1970s

The transcontinental "hippie trail" began in London and ended in Thailand, where the scruffy travelers receive a warmer welcome than in many other countries.

2010

In spite of the political protests that paralyze one of the city's glitziest shopping strips, the number of tourist arrivals shoots up to almost 16 million, proving that tourism is one of the most resilient industries.

1998

The TAT launches its most popular campaign ever, "Amazing Thailand," and showcases the new brand at international travel trade shows around the world the following year.

1987

"Visit Thailand Year" marks the 60th birthday of King Bhumibol Adulyadej, and the number of arrivals surges by 24 percent to 3.4 million.

2016

Tourism contributes 11 percent to the GDP and directly employs more than 4.2 million people, or 11 percent of the total workforce, with many more vendors, restaurateurs, bar owners and taxi drivers benefiting indirectly.

"Short-term gain has all too often been the driving force in the tourism industry, despite the long-term pain that this attitude can lead to. This cautionary view is worth bearing in mind when it comes to safeguarding and nurturing the health of the sector here in Thailand over the coming years."

Kobkarn Wattanavrangkul, Tourism and Sports Minister

FINANCE

A solid financial infrastructure is providing
the foundation for sustainable development

The development of the modern Thai financial sector began in earnest with the establishment of the Bank of Thailand during World War II. The central bank was hastily set up during the Japanese occupation to safeguard the country's financial sovereignty from Japanese attempts to control Thailand's money supply and credit system.

Since then, the Thai banking sector has ramped up its role as financial intermediaries, pooling and transferring monetary resources. With laws and restrictions that aim to strengthen the domestic banking sector, Thai banks have become the center of the kingdom's business community and bolstered its bottom line.

Following the implementation of coherent economic plans that encouraged foreign investment and national development, the Thai economy enjoyed a long period of robust growth in the 1960s. During this time, Thai financial institutions expanded their roles to provide other services, such as clearing and settlement, foreign exchange and risk management. In order to develop the country's capital market as another funding channel for private companies, the Stock Exchange of Thailand (SET) was established in 1975.

Despite boasting a high average GDP growth of 7.8 percent between 1980 and 1996, the Thai economy was beset by various structural problems, especially in the financial sector. As the economic bubble burst, global hedge funds speculated against

the baht. The Bank of Thailand initially (and disastrously) defended the local currency until it almost emptied its foreign reserves, finally succumbing to market forces by floating the baht on July 2, 1997.

The float set off the Asian Financial Crisis and triggered the collapse of the Thai economy. Over six tumultuous months, the baht rose from 25 to the US dollar and peaked at 56 in January 1998. Meanwhile the SET Index plunged to a rocky bottom of 207.31 points in 1998 from its lofty peak of 1753.73 four years earlier.

But in some respects the crisis also proved to be a blessing in disguise. After that debacle, the Thai financial sector underwent a major consolidation. Weak financial institutions were closed down, nationalized, or merged with stronger ones. The surviving banks spent years clearing bad debts, cleaning up their balance sheets and restoring public confidence. All the while, the Bank of Thailand revamped its supervisory role and became more stringent in scrutinizing banks to ensure they followed sound and prudent practices. In addition, the World Bank and International Monetary Fund prescribed some bitter pills in the form of harsh austerity measures.

On the policy front, ever since the year 2000, the Bank of Thailand has adopted a flexible inflation-targeting regime, a popular monetary policy framework used by many central banks all over the world. Its clear inflation target, timely policies and comprehensive communication have helped to keep Thailand's inflation in check over the past decade and a half, which has restored the confidence of both the domestic and global financial communities in the Bank of Thailand.

In 2004, the central bank adopted the four-year Financial Sector Master Plan Phase 1 (FSMP), during which they carried out structural improvements to the sector to enhance its efficiency, strength and access. Key measures under the FSMP Phase

"In the football game of the economy, the central bank acts as center back. In order to safeguard stability during the recent circumstances, we have played a proactive role due to high economic uncertainty. Now that political and economic clarity is greater, the central bank will play a supportive role."

**Former Bank of Thailand governor
Prasarn Trairatvorakul**

Siam Commercial Bank's Chaloem Nakhon branch.

I included the promotion of voluntary mergers to strengthen the financial status and widening the scope of commercial bank business to embrace "Universal Banking," which enables banks to serve all client groups and perform almost all types of financial transactions.

The Bank of Thailand rolled out the FSMP Phase II in 2010, consisting of measures to reduce system-wide operating costs, promote competition and strengthen financial infrastructure. Foreign banks operating in Thailand were also allowed to open more branches or to become subsidiaries. The main purpose of FSMP Phase II is to ensure that Thailand continues to gain strength, improve its efficiency levels and provide better financial access that will enable its financial sector to meet future challenges.

While financial institutions have faced intensified competition from domestic and foreign peers, as well as from non-bank institutions and the capital market, they have also confronted challenges from the increasing globalization of the economic and financial systems. One major example of this was

Thailand's reaction to the global economic crisis in 2008. Lessons learned from the economic collapse of 1997 mitigated the effects of this downturn.

Overall, the continuing implementation of the FSMP over the past several years has substantially improved Thailand's financial infrastructure, ensuring that the system functions smoothly and supports sustainable economic development. In the years to come, innovation and growth are expected to be the financial sector's watchwords, with increased movement toward creative forms of investment, such as crowdfunding.

The emergence of **FinTech** operators in Thailand has also prompted some banks to craft more sophisticated digital payment platforms as a means to prevent losses of traditional sources of revenue from payments, deposits and lending. "Early investment in digital banking will be costly with low return, but it's a requirement for them to fend off any possible major disruption from non-bank fintech operators," Kasem Prunratanamala, head of research at CIMB Securities (Thailand) Co in Bangkok, told *Bloomberg* in March 2017.

FinTech:

An industry comprised of startups and established financial and technology firms that utilize new technology, software and innovation in order to replace or enhance the usage and delivery of financial services.

Challenges Facing the Financial Sector

Debts on the rise Escalating household debt could lead to a greater number of debt defaults, jeopardize financial stability and cause an economic slowdown, if it isn't already. The ratio of debt to income is 121 percent, according to a 2015 McKinsey report, which is on par with levels seen in the US and UK.

Helping SMEs SMEs have a difficult time accessing bank loans because of their size and lack of proven track records.

Loan sharks Low-income earners are incapable of accessing traditional financial services and instead often rely on loan sharks. According to 2011 research funded by the Bill & Melinda Gates Foundation, approximately 27 percent of Thais reported borrowing money, but only 19 percent borrowed from a financial institution and two percent went to an informal private lender.

Financial literacy Many Thais lack basic financial know-how. Heavily influenced by consumerism, many find themselves challenged to provide properly for their children or fall into crisis when the unexpected, such as a health scare, occurs.

Crucial collaborations Close cooperation between the Bank of Thailand and the Ministry of Finance is crucial. The issues surrounding the bank regulating Significant Financial Interests are a case in point.

Investing abroad The Bank of Thailand has in recent years relaxed regulations to encourage Thai companies to diversify their investments through outward Foreign Direct Investment, which will partly help offset baht appreciation pressures from the influx of capital

"I argue that it's appropriate for people to pursue the profit motive in business. If you want to change that, you're going against human nature. But when it comes to setting the rules or creating the institutions, then you should have the general interest at heart, even if it conflicts with your personal interest."

George Soros, hedge fund manager who bet large sums against the baht prior to the Asian Financial Crisis

inflows. That said, uncertainties in global economic and financial conditions could put these outward investments at risk.

Hedging bets Thailand needs to develop more hedging tools for the private sector to manage risks such as those involving the interest and exchange rates, as well as commodity prices. More types of insurance should also be introduced to provide financial cushions when needed.

Fiscal imbalances Thailand needs to find ways to tackle fiscal imbalances, in large part because government revenues tend to remain steady while expenditures rise. This can be especially apparent when

it comes to caring for a rapidly aging society like Thailand's.

State blunders Off-budget government activities such as the rice-pledging scheme can incur losses and erode fiscal stability. Excessive use of such tools may hamper the government's ability to leverage, increase the sovereign credit risk and eventually lead to systemic problems.

Political uncertainty Foreign investors and corporations remain uneasy about the decade-long political strife. Without political follow-through on promises – for example, to improve flood prevention – Thailand's appeal as an investment destination may decline in the face of rising regional competition.

FURTHER READING

- Keynote speech by Dr Prasarn Trairatvorakul, Former Bank of Thailand Governor, at the Application of the SEP to Macro-Economic Management and Business Practices forum, 2016 (available online)
- "The Asian Financial Crisis: What Have We Learned?" by Timothy Lane, *Finance & Development* (magazine of the IMF), 1999
- *The Ascent of Money: A Financial History of the World*, by Niall Ferguson, 2008

TIMELINE

Weathering Economic Turbulence through Checks and Balances

1904

The Book Club becomes the first local commercial bank in Thailand, changing its name to Siam Commercial Bank in 1906.

1940

The Thai National Banking Bureau starts to lay the groundwork for central banking and managing government debts. Two years later it morphs into the Bank of Thailand.

1949

Thailand joins the International Monetary Fund and the World Bank.

1960s

As the Cold War heats up and the Vietnam War explodes, American money pours into Thailand.

1984

The Bank of Thailand adjusts the exchange rate from pegging with the US dollar to a basket of currencies and at the same time devalues the baht by 15 percent to 27 to the US dollar.

1975

The Stock Exchange of Thailand starts trading.

1961

Aided by the World Bank, Thailand begins implementation of its first National Economic Development Plan, which succeeds in promoting economic growth through increased private and foreign investment.

1993

Licenses for Bangkok International Banking facilities are granted as part of financial liberalization attempts.

1997

The baht flotation on July 2 marks the official commencement of the Asian Financial Crisis that eventually wipes out most domestic financial institutions.

2000

The Bank of Thailand adopts the inflation-targeting scheme.

2006

Thailand's central bank imposes draconian capital controls to try to stem rapid appreciation of the baht, forcing offshore investors to keep their money in the country for at least a year or face stiff penalties for early withdrawals.

Gold bars donated by monks to the central bank in 2001.

"If crowdfunding can help innovative start-ups, it could help make Southeast Asia's second-largest economy more competitive.... Thailand needs a start-up generation to help improve the competitiveness. We need to be more innovative and develop our own brands. Global demand is changing."

Vorapol Socatiyanurak, secretary-general of Thailand's Securities and Exchange Commission and member of the National Legislative Assembly

2008

Lessons learned from the 1997 crisis help to mitigate the effects of the global downturn.

2011

Economic disruption from devastating floods causes the central bank to implement the down cycle in policy interest rates for over three years.

SOCK IT TO THEM HARD, NIX.

TRADE

Exports are still the engine of the Thai economy

For centuries, partly due to its advantageous position in Southeast Asia and partly because of its openness to foreign merchants and cultures, Thailand has been the region's gateway. Blessed with abundant natural resources and agricultural goods galore, the kingdom is one of the most economically advanced ASEAN economies. As a founding member of this regional bloc, it boasts a regulatory and trade environment that is amicable to foreign investors.

The Thai economy received a further boost with the implementation of the ASEAN Economic Community (AEC) in 2015, which has eliminated major tariffs and encouraged more interregional trading. Thai investors can now benefit from the stock exchanges within the ASEAN network. Some other AEC perks include various "ASEAN+1" free trade agreements with China, Japan, Korea, Australia, New Zealand and India, which will expand the trade boundaries for Thai goods.

Overall, the Thai economy continues to be export-centric with exports making up more than half of Thailand's GDP, or some US$222 billion in 2016. The country's key export partners are the United States, China and Japan, followed by Australia, the ASEAN countries and the European Union. Topping the list of export items are automobiles, electronics (such as computer hard drives), jewelry and rubber products – as well seafood and rice.

Foreign investment will continue to play a significant role in the Thai economy with investors finding appeal in the country's infrastructure, modern legal framework, many active foreign chambers of commerce and government policies that are free-trade friendly. Since 1995 Thailand has been a member of the World Trade Organization (WTO), which replaced the General Agreement on Tariffs and Trade (GATT), another trade group that Thailand had joined in 1982. The country also has free trade agreements with Australia and New Zealand.

That openness, which has always been a defining Thai trait in terms of culture and economics, whether in times of strife or prosperity, continues

---

THAILAND'S EXPORTS AND IMPORTS

THAI EXPORTS Total: **US$ 222 BILLION** in 2016

THAI IMPORTS Total: **US$ 203 BILLION** in 2016

KEY PARTNERS
- Japan
- Australia
- China
- ASEAN
- USA
- The EU

TOP EXPORTS
- Automobiles
- Electronics
- Jewelry
- Rubber products

KEY PARTNERS
- Japan
- China
- USA
- ASEAN

TOP IMPORTS
- Fuel
- Mechanical & machine parts
- Electronic appliances & parts
- Chemicals
- Jewelry & bullion
- Intergrated circuits

Source: Ministry of Commerce

Situated some 25 kilometers north of Pattaya, Laem Chabang is the kingdom's largest port.

to pay dividends. According to the United Nations Conference on Trade and Development (UNCTAD) World Investment Report 2016, Thailand is a top ten priority destination for foreign investment. Infrastructure is crucial in bolstering such investments. With eleven international airports and numerous deep-sea ports, Thailand is the transport hub of Southeast Asia. That's part of the reason why it was ranked number 46 out of 190 countries in the World Bank's 2017 Doing Business Index, placing third in Southeast Asia, behind only Singapore and Malaysia.

But trade is symbiotic. Thailand also depends heavily on imports. In 2016, those imports totaled some US$203 billion. Fuel topped the list of imports, followed by mechanical and machine parts, electronic appliances and parts, chemicals, jewelry and bullion, and integrated circuits in descending order of demand. The country's biggest import partners are China at 20 percent, Japan at 15 percent and the US with less than a seven-percent share. A report entitled "Asia Development Outlook of 2016: Asia's Potential Growth," prepared by the Asia Development Bank, forecasted that domestic demand should intensify in the years to come.

After the political turmoil of recent years, Thailand's GDP is expected to finally recover somewhat, but will continue to be driven by exports, said Nuntawan

> **"Further economic recovery will depend on the competitiveness of Thai export products and political stability in the years to come."**
> **World Bank, East Asia Pacific Update, April 2015**

Sakuntanaga, director-general of the International Trade Promotion Department in the Commerce Ministry.

However, Thailand's manufacturing industry is still in need of an overhaul. To improve its flagging competitiveness and bolster a sluggish export sector, Thai manufacturers must embrace automation and digitalization, innovate more, and begin pursuing more Original Design Manufacturing (ODM) and Original Brand Manufacturing (OBM).

Analysts estimate that a restructuring of Thailand's export industries may take up to ten years. While major structural changes take place, short-term solutions include exporting more to local markets, such as Cambodia and Myanmar, where consumption is rising. Cleaning up labor practices could also improve trade relations. In 2015, an EU ban on Thai seafood threatened 30 billion baht in exports. Eliminating trafficking is sure to decrease reputational risk, putting Thailand in the good graces of international trade partners once again.

Why Trade Matters to Sustainable Development

Ignites growth Exports have historically driven Thailand's economy and that trend is expected to continue. Altogether, exports make up almost two-thirds of the country's GDP, much of it driven by key sectors such as automobiles and electronics.

Boosts revenue International trade brings in foreign currencies of which 80 percent is the US dollar, one of the strongest currencies in the world.

Adds value Trade boosts the value of the country's natural resources, for example when raw materials like gemstones are turned into Thai-style jewelry.

Provides jobs Export-related industries require a significant labor force and therefore provide employment to millions of Thais in a variety of sectors, from manufacturing to agriculture.

Bargaining chips As a founding member of ASEAN and one of its most economically advanced nations, Thailand is well placed to take the lead in fostering and strengthening inter-regional and global trade relations. Thailand's memberships in the ASEAN Economic Community (AEC) and World Trade Organization (WTO) also increase its bargaining power in trade negotiations and geopolitical disputes.

Positive PR Fostering good relations with export and import partners often has a trickle-down effect that has positive repercussions across sectors like tourism while bolstering the demand for Thai goods.

Challenges Facing Thai Trade

ASEAN trade hub The Thai government wants to make the country into the trade hub of ASEAN to capitalize on the formation of the AEC. On paper, this idea is sound. In practice, Thailand has a long way to go, especially in developing logistics and integrated IT systems to compete with the likes of Singapore and even Malaysia.

Food safety If Thailand wants to make good on its reputation as the "Kitchen of the World," then food safety is of paramount importance.

Fair trade Thailand has legal mechanisms and organizations to support fair trade, such as the Trade Competition Act (1999) and the Office of Thai Trade Competition Commission, to oversee domestic trade, but the laws must be enforced.

Unraveling red tape Boosting cross-border trade with Thailand's neighbors requires solving political disputes and streamlining bureaucratic procedures to modernize the customs department and to provide one-stop services.

More R&D Thailand has begun to talk big about promoting innovation and R&D. Several new innitiatives including Thailand 4.0, Innovation Coupon and Talent Mobility have been launched to help foster cooperation on R&D and encourage innovation-minded business endeavors, but the jury is still out on the results.

Digital economy As the world gets more and more wired, Thai companies need to cash in on the boom in digital retail and the rise of mobile networks. In order to capture a share of the market for online retailing, Thailand must modernize its e-commerce laws in order to assure buyers and sellers that they are trading with fair regulations and platforms.

> *"Trade liberalization is vital to the process of development. Voluntary international exchange widens consumers' range of effective choices and lowers the risk of conflict."*
>
> **James A. Dorn, *Why Freedom Matters***

FURTHER READING
- *Asia Development Outlook of 2014: Fiscal Policy for Inclusive Growth*, prepared by the Asia Development Bank, 2014
- *Thailand: Economy and Politics*, by Chris Baker and Pasuk Phongpaichit, 1995
- Office of the National Economic and Social Development Board, http://eng.nesdb.go.th
- International trade statistics on the Bank of Thailand website, www.bot.or.th/English/Statistics/EconomicAndFinancial/

SPOTLIGHT

The China Factor

As China morphs into a superpower, all the other nations of Asia have been pulled along in the slipstream created by her enormous economic engines. Thailand is no exception. The world's most populous nation is one of the top export markets for Thai goods. While China's middle class flexes its purchasing power muscles and hungers for more products, this staggering sum can only grow.

In 2016, for example, trade between the two countries amounted to over US$68 billion, a leap of around 60 percent over 2010, according to the Ministry of Commerce. In total, around 11 percent of Thailand's exports go to China (second only to the US at 11.25 percent). Some of Thailand's key exports to China include computers, rubber, plastic, components for electronic devices, delivery trucks and refined petroleum. Conversely, China is the largest import partner for Thailand with 20 percent of the kingdom's goods coming from there.

Historically speaking, China and Thailand have enjoyed a robust and mutually beneficial relationship as trading partners. Cultural links and geographical proximity between the two countries led to significant waves of emigration from China to Thailand over hundreds of years. Today, the vast majority of Thailand's 50 richest businessmen are Thai-Chinese (mostly second- or third-generation Chinese). The Thai monarchy has historically enjoyed a mutually beneficial relationship with the Chinese (or Thai-Chinese) merchant class who have generally adopted Thailand as their homeland and generally assimilated quite well.

In 1975, Thailand established diplomatic relations with the People's

Boatload of Thai exports bound for China.

Republic of China. Three years later the countries signed their first trade agreement. As an early part of the ASEAN-China Free Trade Agreement, in 2003 Thailand and China signed a pact that liberalized agricultural markets and cut the tariffs on fruits and vegetables.

The two countries strengthened their fiscal commitments to each other by signing a five-year development plan in 2012 that set specific goals such as expanding trade and investment, boosting tourism and raising bilateral trade to 15 percent a year. "China and Thailand are one family," said the former Chinese Premier Wen Jiabao during a meeting with his Thai counterpart, Yingluck Shinawatra, in Bangkok that year.

With the US cutting military assistance to Thailand and Europe suspending trade negotiations following Thailand's military coup in 2014, the government has made consolidating its ties with China a top priority. More recently, Premier Li Keqiang signed

two memorandums of understanding, one to build a multi-billion dollar rail project and the other to purchase a huge amount of agricultural goods. The railway will link the Chinese city of Kunming with other Southeast Asian countries, while the agricultural deal involves the purchase of two million tons of rice and 200,000 tons of rubber from Thailand.

The launching of the ASEAN Economic Community (AEC) is good news for China and Thailand's economic relations, as the kingdom seeks to secure more investments from the mainland and attract more Chinese tourists.

Developing infrastructure projects is another pressing matter; in the pipeline is a highway that connects southern China with Thailand's largest port, Laem Chabang.

For these reasons, Sino-Thai observers forecast sunny days ahead for relations between the two countries with trade hitting some US$120 billion per year by 2020.

TRANSPORTATION

In the future, cars may take a backseat
to mass transit systems

"Developing countries have been favoring roads over railways for the past 30 years. Sustainability must focus on railways as the worldwide investment in sustainable transportation centers on them. In the next 10 years, 80 percent of investments in public transport in Thailand will go to railways."

Pichet Kunadhamraks, senior civil engineer at the Office of Transport and Traffic Policy and Planning under the Ministry of Transport

In Thailand, the automobile holds dominion over the roads, its dominance the result of two key factors: the car as face-gaining status symbol and a transport system that favors roads. And culturally, in general, Thais, like Americans, love their cars.

When Thailand joined the fast track to economic development in the 1960s and 1970s, American aid money built new roads over former canals and highways through the northeast. As auto prices fell, cars became a more affordable luxury item and a convenient mode of transportation. By the 1990s, with Thailand gaining an international reputation as the "Detroit of Asia" for its car manufacturing plants along the Eastern Seaboard, the roads of Bangkok had become infamously congested. Meanwhile, an aging railway system failed to keep pace with the popularity of road transport. Roads became the favorite mode of not only commuting but transporting goods, with 86 percent of freight sent via these arteries, 12 percent shipped by water and a fraction of a percentage point by air.

However, in the 2000s, the transport network began to shift. First, Bangkok's new, elevated commuter rail, the Skytrain, provided downtown commuters an alternative to driving. The success of the Skytrain encouraged the government to green light plans for 10 more electric railways to encircle the greater Bangkok area, in addition to new commuter trains to be introduced to the capital.

The government also approved plans for double-track lines to be built on all the main routes connecting the different parts of the country as a part of the 10th National Economic and Social Development Plan (2007–2011).

These missing links in Thailand's rail network were long overdue. Although the State Railway of Thailand has been operating for almost 120 years, so far it's only laid some 4,300 kilometers of tracks. More than 90 percent are single tracks and the railway has become notorious for derailments in recent years.

Meanwhile, the military-backed National Council for Peace and Order resolved to work out a plan for the development of national transport infrastructure projects. Transport Minister Arkhom Termpittayapaisith said that Thailand needs continuous investment to get these projects up and running. The budget was set at almost two trillion baht (approximately US$60 billion). More than half of that money will be allocated for railway construction, but the total excludes the cost of double-track, standard-gauge railways and the expansion of Suvarnabhumi International Airport.

The plan consists of five main areas. The first deals with the improvement of existing interprovincial railways, including the replacement of rolling stock, the installation of modern signaling systems, and the construction of double-track railways – both meter and standard gauge – to connect with China via neighboring countries like Laos and Myanmar. The second part consists of buying new buses for the Bangkok metropolitan area and its vicinity, improving roads and bridges, and adding mass transit lines in and around the capital, which will expand the existing Skytrain and subway routes into neighboring provinces. By the time these new lines are up and running, between 2016 and 2022, Bang Sue Station will be the hub connecting them with high-speed railways and buses. With the addition of park-and-ride services, the authorities hope that some 62 percent of commuters will be riding the mass transit system to work and to their leisurely pursuits on weekends – a massive leap over the five percent of commuters who use these services today.

The BTS Skytrain has blazed a new trail for mass transit in Bangkok.

The aviation industry of ASEAN will expand rapidly as the market share of low-cost airlines exceeds 50 percent. The International Air Transport Association predicts that air transportation in Asia and the Pacific will grow faster than that in other regions and is expected to be the source of more than half of all new passengers globally over the next 20 years. Therefore, the capacities of Bangkok's two international airports must be further increased. For starters, the government has set a target to increase the capacity of Suvarnabhumi and Don Mueang airports from serving 45 million passengers per year to 85 million and from 18.5 million to 30 million, respectively, by 2022.

The plan's third pillar supports the building of better highways that will form more efficient linkages between urban warehouses and rural farmlands, and facilitate transfers between tourism hotspots. It also covers the construction of intermodal facilities such as cargo terminals to improve the transport of goods between roads and railways. The fourth and fifth parts of the plan cover the improvement of waterways, including piers and sea ports, while also promoting regional airports and aviation-related industries.

These big ambitions are in danger of stalling due to threats such as political haggling, corruption and red tape. But without a better and more cost-effective transport system that emphasizes railways over roads, and gives commuters better options, Thailand's sustainability and competitiveness will be compromised – not to mention the fact that traffic is likely to remain the number one grumble in the capital, as always.

TRANSPORTATION INFRASTRUCTURE COMPARED TO NEIGHBORS

Infrastructure rankings according to the Global Competitiveness Report 2016-2017 (out of 138 countries)

#2 **SINGAPORE**

#24 **MALAYSIA**

#49 **THAILAND**

Source: Global Competitiveness Report, 2016-2017

COST OF TRANSPORT (per ton per kilometer)

By road
2.12 baht

By rail
0.95 baht

By ship
0.65 baht

By air
0.10 baht

Source: Office of Transport and Traffic Policy and Planning

Why Transportation Matters to Sustainable Development

Hub of choices The growth of industry, agriculture and tourism requires an efficient and balanced transport network that provides commuters and businesses with a wealth of different options.

Staying competitive Transport and logistics have a direct effect on a country's competitiveness. In 2016, logistic costs were equivalent to 14 percent of Thailand's GDP, according to NESDB. Additionally, Thailand ranked 45th in the World Bank's 2016 Logistics Performance Index out of 160 economies, placing it third in Southeast Asia after Singapore (2) and Malaysia (32).

Mass transit networks The construction of 10 mass transit lines covering some 464 kilometers of greater Bangkok will boost the share of mass transit passengers to 60 percent of all commuters, authorities hope, while reducing the number of motorists by up to 40 percent, in order to alleviate the severity of traffic jams, which in turn should improve air quality and reduce fuel consumption.

Quality of life Many countries have created effective mass transit facilities. In Japan, for one, the network of trains is the swiftest and cheapest way to get from Points A through Z. In Holland, a grid of bicycle lanes that link cyclists to parking areas, mass transit systems and a downtown where no cars are permitted makes two-wheeled commuting a breeze. Such examples demonstrate not only a well-designed transportation system, but also some well-coordinated land use planning.

Green sea transport The development of sea ports facilitates a more economical and environmentally friendly mode of transport.

Economic value Mass transit systems unleash new business opportunities around the stations and hubs through which they pass. Highways, railway lines and air routes could help bridge the widening income gap between urban and rural areas.

Gateway to ASEAN Because Thailand is located in the center of Southeast Asia, the country is a potentially strategic hub for railways. In a joint venture with China, new rail links will be built from Bangkok via the Kaeng Khoi district of Saraburi province to the Map Ta Phut area of Rayong province and from Kaeng Khoi via Nakhon Ratchasima province to Nong Khai province, to connect southern China with Laos and Thailand. Beset by delays, it is hoped that the new lines will open by 2018–2020.

"The problem about mass transit development in Thailand is that the projects are planned separately without a vision for connecting them with other kinds of transportation such as passenger buses and vans and motorcycle taxis. This contributes to the escalating costs of urban travel. The government must introduce mass transit services that link with different modes of transportation and they may have to work out systematic subsidies for such services. Such subsidies must be funded with local taxes. Nationwide taxpayers should not be footing the bill for free bus services in Bangkok."

Sumet Ongkittikul, research director for transportation and logistics at the Thailand Development Research Institute in an interview on February 23, 2015

Managing Chronic Traffic Jams

In 2013 the BBC gave Bangkok the first slot in its "Monster Traffic Jams" story. This was not breaking news to any locals, expats or tourists. Traffic congestion has been a chronic complaint for decades. Every time the high tides of traffic seem to be abating somewhat, a new wave of motorists hits the streets to flood the gridlock with yet more vehicles – as was the case in 2011 when the administration of Yingluck Shinawatra introduced a rebate for first-time car buyers, a policy that spurred GDP but ran counter to sustainable development goals. Domestic car sales reached 1.45 million in 2012 and 1.33 million in 2013.

the 20th century: around 16 kilometers per hour.

The rising number of automobiles reflects the growing number of motorists with means – or at least access to credit – who see the car as one of the ultimate status symbols. Of the more than eight million vehicles registered in Bangkok as of 2013, some 57 percent were private vehicles, 37 percent were motorcycles, two percent were trucks and around one percent were taxis. Since the economic boom of the late 1980s and early 1990s every successive government and each new Bangkok governor has had to push the traffic issue to the forefront of their agendas. But

sion with the corrupt traffic cops who let them off the hook for small bribes.

The authorities could take some cues from other countries which have made traffic into a big issue. The United States, for example, has implemented national strategies to improve traffic flows by imposing fees on motorists entering inner city areas, by operating express buses and by promoting strategies to reduce unnecessary trips, as well as staggering working hours. Singapore, on the other hand, operates an electronic road pricing (ERP) system to collect road usage fees during rush hours. In Japan, motorists must prove that they have their own parking space before

During the morning rush hour in Bangkok, a city that can accommodate around two million vehicles in transit, there are an estimated five million vehicles plying the streets at an average speed slower than a horse-drawn carriage at the turn of

more flyovers, more expressways, more traffic cops and the building of the Skytrain and subway have not solved the dilemma. The problem is exacerbated by a lack of private car parks and bad drivers who do not respect the rules of the road in collu-

they can apply for a driver's license.

Perhaps it's time that the Thai government thinks about implementing such measures aimed at changing the behavior and mindsets of automobile drivers by taking aim at their most vulnerable parts: their wallets.

Speedy Trains on the Fast Track?

A high-speed railway project that was shelved during the political conflicts of recent years appears to be getting back on track but not without some speed bumps. The seven new routes, covering some 3,000 kilometers, will be the first standard gauge, double-track lines in Thailand. The wider gauge – 1.435 meters – can accommodate high-speed trains.

In September 2016, after two years of negotiations and more than a dozen ministrial meetings, China agreed to scale back the cost of the first phase of an 873-kilometer high-speed railway line that will link Nong Khai province on the Thai-Lao border with the Map Ta Phut industrial estate on the Eastern Seaboard (with a branch from the Kaeng Khoi district of Saraburi eventually connecting with Bangkok). The cost of the first phase has been reduced to 179 billion baht (US$5.15 billion) from an initial estimate of more than US$16 billion after Thailand said the initial price tag was too high.

The first phase of construction, on a 250-kilometer section of track, began in December of 2016. Thailand will bear the full construction cost, while China is to provide funds for technical systems. If all goes to plan, construction should be completed by the 2019-2020 fiscal year.

Another important route will run south from Nakhon Pathom to Hua Hin on the Gulf of Thailand, and from there onto Prachuap Khiri Khan and Chumphon. Thai transport authorities said high-speed trains will help connect more remote parts of the country and redistribute wealth to the provinces. The catch is that these new routes might not attract many investors, aside from a few strategic lines favored and backed by the Chinese and Japanese, nor much interest from local commuters and

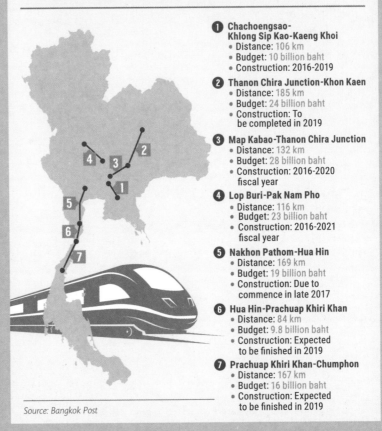

PLANNED DOUBLE-TRACK RAIL LINKS

1 Chachoengsao-Khlong Sip Kao-Kaeng Khoi
- Distance: 106 km
- Budget: 10 billion baht
- Construction: 2016-2019

2 Thanon Chira Junction-Khon Kaen
- Distance: 185 km
- Budget: 24 billion baht
- Construction: To be completed in 2019

3 Map Kabao-Thanon Chira Junction
- Distance: 132 km
- Budget: 28 billion baht
- Construction: 2016-2020 fiscal year

4 Lop Buri-Pak Nam Pho
- Distance: 116 km
- Budget: 23 billion baht
- Construction: 2016-2021 fiscal year

5 Nakhon Pathom-Hua Hin
- Distance: 169 km
- Budget: 19 billion baht
- Construction: Due to commence in late 2017

6 Hua Hin-Prachuap Khiri Khan
- Distance: 84 km
- Budget: 9.8 billion baht
- Construction: Expected to be finished in 2019

7 Prachuap Khiri Khan-Chumphon
- Distance: 167 km
- Budget: 16 billion baht
- Construction: Expected to be finished in 2019

Source: Bangkok Post

travelers. Construction and bidding processes also continue to be beset with delays. Most recently, the Terms of Reference for bidding on five of the projects were scrapped in March 2017 due to alleged irregularities.

In any case, neighboring countries are scrutinizing the situation in Thailand very closely. Because it's the geographical center of the region, Thailand's rail links can determine the future of projects in other ASEAN countries, which means for Thailand, the return on these astronomical investments could be significant as

it positions itself as the hub of ASEAN and a transport bridge between its neighbors and China.

Thailand and Malaysia are set to start talks on the construction of a 1,500-kilometer high-speed railway that would connect Bangkok and Kuala Lumpur. Part of a pan-Asia railway network, the Bangkok-Kuala Lumpur route would be an integral link in a route that eventually runs all the way from Kunming to Singapore, also passing through regional neighbors Cambodia, Myanmar, Laos and Vietnam.

FURTHER READING

- Thailand's Transport Infrastructure Development Strategy 2015–2022 of the Office of Transport and Traffic Policy and Planning (OTP) or www.otp.go.th
- *Guidelines for Transport Development for Sustainable National Development* by the Ministry of Transport, 2012
- *Railway Development in Greater Bangkok* by the Office of Transport and Traffic Policy and Planning, 2014
- *Facilitating the International Railway Network to Promote Thailand as an Economic and Touristic Hub of the Region* by Chalongphob Sussangkarn et al., Thailand Development Research Institute, 2012

REALITY CHECKS

Challenges Facing the Transport Sector

Stops and starts Political instability and corruption stall projects. A case in point is the 106-kilometer-long double-track railway project from Chachoengsao province via Khlong Sib Kao to Saraburi province. The project was supposed to start in 2009 but did not actually get underway until 2016. There is also skepticism over the Transport Infrastructure Development Strategy 2015 to 2022.

Missing links Transport-related organizations often plan their own mass transit projects without consulting each other or linking up with other modes of transport. A prime example is the one-kilometer gap between the Blue Line's Bang Sue station and the Purple Line's Tao Poon station, which has become known as "the missing link." However, the government has announced plans to rectify the issue by mid-2017.

Resourceful staff As of the beginning of 2017, there were six mass transit lines operating in the greater Bangkok area, with several new extensions and lines under construction. In addition, there are standard gauge tracks laid down nationwide. To maximize their potential, these systems will require an army of well-trained personnel dependent on the development of human resources and technology transfers.

Entering ASEAN The AEC will bring about both opportunities and risks. For Thailand, the number of potential consumers will rise tenfold from 68 million to more than 600 million, which means it will also have to stave off competition from neighboring countries marketing the same goods. Without a cost-effective transport system the country's competitiveness will be compromised.

Better logistics A key factor in competitiveness is the shift from low-cost manufacturing to responding faster to customers' demands with more efficient logistics.

Branching out Building transportation networks with neighboring countries, especially through roads and railways, is inevitable. However, expansion is not possible without standard regulations, like insurance services. Without such rules, a given country may not allow vehicles from Thailand to pass through its territory to reach a third-party country.

Unclear separation Organizations responsible for making and regulating transportation policies have yet to be clearly separated to ensure safety, fair fees and efficient maintenance. For example, many countries and states around the world have their own Department of Railways, or Ministry of Rail Transportation, which makes policies and determines investments of rail infrastructure.

Capacity building The aviation industry of ASEAN will expand rapidly as the market share of low-cost airlines exceeds 50 percent, as will the number of arrivals by air. The capacities of Suvarnabhumi International Airport and Don Mueang, the base for low-cost carriers, need to be expanded from serving 45 million passengers per year to 90 million and from 18.5 million to 30 million, respectively, by 2022.

Road deaths Thailand has one of the world's highest rates of road casualties. Improving road safety education, enforcing helmet use for motorcyclists, and putting the brakes on drunk drivers would pave the way for safer roads.

LABOR

Low unemployment masks unrealized potential

Since the implementation of the first National Economic and Social Development Plan in 1961, Thailand has banked on capitalism. Boosting economic growth meant promoting industrial development and exports, as well as attracting foreign investors with a skilled yet affordable workforce.

As Thailand grew rapidly over the next few decades, developing such labor-intensive industries as textiles, garments, footwear and agricultural goods, there was no shortage of workers. Driving a mass migration to the cities and their factories and construction sites were such rural woes as poverty, urban encroachment, and a lack of opportunities both fiscal and educational.

In the aftermath of the financial crisis in 1997 that spread from its epicenter in Bangkok to cause severe repercussions all across Asia, many financial institutions collapsed, businesses went bust, and the unemployment rate spiked, spurring a backlash against migrant workers for supposedly stealing jobs. Only then did Thailand see the need to climb the value chain as it could no longer rely on labor-intensive industries and a stable baht to boost exports and economic growth.

Consequently, the Thai labor market experienced a major structural shift on both the supply and demand sides. The economic turmoil forced a great number of factory workers to return to farming or become self-employed. Meanwhile, some entrepreneurs revised their business models, shifting toward more sophisticated enterprises to cope with the fiercer competition from regional peers bolstered by cheaper wages. Thus, they hoped to hire better-trained workers to occupy these new positions.

Given the need for new skill sets and the increasing price of agricultural commodities, many workers did not take up their old posts even as the economy gradually recovered. The shortage of skilled workers was compounded by the number of young college or vocational students who, rather than fill these positions, opted to pursue the possibilities of higher education. This, in turn, caused an imbalance in the workforce resulting in a dearth of semi-skilled workers willing to toil for menial wages. The shortfall has taken its toll on foreign investment and prevented the Thai economy from growing to its full potential.

At the beginning of 2013, the Yingluck Shinawatra administration implemented a minimum wage hike to a flat rate of 300 baht per day nationwide, a big leap from the previous 171–215 baht (a sliding scale for workers in different parts of the country). Although the controversial decision gave workers a nominal boost in wages, it has failed to keep pace with inflation and, because of a spike in the cost of living, workers may still be worse off than before the pay raise. On the other hand, experts have noted that the raise of wages has outpaced productivity, causing a decline in Thailand's global competitiveness.

Representing another populist policy, the boosting of the starting salaries for civil servants with university degrees to 15,000 baht (and encouraging the private

INFORMALLY EMPLOYED

64%
of Thai workers operate in the informal sector.

0.7%
of the workforce is considered unemployed.

KEY NUMBERS

30%
Roughly the percentage of the labor force with university degrees or higher who joined the informal sector. According to an SCB report in 2014, this figure has risen from 22% a decade earlier.

300
US$9
current minimum daily wage.

1.6 - 3.5 MILLION Migrant Workers
80% of whom are from Myanmar.

Source: Department of Labor, National Statistical Office of Thailand

3D jobs:
"Dirty, dangerous and difficult" or "dirty, dangerous and demeaning" work, frequently undertaken by migrant laborers instead of the local population.

A group of migrant workers on the deck of a fishing boat in Phuket.

sector to do the same) has led students in secondary and vocational schools to aim for bachelor degrees, further adding to the shortage of blue-collar workers.

Another side effect of the minimum wage hike has been the influx of laborers from neighboring countries. Massive numbers of both legal and illegal workers from Myanmar, Cambodia and Laos are now predominantly employed in "**3D" jobs** that are considered beneath the dignity of Thais. At present, more than 1.5 million migrant workers are registered at the Office of Foreign Workers Administration, but the government and labor experts estimate that the actual figure, including illegal workers, is closer to 3.5 million, some 80 percent from Myanmar.

With increasing dependence on migrant workers at home, and negative press abroad about Thailand's human rights violations, the kingdom needs to carefully calibrate policies to ensure that these foreign workers are treated with dignity. Such safeguards are necessary to avoid more stains on the kingdom's international image and to sustain the country's economic growth, which is already under threat from a slumping birth rate and an aging population that may not have the means to retire as early as they had in the past.

These recent issues aside, Thailand has worked hard to keep its people at work. As of 2015, the workforce consisted of 39.2 million people, or almost 71 percent of the working age population. And the unemployment rate has been on the decline over the past decade. Between 2011 and 2016 it averaged 0.69 percent, the second-lowest rate in the region, thanks in no small part to the massive informal sector.

The Informal Sector

A whopping 64 percent of Thai workers make ends meet in the largely unregulated, untaxed and often unaccounted-for informal sector, also referred to as the "grey economy." This is not the criminal underworld but a critical source of jobs to the poor or less educated. From street vendors to motorcycle taxi drivers to freelance creative and Internet entrepreneurs to small mom-and-pop shophouse businesses, Thailand has one of the largest informal economies in the world. In Asia in general, the informal sector is estimated to account for 60 percent of the workforce. Problems posed by the informal sector include the fact that its workers rarely receive security or social benefits, and at worst, are exploited. The sector is also difficult for governments to manage. Skills are not enhanced and economic value is not maximized.

The Long and Arduous Struggle for Equality and Fair Pay

AD 1350–1767

During the Ayudhya period, the *sakdina* or feudal system is established, meaning slaves are considered the property of royalty, aristocrats and high-ranking officials. Peasants also have to spend six to eight months per year working for feudal lords. Prisoners of war are treated as the king's property.

17th–19th centuries

During the early Rattanakosin period, the number of slaves and peasants rises to a third of the population, partly due to several decades of wars against neighboring empires to capture enemy soldiers and convert them into slave labor. During the reign of Rama III (1824-1851), for instance, there are an estimated 46,000 war slaves.

Pre-1932 revolution

The working class becomes the backbone of new, export-driven businesses, such as rubber, sugar, timber and electronics.

1919

Thailand becomes one of the 45 founding members of the International Labor Organization after WWI.

1905

Rama V abolishes slavery in Siam as part of an attempt to modernize the country amid threats of colonization.

1855–1858

The introduction of industrial rice mills gives rise to the working class in Siam. The majority of these workers are Chinese.

1957

First trade union forms under 1956 Labor Act.

1957–1963

The so-called "dark age" of the labor movement during Field Marshal Sarit Thanarat's administration, which implements draconian measures against those disrupting peace and government orders.

1965

Department of Labor is established under the Ministry of Interior.

1997

The Asian Financial Crisis leads to business closures and a surge in the unemployment rate.

1975

The government divides workers into two groups by law. The Labor Relations Act, B.E. 2518, is implemented to govern workers in the private sector, while later in 2000, the State Enterprises Labor Relations Act, B.E. 2543, governs the civil service.

1973

Labor associations co-operate with farmers and students to topple the dictatorship of Field Marshal Thanom Kittikachorn, inspiring a pro-democracy uprising that leads to bloody clashes on October 14 followed by more battles and a right-wing backlash on October 6, 1976, at Thammasat University.

2013

The government led by Yingluck Shinawatra implements a minimum wage hike to a flat rate of 300 baht per day nationwide, a big leap from the previous minimum that applied to workers in different parts of the country.

2014–2016

The US State Department downgrades Thailand to Tier 3, the lowest ranking on its 2014 Trafficking in Persons (TIP) Report, spurring the government to look into the issue more vigorously. In 2016, Thailand is upgraded to Tier 2.

2030

World Bank forecasts demographic change in Thailand that will lead to a severe shortage of workers between the ages of 15 and 64.

"No man needs sympathy because he has to work, because he has a burden to carry. Far and away the best prize that life offers is the chance to work **hard at work worth doing.**"

THEODORE ROOSEVELT, *26th president of the United States of America*

Laborers work high up on the scaffolding of one of Bangkok's many construction sites.

Aspects of Labor in Need of Work

Education mismatch Thai industries are in dire need of skilled scientists, technicians and mechanics. This is the result of an education mismatch, since the majority of university graduates major in social sciences.

Lackluster English skills Unlike surrounding neighbors, Thailand managed to survive Western imperialism in the 18th and 19th centuries to maintain its independence and sovereignty. The downside is that Thais have a comparative disadvantage over neighbors in speaking foreign languages. Throw in weak study habits as well as a rote learning system and the result is that Thai workers are considered to be among the poorest English speakers in all of Southeast Asia.

Double-edged sword The "informal sector" comprising freelancers and those working in small businesses of fewer than five workers has been expanding rapidly of late. This is a double-edged sword that cuts both ways. On the one hand, it has exacerbated the labor shortage in the formal sector. On the other, it has provided a fallback plan for university graduates with degrees in disciplines that possess few career prospects.

Workers' rights wronged Labor protection practices are inadequate in Thailand. As much as 64 percent of the total workforce consists of informal workers, many of whom live in poverty, especially those working the land and living upcountry.

Skills needed Thailand faces impending labor and skills shortages due to an aging population and a workforce lacking proper qualifications in some areas. Deeper emphasis on vocational and industry-specific skills is needed.

AEC worries Deeper regional integration as a result of the AEC could cause the relocation of both investments and industries, as well as a concomitant loss of jobs in Thailand. The key challenge is to combine education and skills development strategies that enhance productivity and competitiveness.

"More attention is urgently needed for these informal sector workers, whose work – by the definition of the National Statistical Office – does not offer any social security or protection."

Thai Health Working Group, "Health Indicators of Thailand's Workforce," 2010

Shoring up human resources Inadequate training programs and human resources activities, especially at small- and medium-sized firms, prevent workers from developing to their full potential.

Migrant workers misused The number of migrant workers is escalating and Thailand's economy has grown dependent on them. They are engaged in almost all activities that require manual labor. With barely any bargaining power, they suffer from unfair wages, low quality of life,

and the absence of a social safety net or any legal protections. In 2014, Thailand was downgraded to Tier 3, the lowest ranking on the annual US "Trafficking in Persons" report detailing abuses of human rights. The demotion was due to the usage of child labor in five different industries, including sugar cane, prawns and

pornography, as well as the exploitation of migrant workers (mostly on fishing boats). In 2016, Thailand was elevated back to Tier 2 to the consternation of some rights monitors.

Late retirement In anticipation of the aging society that is looming due to a 1.5 percent birth rate, Thailand's public and private sectors need to push for policies that promote saving and motivate people to work beyond the traditional retirement age. Employers must also be encouraged to see the value in older workers.

FURTHER READING

- *Thailand: 4.0 and the Future of Work in the Kingdom*, by Will Baxter, International Labour Organization (ILO), 2017
- *ASEAN in Transformation: The Future of Jobs at Risk of Automation*, ILO, 2016
- *Employment Practices and Working Conditions in Thailand's Fishing Sector* by ILO and the Asian Research Center for Migration under the Institute of Asian Studies, Chulalongkorn University, 2013
- Thailand Future Foundation: www.thailandfuturefoundation.org
- *Labor Force Survey in Thailand: 2017*, by the National Statistical Office

SPOTLIGHT

The Impact of the ASEAN Economic Community

The establishment of the ASEAN Economic Community in 2015 was hailed as a major milestone in the regional economic integration agenda of ASEAN. The AEC offers opportunities in the form of a huge market of US$2.6 trillion and more than 622 million people. In the labor sector, there are numerous implications such as the intra-regional movement of specialists in certain fields. However, the regional integration should not bring about any drastic changes to the Thai labor market in the foreseeable future as not all types of workers will be able to move among member nations without restrictions. The AEC blueprint is mainly intended to promote the free flow of goods and services in four different areas: trade and commerce, services, investment and capital. Labor is not a top priority.

ASEAN members have signed Mutual Recognition Arrangements (MRA) that facilitate the movement of qualified workers in seven pro-

fessions, namely, physicians, nurses, dentists, engineers, architects, surveyors and accountants. The MRA for Tourism Professional (MRA-TP), signed in 2012, makes provisions for workers in 32 different tourism professions, from front-desk clerks to maids, travel agents and catering staff, to work anywhere in the region if they are MRA-TP certified. This is the most contentious area. Some locals will certainly face layoffs. Others will lose their jobs to foreigners with stronger foreign language skills.

Even with fewer restrictions due to the AEC, it will still be rather difficult for foreign specialists in the aforementioned seven professions to take jobs from local practitioners due to language barriers, as most patients, clients and co-workers may not be able to communicate fluently in English and miscommunications in these specialized professions could lead to myriad difficulties and dangers in fields like medicine.

It's possible that some local specialists may consider moving abroad but a brain drain is unlikely because most Thais are homebodies, preferring to work here rather than risk living *out there*. Those interested in exploring more lucrative avenues in foreign countries have probably lived abroad anyway.

COMPETITIVENESS

How Thailand measures up in the region
and around the world

The epithet "Teflon Thailand" has long been used to describe the country's ability to slide out of dire situations. For decades its economy remained resilient through coups, floods and paralyzing protests. But perhaps those days are numbered.

In 2014, Asean Confidential, a *Financial Times* research service, calculated that between 2001 and 2005, Thailand's economy grew at an average of six percent a year. By contrast, over the next decade, after divisive political protests and a military coup, it grew an average of 2.3 percent a year. Some of that deceleration had to do with the global economy slowing down and the floods of 2011, but most, says Asean Confidential, resulted from politics. It compared the superior performance of neighboring countries to clinch the argument.

But there is another way of looking at this. As damaging as political turmoil is, it might just be

masking – and probably deepening – the country's more endemic problems. These lie in the heart of Thailand's competitive health.

Thailand's dream of the 1980s and early 1990s to become the fifth "Asian tiger" is now being replaced by the dull reality of getting mired in the "**middle income trap.**" Having maximized the benefits of low-cost labor migration to industrial zones, Thailand looks ill-equipped for the next stage – that of offering higher quality, more productive and innovative manufacturing and services.

During the high-growth period, foreign direct investment (FDI) and low production costs were enough to underwrite the success story behind its Newly Industrialized Country (NIC) status. Governments, including military ones, came and went, but all embraced the expansionary policies of tax privileges and short-term growth. First came labor-intensive industries such as garments and textiles. Soon afterwards a wave of auto industry investments from Japan got the "Detroit of Asia" rolling. Electronic goods were not far behind and in time Thailand won recognition as the single-biggest producer of computer hard drives in the world.

Not even the severe financial crisis of 1997 was able to dent Thailand's progress for long. A 20-plus percent depreciation in the baht against the dollar helped to drive double-digit export growth and open a new chapter in the success story, with a newly transformed and stable financial system underpinned by prudent monetary policies.

But the country's central bank, leading economists and research institutions have warned for years that Thailand has to "move up the value chain," or, in other words, to start producing goods and services of higher quality to replace the declining exports of computer electronics that are fast giving way to the components in smartphones and tablets. To do that, they maintain, Thailand must upgrade its education system and IT infrastructure, while Thai companies

EASE OF DOING BUSINESS

THE WORLD BANK'S EASE OF DOING BUSINESS SURVEY 2017
(out of 190 countries) in comparison with ASEAN countries

#2 SINGAPORE	#99 PHILIPPINES	
#23 MALAYSIA	#131 CAMBODIA	
#46 THAILAND	#139 LAO PDR	
#82 VIETNAM	#170 MYANMAR	

More Thai factories must invest in automation, as well as increase spending on R&D, if the country is to regain its competitive edge.

also need to automate, embrace digitalization and increase spending on R&D. In short, Thailand must be more proactive and innovative. Yet the 2016-2017 World Economic Forum's Global Competitiveness Report shows that Thailand's ranking for innovation fell to 54 (out of 138 countries) from 33 in 2007. At the same time, the education system has failed to address Thailand's poor showing in global performance tables, while the country's need for an infrastructure upgrade to meet the demands of ASEAN "**connectivity**" has been left unaddressed as the political polarization of the last decade appears to have stalled moves to remedy these problems.

To make matters worse, the protracted political turmoil has caused incalculable "brand damage," dramatically advertising the country's weaknesses, such as the arbitrary exercise of power, institutional incompetence, massive corruption and frequently lax law enforcement.

Despite these challenges, there is no doubt that, when it comes to doing business, Thailand trumps most of its neighbors. The World Bank's 2017 Doing Business Report places the kingdom at number 46 among 190 countries, with only Singapore and Malaysia ahead of it. With an overall ranking of 34 in the 2016-2017 Global Competitiveness Index and strong showings in some categories, there is room for cautious optimism.

THE GLOBAL COMPETITIVENESS REPORT

OVERALL GLOBAL RANKING
(out of 138 nations)

#34 THAILAND

#3 USA

#2 SINGAPORE

#1 SWITZERLAND

SPECIFIC RANKINGS

#13 Macroeconomic environment

#54 Innovation

#37 Goods market efficiency

#84 Institutional environment*

*The quality of government, financial, legal and administrative institutions, etc.

Source: Global Competitiveness Report, 2016-2017

Middle income trap:

Some development economists say most countries hit a growth ceiling after a few decades of rapid expansion because they fail to move "up the value chain." This means they continue to compete on low-cost labor and low-quality goods rather than improve the skills and productivity of the country's workforce and infrastructure.

Connectivity:
In economics this refers to linking supply chains (the production, shipment, assembly and distribution of various product components) through better roads, rail, ports and airports as well as more efficient customs procedures.

153

Why Competitiveness Matters to Sustainable Development

Regional supply chains Globalization is also "regionalization." So-called regional supply chains are now the way multinational manufacturing companies work – making one component here and another there, shipping partly finished products across borders so the rest can be done somewhere else. Thailand has to be able to offer attractive options for these companies to win their business, as well as the infrastructure and customs procedures to shift components rapidly and efficiently to neighboring countries in ASEAN and beyond.

Free trade in AEC Trade liberalization hasn't gone away, even if the World Trade Organization is dormant. Bilateral and regional free trade agreements, in particular the ASEAN Economic Community (AEC) that came into being in 2015, continue to open Thailand's economy to strong trade winds of competition.

ASEAN connectivity Thailand has a central role to play in ASEAN's growing economic integration through the AEC and other "connectivity" policies. Its location and resources make it responsible for delivering many of the new railways, roads, ports, pipelines and telecoms linkages on the drawing board, which Thailand's neighbors so urgently need.

Moving up the value chain Thailand can no longer rely on the competitiveness of its low wages. Many of the kingdom's neighbors, in ASEAN and beyond, are becoming newly invigorated competitors for low-cost manufacturers, including garments, footwear and some electronic components, as well as for tourism. Thailand must move up the value chain, offering higher skills and productivity through better education, infrastructure and innovation.

The Thailand brand Global investor confidence and "sentiment" are playing an increasingly important role in deciding where the billions get shifted. While Foreign Direct Investment is still a vital ingredient of a country's growth potential, the growing significance of the ratings-focused corporate and government bond markets means a country's "brand" matters. A decade of political instability has undermined that brand and made global bond investors nervous about the risks involved.

Defining Competitiveness

"Ultimately competitiveness is about raising the prosperity of people, which can be defined as a mix of income, standard of living and quality of life," says Professor Stéphane Garelli, who helps create the annual *World Competitiveness Yearbook* published by the Swiss-based business school IMD (International Institute for Management Development). In a modern economy, competitiveness does not simply ask "How fast can you grow?"

IMD formally defines competitiveness this way:

Competitiveness of Nations is a field of economic theory, which analyzes the facts and policies that shape the ability of a nation to create and maintain an environment that sustains more value creation for its enterprises and more prosperity for its people.

The World Economic Forum, which publishes another annual flagship competitiveness report, defines it as:

The set of institutions, policies and factors that determine the level of productivity of a country. The level of productivity, in turn, sets the level of prosperity that can be earned by an economy.

In the end, competitiveness is about prosperity. But countries only get there through positive interactions between different elements of their state and society. For the IMD, there are three key factors: economic performance, government and business efficiency, and infrastructure. For the WEF, there are 12 pillars of competitiveness, including institutions, infrastructure, macroeconomic environment, health and education, efficient markets, technology and innovation.

The Bank of Thailand: A Model Institution

Central banks play a very visible role in an economy and, at the same time, an almost "spiritual" one – what they say and do affects both the hard-headed decisions of businesses and the more ephemeral sentiment of consumers. And those, in turn, make a considerable contribution to a country's competitiveness.

In this respect, the Bank of Thailand (BOT) has, over the past decade or so, been a star performer, even as other state institutions have been embroiled in the country's divisive politics.

However, it was not always this way. In fact, the BOT played another starring role – of the falling variety – in 1997 when it was at the center of the financial meltdown. A heroic failure by the BOT to defend the baht's "peg" against the dollar ended in the evaporation of the country's reserves and a humiliating 20 percent depreciation of the currency. The economy shrank by 10 percent in 1998 as non-performing loans choked off investment and consumption.

Bank of Thailand headquarters.

"If the country moves toward modern farming and knowledge-based, high value-added service industries, it will be able to escape the 'middle income trap' with sustainability. Two factors really matter: high quality people and investment in research and development. The government has to make these happen."

Somkiat Tangkitvanich, president of Thailand Development Research Institute

There's nothing like learning from your mistakes. By the time the economy was back on its feet, the BOT had put into place procedures not only for its own conduct, but for regulatory and supervisory reforms for the financial system as a whole.

Growth bounced back and by 2003 it peaked at over seven percent. And while politicians have made the occasional barbed comment from time to time, as they do elsewhere, they've kept their distance. The BOT has achieved this despite the fact that it didn't win independence on the execution of monetary policy until March 2008. Even after the coup in May 2014, when the military reshuffled dozens of officials and executives at the top of Thailand's economy, the BOT's distinction was preserved.

The result over the years has been lower inflation, lower-than-average interest rates, a relatively stable currency and sound foreign exchange reserves, even at times of severe cyclical stress and volatility in the global economy.

In July 2014, two months after the coup, former BOT governor Prasarn Trairatvorakul told the Foreign Correspondents' Club of Thailand that the independence of the bank's monetary policy committee was the reason the bank has been able to build up its credibility over the past decade.

"The institutional setup must continue to protect this independence, so that monetary policy can remain a cornerstone of Thailand's macroeconomic stability....The MPC [Monetary Policy Committee) is aware that our operational independence would mean little in the absence of public trust...It is ultimately the very source of our credibility," he said.

Challenges to Improving Competitiveness

Institutions Thailand's institutions are in need of strengthening. Despite scoring a solid 34 in the World Economic Forum Global Competitiveness Index 2016–2017, Thailand ranked an abysmal 84 under the pillar labeled "institutions."

Corruption Thailand's score for Transparency International's 2016 Corruption Perception Index was 35 (where 0 = totally corrupt and 100 = totally clean), falling from a score of 38 in 2015. "Government repression, lack of independent oversight, and the deterioration of rights eroded public confidence in the country," the report noted. By comparison, in 2016 Laos and Myanmar scored 28, while China and India scored 40. For the second year in a row Cambodia was listed as the most corrupt country in Southeast Asia with a score of 21.

Innovation The country's successful development strategy of attracting Foreign Direct Investment into the low-cost manufacturing industry has run its course. In particular, its dominance in the computer disk drive sector has not kept up with changes in global technology where smartphones and other mobile devices are outstripping demand for computers. Spending on R&D is also low compared to similar countries.

Education In all surveys of competitiveness and business–friendly policies, Thailand consistently underperforms at every level in the category of education. The 2015 results of the OECD's triennial Program for International Student Assessment (PISA), which tests 15 year olds, revealed that Thailand placed 54th among the 70 participating countries for mathematics, science and reading. The country's univer-

sities also perform poorly in global tables. Thailand's English-language proficiency continues to prove inadequate in terms of meeting the challenges of a more open professional labor market that has come about with the establishment of the ASEAN Economic Community (AEC).

Infrastructure Thailand needs to maximize its excellent strategic location in ASEAN and its proximity to both China and India. In 2016, the kingdom was on track to spend some US$3 billion on infrastructure development, a figure that is expected to rise to an average of US$9 billion per year until 2020, according to Maybank Kim Eng. The focus will be on constructing new railways, roads and customs posts to establish cross-border trade routes and thus improve Thailand's overall quality of rail infrastructure and import-export facilities.

Funding and credit On the one hand, Thailand scores quite high in competitiveness surveys for "financial market development" and "banking and financial services," including the

costs of capital. On the other hand, access to credit is a big impediment to businesses and, in particular, the development of job-generating small- and medium-sized enterprises. The government's own Competitiveness Report for 2012 complained, "National income per capita remains low, with weak savings and poor investment volume. This reflects the majority of the population's limited access to capital and capability of income generation."

An aging society By 2040, one out of four Thais will be over 60 years old. That will mean that 10 people in the working-age group will have to support at least four elderly persons. This is what is known as a higher dependency ratio. This means the country will have to raise the productivity of the working-age population through the promotion of higher skills, as well as deploy more advanced technology. In addition, Thailand should investigate opportunities for the development of elderly-related businesses, in particular health and medical care services.

FURTHER READING
- *World Economic Forum Global Competitiveness Report 2016–2017*
- *Doing Business 2017: Equal Opportunity for All,* by the World Bank
- *Thailand Economic Monitor: Aging Society and Economy,* by the World Bank, June 2016
- "The Fundamentals and History of Competitiveness," by Stéphane Garelli, *IMD World Competiveness Yearbook,* 2014

TIMELINE

Climbing Up the Ladder of Competition

1961

Thailand's first Economic Development Plan is drafted with support from the World Bank, like many developing countries during the 1960s. The plan identifies FDI as the main engine to drive the Thai economy. Per capita gross national incomes (GNI) is US$101 at the time.

1960s– 1970s

Thailand does well attracting FDI, mainly through the Board of Investment (BOI), created specifically for this purpose in July 1966. A total of about six billion baht pours into Thailand from 1960 to 1982. The aim is partly to substitute locally produced manufactured goods for some imports. Primary farm exports contribute to economic growth.

1984

The Thai baht, pegged against the dollar since 1956, is devalued from 23 against the dollar to 27 baht in order to address a serious current account deficit problem which carried on for more than two decades. The lower baht increases Thailand's export competitiveness.

Early 1980s

Thailand becomes an industrial goods exporter. Under the administration of Prem Tinsulanonda, the government and private sector work more closely together. A key private sector organization, the Joint Public and Private Committee, is set up to convey to the government the needs of businesses.

1970s– 1980s

Thailand gradually shifts to economic growth based on the export of industrial goods aided by both global and local factors.

1985

Industrial countries sign the Plaza Accord, forcing yen appreciation. Thus begins the golden era for Thai FDI as Japanese firms relocate to countries offering cheaper production costs.

Late 1980s to 1996

All sectors boom, from tourism to property, exports and the financial sector. GDP growth averages 10 percent per year. Thailand aspires to become the fifth "tiger economy" of Asia after Japan, Korea, Taiwan and Singapore.

1997

The financial and currency crisis that began in Thailand spreads to other Asian countries. Thailand's economy shrinks by 10.2 percent in 1998, but eventually it remedies those ills by strengthening private sector competitiveness, particularly in the sphere of finance.

"The faster Thailand upgrades its infrastructure, the better its ranking will be."

Professor Arturo Bris, director of the IMD World Competitiveness Center

2005– 2016

Political polarization for more than a decade undermines the country's competitiveness in numerous ways and causes a significant level of "brand damage" that is difficult to quantify.

2016

The kingdom gets the ball rolling on Thailand 4.0, an economic model based on creativity, innovation and high-level services that is intended to move the country toward a value-based economy by reforming its major industries and scaling up the development of new sectors.

CORRUPTION
A deep-rooted menace to Thai society

Whether it's a bribe made to influence an Environmental Impact Assessment study, or a **facilitating payment** accepted to push through a mega-project that might have negative effects on a local community, corruption is one of the biggest impediments to sustainable development.
Its mercenary aim targets short-term gains that benefit those in positions of power. In contrast, sustainable development involves long-term strategies in the interests of all stakeholders.

In Thailand, corruption comes in many forms with a variety of consequences. There are the small bribes, such as the traffic policeman accepting a couple hundred baht in lieu of writing a ticket or a bureaucrat's "processing fee" to ensure certain documents aren't held up. These practices are widespread, deeply entrenched in society and, in certain situations, a necessity to get from Point A to B.

Their prevalence may also explain why many Thais are outwardly tolerant of corruption, at least on this small-scale level. Transparency International's 2014 Corruption Perceptions Index noted that 65 percent of those surveyed believe that corruption is okay if they can reap some benefit for themselves. These common transactions may also explain why 71 percent of Thais perceive the police as "corrupt or extremely corrupt," while 58 percent share that view of public officials and civil servants. However, according to a survey by Chulalongkorn University's Faculty of Economics, this sort of bribe-taking is becoming less common. Thailand's land offices, police stations and customs offices, previously considered the epicenters of such corruption, have all seen declines in such behavior since 1999.

What triggers public outcry and political instability in Thailand is the more lavish and infamous form of corruption commonly referred to as "money politics". This conflict of interest takes place where big money, government and politics intersect and has seriously corroded policymaking, governance and arguably the legitimacy of the entire political establishment.

Reinforced by Thailand's centuries-old patronage system and network of cronyism and nepotism, this corruption of the political establishment includes vote-buying and the outright purchasing of posts within the bureaucracy, impacts public procurement projects, and sees kickbacks and bribes offered to government officials in exchange for special privileges in the form of concessions awarded to particular private companies or criteria specifically designed to favor particular bidders. The size of the bribes paid by businessmen to state officials and politicians account for anywhere from five to 15 percent of the project's value, according to a 2014 survey by the University of the Thai Chamber of Commerce (UTCC). The National Anti-Corruption Commission estimates that in certain years up to 30 percent of the government procurement budget has vanished due to corrupt practices.

At its most systemic level, the intersection of politics and business has seen what is known as wholesale "policy corruption" implemented via legislation usually branded as "populist." These policies are designed to benefit board members, politicians and government officials or their families rather than the general public. Numerous procurement and policy scandals have rocked Thailand, involving everything from the country's infrastructure projects to the purchase of rice, rubber and useless bomb detectors.

Just as prevalent are cases of politicians enriching themselves through purchasing pieces of real estate well below their market value. Not surprisingly, according to Transparency International, Thailand's political parties are second only to the police in terms of being perceived by the public as corrupt, and the kingdom is in the middle of the pack when it comes to its level of public sector corruption as perceived by foreign businessmen.

In a survey of more than a thousand companies in 2013 for the Thai Institute of Directors (IOD), the study broke down the pernicious results of such misconduct. The biggest impact, the respondents

Facilitating payment:

This kind of payment, made to a public official with the explicit intention of speeding up an administrative process, is a euphemism for a bribe.

Thai parliament during a no-confidence vote in Bangkok in 2008. In the past, allegations of corruption have frequently led to instability, coups and new elections.

said, was in weakening Thailand's competitiveness in the region (21 percent believe this). That was followed by a drop in ethics and morals (20 percent), tainting the country's international image (14 percent), putting the brakes on economic growth (11 percent) and posing a threat to good governance (nine percent).

To tackle this multi-pronged vice, Thailand has set up a number of government and independent agencies, such as the Office of the National Anti-Corruption Commission (NACC), the Office of the Public Sector Anti-Corruption Commission (PACC), the Anti-Money Laundering Office (AMLO), the Office of the Auditor General (OAG) and the Office of the Ombudsman Thailand. The Office Information Act, seen as one of the key laws to help stop graft, has also been enacted since 1997. Yet corruption in Thailand remains endemic.

Evidently, the state alone cannot eliminate graft, so the private sector and civil society have increasingly taken on the cause. Fed up with bribery demands, the private sector, for example, has initiated agencies such as the Anti-Corruption Organization of Thailand (ACT) and Thailand's Private Sector Collective Action Coalition Against Corruption (CAC) to monitor and implement measures to ensure transparency of procurement projects while also raising public awareness and promoting actions against corruption.

Average Thais also feel increasingly empowered to voice their dismay, with 72 percent believing that an ordinary person can make a difference in combating corruption, according to Transparency International.

In the coming years, these stakeholders have their work cut out for them if they are to establish a more efficient checks and balances system and the good governance essential for driving Thailand toward a bribe-free society. Doing so is essential if Thailand is sincere in its aim to achieve the lofty goals of the 2030 Agenda for Sustainable Development.

WHO THE PUBLIC SEES AS CORRUPT

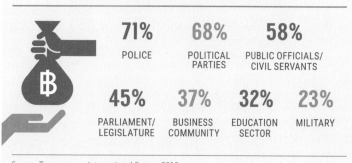

71%	**68%**	**58%**
POLICE	POLITICAL PARTIES	PUBLIC OFFICIALS/ CIVIL SERVANTS

45%	**37%**	**32%**	**23%**
PARLIAMENT/ LEGISLATURE	BUSINESS COMMUNITY	EDUCATION SECTOR	MILITARY

Source: Transparency International Survey, 2013

How Corruption Impedes Sustainable Development

The Securities Exchange Commission protests the amnesty bill of 2013.

Unsustainable system A system of bribery and nepotism, governed by narrow self-interests, is incompatible with the basic political, economic and social agendas of sustainable development, which require an ethical, broad mindset to bring equal benefits to all people, not just a smattering of society's rich and powerful.

It costs lives Corruption, at its worst, facilitates behavior and practices that may lead to sickness, injury or the loss of life, or the destruction of natural resources and ecosystems. By using payments to circumvent building or environmental regulations, for example, a company may ultimately create a hazard that impacts the health or safety of community members.

Heavy toll on the global economy On a macro level, the cost of corruption is said to equal five percent of global GDP (US$2.6 trillion), and it increases the cost of doing business by up to ten percent globally, according to the World Economic Forum and the World Bank, respectively.

Inefficient use of public resources Projects are not allocated to sectors or areas that really need them but to those that offer the best prospects for the personal gain of corrupt officials or politicians.

Distorts competition In many cases, public works are awarded to companies paying the highest bribes – not to those qualified and experienced enough to do the job.

Undermines public trust Endemic corruption causes widespread skepticism about the integrity of politicians and bureaucrats, and it breeds reluctance on the part of the general public to deal with them out of fear they will be squeezed for backhanders. Believing their hard-earned money will be wasted, some people cite the corruption of the political system as an excuse for not paying taxes.

Scares off foreign investment The bribes necessary to push a project through in Thailand can add another five to 15 percent (or more) of the total cost. To foreign investors these kinds of kickbacks are also warning signs that the rule of law means little in Thailand and that business ethics are skewed.

Downgrades standard of living A system infested with graft widens the income gap between the haves and the have-nots. Those without means also find themselves with limited access to public services or recourse to justice in the courts.

Corruption is a bad teacher In survey after survey at both primary and tertiary levels, the majority of students said they do not object to giving bribes or buying votes. Being raised in a country that condones such behavior undermines the larger promotion of an ethical mindset and principles, such as honesty and integrity, that are key building blocks for a sustainable future.

Never completed, the Hopewell rail project was famously hopeless.

FURTHER READING

• *Thailand Corruption Report*, by the Business Anti-Corruption Portal, 2015
• *Thaksin: The Business of Politics in Thailand*, by Pasuk Phongpaichit and Chris Baker, 2005
• "Why Corruption Always Wins in Thailand," by Tulsathit Taptim, *The Nation*, March 18, 2015

SPOTLIGHT

New Legislation to Combat Old Corruption

The generals leading the dozen successful coups carried out in Thailand since 1932 have all said their actions were motivated by the same culprit: corruption. So did the current junta that seized power in 2014 and now rules the country under the National Council for Peace and Order (NCPO).

Department of Special Investigation officials show solidarity in the fight against corruption.

> *"The best suppression tool for corruption is prevention. The saying 'corruption in this life will be punished in the next' should no longer be applicable as an effective fight against corruption."*
>
> **Wasan Phaileeklee, secretary of the Anti-Corruption Reform Panel of the National Reform Council, speaking on the "NRC Blueprint for Change"**

With public outcry against corruption reaching another peak, a new proposal was approved in January 2017 by the military-appointed National Reform Steering Committee under which Thai officials convicted in corruption cases involving more than one billion baht (US$28 million) could face the death penalty. In addition, the hope is that a new law on public service will reduce red tape

and the opportunity for government officials to line their wallets and bank accounts with bribes from citizens or companies.

In Thailand's bureaucracy, with its piles of paperwork that require multiple signatures and stamps, there are ample opportunities for corruption at every turn. Whenever someone needs a new ID or driver's license, to register a house or a business, to transfer a property or open a restaurant, the door of opportunity opens for graft to knock.

Complicated and time consuming, these processes can be streamlined by greasing the palms of civil servants who, it should be noted, are often paid a pittance and expected to

make up the shortfall by taking backhanders. In cases like this, graft drives up overheads, lowers the country's competitiveness levels and deprives the needy of access to essential public services.

To help remedy this chronic ill in the body politic, the government has initiated the Facilitation of Official Permission Granting Act of 2015, which was published in the *Royal Gazette* on January 22, 2015. The law requires government officials and agencies to give clear and precise details regarding the necessary procedures and documents, as well as the specific timeframe, for relevant agencies to consider applications. By doing so, the law will streamline governmental work processes and create a "service mind" among civil servants to deliver faster and better services.

Under the law, government agencies are obliged to inform customers within one week of the scheduled time of their appointment if they cannot grant permission to fulfill the particular request. Government officials are allowed to demand additional documents only once to speed up the processing time and eliminate kickbacks. If the officials fail to comply with the law, service users can file a complaint against them.

The government has also begun creating more one-stop service centers to provide the general public with greater ease of access to a variety of services, while also giving out more detailed information about the necessary application procedures. Such developments, if implemented correctly, if followed and subject to other "ifs", should also aid the private sector by providing them with clear timeframes and costs.

THE THAI
SOCIETY

Education

Health

Family

Public Participation

Inequality

Gender Equality

Conflict

Social issues are the most humane side of sustainable development, since they put real faces on woes like inequality, disease and discrimination that would otherwise remain faceless statistics in reports.

Inevitably, as developing countries like Thailand evolve, robust economic growth will remain a key goal. However, sustainable growth is not attainable without proper attention paid to human values, consumption and concerns. A healthy, well-educated, highly skilled and happy populace is needed to drive the country's progress.

In this respect, Thailand has made tremendous strides. For example, it has developed a solid healthcare system that is now virtually free for all of its citizens. The education sector has also advanced so that every Thai youth of school age enjoys the right to 15 years of schooling, and access to higher education is now more readily available. That

"The human environment and the natural environment deteriorate together; we cannot adequately combat environmental degradation unless we attend to causes related to human and social degradation."

Pope Francis

said, Thailand's rote-based system of learning has led to a lack of critical thinking skills that leaves potential untapped, and hamstrings overall national development.

The family unit lies at the core of any society. In Thailand, as lifestyles change to embrace Western-style consumerism, and as more mothers and fathers leave children in the care of grandparents to seek jobs away from home, the structure of the family unit is changing. Thailand also has the highest rate of teen pregnancy in Southeast Asia after Laos. Meanwhile, many Thai families are menaced by debt, domestic violence, drugs and alcoholism.

However, positive paradigm shifts in gender roles are also being seen here: more women have stepped up to be the heads of households and become primary breadwinners. Thailand now boasts one of the highest rates of women with jobs in ASEAN and one of the highest rates of female executives in the world. Whether through circumstance or choice, women and members of the LGBTIQ community are taking more active roles in the national economy and in public life, combating lingering paternalism.

Yet Thai society as a whole is fractured by inequalities that fuel political strife and exacerbate the rural-urban divide. In particular, returning peace and harmony to the Deep South and reconciling deeply felt political differences across the country are two of the most challenging issues Thailand faces. Finding a sustainable solution will necessitate creating more meaningful, less volatile forms of public participation and dialogue.

As Thailand works toward achieving key 2030 Agenda for Sustainable Development targets such as reducing inequality, promoting inclusiveness and implementing education reforms, so too will it help strengthen the nation's social fabric.

EDUCATION

Reforming education for the 21st century

Student-centered learning:

The principle of creating a learning environment designed to encourage children to discover new skills and knowledge, with teachers facilitating rather than providing front-of-the-classroom teaching.

Thais believe school is a second home for their children. Many parents, however, think that this "home" is in drastic need of serious renovations. And parents are not alone in their concerns. Thailand does not receive high scores in global education rankings, and studies pinpoint the need for further reforms. Indeed, the country is in the midst of a second round of education reforms that runs from 2009 to 2018. The first began with the National Education Act (NEA) in 1999, which followed on the heels of the financial crisis and a new constitution. It boosted enrollments in both schools and universities, and also guaranteed an education for those suffering from disabilities or social deprivation. The second round focuses much more on improving the quality of the education system. Thailand's relatively strong education score on the SDG Index does not make up for the fact that the kingdom must do more to prepare its students.

These misgivings are borne out by other measures – both internal and global – which show that Thailand's students peform poorly in comparison to their international counterparts, particularly in core subjects such as math, sciences and English. In the 2016–2017 Global Competitiveness Report of the World Economic Forum, Thailand's primary education ranked seventh and its higher education fourth out of the nine ASEAN countries included in the report (no data was available for Myanmar). In the latest ranking from 2015 of the Programme for International Student Assessment (PISA), Thailand placed 54th (out of 70 nations) on the performance of 15 year olds in reading, mathematics and science. According to the Times Higher Education University Rankings of 2017, Thailand has only one university among the top 100 in Asia, and one in the world's top 600 (Mahidol University). These rankings cannot be explained away by a lack of financial resources. Thailand has consistently allocated about 20 percent of its national budget – about four percent of annual output – to education, which is among the highest totals in the world. Much of the increased spending in the last decade has addressed the poor pay of public school teachers, once regarded as the system's main drawback. In 2014, the starting salary of a teacher was raised to 15,050 baht per month, up from 7,630–8,340 baht per month during 2009–2012.

However, parents still feel that their children are not getting enough out of their well-financed school system. One obvious manifestation is the high sums parents spend on some 5,000 "cram schools" across the country that provide supplementary education for all ages. Thai students spend hours upon hours each academic year in this kind of intensive, supervised studying environment. Despite their popularity, most of these tutoring centers focus primarily on memorization to prep students for standardized exams. Some experts say that this kind of cramming is indicative of the main shortcoming in Thailand's education system: the lack of **student-centered learning.**

While the student-centered approach has been hailed around the world as the way forward, in Thailand such techniques are still inexplicably shunned in favor of "rote learning" – a grinding

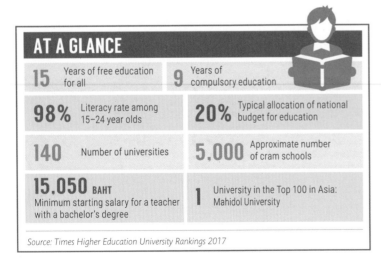

AT A GLANCE

15 Years of free education for all		**9** Years of compulsory education	
98% Literacy rate among 15–24 year olds		**20%** Typical allocation of national budget for education	
140 Number of universities		**5,000** Approximate number of cram schools	
15,050 BAHT Minimum starting salary for a teacher with a bachelor's degree		**1** University in the Top 100 in Asia: Mahidol University	

Source: Times Higher Education University Rankings 2017

Some 35,000 students sit for an exam hoping to gain admission to Srinakharinwirot University.

and stultifying process of drilling, memorizing and copying that leads to passive students who lack critical thinking skills. It's not that the authorities are unaware of the issue. As far back as the year 2000, the country officially adopted student-centered learning, at least in theory. However, more than 15 years later, it is still a rarity in most Thai schools.

According to Thitinan Pongsudhirak, a political scientist at Chulalongkorn University, "The mindset is from the nation-building and Cold War period to produce obedient and nationalistic citizens, which does not fit 21st-century needs. It is hierarchical, top-down, with a systematic lack of critical thinking." Thailand's dated assessment system only serves to reinforce this backward mentality. The multiple-choice Ordinary National Educational Test (O-NET) and Advanced National Educational Test (A-NET) run by the National Institute of Education Testing Services have been roundly criticized, mostly because they evaluate memory rather than thinking, but also because of the narrow, parochial and moralistic nature of some questions and the "correct" answers. Such exams also influence teaching styles, reinforcing the tendency of teachers to fall back on rote learning. This combination of memorization-based exams and a lecturing style of instruction tends to produce passive, uncreative students.

When examining the Thai education system, the onus has often fallen on the teachers, who, to be fair, frequently fail to impress. In 2010, the Office of the Basic Education Commission conducted the first national exam to test secondary teachers in their own subjects. The result was that 88 percent of computer science teachers, 86 percent of biology teachers, 84 percent of math teachers, 71 percent of physics teachers and 64 percent of chemistry teachers failed their own subjects. Thus, while Thailand has met the Millennium Development Goal of universal primary education for all, UNDP notes that "the greatest challenge lies in improving quality."

The harsh truth is that Thailand's entire education system needs an overhaul. Moving forward, the kingdom must prioritize teaching **21st-century skills**, and find ways to enhance vocational training and encourage enrollment in such classes. Partnerships between the private sector and Thailand's top universities and vocational institutions present one opportunity for advancement on this front.

Thailand also needs to embolden its teachers to be more creative, innovative and engaging, and amp up the quality of teacher training initiatives. Accountability is also key. As TDRI's Somkiat Tangkitvanich and Supanut Sasiwuttiwat argue: "Effective education reform must incorporate the creation of an accountability system." This, it is said, is vital to the success of other reforms. Such a system should reward teachers and administrators according to student performance. It's also high time that the book on rote learning be shelved for good.

21st-century skills:

A range of skills, knowledge, habits and character traits necessary to compete in the modern world. They include critical thinking, problem solving, synthesizing information, research skills, creativity, self-direction, innovation, digital skills, financial literacy and civic/ethical literacy.

The Elephant in the Classroom: Implementing Reform

Over a period of about 40 years, King Bhumibol Adulyadej made hundreds of speeches about education. A biography called *King Bhumibol Adulyadej: A Life's Work* observed that the monarch's thinking on education included, among others, the following themes: knowledge from books must always be tested by application; the ability to analyze what has been learned and relate it to reality is more important than memorizing facts; education must include ethics and mindfulness. Thailand has not put into place an education system in line with His Majesty's views and the needs of a modern, competitive economy.

This is not for want of loudly declared intentions to reform the education system. Governments and political parties, leading civil servants, intellectuals and think tanks have repeatedly trumpeted the need for change. Right now Thailand is in what is called the second decade of education reform, which runs until 2018. The first began with the National Education Act (NEA) in 1999. At the time, a respected academic and member of various education commissions, Dr Prawase Wasi, maintained that continuing the traditions of Thai education was "pushing the country into a national disaster."

Yet the consensus regarding the past 15 years is that progress has been slow – and the reality is in keeping with the wider public view that changes have largely failed to fulfill the promise of the NEA.

With the coming to power of the National Council for Peace and Order under General Prayuth Chan-ocha, some educationalists had hoped that the heavy focus on reforming all Thailand's institutions might rejuvenate the effort to transform the country's education system. So far, though, the emphasis has been mostly on the moral side.

"You should not think that you are studying in order to pass an entrance exam, because our existence doesn't depend on whether you marked down the right multiple choice answer on your exam. Our existence depends on working and analyzing various problems."

King Bhumibol Adulyadej

In June 2014, in his nightly address to the nation, the general who became prime minister said the Ministry of Education should include in the curriculum subjects that help reinforce the values of "Being Thai": national pride and upholding the institution of the monarchy. "The purpose is to instill discipline, strengthen the physical and mental state, and reinforce conscience and social responsibility," he said.

Further reforms of the education system are to be overseen by a new "super board" that was established in 2015. Two months after the establishment of the board, Permanent Secretary for Education Suthasri Wongsamarn stated that national reconciliation was an important mission of the Ministry of Education.

However, the effort must begin with unity in the family before spreading to the community and the nation. Members of society should take pride in Thai culture and history, she said. Additional curriculum subjects would be introduced in order to raise awareness about the identity of the Thai people and nation, including history, civic duties and moral education.

On the plus side, there have been positive developments as well. In June 2016, Prayuth announced that free basic education was being expanded from 12 to 15 years in order to cover kindergarten levels as well primary and secondary. The new 15-year state-subsidized plan provides free education through the Mathayom 6 (Grade 12) level, or the equivalent in vocational schools. It also covers special education for students with disabilities or special needs, and "welfare education" for underprivileged children.

Due to the country's chronic shortage of skilled workers, subsequent governments over the past decade have tried to enhance and expand vocational education and to encourage more students to enroll in these classes. However, to date, these efforts have not produced the level of results desired.

For its part, the current government increased the budget of the Office of the Vocational Education Commission (OVEC) by about a third in 2015. The OVEC has also developed a new curriculum at a higher-level certificate of vocational education to teach more 21st-century skills and try to meet the evolving challenges of the ASEAN Economic Community (AEC), which came into being at the end of 2015.

What is clear is that the pressure to turn out more skilled vocational graduates is mounting. A 2016 survey by the Quality Learning Foundation (QLF) determined that Thailand needs to add as many as two million vocational skilled labourers over the next five years.

QLF Executive Assistant Manager Patanapong Sukmadan said in a December 2016 article in *The Nation*: "It's necessary to boost vocational students' skills including language. If we don't start now, more Thai youths will be unemployed. The gap between formal and vocational education must be narrowed...."

Patanapong also said that the QLF and the Office of Basic Education Commission had organised pilot vocational education programs in ten provinces in an effort to help prepare a labour force in line with the aims of Thailand 4.0's area-based educational reforms.

Universities are Taking Sustainability Education to the Next Level

As the concept of sustainable development has gained notoriety, higher education institutions in Thailand have taken notice and designed curricula and degrees to prepare individuals for careers in this field.

The most notable institution is the Sasin Center for Sustainability Management (SCSM). Established in 2011, SCSM is a collaboration between the Sasin Graduate Institute of Business Administration of Chulalongkorn University and the Corporate Responsibility & Ethics Association for Thai Enterprise. The program uses the core principles of sustainable development, corporate social responsibility and the Sufficiency Economy Philosophy to groom "future leaders" in sustainability management. In 2016, SCSM had 130 MBA and EMBA students. SCSM also offers sustainability management consulting services, training and workshops to the public and private sectors, and to civil society organizations.

The Asian Institute of Technology's School of Environment, Resources and Development offers a master's degree in Climate Change and Sustainable Development. The program is designed to develop professionals who can contribute to addressing issues related to climate change impacts and the sustainable management of resources.

Srinakharinwirot University's International College for Sustainability Studies, which was established in 2005, offers bachelor's degrees in Sustainable Tourism Ecotourism, Recreation and Hospitality Management. The courses are designed to expand students' perspectives on contemporary issues in tourism management by teaching them about the importance of sustainability, environmental conservation, heritage preservation and cross-cultural communication.

More recently, Mahidol University in 2016 announced that it plans to establish a Mahidol Center for Sustainable Development. Mahidol's Faculty of Science also has a Center of Sustainable Energy and Green Materials which was established in 2006 to advance research and development of alternative energy technologies.

The organization PRO-Green also offers support to academics who wish to carry out research on a range of topics related to the Green Economy, as well as students who want to pursue degrees in this area.

Challenges Facing the Education System

Huge budget Thailand spends more of its national budget (20 percent) on public education than any other item. This is relatively high by global standards, yet the quality of education does not meet those standards, reflecting a rather poor return on investment.

Slow to evolve Despite frequent announcements of education reform over the years, including a desire to shift to more child-centered learning methods, little has been achieved because of bureaucratic inertia, weak implementation, lack of accountability and policy discontinuity resulting from unstable politics. In addition, a highly centralized, bureaucratic national testing system reinforces rote learning and standardization. Pass rates are even substantially lower than those for global testing schemes, which suggests there is something amiss with the tests themselves rather than the students taking them.

Teaching creativity At times, inadequately qualified teachers wedded to traditional top-down teaching methods preside over passive pupils who are discouraged from questioning or thinking for themselves. Without more creative and innovative graduates, Thailand is likely to remain trapped in the middle-income zone, unable to meet the challenges of an increasingly competitive and global "knowledge economy."

Retention rate While Thailand has achieved the UN's Millennium Development Goal of universal primary education, on a national level, only 75 percent of children are actually entering Grade 1 at age six. In the northeast, that figure dips to 65 percent. The "retention rate" of Thai students over the 12 years of basic education has also been declining.

Only 60.7 percent of those who entered Grade 1 in 2001 were still in school in 2012.

Unqualified graduates Despite higher enrollments, Thai universities are still turning out graduates in social sciences and the arts who struggle to find jobs. Meanwhile, the sciences are being relatively neglected to the detriment of employers who cannot fill vacancies that require such skills. In 2011, UNESCO reported that Thais with tertiary qualifications lacked the following: "communication skills, computer and ICT-using abilities, management, calculation skills, problem solving, teamwork, responsibility, honesty, tolerance, discipline, punctuality and leadership."

Learning more English English proficiency is near the bottom of ASEAN's class and shows no signs of improving, which is a serious impediment to Thailand's competitiveness.

Vocational training Successive attempts over the past decade to promote and improve the quality of vocational training at upper secondary and tertiary levels have delivered weak outcomes leading to an undersupply of skilled vocational workers. One reason for this is that high school graduates are increasingly opting to attend universities instead of vocational institutions. In fact, the Office of the Vocational Education Commission said in 2017 that some 59 percent of private vocational schools in Thailand are in critical condition and have fewer than 500 students.

Learning for life As Thailand grapples with the challenges of reform in 11 areas of public life, education is in danger of being elbowed out by other pressing issues. Educationalists fear that curriculum

changes intended to improve public morals and military-like rigor are taking precedence over the kind of reforms necessary to promote critical thinking and life-long learning.

Deep divide Education provisions and outcomes are still very unequal across the country, deepening the urban–rural divide that underlies much of the country's political instability. Meanwhile, expensive and elitist private and international schools turn out better-qualified young people disconnected from the majority of their compatriots.

Accountability Thailand needs more accountability on both sides of the teacher's desk. Experts argue that remuneration for teachers and headmasters should be linked to improvements in students' learning outcomes. Meanwhile, students should be held to higher standards, disciplined for cheating and plagiarism, and made to repeat a grade if warranted by an inadequate academic performance.

Greying teachers Thailand's aging demographic means that 40 percent of teachers will retire in the next ten years. Replacing them with high-quality graduates will prove difficult, especially outside urban areas.

FURTHER READING

- "Education in Thailand: Changing Times?" by Daniel Maxwell, 2014
- "Revamping the Thai Education System: Quality for All," by Somkiat Tangkitvanich and Supanut Sasiwuttiwat, *TDRI Quarterly Review*, June 2012
- "A Decade of Education Reform in Thailand: Broken Promise or Impossible Dream?" by Philip Hallinger and Moosung Lee, *Cambridge Journal of Education*, June 2011
- "Education Reform in Thailand," by William Stanley, Stamford International University, 2012
- "Thailand's Small School Challenge and Options for Quality Education," by Dilaka Lathapipat, World Bank, 2016

TIMELINE

The Development of Education in Thailand

13th–14th centuries

During the Sukhothai period, "schools" are private houses, temples, philosophy institutes and royal courts. Boys study military subjects and martial arts, while girls study embroidery and etiquette.

14th–18th centuries

The first textbook in the Thai language, *Chindamanee*, is written by a monk, Phra Horatibodi, during the Ayudhya era when missionary schools are also founded.

Mid-1800s

Thai-language textbooks are printed for the first time.

1871

An English-style school is set up in 1871 at the palace during Rama V's reign. The school prepares scions of royalty to study abroad.

1916

Chulalongkorn University is founded. Its first four faculties are Medicine, Political Science, Arts and Engineering.

1898

The first education plan is launched, designating pre-school, primary, secondary, technical and higher education systems.

1887

Rama V establishes the Ministry of Education.

1884

Rama V believes the children of commoners should be educated as part of modernization. The Wat Mahanaparam School is started with royal support.

1874

The first school for girls, Kula Satree Wang Lang, set up by an American missionary, marks the beginning of education for females in the country. Later it's renamed the Wattana Witaya Academy.

1932

Constitutional monarchy supplants absolute monarchy. The First National Education Program decrees schooling for every Thai regardless of sex, social background or physical condition.

1934

The University of Moral and Political Science is founded as the country's first open university. The first two faculties are Law and Accounting. It is later renamed Thammasat University.

1943

Kasetsart University (Agricultural Studies), Mahidol University (Medicine) and Silpakorn University (Fine Arts) are founded.

1951–1957

Thailand's first international schools, such as the International School Bangkok (ISB), Ruam Rudee International School and Chiang Mai Children's Centre, open.

1999

The National Education Act is created to reform education to help Thailand overcome the effects of the 1997 financial crisis by decentralizing the education system. This marks the "First Decade of Education Reform."

1970s–1980s

Reforms aim to create more up-to-date curricula and equality in education following the student uprising and the fall of the military regime in 1973.

1971

Ramkhamhaeng University is established as the kingdom's first open-admission public university.

"Teachers and principals should be made more accountable to students and parents by linking their remuneration to improvements in students' learning outcomes. This should be coupled with enhancing 21st-century skills among Thailand's next generation."

Dr Somkiat Tangkitvanich, president of TDRI

2002–2006

The 9th Social and Economic Development Plan emphasizes the role of higher education in pursuit of a society that is "knowledge-based."

2007

The Education Ministry makes the Sufficiency Economy Philosophy one of its goals and later integrates the philosophy into the national curriculum.

2016

The government expands free basic education to 15 years in order to cover kindergarten levels as well primary and secondary.

HEALTH

Free coverage gives healthcare system further booster shot

In 1930 the average life expectancy in Thailand was a mere 31 years. By the 1950s it had risen by some two decades, and as of 2016 it stood at 75 years. That leap in longevity speaks volumes for the vast improvement of the kingdom's healthcare system.

Thailand had achieved most of its health-related Millennium Development Goals by 2004, well in advance of the 2015 deadline laid down by the United Nations in 2000. Among these triumphs is Thailand's positive maternal mortality and neonatal mortality rates; infant vaccination rate (99 percent) and healthy life expectancy at birth; the elimination of malaria in all but the most far-flung frontiers; and the cutting of new HIV infections by more than 80 percent since the peak of the pandemic in 1991. Furthermore, almost 98 percent of the populace now has access to improved water and 93 percent enjoy proper sanitation.

Such headway is not surprising given Thailand's long history of traditional medicine based on local wisdom and raw materials and its more recent track record of facing down threats and improving its facilities and professional know-how. As far back as the 1800s, Thailand's kings were supportive of Christian missionary-led health clinics and services, so much so that all missionaries were referred to as "doctor." In the 1900s, the royal family continued to support the expansion of the medical system, with King Bhumibol's own father, Prince Mahidol, leading

the way. In the 20th century, as Thailand's medical system modernized and global breakthroughs in prevention and treatment were adopted, outbreaks of polio, smallpox, cholera, yaws and other deadly diseases, as well as the scourge of malaria, were largely eliminated as rural and urban menaces.

In addition, Thailand has become a global center of wellness spas and holistic treatments, such as Thai massage, and is increasingly investing in developing its traditional practices as well as age-old understanding of herbal ingredients into modern products.

Still, 21st-century Thais face a slew of new threats including obesity, substance abuse, poor road safety and the health ramifications of living in an increasingly industrialized environment. In addition, according to the chapter on Thai healthcare in the book *Sufficiency Thinking: Thailand's Gift to an Unsustainable World*, "a shortage and maldistribution of health personnel, together with skilled health professionals shifting from public to private hospitals and from rural to urban areas, have contributed to inequalities in health outcomes between rich and poor." Indeed, an estimated 50 percent of all physicians practice in and around Bangkok.

Until the year 2000 almost one-third of all Thais had no healthcare coverage. In 2002, the **Universal Coverage Scheme or UCS** (popularly known as "The 30 Baht Health Scheme") was launched under the National Health Security Act. The clever nickname derived from the fact that patients had to only pay 30 baht for administration fees, no matter the prescription or operation, for each visit or admission to a hospital or clinic (in 2007, the co-payment was abolished and the UCS became free). The new card consolidated the Low Income Health Care Scheme and the Voluntary Health Card plan and incorporated 30 percent of the uninsured into the UCS.

Since the plan was hatched, almost the entire population has been covered by one of the three big health insurance policies: the Civil Servant Medical

LEADING CAUSES OF DEATH IN THAILAND

19%	Cancer		4%	Chronic pulmonary diseases
12%	Heart disease		4%	Diabetes
10%	Strokes		4%	Road accidents
9%	Respiratory infections		2%	Kidney disease
4%	HIV / AIDS		2%	Cirrhosis

Source: Public Health Ministry, 2012

Free daily aerobics programs at dusk, like this one in Yala, are part of a national health campaign.

Benefit Scheme for civil servants (7.82 percent), the Social Security Scheme for company employees (16.6 percent), and the UCS for the rest of the primarily rural populace. The latter scheme also strove for a more egalitarian approach to public health, serving both the needs of the poor, who could not afford treatments, and helping those with means with costly treatments like chemotherapy for cancer. Thanks to the UCS, the household costs for taking care of family members stricken with catastrophic illnesses have been steadily declining, from a total of 5.7 percent in 2000 to only 3.3 percent in 2009.

However, one of the most severe side effects of the triple-pronged plan is that the number of outpatient visits has spiked, thereby putting strain on the system that taxpayers must also shoulder. To cope with that increase, the government has had to boost its spending to support these programs from 56 percent of total health expenditures in 2001 to 75 percent in 2010. For 2015, the total budget for the UCS amounted to almost 143 billion baht to cover around 49 million Thais, inclusive of antiretroviral medicines for HIV/AIDS patients and renal treatments like dialysis.

The UCS is far from perfect. There is a disparity between the facilities in metropolitan areas and the countryside, just as there is a gulf between how the poor, the middle class and the wealthy are treated by medical staff. That said, in tandem with the UCS has come a new way of looking at health. The existing system has generally been passive: when people get sick they go to the hospital. However, in the new millennium a way of thinking that advocates prevention has entered the public discourse and the mindsets of policy makers.

Since then, "health" has been redefined in a broader context, covering physical, mental, social and even spiritual wellbeing. No longer just a matter for doctors and patients, health is starting to be recognized as an intrinsic part of human and social development as well as a fundamental right.

Universal Coverage Scheme (UCS):

This term is used to define a healthcare system that ensures all people obtain the treatments and services they need without suffering any undue financial hardship to pay for them.

HEALTHCARE COVERAGE

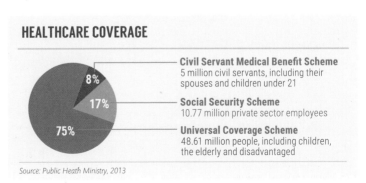

Civil Servant Medical Benefit Scheme
5 million civil servants, including their spouses and children under 21

Social Security Scheme
10.77 million private sector employees

Universal Coverage Scheme
48.61 million people, including children, the elderly and disadvantaged

Source: Public Heath Ministry, 2013

Strengths and Weaknesses of the Thai Healthcare System

STRENGTHS

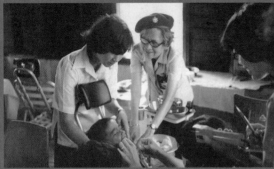

The late Princess Mother assists volunteer dentists.

- Thailand had already achieved most of its health-related Millennium Development Goals (MDGs) by 2004, well in advance of the 2015 deadline. This prompted the country to pursue an MDG-plus strategy, with expanded targets like further reducing the rates of maternal mortality, HIV infections and malaria at regional levels, especially among the northern hill tribes and southern Muslim communities.

- The UCS covers almost the entire population.

- Strong technical capacities ensure that healthcare workers meet a high standard.

- The National Health Assembly (NHA) is an effective platform for developing policies through the exchange of scientific data and knowledge transfers with different sectors.

- The UCS also covers anti-retroviral treatments for people living with HIV/AIDS, and has reduced the number of new HIV infections by more than 80 percent. It also fully covers dialysis and kidney transplants for patients with chronic renal failure.

- Bangkok's best private hospitals are so highly regarded and relatively affordable that they receive so-called "medical tourists" from around the world.

- The monarchy has a long-standing tradition of promoting nutrition, developing health projects and mobilizing funds to support public health.

WEAKNESSES

- Coordination between health-related agencies is still lacking.

- Inefficient management and integration of medical information persists.

- There is a lack of focus on prevention of illnesses and chronic diseases through family medicine.

- Spiritual well-being and traditional treatments have been pushed out of the mainstream.

- There is a shortage and lack of even distribution across different regions of health resources (such as hospital beds) and personnel (such as nurses, dentists and pharmacists), especially in the more remote areas.

- A significant difference remains between the services and equipment of expensive private hospitals and that of cheaper but overcrowded public facilities.

- Increased focus on profits by private hospitals and doctors, who can earn more by working in specialty areas or by focusing on medical tourists. Doctor-patient relationships have also deteriorated with doctors either too rushed or too profit-driven to communicate with and assess patients effectively.

- The discrepancy between the three schemes (the CSMBS, SSS and UCS) is a major source of inequity. Moreover, the CSMBS is inefficient as it pays on a fees-for-service basis. The overuse of medicine and diagnostics means that it costs four times more per capita than the UCS.

Oversized teeth help children learn to brush in Korat.

TIMELINE

From Eradicating Smallpox to the Rise of "Mister Condom"

1835

American missionary Dan Beach Bradley arrives. He performs the first surgery and amputation in Siam and introduces a vaccine for smallpox.

1888

Siriraj Hospital, the kingdom's first hospital, opens.

1893

The Bangkok Nursing Hospital, the first private facility and still operating today, is established.

1914

In honor of their father, Chulalongkorn Hospital is built with donations from King Vajiravudh, or Rama VI, and his brothers and sisters, along with funding from the Thai Red Cross Society.

1962

Smallpox is eradicated after successful vaccination campaigns.

1946

In the wake of Rama VIII's still-unsolved death, Thai doctors first experiment with forensics.

1942

The Ministry of Public Health is established.

1929

Prince Mahidol, who obtained a medical degree from Harvard University, passes away.

1927

The first use of an X-ray machine in Thailand.

1973

Doctors at Siriraj Hospital successfully perform the first kidney transplant.

1987

The first HIV/AIDS case is reported. As a countermeasure, Mechai Viravaidya, or "Mister Condom," launches his safe-sex campaign by giving away millions of condoms while disseminating information about the pandemic to Thais and foreigners.

2002

The National Health Insurance Bill is endorsed, ushering in the 30 baht healthcare scheme.

2007

The endorsement of the National Health Act marks a major turning point in the healthcare system by taking a multi-stakeholder approach that stresses a more holistic approach under the motto, "All for Health and Health for All."

2016

Thailand becomes the first country in Asia to effectively eliminate mother-to-child transmission of HIV and syphilis, according to criteria devised by WHO.

"Good physical health is a factor supporting the economic progress and social security of the country because it leads to good mental health. Physical and mental fitness enable the individual to effectively serve the nation while refraining from imposing burdens on the nation."

King Bhumibol Adulyadej

It's Time to Get Healthy, Thailand

With a population that is short on free time and increasingly engrossed in sedentary activities, obesity has emerged as a growing public health threat in Thailand. Indeed, with an overall obesity prevalence of 32.2 percent, Thailand ranks second in the ASEAN region — behind Malaysia's 44 percent — with the highest number of obese citizens, according to the latest World Health Organization data. But it's not just adults. Thailand's Multiple Indicator Cluster Survey 2012 found that 10.9 percent of children under age five are overweight in Thailand.

The key catalysts for Thailand's bursting beltlines are a decrease in physical activity and heavier consumption of processed foods, sugar and soft drinks. It may come as no surprise that Bangkokians are the most at risk of becoming obese. In today's fast-paced Thai society, many adults work six days a week and have long commutes.

As a result, parents have less time to prepare meals at home and are relying more on fast food options and convenience stores like 7-Eleven to feed themselves and their kids. And the vast majority of what's on offer — processed meats, synthetic foods, boxed fruit juices, microwavable meals, candy, etc, — contain excess sugar, high fructose corn syrup, trans-fats and sodium. Far from nutritious, the lack of a balanced diet means that children aren't getting the nutrients they need to fully develop. A recent study by the Southeast Asia Nutrition Survey concluded that due to a lack of exercise and poor nutrition, in the next decade an increasing number of Thai children are likely to be overweight, shorter and have lower IQs.

While Thailand is globally renowned for its cuisine, these days many of the staple dishes consumed by Thais aren't all that healthy either. Many are fried and heavy on oil. Fish sauce, shrimp paste and curry paste, which all figure prominently in Thai food, are also all high in sodium. Just a half cup of coconut milk — a key ingredient in many soups, curries and desserts — contains over 200 calories and more than a day's recommended allowance of saturated fat.

Last but certainly not least, there's the sugar predicament. It would not be unfair to say that for a typical Thai person, their Achilles heel is actually their "sweet tooth."

According to the Ministry of Health, the average Thai consumes 30 kilograms of sugar per year, which amounts to over three times the maximum recommended intake of 25,000 milligrams per day. These days, even in traditional cooking a hefty dose of refined sugar is added to the famous *pad thai* where previously cooks would have opted for tamarind paste as a sweetener. Not to mention the sugar-laced syrups and condensed milks heaped upon desserts and into coffee drinks and shakes.

And all that sugar consumption has consequences. According to WHO, adults who consume less sugar have lower body weights, while evidence shows that increasing the amount of sugar in a person's diet is associated with an increase in weight. Excess sugar consumption can also lead to metabolic syndrome (pre-diabetes) and even Type 2 diabetes, a growing problem in the kingdom. In fact, the number of Thais with diabetes rose from 1.5 million in the year 2000 to 3.2 million in 2013. That figure is expected to reach some 4.3 million by 2035.

But Thailand's obesity problem isn't just about diet. The rise of personal smart phones, laptops and tablets mean that fewer children and young adults are using their spare time for physical activities like recreational sports. The prevalence of relatively cheap taxis, motorcycle taxis, public transport - and the fact that so many people own their own cars and motorbikes - also means that fewer Thais are willing to walk from Point A to Point B, even for short distances. In Bangkok, the dearth of green spaces and public parks doesn't help matters either.

While government agencies could certainly do a more thorough job of raising public awareness about the benefits of a proper diet and regular exercise, the choice of whether or not to eat right and lead a healthy lifestyle ultimately falls on the shoulders of the individual.

OVERWEIGHT POPULATIONS IN SOUTHEAST ASIA

Overweight prevalence (%) for adults of both sexes (BMO of > 25kg/m2)

Country	%	Country	%	Country	%
Vietnam	10.2	Myanmar	18.4	Singapore	30.2
Cambodia	12.1	Indonesia	21	Thailand	32.2
Laos	13.3	Philippines	26.5	Malaysia	44.2

Source: WHO Noncommunicable Diseases Country Profiles, 2011

FURTHER READING

- *The Kingdom of Thailand Health System Review,* by the Asia Pacific Observatory on Health Systems and Policies, 2015
- *ePatient 2015: 15 Surprising Trends Changing Health Care,* by Rohit Bhargava and Fard Johnmar, 2015

REALITY CHECKS

Challenges Facing the Healthcare System

Teenage pregnancies Thailand's teenage pregnancy rate is the highest in Southeast Asia after Laos. Of every 1,000 teens aged 15 to 19, about 60 are mothers – one of the highest rates in the world and more than 20 times higher than Singapore's, according to the SDG Index.

Sedentary lifestyles With more Thais working in offices and spending idle time parked in front of computers or using their smartphones, fast food and inactivity have become the prime culprits in a gargantuan increase in waistlines. More than one-third of Thais aged above 15 are fighting obesity today, which has also led to an upsurge in diabetes. A lack of urban green spaces can also make it inconvenient for city-dwellers to lead a more active lifestyle.

Food security Feeding a growing population poses many problems, such as the quality and quantity of food, widespread access to nutritious food sources, the proper use of such resources and their long-term sustainability. Reforms of the country's agricultural sector and food production systems are needed to ensure food security remains a priority for future generations.

Bangkok magnet The concentration of health personnel in Bangkok and other urban areas in the central region remains high, especially in the private sector, leaving other areas like the northeast region lacking in qualified staff. Physician density per 1,000 people was only 0.4 across the country, according to the SDG Index.

Smoke signals In 2015, the National Statistical Office revealed that the number of Thais over 15 years of age who smoke had risen to 11.4 million in 2014 representing a 21-percent increase over the year prior. The study showed that the average age of new smokers dropped to 15.6 years of age from 16.8 in 2007. Some attribute this rise to the efforts of tobacco companies to counteract government control measures through marketing strategies that violate the law.

Substance abuse In 2014 Thailand ranked first among ASEAN countries in alcohol consumption. Meanwhile, around 1.4 percent of Thais are addicted to amphetamine-type stimulants such as *yaba*, one of the highest rates in the world.

Overmedicating About 38,000 Thais die from antibiotic-resistant bacteria annually because of over-prescription and the prevalence of antibiotics in food and the country's water supply, according to research conducted by Chulalongkorn University's Drug System Monitoring Mechanism Development Center.

Mental health Only about one million Thais undergo regular mental health treatment, while the Department of Mental Health estimates that as many as 20 percent of Thais suffer from some form of mental illness including psychosis, anxiety disorders and depression.

Road fatalities Thailand has the dubious distinction of ranking second in global traffic fatalities after Libya, with more than 24,000 road deaths per year and a mortality rate of 36.2 per 100,000, according to the World Health Organization.

Malnutrition The SDG Index singles out stunting, undernourishment and wasting as areas that Thailand needs to address to achieve Goal 2 of the 2030 Agenda for Sustainable Development. According to Thailand's Multiple Indicator Cluster Survey (MICS) 2012, carried out by the National Statistical Office with assistance form UNICEF, one child in six under the age of five is stunted. In addition, nearly one in ten children under age five is underweight and 6.7 percent suffer from the acute level of undernourishment known as wasting.

Aging society Around eight million people (13 percent of the population) are over age 60 in Thailand. With a younger generation that is less equipped to take care of them, this puts added strain on both families and the country's healthcare system.

Spiritual well-being As Thailand continues to modernize, Thais of all ages are seeing their free time dwindle. Overtime hours, afterschool classes and long commutes often leave little time to cultivate spiritual wellbeing, particularly for those living in congested urban areas.

Good governance As with many parts of the public and private sectors, transparency and good governance are perennial challenges for the Ministry of Health.

FAMILY

The most binding of social ties is loosening with the times

Thai families are famously close-knit with an extended family often living under the same roof or in the same compound. As a **matrilocal** culture, this means that the husband tends to live with his wife's family. For a daughter's hand in marriage, a potential groom will still often be expected to pay a dowry, which is sometimes but not always given to the couple so they can begin their new life together. These matriarchal aspects of traditional Thai culture are counter-balanced against the burdens that women must shoulder in terms of domestic chores, raising children and tending to ailing parents.

The structure of family units is evolving in Thailand with an increase in [

REASONS

Why Family Matters

Loving care Human infants remain helpless longer than the offspring of any other creature. They also require much more care and attention to develop both emotionally and intellectu-ally. Individuals who have suffered family neglect often struggle with negative emotional, psychological and behavioral effects.

Pillar of strength Family is the primary source of support for individuals and the last resort when emergencies arise.

Social tool As a social tool that helps to control biological impulses, the family unit can encourage individuals to limit sexual intercourse to one partner.

Financial necessity Family-run businesses and family farms are still crucial to Thailand's economy and social structure. Families provide individuals with support networks and loans during times of unemployment, debt and financial crises.

Common denominator The family unit is the connective tissue that binds us to one another and is the most common denominator of the human race. Families are where people find their strongest sense of identity, community and belonging.

But the forces of globalization and consumer culture have exerted a tremendous influence over Thai family life in recent years, especially in rural areas. To make a living and escape the tedium of farming life, more parents have been migrating to urban areas or moving abroad in search of well-paid jobs, leaving children behind in the care of grandparents or other relatives.

This mass migration has resulted in a steadily rising divorce rate in Thailand. According to the Ministry of Interior, the number of family members in a household dropped from 5.2 in 1980 to 3.1 in 2015. Simultaneously, the number of nuclear families living on their own has fallen from 63.1 percent in 1987 to 49 percent in 2016, as the rising cost of living has caused young couples to live with their parents whom they also look after.

However, these changes may have also reduced domestic friction. According to the Report on Gender-based Violence against Women and Girls Indicators issued in December 2011 by the Office of Women's Affairs and Family Development, the reported cases of violence committed by a

...useholds and single mothers raising children on their own.

FAMILY STATISTICS

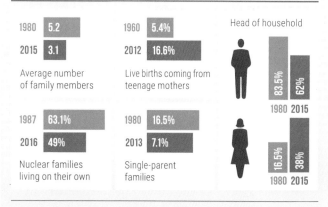

1980	**5.2**	1960	**5.4%**
2015	**3.1**	2012	**16.6%**

Average number of family members

Live births coming from teenage mothers

1987	**63.1%**	1980	**16.5%**
2016	**49%**	2013	**7.1%**

Nuclear families living on their own

Single-parent families

Head of household

83.5% 62%

1980 2015

16.5% 38%

1980 2015

Sources: Ministry of the Interior, National Statistical Office of Thailand, UNFPA

family member dropped from 45 percent of all violent incidents reported in 2005 to 34 percent in 2009. The most recent figures from the Thailand Domestic Violence Information Center revealed that the reported incidents of domestic violence have decreased from 1,076 in 2011 to 855 in 2013. (However, the number of unreported altercations is thought to be much higher.)

Increasing teenage pregnancy in Thailand is another issue affecting families. Some observers contend that the rising teenage pregnancy rate is the result of looser family ties, increased exposure to graphic content through the Internet and media, and insufficient sex education. In 2013, out of every 1,000 births to women aged 10-49, 4.6 were to adolescents girls aged 10-14 and 163.2 were to adolescents aged 15-19, according to the Ministry of Public Health. In fact, after neighboring Laos, Thailand has the highest rate of teenage pregnancies in Southeast Asia.

The Ministry of Social Development and Human Security has launched a number of campaigns to discourage teenage pregnancies and support single mothers. But the fissures run deeper than that. They are symptomatic of a greater divide; in a digital era defined by mass telecommunications the distance between family members is at once closer and farther than ever. How to get parents and their children to spend more quality time together was the quandary the government faced when developing such campaigns as *Krob Krua Ob Oon* ("Family Warmth"). In rural villages, all over the country, the authorities have encouraged blood relations to get together to participate in games designed to promote family values and loving households.

For all these well-intentioned efforts, how much can any government legislate what goes on in the homes of its citizens? Simply put, the family traditions of yesterday cannot keep pace with the high-speed world of today. The signs of change are everywhere. Even in traditional family units, women are increasingly emerging as the head of the household, up from 16.5 percent in 1980 to 38 percent of all families in 2015. More and more Thais are also delaying marriage and childbearing in favor of pursuing careers or exploring non-traditional lifestyles, such as coupling without children, single-parenting, or living alone. As a result, Thailand's birthrate has fallen to an average of 1.5 children per women. In 2015, Thai mothers gave birth to some 704,000 children, a decrease of almost 80,000 births versus 2013. For those who do choose to have children, often enough it is the grandparents who end up being responsible for raising their own children's brood. Yet despite the evolving nature of the typical Thai family, the devotion to family life in whatever form, and the rearing of children with whatever arrangement of guardians, is likely to remain an intrinsic element of Thai society.

Matrilocal:

Derived from anthropology, the term denotes a culture where the husband goes to live with or near the wife's family.

When the Old Raise the Young

Since the dawn of the 1960s when Thailand implemented the first National Economic Development Plan, a growing demand for labor in the industrial sector has encouraged villagers, especially in the vast and impoverished northeast region, to migrate in search of more lucrative opportunities.

This has led to a boost in the household incomes of pastoral villages, but the downside has been the slow disintegration of the family unit. With more and more parents toiling in industrial estates or in urban factories and construction sites, grandparents stay in the villages to take care of the young.

As the Multiple Indicator Cluster Survey (MICS) conducted in 2012 by the National Statistical Office and supported by the United Nations Children's Fund (UNICEF) noted, around 21 percent, or three million, of the kingdom's children do not live with either of their parents due to internal migration. In the northeastern region of Isaan, that figure rises to almost one-third of the young.

In such households, where both parents are migrant laborers, some 85 percent of the children are looked after by grandparents, while aunts, siblings and other relatives account for the other guardians. The average age of the caregivers in these households is 58, while the oldest is 87, according to a research paper on "Children Living Apart from Parents Due to Internal Migration" conducted in 2012 by the Institute for Population and Social Research at Mahidol University.

One theoretical downside of this upwardly mobile trend is that children, who need consistency and strong parental figures during their formative years, are at a disadvantage in their intellectual development. The problem is exacerbated

"Families are earning more but living apart."

Nipon Poapongsakorn, distinguished fellow at TDRI

by the ebb and flow of money from their parents, or so goes one theory. But the truth and trouble are that comprehensive studies tracking these young people as they move from childhood to adulthood are almost nonexistent. In shorter-term studies, however, examining their health, academic performances and overall life satisfaction, children raised by relatives and those reared by their real parents did not differ significantly.

Just the same, one must also consider the effects of parenting on aging caregivers. Research papers published by Khon Kaen University's Graduate School in 2008 and 2009 indicate that child-rearing has become a burden

on grandparents, who are often poor in health and cash. So these seniors – especially the widowers – tend to suffer a dramatic decrease in their quality of life.

The researchers also pointed out that the older guardians have a tendency to spoil their grandkids in order to compensate for their lack of parental contact or – stressed out by money and health woes – take an iron-fisted approach to parenting. Another major concern, the papers noted, is that with limited knowledge of the modern world, and with outdated methods of rearing the young, the elderly may not be able to adequately supervise or prepare their charges for the future in a rapidly changing world.

FURTHER READING

- *Family Institution: Development and Changes*, by Sirirat Adsakul, 2010
- *The Family in Flux in Southeast Asia*, by Yoko Hayami, et al, 2012
- *Globalization and Families*, by Bahira Sherif Trask, 2010
- *Parenting for the Digital Age*, by Bill Ratner, 2014
- *Tech Savvy Parenting*, by Brian Houseman, 2014

SPOTLIGHT

Tackling Teenage Pregnancy

More than one million babies have been born to teenage mothers in Thailand over the past 15 years, with a 31 percent increase from the year 2000 to 2014. The average age of pregnant teenagers continues to fall year on year, and despite Thailand's rapidly aging population, the country's teenage birth rate remains high.

In response, the government's new Prevention and Solution of Adolescent Pregnancy Problem Act, inacted in 2016, aims to decrease the number of teen pregnancies in ten years' time and provide better support for the kingdom's hundreds of thousands of teenage mothers.

Developed with support from the United Nations Population Fund (UNFPA), the new law states that all young people between 10 and 19 years of age must have access to services and information on sexual and reproductive health, both at school and at work, and that pregnant teenagers must be able to access proper care and advice, and are encouraged and allowed to remain in school. The law also calls for vocational training to be made available to teen mothers to help them find jobs.

"We can't stop teenagers having sex, but we can help them make sex safe," said Dr. Jetn Sirathranont, Chairman of the Committee on Public Health in the National Legislative Assembly. "Comprehensive sexuality education is vital for children and adolescents so they know how to live and have healthy sexual relations. Every student in every school will have the right to this thanks to this act."

Families in Thailand often force their pregnant daughters to leave school and even their homes due to perceived shame. For these teens, returning to school or work after giving birth can also be problematic and stressful due to social stigma.

Yoriko Yasukawa, UNFPA's Regional Director for Asia-Pacific and Thailand Country Director, said that tackling the stigma, attitudes

> *"Comprehensive sexuality education is vital for children and adolescents so they know how to live and have healthy sexual relations."*
>
> **Dr. Jetn Sirathranont, Chairman of Thailand's Committee on Public Health**

and barriers that teenage mothers face in terms of acquiring education and jobs are essential to ensuring the success of the new law.

"It's a huge step forward," said Yasukawa, adding that the act ensures that "sexual and reproductive health services, education and information are delivered in a respectful, youth-friendly and confidential way."

These advances in sex education are imperative given that topics like abortion have typically been taboo in the classroom setting. Abortion is technically illegal in Thailand except in cases of rape or incest, to save a woman's life or preserve her physical or mental health, or if the woman is under 15 years of age. However, it is increasingly common for exceptions based on potential psychological harm to include women who are emotionally distressed at the prospect of having children they cannot afford to raise. This interpretation of the law is significant given that some 70 percent of women seeking an abortion cited "economic, social or family reasons," according to a health department survey that spanned 13 provinces in 2014. That same study also found that one in four women seeking abortions were students.

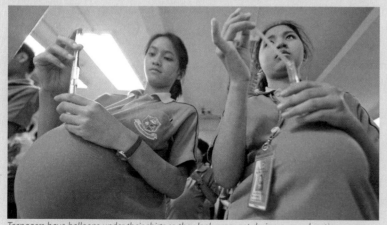
Teenagers have balloons under their shirts so they look pregnant during a sex education program

PUBLIC PARTICIPATION

Bringing power to the people in fits and starts

Citizen journalism:

Citizens take on the role of journalists who report, photograph and analyze the news. The downside is that the quality of these reports can be shoddy, while the facts are flimsy.

The overall purpose of public participation is to involve those who will be affected by a decision in the actual decision-making process. That decision could be as small as the selection of a new school principal or as momentous as a new constitution. According to the International Association of Public Participation, public participation is an active, dynamic process with five key pillars: inform, consult, collaborate, involve and empower. By involving the public, more sustainable decisions are made.

Since Thailand's transition from absolute to constitutional monarchy in 1932, the country has had 19 constitutions and charters. Despite the different socio-political contexts of each constitution's design, one principle that all 19 charters have in common is that sovereignty belongs to the Thai people. But until the 1997 constitution, known as the People's Charter, the role of Thai citizens in the decision-making process had been fairly limited, with the country not allowing its citizens to participate in the political processes in ways other than through representatives. During the 65 years from 1932 to 1997, in which Thailand was ruled mostly by unelected leaders, other mediums used by the Thai people to participate in policymaking included street demonstrations and grassroots movements, e.g., the peace movement against Korean War in 1952, the Farmers Federation of Thailand movement from 1974-1992, the demonstrations against military dictatorship in 1973, 1976, and 1992, and the Assembly of the Poor movement advocating for equity and fairness, popular participation, and self-determination for all Thai people who were affected by large-scale development projects.

But little real power was vested in the hands of the people until the groundbreaking, citizen-led constitution of 1997 enshrined this right. This constitution contained a strong bill of rights and liberties with support from a range of civil society groups. This charter represented a significant shift toward a more open and democratic society. It provided a legal basis for increasing the individual's degree of public participation, from the right to access information and protecting freedom of speech to influencing public policy, local resource management, administrative decentralization and referendums.

If it all looked good on paper, the reality was something of a cautionary tale. A study conducted by King Prajadhipok's Institute (KPI), an independent public organization under the supervision of the National Assembly, discovered that of 16 legislative proposals submitted by eligible voters in the period from 1997 to 2006 none were approved by the National Assembly. On environmental matters, the civic groups fared better, stopping the construction of Hin Krut Power Plant in 1998 due to concerns about its environmental and health impacts. In 2005, a consumer rights group opposed the privatization of the Electricity Generating Authority of Thailand (EGAT). They took the case to the Supreme Administrative Court, which granted an injunction to stop the listing.

FREEDOM IN THAILAND		(1 = BEST, 7 = WORST)		
	STATUS	FREEDOM RATING	CIVIL LIBERTY	POLITICAL RIGHTS
2017	NOT FREE	5.5	5	6
2013	PARTLY FREE	4.0	4	4
2010	PARTLY FREE	4.5	4	5

Sources: Freedom House, www.freedomhouse.org

Anti-government protests gather steam during the "Bangkok Shutdown" street protests of early 2014 before the military takeover in May.

Successes like these have been few and far between. For the majority of Thais, street protests are the favorite form of public participation. When political rivals reach a deadlock, the military steps in to resolve the situation and seize power.

That was the case in 2006 when the military staged a coup and abrogated the constitution. Despite criticism of the drafting process for a new charter, it was the first time the government organized a nationwide referendum for the public to approve or deny the entire draft. The turnout was around 58 percent.

The 2007 constitution sought to expand rights, liberties, public participation and decentralization with clear and specific provisions. Unfortunately, the new constitution did little to bring about any political reconciliation in a deeply divided land. The deeper those cracks, the more voters demanded their say at the ballot box. By the elections of 2011, three-quarters of all Thais voted. The election for Bangkok's governor saw a similar surge of interest, peaking at almost 64 percent of the electorate in 2013.

On a grassroots level, community groups are mobilizing on a scale never seen before and using **citizen journalism** and **crowdsourcing** to make sure their voices are heard. The Community Organization Council Act of 2008 established a large number of community councils to promote networking among like-minded groups. For them, the most important aspect of public participation is decentralization. While the country's national politics have been mired in a series of crises, civil society, local government officials, and scholars at the provincial and district levels have tried to take a more direct role in local administration. However, in 2014, the military seized power again in a bloodless coup to end six months of political turmoil; it was the nation's 19th successful coup in 83 years.

In summary, while elections have been held for decades and Thailand has long had an active media, it has been about 20 years since true public participation has been enjoyed by the people. Time is needed for both the authorities and Thai citizens to learn how to constructively participate in public policy formulation and the decision-making process. All sectors in Thai society are still green, but keen to learn how to implement the participation process properly and effectively. The participation process requires reciprocity, communication, conciliation, and respect. This process, if implemented correctly, might reduce any future conflicts.

Crowdsourcing:

The practice of obtaining services, ideas, content or financial contributions from a large group of people – mostly online – rather than resorting to traditional businesses or governments for funding.

Why Public Participation Matters

Empowering individuals to become active citizens
Public participation helps people exercise their democratic rights while encouraging them to think beyond their own self-interests. As people develop knowledge and understanding of crucial issues, they are able to make well-informed decisions about the services and public policies that affect their quality of life, which in turn serves the greater good by strengthening communities and democracy.

Everyone has a say Public participation is a dialogue that seeks to involve everyone in the decision-making process, regardless of political affiliations, race, social status or wealth. It helps to break down social hierarchies and barriers.

The potential to reduce conflicts Public participation provides opportunities – and a forum – for groups with divergent interests to express their needs and concerns without having to be adversarial. This can reduce tensions, though it's certainly not a panacea.

Fostering cooperation; bolstering trust Where public participation in the decision-making process is non-existent or ineffective, the public's suspicion of policies or development projects runs high while trust remains low. Full public participation boosts those levels of trust necessary for the state to maintain its legitimacy.

Improved quality of the decision-making process
Increased public participation allows the citizenry to inform the government of its preferences and concerns, as well as apprising them of other alternatives in developing projects or policies.

Accountability and transparency The more active the public is in these processes, the more accountable civil servants must be, which reduces corruption and improves efficiency.

Staff count votes during 2012 elections.

FREEDOM IN THAILAND

OUT OF
180 COUNTRIES,
THAILAND RANKED **136**TH
FOR PRESS FREEDOM

Source: 2016 Reporters Without Borders survey

"...political and social participation has intrinsic value for human life and well-being. To be prevented from participation in the political life of the community is a major deprivation."

Amartya Sen, economist and philosopher

Watchdogs Unleashed in Cyberspace

Only a well-informed people can truly participate in public policy formulation and make informed decisions. Typically the media plays a crucial role in this process: raising awareness by spotlighting important issues and providing objective reports. For any democracy, or aspiring democracy, it can be said that a free and fair press is of paramount importance.

Unfortunately, the media in Thailand can be rather weak in terms of objectivity – prejudice, favortism and self-censorship, as well as political and business influences, have all placed parameters on what can and cannot be reported. In addition, there are legislative restrictions on media. The *lèse majesté* law, criminal defamation, Computer Crime Act (CCA) and the regulations imposed by the junta will continue to put a serious dent in Thailand's reputation regarding media freedom.

As a repercussion of this, in recent years alternative media sources have emerged and become important tools for people to share opinions and voice their concerns to policy makers. Through alternative media outlets, watchdogs and blogs such as *Isaranews*, *Siam Intelligent Unit*, *Prachatai*, ThaiPublica, Deep South Watch and iLaw, along with social media like Facebook and Twitter, the media landscape has diversified dramatically.

These outlets allow common people to bypass the gatekeepers of traditional, mainstream media and share the information and perspectives citizens deem important. These platforms encourage a more open media environment and provide channels for the public to access information, express their wishes and opinions, participate in decision-making, exercise their right of supervision, and make choices that benefit their wellbeing.

According to a 2016 report by We Are Social, Thailand has 38 million Internet users out of a population of 68 million people. A recent survey conducted by the Electronic Transaction Development Agency (ETDA) revealed that Thais spend seven hours a day on average chatting online, catching up on news, watching shows, posting photos and searching for information.

While online media has become a key source of information for the Thai people and a tool for greater engagement of citizens in public policy decision-making, people have been left with an enormous task of verifying information by themselves from numerous forums and platforms. Online media face a number of obstacles to providing important information to the public, including the CCA which is designed to control online news portals. The CCA allows the state to take action against Internet users deemed to be sharing information considered detrimental to national security. Many rights activists consider the CCA to be a serious violation of freedom of speech.

Thai netizens have also used social media for crowdsourcing and submitting petitions. A group of advocates from the Thai Netizen Network, a nonprofit advocacy group promoting online privacy and Internet freedom, launched an online petition through the website Change.org, submitting almost 21,000 names of people opposed to the 10 digital economy laws.

Now, more than ever, the decision-making process depends upon two-way communication and information disclosure. Where access to information has been blocked, and where information has proven unreliable or its release has been unnecessarily delayed, the risk of social conflict will grow.

Without more open information, more responsive communications through all available media, and more meaningful public participation, any development or reform cannot be achieved and sustained.

A new generation of tech-savvy activists are taking their causes to cyberspace.

Toward a Just, Inclusive Society

Promoting the rule of law, supporting a free media and civil society, safeguarding the independence of the judiciary and ensuring equal access to justice are key pillars of sustainable development. These precepts not only foster a healthy society, but also create the foundation needed for sustainable development to take root and flourish.

Notably, according to a 2015 World Justice Project report, the top four countries in terms of rule of law — Sweden, Norway, Finland and Denmark — are the top four countries in terms of sustainable development as ranked by the SDG Index. Thailand was 56th out of 201 countries ranked by the 2015 World Justice Project report and 55th in the SDG Index. Therefore, a clear correlation between rule of law and sustainable development appears to exist.

With that in mind, Thailand has much to improve. In recent years, in an attempt to mitigate turmoil in Thai society, press freedoms and free speech have been curbed significantly, while journalists and activists have faced state-sponsored intimidation. Under the current military government, critics face the prospect of being sent to "re-education camps" if they step out of line on more than one occasion. According to iLaw, since the 2014 coup, hundreds of people have been summoned by the junta for attitude adjustment.

Rights monitors note that Thailand's courts and laws have frequently been manipulated by powerful individuals, companies and political actors as a means to further their aims. Corruption-related prosecution has regularly been wielded as a political weapon by various administrations since the mid-2000s, and laws like lèse majesté, criminal defamation, sedition and the Computer Crimes Act are increasingly employed to silence dissenting voices, settle personal vendettas, eliminate political opponents or simply to make an example out of someone. Journalists, civil society activists, politicians and government critics are frequently targeted by such lawsuits. While this phenomenon is not endemic to Thailand alone, it clearly contradicts the aim of building an inclusive society that welcomes debate among parties with differing views.

In particular, Thailand's lèse-majesté law, which makes it a crime to insult the royal family, has seen a significant spike in use during the latest period of political upheaval. As outlined by Section 112 of the Criminal Code, "Whoever defames, insults, or threatens the king, the queen, the heir-apparent, or the regent, shall be punished with imprisonment of three to fifteen years." While it is common for countries with constitutional monarchies to have such laws, in Thailand their application is sometimes perverted by individuals and political actors with ulterior motives.

Since the May 2014 coup d'état, the government has aggressively pursued lèse majesté cases, charging at least 82 people under Section 112, according to iLaw. Some 64 people have also been charged with sedition (Section 116) since the coup.

> "Moving forward towards a prosperous new normal requires that we fundamentally change our way of thinking, attitudes and mindsets to embrace openness, a diversity of views, as well as values that support societal change."
> **Anand Panyarachun, former prime minister of Thailand**

Even outside Thailand's political sphere, the weak implementation of rule of law has had a detrimental effect on the distribution of justice. Private companies that find themselves facing strong opposition from local communities may try to bend the judicial system to their will by filing defamation lawsuits. However, matters can sometimes escalate far beyond mere legal action.

Thailand was ranked the eighth most dangerous country in which to defend land and environmental rights, according to a 2014 report by Global Witness. Some 60 activists who led opposition to coal plants, toxic waste dumping, land grabbing or illegal logging have been killed or have disappeared under dubious circumstances in the past two decades, according to a May 2016 report in The New York Times.

Because many of these incidents occur in rural areas and are linked to well-connected individuals or high-profile business interests, these killings are unlikely to receive much media coverage and investigations tend to produce few meaningful convictions. Freedom House's 2015 country report on Thailand notes that "even in cases where perpetrators are prosecuted, there is a perception of impunity for the ultimate sponsors of the violence."

Making Thailand more just and inclusive won't be easy. But it is imperative to strive toward this in order to achieve a sustainable societal model. As former prime minister Anand Panyarachun said in a 2016 speech to the Foreign Correspondents Club of Thailand, "moving forward towards a prosperous new normal requires that we fundamentally change our way of thinking, attitudes and mindsets to embrace openness, a diversity of views, as well as values that support societal change."

FURTHER READING

- *The Public Participation Handbook,* by James L. Creighton, 2005
- *Participatory Democracy,* by Borwornsak Uwanno, 2005
- "Cyber Social Networks and Social Movements," by Soran Shangapour, Seidawan Hosseini and Hashem Hashemnejad, *Global Journal of Human Social Science,* Vol. 11, Issue 1, Version 1.0, 2011
- *Truth on Trial in Thailand: Defamation, Treason, and Lèse-Majesté,* by David Streckfuss, 2011

REALITY CHECKS

Challenges Facing Public Participation

Perception of bias Television is by far the most popular medium in Thailand. The majority of Thais rely on television as their primary source of news. The National Broadcasting and Telecommunications Commission (NBTC) acts as an "independent regulator"; the regulating process, however, is politicized as major television stations and radio frequencies are still owned and controlled by the military and several government agencies.

Democratic values Some observers say Thais possess only a partial understanding of democracy and of the roles, rights, liberties, duties and responsibilities of citizens in a participatory democracy. Democracy is not a system per se, but a set of norms and values that serve as a bedrock of the system. More must be done to educate the general public on these matters.

Social status The traditional Thai structure of society divides people into social classes in regard to carefully nuanced distinctions of age, education, ethnicity, occupation, wealth and proximity to power. An unquestioning deference to authority, especially politicians, high-ranking military men and the wealthy elite, may not always engender the dialogue that is a prerequisite to productive public participation.

Centralized administration The balance of power often tips in favor of the upper echelons in Bangkok who make the big decisions about public policies and infrastructure projects so crucial to the development of the provinces.

Rule of law The rule of law is essential to lay down the limits of political interference in decision-making processes and to encourage responsible policymaking.

Freedom of speech Thailand has had an uneasy relationship with freedom of speech. Military-led governments have not welcomed criticism or investigative reporting. Elected politicians have threatened lawsuits against their detractors, and the *lèse majesté* law is abused. As a result, self-censorship in the media is common. The current junta has imposed a number of restrictions on the press and social media users, scolding some for expressing their views and detaining others.

Nepotism common Preferential treatment given to friends, relatives, or business partners is common in Thai culture, undermining those who might succeed through merit.

Access to information While the 1997 Official Information Act has been widely accepted as a useful tool, it has also created challenges for the traditional bureaucratic system. There are many obstacles to overcome, such as harassment and intimidation, when the general public tries to get information or access to the contracts for mega-development projects. Sometimes people have had to take cases to court because officials rejected their requests for disclosure. Instead of helping public participation the law may become a hindrance and legal tool to delay or deny access.

Male dominance In 2011, women made up 15 percent of MPs, 16 percent of senators, and 17 percent of senior civil service positions despite outnumbering men as civil servants, according to the UN Women Thailand Country Program.

INEQUALITY

Vast divides separate Thailand's haves and have-nots

The issue of inequality is shared among countries rich and poor, with topics like income inequality, rural-urban divides and the unequal distribution of public resources inciting heated debate. Thailand is very enmeshed in this contemporary conundrum, though with some very Thai-specific features. While some of the kingdom's richest tycoons are themselves self-made billionaires, their rags-to-riches stories are outliers that belie the pervading sense that Thailand is not distributing its opportunities fairly. Yes, it's true that economic growth has helped to lift millions of Thais out of poverty. But as of 2017, the distribution of wealth and resources remains highly imbalanced when compared to other upper-middle-income countries.

Once an absolute monarchy, where the king essentially owned or controlled the majority of capital, Thailand is still a place where power and wealth remain remote to the vast majority of the population and the sense of participating in a civil society is weak. Many of the socioeconomic problems of today are a direct result of a liberal development philosophy that has favored specific groups and geographical areas over others. "This uneven developmental process has not given sufficient attention to conflicts of interest between capitalists and workers, urban and rural populations, and landlords and farmers," wrote analysts Phallapa Petison, Werapong Prapha and Veerathai Santiprabhob in their paper "Community Sufficiency

Gini coefficient:

Devised by the Italian statistician Corrado Gini, the Gini coefficient measures the extent to which the distribution of income among individuals or households within an economy deviates from a perfectly equal system distribution. A low Gini coefficient indicates a more equal distribution, with 0 corresponding to complete equality, while a higher coefficient indicates more income disparity.

> **"Many of the economic and social problems we currently face, including the simmering political tensions and sporadic clashes we have suffered in the past decade, can be traced back to the injustice and inequality inherent in our society."**
>
> **Anand Panyarachun, former prime minister of Thailand**

in Nan Province." As a result, "Thai society has become divisive and fragile," they added.

While Thailand certainly has expanded its middle class, decision-making and resources are still concentrated in the capital, and the divides between Bangkok and the rest of Thailand, between policy-makers and communities, and between the owners of capital and the vast workforce doing their bidding have not been meaningfully bridged. After decades of imbalanced growth, some critics now frame Thailand not as a growing democracy, but as an oligarchy run by a network of elites comprised of the monarchy, top companies, state-owned enterprises and other powerful families.

It should come as no surprise that many Thai leaders and observers see the current situation as increasingly untenable. "For development to be sustainable, the fruits of economic growth must be spread widely and fairly to foster social cohesion and continued economic and political legitimacy," said former prime minister Anand Panyarachun in a 2016 speech to the Foreign Correspondents Club of Thailand. "Many of the economic and social problems we currently face, including the simmering political tensions and sporadic clashes we have suffered in the past decade, can be traced back to the injustice and inequality inherent in our society," the former prime minister said.

What are some of the driving factors for inequality in Thailand? To begin, the kingdom suffers severe income inequality, which is reflected in its score as calculated by the **Gini coefficient**. This internationally accepted measure calculates the extent to which the distribution of income among individuals or households within an economy deviates from a perfectly equal system distribution. A low Gini coefficient indicates a more equal distribution, with 0 corresponding to complete equality, while a higher coefficient indicates more income disparity. By this internationally recognized standard, Thailand's level of income inequality has fallen from

A worker walks past a luxury condo ad in Bangkok. Due to various factors, many Thais are shut out from the opportunities and prosperity found in Bangkok.

0.52 in 2000 to 0.47 in 2013. This figure, however, is still higher than Indonesia, Laos, Vietnam and Cambodia and means Thailand has among the highest income inequality levels in Southeast Asia. This fact and others led Thailand to score below its region's average on Goal 10 (Reduced Inequalities), according to the SDG Index.

Thailand also features a radically unequal distribution of land. The top ten percent of all landholders (roughly 1.5 million individuals and juristic organizations) hold more than 60 percent of all land, while the bottom ten percent hold a mere 0.7 percent, according to data from the Land Development Department. The largest landholder, Thai Beverage chairman Charoen Sirivadhanabhakdi, reportedly holds around 101,000 hectares.

Meanwhile, in 2013 members of parliament reported holding land that, on average, was worth almost 31 million baht (about US$900,000). These numbers speak to some of the vast inequalities at play in Thailand and the need for more effective measures to better balance land and wealth distribution – something the kingdom continues to struggle to enshrine in its laws.

Since the 1990s, a slew of regressive tax measures have fueled an environment whereby Thailand's rich can gain from the tax system. "One cause of the great inequality in wealth distribution is that Thai governments have never seriously taxed land or assets, but instead have often offered tax benefits that actively contribute to greater inequality," wrote Sarinee Achavanuntakul, Nathasit Rakkiattiwong and Wanicha Direkudomsak in their paper "Inequality, the Capital Market and Political Stocks." In Thailand, the ratio of tax to GDP is around 17 percent. In many other middle-income countries with similar levels of development, the ratio is much higher: 25 percent in Venezuela and 32 percent in Turkey.

To date, even sincere government efforts to address tax reforms or land and wealth redistribution through policy changes have gained little ground. For example, in 2011 the National Reform Commission proposed wide-ranging land reforms including a progressive tax on people owning large tracts of land, a cap on private land ownership, a proposal to allocate land to a million poor families and a system to provide legal assistance to villagers facing encroachment charges. But six years on, the success of these proposals is still uncertain.

The Alarming Acceleration of Household Debt

A growing concern in Thailand is that the poor and middle class are saddled with significant household debt, which surged to 84.2 percent of GDP in 2015, up from 54.6 percent in 2007.

In recent years, a slowdown in GDP growth and government initatives to spur domestic spending such as tax incentives for first-home buyers and first-car buyers have played significant roles in pushing up the debt level of Thai households. Some 36.7 percent of that debt was owing to the purchase of land and residential units while around the same amount was spent on household consumption. The remainder was spent on business, farming, education and other expenditures.

According to a survey by the National Statistical Office, average household debt leapt to 156,770 baht in 2015 from 116,681 baht in 2007. This distressing acceleration of household debt threatens future liquidity and the debt serviceability of the household sector, and raises concerns about non-performing loans at financial institutions. The continuing increase in household debt could also create more adverse effects as it impinges on purchasing power, limits consumption to the bare necessities and drags down the overall standard of living. Although there have been no imminent signs that the loan quality at commercial banks is deteriorating while household debt grows at a slower pace, the Bank of Thailand has periodically expressed its concerns over these high figures. The situation needs to be closely monitored by financial institutions and regulators who must take steps to ensure that household debt does not go through the roof. There is no need to be too alarmist; precautions can be taken.

In order to strengthen the financial position of Thai households and increase their immunity to future economic risks, it is essential to promote fiscal discipline by reducing excess spending while at the same time encouraging people to save money, stop borrowing and take out insurance policies to cope with emergencies. Shoring up the finances of the household sector would enhance overall economic stability, too. That would keep the wheels of the economy turning by ensuring that the public's purchasing power is not impeded, thereby mitigating the risk of yet another recession.

BY THE NUMBERS

RISE OF AVERAGE HOUSEHOLD REVENUE

18,660 baht per month	26,915 baht per month
2007	2015

RISE OF AVERAGE HOUSEHOLD DEBT

116,681 baht per month	156,770 baht per month
2007	2015

TOTAL HOUSEHOLD DEBT (% of GDP)

54.6%	84.2%
2007	2015

RISE OF AVERAGE HOUSEHOLD EXPENDITURES

14,500 baht per month	21,157 baht per month
2007	2015

DEBT BREAKDOWN

32.4% purchase of land or residential units

41.3% household consumption

26.2% business, farming, education and others

Source: National Statistical Office, 2015

Uneven Distribution of Opportunities

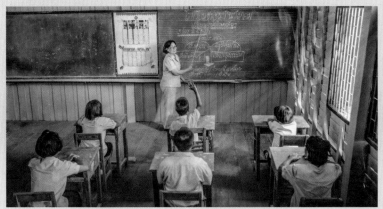

Rural schools tend to have far fewer resources than their urban counterparts.

The acute concentration of wealth, income and decision-making in Bangkok has spawned an urban-rural divide in terms of access to resources, services and opportunities at all levels. Poorly-educated migrant laborers from rural regions who flock to cities in search of work and a higher standard of living often find themselves toiling in low-skill, high-pressure jobs and living in shanty-towns. For these laborers it's a vicious cycle that leaves them trapped in a spiral of downward mobility.

Duangmanee Laovakul, an Assistant Professor in the Faculty of Economics at Thammasat University, said "one reason for inequality in Thailand is the under-supply of public goods and services" reflecting a long-held complaint of those outside Bangkok that resources are too centralized. Indeed, in provincial Thailand everything from educational resources to physical infrastructure to hospital equipment to public funding is lacking when compared to the country's thriving capital.

"There are too few good schools, inadequate public transport facilities, and no comprehensive provision for old age even though the society is rapidly ageing," Duangmanee wrote in her paper "Concentration of Land and Other Wealth in Thailand." Thus, for the poorest of the poor, especially those living outside Bangkok, the pathway to a decent or higher education, a quality job, justice or simply a bank loan can be difficult to imagine, let alone access. Darker skin, family name, birthplace, gender and other indicators outside of an individual's control are often implicitly considered during, for example, job hiring. This, in turn, can create a "cycle of deprivation" in which a family's lack of opportunities are transmitted from parents to children and onto grandchildren, ensuring that future generations remain trapped in poverty.

For such people, access to quality education is a prime example. Principals of elite schools are known to expect fees, donations or outright bribes to admit students and acceptance is largely based on one's ability to pay tuition. Children of the elite can afford to attend private Catholic or international schools, hire tutors or travel abroad to learn foreign languages or other vital skills, and therefore have a significantly better chance to follow in their parents' footsteps. Meanwhile, rural children who primarily attend under-funded public schools lack qualified teachers and other essential education resources. The simple fact is that poor-quality education during primary and secondary years hinders the ability of a child to later pursue higher education and unleash his or her potential.

"Youth from socially and economically disadvantaged households have fewer opportunities to enter tertiary education than youth from privileged households," creating a pattern of lower wages when the former enters the labor force, said Dilaka Lathapipat, a human development economist at the World Bank, in the paper "Inequality in Education and Wages."

Thailand needs to tackle the "challenging task of overcoming wealth-related inequality in college preparedness from an early age by

> *"There are too few good schools, inadequate public transport facilities, and no comprehensive provision for old age even though the society is rapidly ageing."*
> **Duangmanee Laovakul, assistant professor at Thammasat University**

providing good child care facilities in poor communities and eliminating the huge disparity in the quality of basic education provided by resource-poor and resource-rich schools," Dilaka said in a 2012 paper, "The Influence of Family Wealth on the Educational Attainments of Youths in Thailand." The most direct way to accomplish this, he said, is to consolidate rural schools and then equip "hub schools" with better resources and an adequate number of trained teachers.

Unofficial Oligarchy?

The constant protection and advancement of Thailand's wealthy, whether through favoritism in the form of tax breaks or cronyism, has led to a contentious debate about the structure of Thai society as a whole. As Pasuk Phongpaichit and Chris Baker wrote in their 2016 book *Unequal Thailand: Aspects of Income, Wealth and Power*, "informal networks and coalitions of the few play major roles in the distribution of power and economic benefits..." They add that "these economic inequalities underlie inequalities of power, social position and access to resources of all kinds."

The Director of the Asia Research Centre at Murdoch University, Kevin Hewison, argues a similar premise in a 2015 article in the *Kyoto Review of Southeast Asia*: "The inequality of conditions in Thailand is the fundamental fact from which all others are derived. Economic and political inequalities in Thailand are mutually reinforcing conditions that have resulted from the ways in which the gains of rapid economic growth have been captured by elites."

Many Thailand observers would agree and argue that dating back to the era of absolute monarchy through to present day, the "fruits of economic growth" continue to be shared only among a few. Instead of building an accountable and transparent meritocracy based on the rule of law, these critics argue that the country's elite have used their elevated status to stack the odds in their favor and protect the various apparatus that further their financial ambitions and solidify their power base. "Relatively low incomes, skewed ownership and the siphoning of income to the already rich indicate a long-standing pattern of exploitation," Hewison said.

According to this criticism, in many ways, maintaining "the gap"

> *"The inequality of conditions in Thailand is the fundamental fact from which all others are derived. Economic and political inequalities in Thailand are mutually reinforcing conditions that have resulted from the ways in which the gains of rapid economic growth have been captured by elites."*
>
> **Kevin Hewison, Director of the Asia Research Centre at Murdoch University**

has become almost as important as the profits, monopolies and alliances it helps to engender. Pasuk and Baker argue that mechanisms have been put into place by elites to safeguard the very existence of Thailand's equality gap. "The persistence of economic inequality is a function of the strength of oligarchy. The rule of the few is found not only at the national level but also in the operation of institutions at all levels of society. The few rule and prosper by

cultivating and defending privileges and monopolies of various kinds, and by opposing extension of the rule of law which might form the foundation for greater equality," Pasuk and Baker write. As related to Target 10.2 of the 2030 Agenda for Sustainable Development, which promotes the "political inclusion of all," the feeling that the powerful public and private sectors do not properly listen to the concerns of community stakeholders has fueled the sense of inequality in Thai society.

As an example, Nopanun Wannathepsakul points to the hybrid semi-public, semi-private nature of PTT and the Electricity Generating Authority of Thailand. "These massive hybrid organizations in the energy sector have been created by a 'network bureaucracy,' which commands great official power. This network spreads across all agencies involved with the energy sector... The hybrid nature of these corporate groups and the power of this 'network bureaucracy' have contributed significantly to the rapid expansion of these corporations," Nopanun wrote in the 2016 paper "Network Bureaucracy and Public-Private Firms in Thailand's Energy Sector."

At the same time, Nopanun says some executives in these organizations "hold or once held high public office; they are chairpersons or directors on several boards; at the same time they have high positions in public agencies that oversee these organizations.... Their remuneration from these multiple posts is generous."

The influence of connections, nepotism and the pure greed of those in power has not gone unnoticed by the average Thai, with the result that inequality, and the structures that maintain it, are increasingly being challenged through political protests.

FURTHER READING

- *Unequal Thailand: Aspects of Income, Wealth and Power,* by Pasuk Phongpaichit and Chris Baker, 2016
- "Inequality and Politics in Thailand," by Kevin Hewison, *Kyoto Review of Southeast Asia,* 2016
- *Property Tax in Thailand: An Assessment and Policy Implications,* by Duangmanee Laovakul, 2016
- "Democratic Governance: Striving for Thailand's New Normal," a speech by Anand Panyarachun to the FCCT in March 2016, available online

REALITY CHECKS

Challenges to Closing the Equality Gap

Tax and land reforms The implementation of meaningful taxes on land, wealth and capital gains is long overdue in Thailand. In order to reduce inequalities, the state must reform its tax laws and pass legislation based on the principles of fair taxation and ability to pay. It is well known that Thailand needs to address the issue of land distribution. In 2011, the National Reform Commission proposed wide-ranging land reforms including a progressive tax on people owning large tracts of land, a cap on private land ownership, a proposal to allocate land to a million poor families and a system to provide legal assistance to villagers facing encroachment charges. So far, however, few changes have been implemented. A proposed inheritance tax of five percent has been revised by the National Legislative Assembly (NLA), which raised the minimum value subject to inheritance tax to 100 million baht. Meanwhile, a tax bill on new land and buildings, which had received the green light from the cabinet, was altered to be applied to properties worth 50 million baht and over.

Distribution of opportunities and resources Thailand's vast rural-urban divide is exacerbated by unequal distribution of governance power, wealth and income with the lion's share of resources and opportunities going to those in urban centers like Bangkok. To Leave No One Behind on its path to sustainability,

Thailand must work harder to share the benefits of its development and prosperity more evenly.

Teachers for rural schools
Inequality in education is fueled by uneven distribution of resources and teachers across Thailand's more than 30,000 schools. In the worst scenarios, a single teacher might have to teach every subject to one or more grades, and also lacks adequate physical resources to help them with this difficult task. There are also serious discrepancies between the qualifications of teachers in Bangkok and those in rural schools. For example, according to World Bank data, the share of teachers with a graduate degree in Bangkok is 19.7 percent while in Mae Hong Son, Thailand's poorest province for several years running, the share is just 8.7 percent. Likewise, the average

years of experience for a teacher in Bangkok is 25.3 while in Mae Hong Son it is just 10.8 years. Tellingly, the World Bank has determined that the single most important factor that contributes to education inequality in Thailand is the number of teachers per classroom. In Bangkok, that ratio stands at 1.61 while in Mae Hong Son it is only 0.71 teacher per classroom.

Rule of law versus impunity Thai society is still very much governed by a hierarchy of power and position. From politicians to tycoons to village headmen, those seen as having elevated status are treated with deference. To challenge that hierarchy in a brazen manner can provoke serious legal action initiated by the more well-connected individual, or even violence. Government officials, members of the security forces and wealthy Thais have long enjoyed impunity for their crimes, even in cases of deadly offences. Corruption, in particular, has become a shortcut for politicians, officials and others to enrich themselves, representing a constant bleeding away of funds and resources that otherwise could have been applied to uplifting the larger populace. The reason is Thailand's weak rule of law, and a culture of impunity that often enough sees the victims prosecuted, instead of the perpetrators. This obviously has a detrimental effect on the distribution of justice, but also impedes sustainable development and the fostering of an inclusive, egalitarian society.

GENDER EQUALITY

The long road to a fairer society

Equality between the sexes is a human right that the United Nations has highlighted as one of its 17 Sustainable Development Goals (SDGs). By achieving this equilibrium, the lives of half the world's population, including more than 34 million women and girls in Thailand, could be vastly improved.

Once defined by the old proverb "the husband is the forelegs of the elephant and the wife the hind legs," implying that men are leaders and women followers, Thai women have made enormous strides in the past few decades. Perhaps no other figure illustrates this as well as Thailand's ranking in the world's top 10 countries with the highest number of female executives. Few other nations can also boast such a high workforce participation rate for women: 64.3 percent of females aged 15 and above are employed.

In other crucial respects, Thai females enjoy quite a few essential freedoms and protections that women in many other developing countries can only dream of. Married women are free to adopt titles or family names, according to their preference. The rights of women to file for divorce and to gain custody of their children and assets have also been recognized.

Over the years, the state has identified other benchmarks of gender equality. As part of the 10th National Economic and Social Development Plan (2007–2011), the Women's Development Plan cemented five key pillars to promote the advancement of women and gender equality: mobilizing all stakeholders to advocate gender equality; enhancing female participation in the policy-making process; improving healthcare services; strengthening women's rights to human security; and fostering more economic participation.

However, for all these policies and successes, the gender gap remains as deep-seated as any cultural prejudice. Still seen as caregivers in many ways, Thai women often bear a disproportionate amount of the household chores, from cooking and cleaning to child-rearing. Many do so while working full-time. Thai women, like those in other Asian countries, still earn less than men in every major industry sector. Women are also far more likely to be employed part-time or on a temporary basis. For these reasons and others, in 2015 Thailand ranked 87th out of 179 countries in UNDP's Gender Inequality Index.

Nowhere is the disparity between the sexes revealed in more shocking fashion than in the statistics of sexual abuse and domestic violence. In 2013, almost 32,000 women and girls were battered and/or sexually assaulted, amounting to 87 cases per day, according to the Public Health Ministry's One Stop Crisis Center Report that year. As a result, Thailand was ranked 36th among 75 countries in acts of physical violence committed against females, and seventh out of 71 countries for sexual assault, according to the Thailand Institute of Justice.

On the other side of the gender gap, Thai society is well known for its acceptance of homosexuality and gender diversity including kathoey ("transgendered"). Thailand has an estimated 600,000 gay men, according to the Ministry of Pubic Health. The country has also become a top destination for gender reassignment surgery – a distinction that has

ABUSE OF WOMEN IN THAILAND

31.866	Women and girls physically and sexually assaulted
87	Number of females assaulted every day
186	Deaths due to domestic violence
4.000	Rape cases reported to police
2.400	Arrests made by police

Sources: Thailand Institute of Justice (TIJ) and UN Entity for Gender Equality and the Empowerment of Women (UNWOMEN), 2013

Thanks to the national policy on eradicating gender disparity at all levels of education, both girls and boys are entitled to 15 years of free education. In areas like increasing literacy these policies have earned top marks. In 1994, nearly two-thirds of the illiterate population were women. Today, among 15 to 24 year olds, the rate of literacy is almost identical for both genders at around 98 percent, a significant achievement.

Women participate in a self-help group in Buriram province.

lent a veneer of acceptance to those in the LGBTIQ community. But for all the superficial tolerance, members of the LGBTIQ community still face discrimination in the workplace and are frequently depicted through negative stereotypes in the media. Being accepted by their own flesh and blood is often the biggest struggle they face in this family-centric nation.

The protection of women's rights and those of the LGBTIQ community are deeply intertwined. Both remain vulnerable and marginalized. But with more female and LGBTIQ representation in the corridors of political power and less violence against them, the massive gains that Thai women have made in fields like education and economics could be further bolstered to bring about a much fairer society. With **gender mainstreaming** becoming the norm among public policymakers, it is time to celebrate rather than bemoan the differences between the sexes, bringing them closer together in a marriage of equals who are allowed to express themselves freely.

Gender mainstreaming:

A public policy approach that promotes gender equality by evaluating the impacts that proposed legislation and programs will have on both men and women.

VIOLENCE AGAINST WOMEN IN THE REGION

● Last 12 months ■ Lifetime

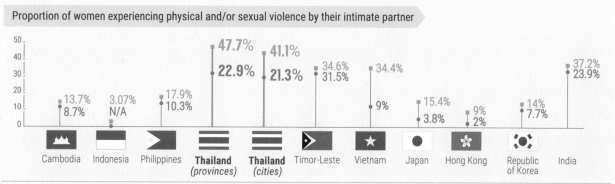

Proportion of women experiencing physical and/or sexual violence by their intimate partner

Cambodia	Indonesia	Philippines	**Thailand** *(provinces)*	**Thailand** *(cities)*	Timor-Leste	Vietnam	Japan	Hong Kong	Republic of Korea	India

- Cambodia: 13.7% / 8.7%
- Indonesia: 3.07% / N/A
- Philippines: 17.9% / 10.3%
- Thailand (provinces): 47.7% / 22.9%
- Thailand (cities): 41.1% / 21.3%
- Timor-Leste: 34.6% / 31.5%
- Vietnam: 34.4% / 9%
- Japan: 15.4% / 3.8%
- Hong Kong: 9% / 2%
- Republic of Korea: 14% / 7.7%
- India: 37.2% / 23.9%

Source: World Bank Development Report, 2012

Why Gender Equality Matters to Sustainable Development

Productivity boost A greater degree of gender equality in the workplace can have a positive effect on productivity. A World Development Report from 2012 found that the output per worker in East Asia and the Asia Pacific could be boosted from seven to 18 percent with the addition of more female entrepreneurs and staffers.

Healthier children Promoting gender equality is an investment in the next generation. Healthier, better-educated mothers have healthier, better-educated children, which has a direct effect on the offspring's wellbeing and future prospects. In fact, these consequences begin in the womb, as a mother's health and nutrition strongly affect the child's physical and cognitive development, the World Bank said in a 2012 report.

Empowered voices Strengthening women's voices can enhance the quality of developmental decision-making. In Hong Kong, for example, helping female migrants toiling as maids to stand up for their rights over the last two decades has contributed to the enactment of laws that provide them with some of the most comprehensive legal protections in the world for domestic workers.

Mothering nature Women are responsible for half of the world's food production overall and up to 60–80 percent in developing countries. Women also have an important role in establishing sustainable use of resources such as in small-scale fishing communities. In these roles, their extensive knowledge of natural resource and ecosystem management and their contribution to environmental sustainability should be taken into account.

Justice for all The protection of women's rights and those of the LGBTIQ community, which are deeply intertwined, can help minimize social prejudice and promote human rights among individuals in marginal and vulnerable communities.

Landmark Legal Moments for Gender Equality in Thailand

- The 1974 constitution was the first to mention gender equality (Section 28).

- Thailand ratified the Convention on the Elimination of All Forms of Discrimination Against Women (CEDAW) in 1985 and its Optional Protocol in 2000.

- Thailand's maternity leave law, implemented in 1993 as part of the Labour Act, allows 90 days of leave. For the first half, mothers receive full salaries. If they choose to take an additional 45 days of leave, they can do so at 50 percent pay. Those working in the informal sector receive no such benefits.

- Thailand endorsed the Beijing Platform for Action in 1995 during the Fourth World Conference on Women. The declaration stated 12 key areas where urgent action was needed to ensure gender equality.

- In 2000, Thailand embraced the Millennium Development Goals (MDGs). Significant efforts have been made to integrate these international principles and instruments into policy and programing frameworks, as evidenced by the Constitution B.E. 2550 (AD 2007), which contains provisions for anti-sex discrimination and gender equality. The Domestic Violence Victim Protection Act B.E. 2550 was also established.

- The Gender Equality Act came into effect in September 2015. The law aims to eliminate discrimination among the sexes and is the first Thai law to contain language explicitly recognizing gender diversity. Additionally, a special committee has been set up to promote parity and mediate on cases of gender discrimination among the sexes, including "any act or failure to act which segregates, obstructs or limit any rights, whether directly or indirectly, without legitimacy because that person is male or is female or has a sexual expression different from that person's original sex." Offenders could face as much as a six-month jail term or a fine of up to 20,000 baht.

The Rise of Female Executives

Few figures illustrate the financial betterment of women's lives in Thailand more positively than the huge number of female executives and entrepreneurs. In 2012, Thailand topped the global list of women running their own companies, according to the US-based Global Entrepreneurship Monitor (GEM) survey, which also said that 12 women started or ran their own businesses for every 10 men. Although the rate has dropped a little in recent years, it remains remarkably high, with Thai women accounting for about one-third of the board members of companies registered with the Ministry of Commerce.

To commemorate International Women's Day on March 8 every year, Grant Thornton, a London-based professional service network, releases its annual International Business Report, which includes a survey of women occupying executive roles. In 2016, Thailand came in sixth with 37 percent, well above the global average of 24 percent, and only slightly behind ASEAN leader the Philippines (39 percent).

Sumalee Chokdeeanant, an assurance partner at Grant Thornton Thailand, said, "In Thailand and in many places in Southeast Asia, it is not unusual for women to be in senior executive roles. Asia is a strongly family-orientated society and the female has always played a key role in the household, especially governing finances. Over time, as business has grown, this has simply become the norm for us in Thailand. This is evidenced by a continuing growth trend of women in executive roles here." She added that the integration of the ASEAN Economic Community means more golden opportunities for women to take on senior managerial roles in the years to come.

In the domestic labor market, female participation has been on the rise for many years, thanks in no small part to the enactment of the Labor Protection Law of 1998. That law stipulated gender equality in employment, health security, work safety and the prohibition of sexual harassment in the workplace. This legislation, in tandem with a raft of governmental measures to boost the education levels of girls and their access to tertiary institutes, has resulted in what the UN's Gender Equality Index has called one of the highest "labor force participation rates" for women in the world: a little over 64 percent. Thai females also account for about one-third of the board members of companies registered with the Ministry of Commerce.

For Thailand and other nations, the more balanced the workplace and the executive boards are between men and women, the more well informed are the decisions that form the axis upon which these businesses pivot. Francesca Lagerberg, global leader for tax services at Grant Thornton, said: "That greater diversity in decision-making produces better outcomes is no longer up for debate. For businesses, better decisions mean stronger growth, so it is in their interests to facilitate the path of women from the classroom to the boardroom."

KASIKORNBANK President Kattiya Indaravijaya speaks during an interview.

> **"It's still a male dominated world. Women are playing major roles in the business world, but there is still more to come."**
>
> **Kattiya Indaravijaya, one of KASIKORNBANK's four presidents**

According to Kattiya Indaravijaya, one of KASIKORNBANK's four presidents, all of Thailand's key sectors still need to do more to encourage mentoring, access to childcare and opportunities for the advancement of women. "It's still a male dominated world. Women are playing major roles in the business world, but there is still more to come," she told *Bloomberg* in 2017.

Providing "SUPPORT" for Women

Gender inequality heavily impacts the nation's most vulnerable communities – such as rural people, impoverished neighborhoods and ethnic or religious minorities – because they often don't enjoy the same access to healthcare, education, technologies, social networks and other resources that the wealthy, the privileged and the urban do. This fact has not escaped the notice of the royal family, who has

crafts. These products are in turn sold in specialized SUPPORT Foundation stores throughout the country, linking the marginalized or remote groups to a direct market. Queen Sirikit has also promoted their work and their handicrafts by wearing them and bringing them with her on trips abroad.

While this provides an additional income stream to indigent families (many of them headed by women

reliance. With these new skills, marginalized women are better equipped to determine their own future. Those who produce exceptionally fine work also have the opportunity to become SUPPORT instructors.

The royal family has started numerous such projects aimed at poverty alleviation, women's empowerment and social justice, not least of which are Princess Maha Chakri Sirindhorn's Sai Jai Thai Foundation and Phufa stores. While the Sai Jai Thai Foundation targets military personnel, police officers and civilians who were hurt in the line of duty, the Phufa stores aim to raise rural people out of poverty. Although not specifically focused on gender equality, both of these projects address issues of social justice, which feeds into the advancement of all marginalized groups – including women.

Queen Sirikit (right) founded the SUPPORT Foundation in 1976

established several foundations and organizations to include marginalized communities in the national goal of achieving sustainable development.

One long-standing example is the "Foundation for the Promotion of Supplementary Occupations and Related Techniques of Her Majesty Queen Sirikit of Thailand," also known by its shortened form, the SUPPORT Foundation. Established in 1976 by Queen Sirikit, the foundation is guided by the principle of fostering greater self-reliance and provides supplemental employment to mostly low-income rural women and teenagers by training them in the production of traditional Thai arts and

who, without this opportunity, may have taken up riskier employment involving migration to urban areas or the sex trade), the program has had the added benefit of preserving a rich national heritage of arts, crafts and traditional vocations that are important to the Thai sense of identity.

Many of the trainees participating in SUPPORT are, in fact, single mothers, female heads of households and teenagers. Through the program, these women are not only provided with tangible, income-earning skills in such folk crafts as silk weaving, silver and gold smithing and basketry, but they also gain a sense of pride, empowerment and self-

Established in 1975, Sai Jai Thai offers both a stipend and vocational training in leather and glass artisanship to the handicapped to compensate veterans for their sacrifice and to encourage livelihood development. Sai Jai Thai crafts are sold in gift shops in high-traffic locations such as Suvarnabhumi Airport.

On the other hand, the Phufa stores are high-quality handicraft shops that carry region-specific goods sourced from rural and impoverished communities. Similar to the queen's SUPPORT Foundation, the Phufa stores offer marginalized communities a distribution channel and encourage the use of local resources and fair trade to give vulnerable communities employment opportunities and greater self-reliance.

Such programs go a long way toward helping to level the social playing field, addressing economic inequality and gender issues, as well as handing back more power to the people.

FURTHER READING

- *Gender Equality and Human Rights*, by Sandra Fredman and Beth Goldblatt, UN Women, 2015
- *Being LGBTI in Asia*, by UNDP, 2016
- *Gender Mainstreaming in Environment and Sustainable Development Projects: A Perspective from the Asia-Pacific Region*, by UNDP, 2015
- *Driving the Gender-Responsive Implementation of the 2030 Agenda for Sustainable Development*, by UN Women, 2016

REALITY CHECKS

Challenges to Reaching Gender Equality

Nong Toom is a famous transgender boxer.

Thai women are increasingly taking the initiative to earn income through small businesses.

Social stigma Discussions of sexuality in general are still taboo in Thailand. Sex education in Thai schools is addressed from a biological perspective with little discussion of sexual wellbeing or health. The current textbooks still describe homosexuality as an illness or aberration.

Violence on the rise In a 2014 report by the Thailand Institute of Justice, Thailand ranked 36th out of 75 countries in the number of acts of physical violence against women reported. It also ranked seventh out of 71 countries for sexual assaults.

Political representation Following the 2014 coup, the proportion of seats in parliament held by women fell from 15.8 percent in 2013 to 6.1 percent

in 2014-2016. This is well below the global average of around 23 percent. LGBTIQ representatives are also needed to ensure gender equality.

Teaching new values To change the social norms that perpetuate gender inequality, first it is necessary to change the education system that fosters and reinforces them. Reforming school curricula, revising learning materials including accurate information about LGBTIQ issues and encouraging a more open, gender diversity-friendly environment will address such pitfalls and inspire a greater sense of fair play.

Better law enforcement While many laws are on the books to combat trafficking in women and

children, they are not adequately enforced or are even circumvented by complicit parties. Consequently, Thailand has come under fire from many international governments and human rights organizations.

Same-sex marriage legislation Until the passage of the Civil Partnership Registration Bill, same-sex couples will not have the right to marry or enjoy employee benefits and other basic rights.

Sexual harassment Although sexually harassing employees is deemed unlawful, prosecuting offenders is difficult due to a vague legal definition, which depends on the interpretation of the judges, the police and those involved.

CONFLICT

Sustainable development requires dialogue and peace

While some conflicts may ultimately give rise to a brighter future for a country, violent unrest and social turmoil are typically antithetical to sustainable development. Whether born from political or religious divisions or another cause, the impacts are largely the same in any nation: physical loss of life and limb that destroys families and foments distrust and hostility, environmental degradation, the devastation of communities, economic loss, infrastructure damage and the derailment of positive policy initiatives. Some conflicts feed off the environment itself, with groups pillaging property, such as antiquities, or plundering natural resources in order to raise money or publicize their campaigns.

Although it has been hundreds of years since Thailand became truly engulfed in war, the kingdom has not been immune to serious unrest. Since the abolition of the absolute monarchy in 1932, the kingdom has been convulsed by 12 successful coups and many more failed attempts. While the periods of military rule that followed have brought stretches of development, security and peace to the country, they have also on occasion – in the 1970s, 1990s and 2000s – led to violent unrest.

Currently Thailand suffers from two key conflicts: one between political factions known as the **red shirts** and **yellow shirts**, which has resulted in mass street protests over the past ten years that have paralyzed large sections of Bangkok and triggered tense protests across the country. The other conflict consists of an insurgency in the three Deep South provinces of Yala, Pattani and Narathiwat, where a large percentage of the region's population is of Malay descent and most residents from the older generation still speak Patani Malay (*Yawi* in Thai) as their first language.

The situation in the provinces of Thailand's Deep South is a primarily localized conflict which involves operations of limited scale, but has claimed more than 6,700 lives since the last flare up in 2004, according to Deep South Watch, which is connected to the Prince of Songkla University in the province of Pattani. In August 2016, a spate of bombings rocked tourist destinations from Hua Hin to Phuket, killing four and injuring more than 35. Although no group claimed responsibility for the attacks, authorities were quick to point the finger at southern Muslim insurgents. As historian and expert on the issue Duncan McCargo notes, the causes of the conflict are "complex and often opaque. Alternative explanations include questions of identity, historical injustice, economic inequality and discrimination, unequal power relations, and networks of criminality involving local politicians and members of the security forces."

Although both of these conflicts are generally contained to small groups, they have impacted Thailand's image. The country ranked 125 out of 163 countries in the 2016 Global Peace Index (GPI). This

THE PILLARS OF PEACE

A visual representation of the factors comprising positive peace. All eight factors are highly interconnected and interact in varied and complex ways.

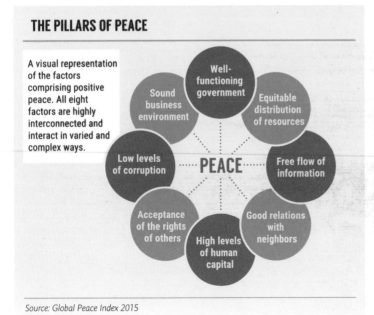

Source: Global Peace Index 2015

A soldier helps schoolchildren off a bus in the Deep South. Militants have targeted schools, which they see as representatives of the state.

index gauges global peace through 23 qualitative and quantitative indicators under three broad themes: the level of safety and security in society, the extent of domestic or international conflict and the degree of militarization. Only the Philippines and North Korea fared worse than Thailand in the Asia-Pacific region. The report noted that while Thailand had the second-highest absolute improvement in the 2016 GPI, its principal gains were primarily due to improved relations with Cambodia, with whom its relationship had been a "source of friction in the past." The report also noted that relative stability in the kingdom had "come at the cost of an erosion of Thailand's democratic institutions as it does not appear likely that the military will relinquish power anytime soon." Meanwhile, the financial cost of Thailand's conflicts was estimated at some US\$70 billion, which is \$1,033 per Thai, or seven percent of GDP. It's interesting to note that the 12 "most peaceful" countries are all stable democracies.

While Thailand's two key conflicts fill the newspapers and can give the appearance of a country in constant turmoil, the kingdom has remained incredibly resilient in many respects. In 2016 it ranked as the world's 11th-most-popular tourist destination, and although no stranger to slumps, it has maintained its status as the region's second-strongest economy.

The corrosive effects of Thailand's conflicts are thus slightly harder to identify and the greatest loss is likely to be found in the unknown opportunity costs suffered over the long-term. In other words, a lack of policy and political continuity over the past decade in particular may have created an era of lost opportunities for the country during a time of increasing global competitiveness.

In Thailand, conflict in the Deep South and polarization between the so-called red shirts and yellow shirts has led to a breakdown of trust and dialogue. The ongoing cycle of unrest also means that Thai political institutions are forever channeling their energy into negotiating ways out of new crises, rather than into more long-term strategies of development. As a whole, detrimental forces such as these threaten to derail Thailand's best efforts to chart a sustainable pathway.

"Dealing with the symptoms will no longer be adequate to heal the gaping wound in our country. We need to deal with the bacteria and the viruses that lie at the roots of our national malaise", long-time politician Surin Pitsuwan wrote in a June 2016 *Bangkok Post* editorial. "It is high time that the Thai bureaucratic system be overhauled before a rebellion of the periphery."

Red shirts:

Broadly speaking, most red shirts hail from the north and northeast. They may be backers of former prime minister Thaksin Shinawatra or spurred by their perspectives on inequality or see the movement as a vehicle to achieve other objectives such as a redistribution of power or wealth.

Yellow shirts:

They tend to be distrustful of Thaksin Shinawatra and his network and the aims of the red shirt leaders; they are generally comprised of supporters of the Democrat Party, royalists and the Bangkok middle class. They may also have benign objectives such as rooting out corruption and inequality, which seemingly overlap with the causes of the red shirts.

How Conflict Impedes Sustainable Development

Foreign investment Political upheavals and uncertainty create doubt among foreign investors searching for stable markets. In the first nine months of 2014, beset by street protests, Thailand's economy grew by only 0.2 percent, though complete economic recoveries in Thailand after such crises are common.

Post-traumatic stress The majority of those killed in southern Thailand have been civilians, including many children, teachers and religious figures. The psychological toll this takes on communities and families there can be significant, with long-term effects on a generation born into a culture of fear and violence.

Harming tourism Every time more protests flare up, or a coup is carried out, tourism takes a hit. As the country's biggest earner of foreign exchange, any such downturn in arrivals impacts the economy.

Education under threat In southern Thailand, insurgents raze schools or other institutions deemed to be sponsored by the state. Teachers have been targeted and murdered by insurgents, leading to fear among students and a shortage of qualified educators. In Bangkok, as a result of political protests, on several occasions the city has ordered schools across the capital to close, affecting tens of thousands of students.

Lives and businesses interrupted Protests have forced the shuttering of shops in the center of Bangkok and triggered the closure of the main international airport at one time. SMEs, which are large contributors to GDP and employment, in particular bear the brunt of such shutdowns, as they are less able to absorb the downturn. More than a decade of violence in the Deep South has likely stunted some of the potential for economic growth there, for example in the tourism sector.

Political instability Constant changes of government and uncertainty make it more challenging for politicians to implement effective, long-term policies, and more difficult for the private sector and international partners to compete for government contracts.

Rights suppressed Any such conflicts are a convenient excuse for the powers-that-be to crack down on press freedom and personal liberties. After the 2014 coup, the junta imposed martial law to keep dissent and journalists in check. Down in the southernmost provinces, the military and law enforcement officials have resorted to detentions, warranted or not. Such a climate of repression may exacerbate the anger of some locals in the Deep South who then commit atrocities, or carry out counterattacks, in a cycle of violence that has been ongoing for more than a decade now.

A popular joss house in Pattani province suffered a drop in visitors after the southern insurgency resumed in 2004.

Keys to Bridging Divisions

Fresh elections Though elections invite more political squabbles to surface and protests to occur, they are a necessary prerequisite to re-establishing democracy.

Voters at the polls during the 2014 general elections.

Sincere dialogue Whether between red shirts and yellow shirts, security forces and militants, or community stakeholders and Bangkok leaders, sincere dialogue absent of self-interest and personal agendas and bias is the only way inclusive and sustainable solutions will be found.

A free, unbiased press Thailand ranked number 136 out of 180 countries in the 2016 Reporters Without Borders World Press Freedom Index. A free and objective press can provide a platform for airing the views of opposing sides on national issues and in doing so it can help to heal these rifts. At the same time, media bias remains an issue that must be resolved.

On the fringes Malay-Muslims in the Deep South have long complained that they feel detached from mainstream Thai society and culture. Recognizing that the issue is a cultural one as well as a security one could undermine local support for separat-

ist groups and help pave the way for constructive dialogue and, eventually, peace in southern Thailand.

Bold ideas The problems in the Deep South are complicated and require long-term involvement and perhaps the implementation of bold ideas. Among their many recommendations, the National Reconciliation Commission, chaired by former prime minister Anand Panyarachun, recommended promoting dialogue with militant groups; creating a regional development council; establishing a fund for reconciliation and healing; devising procedures to deal quickly with complaints against government officials in the region; and creating a new agency to oversee the administration of the area. In general, the commission encouraged authorities to bridge the gap between the state and the region by recognizing the region's unique identity as well as social and economic injustice.

Human wrongs Security forces in the Deep South and the insurgents must stop committing abuses of

human rights, which serve only to further fan the flames of conflict.

Redistributing wealth A disproportionate amount of government revenue is spent in Bangkok. Decentralization would help to develop the poorer regions of the country and address the lingering resentment toward the capital's elite that has galvanized the red shirts.

Rule of law Implementing a culture that respects the rule of law would deter protesters from resorting to illegal means to achieve their goals and demonstrate that negotiation, rather than brinkmanship and violence, is the way forward.

Investigate disappearances As various NGOs have suggested, the state could set up a special commission to investigate all the mysterious disappearances in the southern provinces, some of which are suspected to be the result of kidnappings by state officials, such as the apparent forced disappearance of human rights lawyer Somchai Neelaphaijit in 2004.

SPOTLIGHT

Reds versus Yellows

The conflict between the red and yellow shirts is often traced back to 2006 when Prime Minister Thaksin Shinawatra was deposed in a coup. His supporters, hailing largely from the north and northeast of Thailand, long-time underdogs in the political arena, donned red and took to the streets to protest his ousting. Their rivals, dressed in yellow, the color of royalty in Thailand, were seen by the red shirts as aligned with the Thai elite, which had once dominated the government and fiercely opposed Thaksin's regime.

The events of 2006, with mass protests bringing Bangkok to a standstill, were to play out time and again over the coming years, with the yellow shirts occupying Government House for five months in 2008 and then triggering the closure of Suvarnabhumi International Airport for eight days by staging protests there later that year.

In 2010, the red shirts resorted to the same tactic, squatting in the area around the Phan Fah Bridge before moving to Lumphini Park and the glitzy Ratchaprasong shopping district of Siam Square, which they occupied for months, closing down all the malls and shops in the area. After the

"Yellow shirts" wave clappers during a 2008 demonstration in Bangkok.

ruling Democrat Party ordered the army to storm the barricades, the red shirts responded by launching dozens of arson attacks across the city. All in all, 91 people lost their lives in the conflagrations of 2010 and many more were injured.

No matter which color-coded partisans started the protest, the results were often the same: shops and schools closed en masse, negative news reports circled the globe, and tourist arrivals declined. The protesters' penchant for targeting Bangkok's biggest thoroughfares and busiest districts, to pressure parties into making concessions so as not to lose the support of business owners, has worsened the economic damage.

The real costs of these conflicts in terms of the country's future development and its international image are rarely considered by the protesters, who are largely focused on ousting one party in favor of another.

Perhaps the word most reiterated in political circles and the media in recent years is reconciliation. The meaning is clear enough, but how to achieve such a truce without military intervention or an authoritarian government remains an elusive goal.

Red shirts gather en masse for a rally.

FURTHER READING

- Deep South Watch, think tank and website in Thai, English and Malay, www.deepsouthwatch.org
- *Red vs Yellow: Volume 1: Thailand's Crisis of Identity,* by Nick Nostitz, 2011
- "Thailand's Good Coup: The Fall of Thaksin, the Military and Democracy," edited by Michael K. Connors and Kevin Hewison, special issue of *Journal of Contemporary Asia* 38, no. 1, 2008
- *Good Coup Gone Bad: Thailand's Political Developments Since Thaksin's Downfall,* edited by Pavin Chachavalpongpun, Institute of Southeast Asian Studies, 2014

SPOTLIGHT

Deep Trouble in the Deep South

The region often known as Patani had only a loose tributary relationship to Thai rulers until King Chulalongkorn began to pacify the outer-lying regions of his kingdom in order to consolidate national borders. Attempts to bring the provinces under Bangkok rule during a time of Western encroachment and imperialism culminated with the Anglo-Siamese Treaty of 1909, in which the British recognized Siam's sovereignty over the area.

For the next few decades, the population was mostly left alone until 1934, when Field Marshal Pibulsongkram instituted a process of "Thaification" of the southern provinces, forcing Malays to adopt the customs and language of the dominant Thai ethnic group, effectively denying them any status as a separate ethnic minority. Attempts by Haji Sulong and other local leaders to gain better treatment for the region after World War II ended in failure and, ultimately, Haji Sulong's arrest and disappearance.

But it was not until the late 1960s that a more formal rebellion movement was formed and flashpoints lit up the south as the Patani United Liberation Organization (PULO) attacked state institutions with the goal of secession.

The conflict periodically flared up from the 1960s until 2001, when shadowy insurgent groups, radicalized by hardline Islamic teachers, began expanding their targets to include attacks on police stations and army barracks. The violence escalated significantly on January 4, 2004, when rebels set fire to police stations and schools and made off with weapons from a local army arsenal.

The military exacted a harsh revenge. After launching a series of attacks in the three provinces that left dozens of militants dead, 32 of them, armed with only knives and a single gun, holed up in the 18th-century holy site of Krue Se Mosque

The ancient Krue Se Mosque in Pattani was the sight of a massacre in 2004.

in April 2004. After a seven-hour standoff with the military, heavily armed soldiers stormed in to shoot all of them dead.

"The Tak Bai Massacre" followed in October 2004. After demonstrations outside the police station in the town of Tak Bai in Narathiwat province, hundreds of protesters, their arms tied, were stacked on top of each other in trucks. During the five-hour ride to the army camp in Pattani province, 78 of them died from suffocation and other causes.

These incidents, as well as the still-unsolved disappearance of the Thai-Muslim lawyer and human rights activist Somchai Neelaphajit also in 2004, triggered a dramatic upsurge

in the violence and only further entrenched distrust. Militants, some of whom now proclaimed a radical goal of Islamicization of the southern provinces, began targeting any entity deemed to be a collaborator with the state: government officials, schools, teachers, monks and the families of teachers.

More than 6,700 people have been killed since 2004, according to Deep South Watch, both from insurgent attacks and as a result of military responses to the violence and unrest. No clear solution is in sight.

But it's difficult to point fingers. Both sides are guilty of committing atrocities and the causes of the violence are varied and complicated. As a statement released by the Thailand office of the international NGO Human Rights Watch pointed out, "The cycle of human rights abuses and impunity contributes to an atmosphere in which Thai security personnel show little regard for human rights and secessionist insurgents have committed numerous atrocities."

THE THAI
CULTURE

Monarchy

Religion

Heritage

Although it is not one of the traditional three pillars of sustainable development, culture is increasingly seen as an intrinsic part, since it forms – through religion, history, heritage and more – the psyche of any nation. And in Thailand, so much of the country's culture dovetails wonderfully with the principles of sustainable development.

Ninety-five percent of Thais are adherents of Buddhism, which encourages compassion, holds nature sacred, and illuminates its followers about cause and effect, and the interdependence of all living things. Buddhism also promotes the philosophy of moderation, or the "middle way," which preceded the Western-style consumerist culture that has become more prevalent in Thailand. The religion may yet serve as an antidote to capitalism's grosser excesses.

For centuries, Thailand's other cultural pillar, its monarchy, has steered the kingdom's development and united the country. The late King Bhumibol Adulyadej displayed his foresight by implementing many royally initiated projects on everything from farming and community development to water management. Furthermore, His Majesty's formulation of the Sufficiency Economy Philosophy offers a home-grown spin on many of the practices and values that form the basis of sustainable development thinking around the world.

Thailand also possesses a rich history of cultural diversity and is famous for successfully absorbing outside influences and ideas as well as peoples. Both Buddhism and the Thai people are famous for such accom-modation. All of these cultural traits provide a platform for sustainable development to thrive here.

The pillars of Thailand's culture, however, are not immune to change and the traditional principles they embody are perhaps no longer as potent. While the grandiose temples at the protected ancient cities of Ayudhya and Sukhothai may continue to stand tall as tourist attractions and the embodiment of national pride, Thai culture itself has rapidly evolved to include new global modes, technology and values. With the pull of the future as strong as the influence of the past, tensions are appearing. How well Thailand's past values and beliefs are reconciled with new ideas will continue to define Thailand's future development path and its identity.

MONARCHY

An ancient institution navigates the modern world

Thailand's monarchy is one of the most enduring of all monarchies in a region once full of them. From the 13th century until 1932, when absolute monarchy was replaced with constitutional monarchy, Siamese kings ruled. And from 1932 until today, they have continued to reign.

This span of more than 750 years, covering dynasties both short-lived and legendary, is remarkable for not only its length but also for its ongoing impact – few countries in the world today remain so extraordinarily dedicated to and defined by their royalty.

Traditional systems of patronage and hierarchy combined with Buddhist teachings of righteous kingship mean the monarchy has long been paramount in this largely Buddhist nation, not only in terms of mystique and symbolic power, but also in terms of leading and developing the nation. The early kings of the current Chakri dynasty were preoccupied with re-establishing sovereignty, fending off threats from near and abroad, and

stabilizing the kingdom they had based in Bangkok after the Burmese sacked Ayudhya in 1776.

The fourth king, Mongkut (Rama IV), is credited with beginning the modernization process taken over by his son, Chulalongkorn (or Rama V), toward the end of the 19th century. King Chulalongkorn led a massive reformation and expansion of the government bureaucracy sometimes known as the "Revolution from Above." Centralizing power in Bangkok, he and his court formed the foundations of the current nation-state: slavery was abolished, the judiciary and legal code were modernized, roads and railroads were built, a standing army was created, borders were consolidated and revenue gathering systems were streamlined.

The ninth king, Bhumibol Adulyadej (or Rama IX), provided the crown with fresh relevance during the constitutional era. While he recognized the importance of the monarchy's symbolism to the country's identity and people, he also transformed the monarchy into an agent of development. Shocked by the hardships he discovered among the people upcountry, the king became passionate about his own brand of sustainable development. His advice – on everything from protecting forests and watersheds to improving soil quality – as well as his localized approach, are echoed by the development practices of the UN today. Through NGO-style institutions such as his Chaipattana Foundation and more than 4,400 royally initiated projects, King Bhumibol found ways to circumvent red tape in the name of more long-lasting solutions to the social, economic and environmental problems Thailand has faced.

Today, however, in an age of new and interactive media, and with the death of His Majesty King Bhumibol in October 2016, the future role of the monarchy has become a topic of increasing discussion. How will it adapt – at what pace and to what degree – to the new challenges facing Thailand?

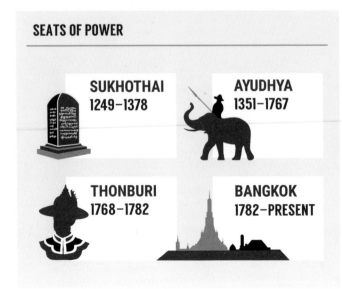

SEATS OF POWER

SUKHOTHAI
1249–1378

AYUDHYA
1351–1767

THONBURI
1768–1782

BANGKOK
1782–PRESENT

King Bhumibol Adulyadej and Queen Sirikit (center, seated) sit with their guests at the Ananta Samakhom Throne Hall in Bangkok on June 12, 2006, before the royal barge ceremony the next day to mark the king's Diamond Jubilee. Monarchs and dignitaries from around the world joined this momentous occasion.

REASONS

Why the Monarchy Matters to Sustainable Development

Public works The rulers of the Chakri dynasty have frequently been progressive on issues involving religious freedom, public health, education and national development. Rama V (King Chulalongkorn), for example, decreed that the public could practice any faith, despite an overwhelming Buddhist majority in his kingdom. King Bhumibol traveled through the country to initiate irrigation projects and agricultural development, as well as the expansion of healthcare and education to rural areas.

Diplomacy and leadership Through private audiences with the prime minister, cabinet ministers, international leaders and ambassadors, the monarch acts as a key voice on national affairs. Skillful reforms, negotiation and maneuvering likely saved Siam from being colonized over a century ago. King Bhumibol's world tours in the 1960s engendered goodwill toward Thailand during the Cold War.

Moral power The unofficial social contract for a Thai monarch binds him to observe the "Ten Principles of Kingship" (*dosapit rajadhamma*), a royal code of governance to ensure he acts in the larger interests of his people. Constitutionally, the king also possesses the royal prerogatives "to be consulted, to warn and to encourage." Thus the king can act as both a moral compass and as a check on greed and power. King Bhumibol's birthday addresses to the public, for example, often offered advice on social issues.

Bearer of tradition, culture and national identity
Borrowing from Buddhist, Hindu and Brahman traditions, many rituals and symbols surrounding the monarchy date back hundreds of years. The ceremonies, regalia and iconography of the monarchy and Royal House of Chakri, not to mention its mystique and other intangible qualities, play a vital role in forming Thai culture and projecting the Thai national identity both at home and abroad.

Stability over chaos Thais refer to the monarch by many names, including "God Upon Our Head" (*phra chao yu hua*) or the *dhammaraja*, a righteous ruler who promotes justice, virtue, wisdom and disperses the *dhamma* to the people. These names are born out of the Buddhist ideal of kingship and speak to the monarch's influential role in creating stability for his people. As historian Frank Reynolds put it, the main function of the king has traditionally been "to constitute the central pinnacle – the bond between the divine and the human, around and below which the Thai civil order took form."

Challenges Facing the Monarchy

Public image Like any institution, the monarchy has seen ups and downs in its public image over the centuries. In the 1920s, during a period of financial struggle and growing republican sentiment, Rama VI actively engaged his critics through a newspaper column. Nowadays, the unchecked dissemination of information, news and gossip over the Internet forms a communications nightmare for the palace and government, which almost never directly engages the chatter. The latter's Computer Crime Act has done little to stem the tide of talk. Due to the *lèse-majesté* law and other sensitivities the media is reluctant or unable to report on royal affairs properly. Without a credible filter sorting fact from the fiction, innuendo and misinformation run rampant.

Financial transparency The wealth of the monarchy consists of land, corporate investments and other assets worth upwards of US$37 billion, according to *King Bhumibol Adulyadej: A Life's Work*. This wealth is managed by the Crown Property Bureau (CPB), which has been instrumental in developing the country's infrastructure and remains one of the largest investors in the Thai economy to this day. Unlike other monarchies around the world, which are funded almost totally by allocations from the national budget, the Thai monarchy's day-to-day expenses are mostly covered by the CPB. However, the CPB is under no obligation to provide the public with details of its accounts or activities. Even though the CPB's wealth has been frequently used to fund public works, critics argue the monarchy should abide by modern-day standards of transparency.

Above politics A common phrase in Thailand describes the monarch as "above politics." The reality, however, is that he exercises his constitutional power through royal assent, signing new legislation into law. The monarchy's critics have argued that the

Bhudthan Thom Throne in the Grand Palace.

palace – or its institutions such as the Privy Council, which consists of retired members of the civil service and military – of influencing political outcomes through their clout, prestige and popularity. This perception is one reason the institution has been brought down into the political fray.

Balancing act Monarchies around the world, including Thailand's absolute monarchy in 1932, have been replaced by new systems of governments. The new dynamics challenge a monarchy to create and maintain its role. In addition, the military and political factions have frequently cited their fealty to the monarchy as

justification for their actions. In all likelihood, the monarchy will continue to have to navigate national crises and a tumultuous political landscape featuring a powerful military, a dynamic that inevitably shapes public perception of the monarchy itself.

Article 112 Although the law is common in countries with monarchies, Thailand's *lèse majesté* law, sometimes referred to as Article 112, is one of the most punitive in the world. He or she who "defames, insults or threatens" the king, queen or heir apparent or regent" can receive a sentence of up to 15 years in prison, with a minimum of three years. Uniquely, the law allows anyone to file a complaint. Sadly, the law has become a political tool abused by many. Owing to the prestige of the monarchy, the police and the courts are reluctant to deem that complaints have no merit. Moreover, media are unable to report freely on cases or censor themselves for fear of repeating the offense. The international community frequently hears of people being jailed for comments that would be protected by free speech laws in their countries. Because it is abused by others, Article 112 can make the monarchy and Thailand look out of step with international free speech norms.

Smooth transitions Successions prior to the end of absolute monarchy in 1932 were often contentious, owing to rival claims and the jockeying of factions. One reason for this tension was that royal succession did not follow the system of primogeniture common to other monarchies in which the oldest male heir automatically accedes to the throne. Under the current Palace Law of Succession, the existing monarch names his or her own heir.

FURTHER READING

- *The Lords of Life* by Chula Chakrabongse, 1960
- "The Old Siamese Conception of the Monarchy," by Prince Dhani Nivat, *Journal of the Siam Society*, 1947
- *Kings, Country and Constitutions: Thailand's Political Development 1932–2000*, by Kobkua Suwannathat-Pian, 2003.
- *The King of Thailand in World Focus*, edited by Denis D. Gray and Dominic Faulder, 2008
- *King Bhumibol Adulyadej: A Life's Work*, edited by Nicholas Grossman and Dominic Faulder, 2011

TIMELINE

A History of Kings

12th century

Buddhist revival spreads east from Sri Lanka and is embraced by the Tai people.

1249–1378

Steeped in Buddhist traditions, the Sukhothai kingdom, located in central modern-day Thailand, flourishes.

16th and 17th centuries

Golden age of the kingdom of Ayudhya, which thrives as a trading center for all of Southeast Asia and welcomes settlers from Europe, India, Persia and Arabia.

1767

The Burmese sack Ayudhya, razing it to the ground. A new seat of power founded by King Taksin later emerges downriver in Thonburi.

1873

Rama V (King Chulalongkorn) comes of age and assumes power. Over the next few decades, he implements an ambitious program of reform, centralizes power, overhauls the bureaucracy and revenue system, consolidates borders, modernizes infrastructure and successfully completes a "Revolution from Above."

1851

The progressive Rama IV (King Mongkut) accedes to the throne and begins to contend with rising Western influence in the region.

1782

Royal House of Chakri founded. Its first ruler, Rama I, moves the new capital across the Chao Phraya River to modern-day Bangkok.

1925

The reign of Rama VI (King Vajiravudh) ends with the monarchy in a weakened position as support for absolutism begins to crumble. His successor Rama VII (King Prajadhipok) considers the potential of constitutional monarchy.

1932

Thailand's absolute monarchy is overthrown by a small clique of military men and civilians. The country's first constitution is promulgated with the king remaining the head of state.

1935

Rama VII abdicates while in Europe and lives the rest of his life there.

1946

Rama VIII (King Ananda Mahidol) is found dead in his bed at the age of 20 from a gunshot wound. The death remains a mystery. His 18-year-old brother, Prince Bhumibol Adulyadej, becomes king.

1950

Rama IX (King Bhumibol Adulyadej) returns from Europe to Bangkok for his official coronation ceremony, marriage and the cremation of his brother.

1973

A student uprising calls for the overthrow of the military-led government. Rama IX intervenes in the crisis to try to dissolve tensions. After violence breaks out, the king names a new prime minister.

2016

King Bhumibol Adulyadej dies at age 88 on October 13. His son, Maha Vajiralongkorn, is declared king on the first of December.

The king and queen after marrying in 1950.

> *"I will rule the land righteously for the benefit and happiness of the people."*
> **King Bhumibol Adulyadej at his coronation, May 5, 1950**

Sustainable Development in Words and Deeds

One of the central tenets of sustainable development is the idea of leaving the world a better place for future generations. Achieving that goal does not require the current generation to sacrifice, but asks that they should protect key resources and create stronger systems that will lead society toward future success. Thus, at the heart of sustainable development are public participation, careful analyses of challenges, and long-term planning and solutions.

In words and deeds, King Bhumibol Adulyadej embodied these core values of sustainable development from the moment he began pursuing his own development projects in the 1950s, long before the term "sustainable development" was coined. The issues he tackled – water security and management, soil quality, crop yields and farming techniques, forest conservation, education and healthcare – remain at the core of development agendas today.

The king briefs students on a project devoted to water management near Hua Hin.

The king (left), officials and villagers at a poppy field in the Golden Triangle.

Against the grain of large-scale modernization, the late king promoted site-specific measures that were sensitive to the environment and the socio-economic conditions of the people. Today, there are over 4,400 such royally initiated projects throughout the country aimed at creating sustainable development.

King Bhumibol once commented that "our existence depends on working and analyzing various problems." For a monarch who could have remained removed from his people, he became famous for his closeness to them. Taking up their petitions and rather than relying on the reports of government officials, he preferred to personally study the issue by poring over research materials and then going on a field visit. Often driving himself to the location in question, once he arrived he would walk through the site to get a feel for

the place and also to show moral support to the locals, whose spirits were often lifted by the interest he showed in their problems. A team that accompanied the king would collect information at the site and also help assess the problem. The king then held a public hearing on the spot – an approach which came to be known as "rapid rural appraisal."

After the villagers and local officials agreed on a course of action, King Bhumibol would fund the development project through his Chaipattana Foundation, which means "Victory through Development."

By the time of an interview with BBC correspondent David Lomax during the making of the documentary *Soul of a Nation* in 1979, many of the king's projects were meeting with success. The most famous was his opium eradication scheme known as The Royal Project in the "Golden

Triangle." But the king said the job was far from over. Instead, he predicted it would take three decades to eradicate opium from Thailand.

Moreover, he recognized that the elimination of the opium trade was in itself not the goal. "The other task is to give these people a better way of life," he told Lomax. "So this will continue even if – and I don't think it is in the very near future – opium is eradicated like smallpox has been. We have to continue the program for a very long time, so that we will give these people a better life and also so that everyone will benefit from it."

The words were typical of the king's approach. On another occasion, regarding the country's trouble in managing its water resources, he observed: "If we take proper care of

the environment, there will be water for many hundred years. By that time our descendants might be thinking of some new ways to solve the problems." When the king spoke about education, he described it as "a never-ending process." One of his most ambitious educational projects, a massive Thai-language encyclopedia for students, was intended to "transfer knowledge, culture and ethics for the next generation."

When he considered the health of his people, he did not offer short-term remedies and cure-alls. Instead, he looked for practical solutions. The Japanese emperor Akihito, when he was crown prince, offered the king a species of freshwater fish known as *Tilapia nilotica*. The king bred it under the Thai name *pla nil* and introduced

it to the country as a popular, new source of protein. After his visit to a Danish milk plant in 1960, the monarch returned to Thailand and helped introduce more dairy into the diet of Thais by opening a dairy plant.

The king's experiences and ideas were ultimately formalized in the late 1990s into the king's own development philosophy known as the "Sufficiency Economy Philosophy."

"The kings of prior reigns had ruled the land. But HM the King of this present reign has ruled the hearts of the people..."
M.R. Kukrit Pramoj, 13th prime minister of Thailand

When the UN Honored the King

In May 2006, the UN presented its first UNDP Human Development Lifetime Achievement Award to King Bhumibol. UN Secretary-General Kofi Annan traveled to Klai Kangwon Palace in Hua Hin, where the king was staying at the time, to make the presentation. The award was timed to mark the 60th anniversary of King Bhumibol's accession to the throne, and honored his Sufficiency Economy Philosophy and his promotion of sustainable production and consumption. "As a visionary thinker, Your Majesty has played an invaluable role in shaping the global development dialogue," Annan said at the presentation. "Your Majesty's Sufficiency Economy Philosophy, emphasizing moderation, responsible consumption, and resilience to external shocks, is of great relevance worldwide during these times of rapid globalization. It reinforces the United Nations' efforts to promote a people-centered and sustainable path of development."

In the *International Herald Tribune*, former prime minister Anand Panyarachun wrote at the time of the award, "We are in desperate need of technological solutions to our energy problems, a more equitable distribution of wealth, a level playing field for international trade and more generous development aid to poor countries. But this will not be enough. A more profound transformation of our societies, our values and the way we consume is needed. The king's philosophy of 'Sufficiency Economy' offers just that – a more balanced, holistic and sustainable path of development and an alternative to the clearly unsustainable road the world is currently traveling down."

RELIGION

The spiritual and ethical roots of sustainable development

On the tricolored Thai flag of red, white and blue, where red represents the nation and blue the monarchy, Buddhism's purity is symbolized by the color white. Long before a national consciousness even existed, the religion helped form the values, heart and soul of Thai communities. For centuries, monks were among the most respected leaders and most well-educated people, especially in the countryside, where they exerted an enormous influence in their villages. The precepts of karma and compassion and observing the eightfold path of virtues have long served as a moral compass for most Thais.

But Buddhism has two distinct components in Thailand: as it is practiced by monks, who follow myriad rules, and as it is practiced by laypeople, who are not required to follow as many of those tenets. For monks, by immersing themselves in the dharma, or teachings of the Buddha, they hope to attain a greater sense of enlightenment and ethical behavior. Oftentimes they take on professorial roles in their communities, spreading the wisdom of the Lord Buddha to the devout while also serving as advisors to them in many different capacities.

For laypeople, from the cradle to the crematorium, Buddhism is an integral part of their lives easily observed in everything from giving alms to monks during their daily rounds or having saffron-robed holy men chant at weddings and funerals. The religion also plays a significant role in Thai culture. On the calendar are national holidays like Visakha Bucha Day, which commemorates the birth, enlightenment and death of the Buddha. Many parts of the country also celebrate Buddhist Lent, which marks the Buddha's first sermon and the annual three-month retreat for monks during the rainy season, in spectacularly different ways and add Buddhist touches to their own provincial festivals.

The compassion and tolerance that the Buddha exuded is evident in the fact that Thais, and all foreign residents in the kingdom, are free to choose any faith they wish to follow. There is no official state religion, though some 94 percent of Thais still identify themselves as Buddhists. The other 5 percent are made up of Muslims, mostly in the southern provinces of Yala, Pattani and Narathiwat, with scatterings of Christians across the central region and animists in the hill tribes up north.

Though Buddhism in Southeast Asia is mostly of the Theravada school (a fundamentalist interpretation of the original teachings), the Thai strain has evolved to include parts of different creeds. A widespread belief in spirits comes from animism. Some of the different deities like Brahma (the god of creation) and Ganesha (the elephant-headed god of knowledge) and that bird-like symbol of Thai royalty, the Garuda, are from Hinduism, which came to Thailand via the Khmer empire. Chinese deities like Guan Yin, the Goddess of Mercy, are also worshipped, and the yin/yang symbol of Taoism is a common sight.

There are more than 37,000 Buddhist temples and around 355,000 monks and novice monks. During the three-month rainy season retreat known as Buddhist Lent, the number of monks and novices swells by some 30 percent. Boys can become

"The way to extinguish desire is to become a giver. Give regularly. The act of giving and the extinguishing of desire: they always happen simultaneously. Giving is for reducing greed in one's mind, without hoping for any better thing in return."

Somdet Phra Nyanasamvara, the late Supreme Patriarch

Monks pray before morning meal at Maha Chulalongkorn's Buddhist University's Nong Khai campus.

novices from the age of 7, but men must wait until they turn 20 to ordain as monks. Then they are bound to follow 227 precepts, whereas laypeople only have to refrain from such things as lying, stealing, cheating and drunkenness.

In Sanskrit, Buddha means "awakened." By that he meant awakened to reality as it is – not the delusions of ego, anger and lust which are mental constructs but the ephemeral nature of feelings and the inescapable life cycle of all sentient beings: birth, old age, suffering and death (valuable lessons that every Thai school kid must still learn by heart). These are also the cycles of nature. Throughout the Buddha's teachings, the natural world is of primary importance; it's often used as a metaphor or touchstone. In one famous sermon, the Buddha simply sat and held up a lotus blossom until it whithered to demonstrate life's impermanence. Because of such associations, the flower is used in religious rituals and has a host of different meanings for Thai people, some secular and some spiritual.

By respecting nature and all sentient beings, no matter how minuscule, the Buddha also sought to emphasize the interdependence of all life, the forests and oceans, the animals and people, to produce a body of teachings, ethics and practices that encouraged **mindfulness**, which remains at the roots of ethical behavior in Thailand.

THE KINGDOM'S RELIGIONS

94% Buddhism

1% Christianity

5% Muslim

Less than **1%** Hill tribe animism and other

Source: CIA World Factbook

Mindfulness:

The state of active, open attention on the present. When you're mindful, you observe your thoughts and feelings from a distance, without judging them good or bad. Instead of letting your life pass you by, mindfulness means living in the moment and awakening to experience.

REASONS

Why Buddhism Matters to Sustainable Development

Interdependence and deep ecology The Buddhist belief in cause and effect and the interdependence of all human beings with nature speaks to the most vital principles underlying sustainable development. As with the world's other great faiths such as Christianity and Islam, Buddhism preaches the virtue of empathy and cultivating compassion for all living creatures.

Key principles The dharma, which teaches that worldly consumption and pursuits are fundamentally unsatisfying, can act as a healthy check on unbridled consumption and create a mindset compatible with the goals of sustainable development. Moreover, the dharma is the moral compass of the vast majority of Thai people. Through these Buddhist teachings, they learn about discipline, wisdom, and selflessness.

Sense of place Buddhism's concepts of rebirth and karma help Thais make sense of the world and their place in it. Classic Buddhist texts encourage ethical behavior by presenting a hierarchical universe in which bad behavior will be punished with purgatory while good behavior will be rewarded now and then.

Trove of culture Buddhism is a bastion of Thai customs, tradition and culture, from rituals at home and in the temple to festivals and weddings to the daily alms collections across the country each morning. It is at the heart of the Thai national identity.

Higher learning Many youth from disadvantaged backgrounds would never have a chance to attend institutions of higher learning outside of Buddhist universities.

BUDDHISM BY THE NUMBERS

There are
37,713
temples

There are
355,295
monks and novices

During Buddhist Lent, the number of monks and novices increases
by about **30%**

Monks must follow
227
precepts

Facts You Need to Know about Religion in Thailand

- Ever since the days of the Ayudhya kingdom, circa the 16th century, both Thais and foreigners have been free to practice their own religions.

- The cornerstone of Buddhism is self-reliance, as denoted by the Buddha's famous last words to his disciples, "Work out your own salvation with diligence and heedfulness."

- The hill tribes of the north live in harmony with nature as they worship her healing and life-giving properties in the form of spirits which dwell in trees, fields and rivers.

- All the great religions preach that life is full of suffering. Only the Buddha taught how to eliminate that suffering through the doctrine of non-attachment to delusions like the ego and the removing of constantly shifting cravings through meditation and contemplation.

FURTHER READING

- *Handbook for Humankind*, by Bhuddadhasa Bhikku, 2005 (English edition)
- "The Buddhist Attitude Towards Nature," by Lily de Silva, Access to Insight, 2013, www.accesstoinsight.org
- BuddhaNet, the Worldwide Buddhist Information and Education Network, www.buddhanet.net
- *Socially Engaged Buddhism*, by Sulak Sivaraksa, 2005
- *Transforming Thai Culture: From Temple Drums to Mobile Phones*, by William Klausner, 2004

TIMELINE

Rises to Prominence, Falls from Grace and Sacred Dates

228 BC

Buddhism's great propagator in India was King Ashoka the Great. Once a barbaric warmonger, he warmed to the Buddha's brand of pacifism and made it the national religion in India, while building 84,000 stupas and encouraging Buddhist missionaries to disseminate the teachings across Southeast Asia, so it reached present-day Thailand in 228 BC. Asoke Road in Bangkok is named after him.

AD 1238

After the first Siamese kingdom of Sukhothai ("Dawn of Happiness") is established, Theravada Buddhism becomes the state religion. Buddha images crafted in this style are commonly regarded as superior in beauty and workmanship.

1767

With the sacking and burning of Ayudhya by the Burmese many invaluable, palmleaf manuscripts of Buddhist teachings are torched. The invaders also decapitate many Buddha images, not only to loot the heads, but to crush the spirit of the Siamese.

16th century

The Portuguese, the first Westerners to live in Ayudhya, build Catholic churches.

AD 1351

After the founding of the Ayudhya kingdom, the monarchs of this dynasty become renowned as temple-builders, so much so that foreign visitors call the capital a "city of gold" that makes Paris and London look drab by comparison.

1833

Prince Mongkut, a monk for 27 years before he was crowned as King Mongkut, founded the Dhammayut Sect, to reform Buddhism in accordance with its Pali roots. But this move also results in a Bangkok-centered *sangha* (monkhood) that usurps power from village temples.

1782

Rama I establishes the still-reigning Chakri Dynasty in the new capital of Bangkok. After receiving copies of the Tipitaka (Buddhist scriptures) from Sri Lanka he convenes a council tasked with the feat of standardizing the Thai translation before sending them to temples in the country.

1900

The Dhammayut sect produces two widely revered monks, Ajarn Mun and Ajarn Sao, who revive the forest meditation tradition.

1878

King Chulalongkorn (Rama V) issues The Edict of Religious Toleration, which affirms the right of Christians or any religious adherents to practice their faith in the kingdom.

1975

One of Ajarn Mun's most devoted disciples, Ajarn Chah, also becomes a meditation teacher and in this year he founds a forest temple called Wat Pah Nanachat. That monastery has now sprouted 10 other branches in England, Australia and Europe.

1994

Phra Yantra, who claims he was Genghis Khan in a previous life, is officially defrocked for a variety of offenses. His fall from grace is one of many scandals to rock the Thai monkhood in the 1990s.

Late 1990s

BuddhaNet and many other websites devoted to the teachings of the Lord Buddha go online in addition to digital versions of the scriptures.

2003

The ongoing conflict in Thailand's deep south flares up again, in general pitting Muslims against Buddhists. Many monks have been killed.

2008

In spite of monks protesting once again – and even sleeping in coffins outside the parliament buildings – to make Buddhism the country's official religion, the push fails.

Challenges Facing Buddhism in the Material World

Disgraced monks Since the 1990s, a number of high and mighty monks have been reduced to lowly criminals fleeing the country on charges of sexual misconduct or caught accumulating huge amounts of cash, cars and luxury goods.

Losing their religion Ordaining as a monk, once a duty for every young man, has lost some of its prestige.

Commercial concerns Many supplicants come to pray for material riches and business success at shrines and temples rather than focus on the Buddha's teachings.

Influence of idolatry Thai Buddhism remains infused with Hinduism and myriad superstitions, such as the belief in the magical power of amulets and tattoos, which critics say have perverted the religion.

Redefining merit The concept of "merit" needs to be redefined. Doing good deeds for society can be as valuable, if not more so, than donating money to temples.

Fringe elements The hardline supporters, who are against abortion and other progressive causes, have alienated moderates, while some

abbots have taken leading roles in political protests when critics say they should not intervene in state politics.

The business of Buddhism Some temples have accrued enormous wealth through donations, and there is little transparency or oversight regarding temple assets. This has led many to worry that some temples and monks are acting at odds with Buddhist teachings.

Socially disengaged Some observers recommend that Buddhism's relevance to the modern world should be promoted vis-à-vis the "socially engaged Buddhism" of Thai thinker and theologian Sulak Sivaraksa.

Material world Buddhism's core values of non-violence and non-attachment to material objects are more and more threatened by a consumerist society that advocates the latter and a mass media that glorifies the former.

Gender gap The fact that women are not granted the opportunity to ordain as monks in Thailand is a contentious issues for some, though they can wear the white robe of a Buddhist nun, who has fewer rules to follow.

"Religion is at the heart of social change, and social change is the essence of religion."

Sulak Sivaraksa, *Seeds of Peace: A Buddhist Vision for Reforming Society: Collected Articles*

Forest Tradition Takes Root

Ajahn Chah practices meditation in the forest.

Before Ajahn Chah became world-renowned for establishing Wat Nong Pah Pong in northeastern Thailand and popularizing the Forest Tradition on an international level, as a young monk he wandered the countryside grappling with the question of how to apply Buddhist teachings on morality, meditation and wisdom. His father's death, just five years after he had received bhikkhu ordination, was a stark reminder about the fragility of human life. And so, practicing in the austere Forest Tradition and using reflections on mortality to explore the meaning of life, for the better part of a decade he ventured deep into remote forests, caves and cremation grounds in search of ideal places to develop meditation practice.

Then, in 1954, he was invited to return to his home village in Ubon Ratchathani province. Nearby, under a canopy of thick forest, he established Wat Nong Pah Pong. In this quiet spot, rumored to be inhabited only by cobras, tigers and ghosts, disciples flocked to Ajahn Chah despite hardships such as malaria and a lack of shelter. As the years passed, he gained notoriety as an influential teacher in the Forest Tradition.

The tutelage of Ajahn Chah was said to be harsh and challenging, with the aim of pushing monks to test the limits of their endurance so they would develop patience and

fortitude. They practiced moderation, and learned to exist in harmony with the ecosystems in which they lived. The emphasis was always on "letting go," on surrendering to the way things are. According to Ajahn Chah, "Mindfulness is life. Whenever we don't have mindfulness, when we are heedless, it's as if we are dead."

His followers were expected to maintain disciplined observance of the Vinaya, the rules of the sangha. The Forest Tradition is particularly strict regarding food. Adherents observe the "one eaters practice," whereby they consume only one meal in the morning. It also emphasizes the realization of enlightenment as the focus of monastic life.

Implicit in the Forest Tradition is the concept of sustainability. During the last half of the 20th century, rampant logging and land clearing wiped out the vast majority of

environment. Likewise, they have been integral in the replanting of woodland areas and are credited with the resurgence of forests in areas around their monasteries, which in essence have become forest sanctuaries.

In 1967, Ajahn Chah took on his first Western disciple, an American monk named Venerable Sumedho who had been practicing intensive meditation near the Lao border. Over the next few years Wat Nong Pah Pong drew more Western monks. In 1975, in an unprecedented move, Ajahn Chah founded Wat Pah Nanachat, now known commonly as the International Forest Monastery, with Ajahn Sumedho as the abbot. It would be the first monastery in Thailand with the aim of training English-speaking Westerners in the monastic Vinaya, as well as the first run by a Westerner.

A small house for studying Dhamma inside Wat Pah Nanachat in Ubon Ratchathani province.

Thailand's rainforests. Due to the location of their forest monasteries, Ajahn Chah's disciples frequently found themselves on the frontlines in the battle against timber companies and plantations. As the forest is a crucial factor in allowing devoted adherents to follow in the footsteps of Buddha, monks of the Forest Tradition and local lay people work tirelessly to preserve their natural

Today Wat Nong Pah Pong has more than 250 branches across Thailand, and more than 15 associated monasteries and 10 lay practice centers around the globe including the Abhayagiri monastery (Fearless Mountain) in northern California and the Cittaviveka monastery in the United Kingdom, the latter of which was the first Ajahn Chah monastery to be established outside of Thailand.

HERITAGE
Cultural achievements with lasting value

Thailand, with a more than 700-year-old history played out between great empires and spanning a succession of royal dynasties, has produced one of the world's richest and most varied cultures. Extending from the former Lanna kingdom of the north through the ancient capitals of Sukhothai and Ayudhya to the Khmer-style temples of Lop Buri and the northeast and down to southern lands once ruled by rajahs lies a still very visible layer of Thailand's ancient past.

> *"Historical sites are our nation's prestige. Even a single block of old bricks is valuable to preserve. With no Sukhothai, Ayudhya and Bangkok, Thailand is meaningless."*
> **King Bhumibol Adulyadej**

Then add on centuries of openness to trade and new peoples, from the millions of Chinese moved by revolution and hardship to Thailand's plains, ports and cities to the steady stream of Western and Asian capitalists, soldiers, traders, speculators and wanderers. With so many influences, it is no wonder that Thailand's heritage can appear complex, deep and downright confounding.

The remnants of this past are not simply experienced through the beautiful Buddha images and temples that attract tourists today, but can also be witnessed in the rituals maintained by its monarchy, heard in the songs of its musical forms, or literally tasted in the contrasting flavors of a curry from one of the world's most beloved cuisines.

To preserve such heritage in the face of globalization can be a matter of both angst and debate in Thailand, and a matter of complexity beyond any single body. It is, indeed, a paradox of sorts since Thailand's heritage itself was born out of change, and an acceptance of it. In all likelihood, Thailand's **tangible cultural heritage** and intangible cultural heritage will face continued pressure, if not from a new generation arguably less interested in the past, then from business magnates looking to bulldoze historic districts in favor of generic malls, condos and infrastructure projects. Amidst the rush for development, in a country where old can be seen as old-fashioned and new as synonymous with better, locals have had to fight to preserve their districts whether in cities or the countryside.

THAILAND'S UNESCO WORLD HERITAGE SITES

CULTURAL

1991 The ancient city of SUKHOTHAI and associated historic towns in KAMPHAENG PHET provinces

1991 The ancient city of AYUDHYA

1992 BAN CHIANG Archaeological Site

NATURAL

1991 THUNGYAI-HUAI KHA KHAENG Wildlife Sanctuaries

2005 DONG PHAYAYEN-KHAO YAI Forest Complex

A mural from the Temple of the Emerald Buddha in Bangkok's Grand Palace. It depicts Hanuman, a key character from the national epic, the 'Ramakien'.

Government policies have helped. In a landmark case, the Fine Arts Department designated 20 percent of the island on which the historic old city of Ayudhya is situated as a National Historical Park in 1976. Recognition from international bodies such as UNESCO has also helped keep major cultural landmarks from falling prey to developers, while at the local level people have banded together to use the amplification offered by social media to ensure their voices are heard. This is increasing awareness around the need for more people-centered development that also respects cultural heritage.

What does any of this have to do with sustainable development? In 2013, UNESCO released a report on why cultural heritage should form a key component of the UN's post-2015 development agenda. One of the main arguments was that activities and output related to culture are "green by design," as they have developed over generations and embody holistic, sustainable responses to specific environments.

As for **intangible cultural heritage**, which refers to traditions and cultural practices, UNESCO stated, "While fragile, intangible cultural heritage is an important factor in maintaining cultural diversity in the face of growing globalization, an understanding of the intangible cultural heritage of different communities helps with intercultural dialogue, and encourages mutual respect for other ways of life."

Why Heritage Matters

Culture leads to innovation Human development thrives on creative expression and cultural heritage as a means of emotional and psychological catharsis, intellectual stimulation and the exploration, celebration and transformation of the human condition.

Economic value Thailand's rich heritage is one of the main drivers of the tourism industry, the country's top foreign exchange earner. In areas outside of Bangkok rich in heritage but struggling, culture-related initiatives can revitalize economies, as long as benefits to the community and long-term durability are given precedence.

Unity through understanding Promoting cultural exchanges in Thailand's different regions could help defuse potential tensions by fostering tolerance and respect among communities.

Windows into the past Heritage offers a window into the past through which future generations can gain knowledge as well as insight about their ancestors, roots and place in history.

Priceless legacy Heritage represents shared traditions, preserving a collective memory that binds people together. These are ecosystems of the spirit that add value to people's lives and cannot be assessed in monetary terms.

Evolve or Perish

"I don't see the point of preserving something so that it becomes dead. If it must evolve, we must allow it to evolve," said M.R. Chakrarot Chitrabongs.

It's perhaps a surprising sentiment to hear from not only the former secretary general of the National Culture Commission and former permanent secretary of the Ministry of Culture, but also a man with a pedigree that could justify his claim that he has Thailand's heritage "in his veins." Both of M.R. Chakrarot's grandfathers are legendary figures in Thailand: Prince Damrong Rajanubhab (1862–1943), often called the "Father of Thai History," and Prince Narisara Nuvadtivongs (1863–1947), renowned for his contributions to Thai arts.

However, despite such roots, M.R. Chakrarot does not hold traditional Thai culture to be sacrosanct and to be preserved at all costs. For him, culture should remain relevant and the primary defining trait of Thai culture, M.R. Chakrarot said, has

always been its ability to adapt and assimilate outside influences.

"I have no fear for Thai culture because if you study and understand Thai culture, it is a culture that has evolved from centuries of accep-

> "I have no fear for Thai culture because if you study and understand Thai culture, it is a culture that has evolved from centuries of acceptance of foreign cultures."
>
> **M.R. Chakrarot Chitrabongs, former permanent secretary of the Ministry of Culture**

tance of foreign cultures. Very little of Thai culture today is our own, except one might say the language, but even with that there were influences," he explained. "Many of our customs can have their origins traced to ethnic minorities in the land or sometimes way outside of the land."

In this respect, he echoed the opinions of his grandfather, Prince Damrong, who said: "When [Thais] saw some good feature in the culture of other people, if it was not in conflict with their own interests, they did not hesitate to borrow it and adapt it to their own requirements."

M.R. Chakrarot said that while change is inevitable and outside influences are constant, for culture it is important that a balance be maintained between humanity's three main spheres of influence: materialism, society and nature. By striking this balance, culture can grow out of the rich heritage of the past yet be redefined and modernized by future generations.

Dancers with the Thai country music superstar Nok Noi Urai Phon exit a stage in Bangkok.

Movements in the Arts, Landmarks in Heritage

1495 BC to 900 BC

Though the date range of the Bronze Age site, Ban Chiang, in Udon Thani province, is still in dispute, the area contains evidence of one of the most important and advanced prehistoric settlements discovered in Southeast Asia.

1926

The National Museum opens with Prince Damrong classifying the stylistic periods for religious architecture, along with ceramics and sculptures, into eight stylistic periods in a chronological sequence named after historic polities: Dvaravati, Srivijaya, Lopburi, Chiang Saen, Sukhothai, Uthong, Ayudhya and Rattanakosin (Bangkok).

1943

Italian sculptor Silpa Bhirasri (1883–1962), born Corrado Feroci, founds Silpakorn University, the country's primary art school. His arrival in 1923 at the invitation of Rama VI reflects an embrace of Western styles as well as increased patronage in the arts. Bhirasri's sculptures such as Bangkok's Democracy Monument remain iconic.

1976

The Fine Arts Department, spurred by the activism of Sumet Jumsai, designates 20 percent of the island on which the historic old city of Ayudhya is situated to be protected as National Historical Park land, ending decades of plundering. In 1988, another conservation victory is scored when the "Phra Narai Lintel," which had gone missing in the 1960s before turning up at an American museum a decade later, is returned to Phanom Rung Historical Park in Buriram province.

2000

Montien Boonma dies prematurely of cancer at the age of 47 only a few years after his wife succumbed to the same illness. His installation works, such as "Lotus Sound" and "Sala of the Mind," reflect his deep Buddhist beliefs and incorporate Thai materials such as rice sacks, herbs and buffalo horns. In death he has experienced a massive renaissance with retrospectives of his work held all over the world.

1249–1767

Regarded as the cradle of Thai civilization, the Sukhothai kingdom, which lasted from 1249 to 1378, is famous for its ceramics and Buddha images. King Ramkhamhaeng is credited with developing the Thai alphabet based on Khmer-Sanskrit models, and *The Story of the Three Planes of Existence, a Buddhist Cosmology*, remains the oldest surviving work of Thai literature. Indic and Khmer traditions from the Angkor Empire informed the court arts and traditions of the Ayudhya kingdom that followed Sukhothai. In 1767, it is sacked by the Burmese, and many cultural treasures are destroyed.

Late 1700s

Thai culture rises from the ashes of Ayudhya as a new dynasty begins. Many pillars of the country's cultural legacy date to this time: the construction of Wat Arun as well as the Grand Palace and the Temple of the Emerald Buddha; the adaptation of the *Ramakien* as Thailand's national epic; and the epic poems of Sunthorn Phu (1786–1855).

1952

Prime Minister Plaek Pibulsongkram establishes the Ministry of Culture. From this point on there will always be a government institution charged with overseeing Thailand's cultural policies.

1954

Suraphol Sombatcharoen (1930–1968) releases *Nam Ta Lao Vieng*, one of the first hits in a musical genre that would come to be known as *luk thung* (literally, "children of the fields") in the 1960s. The melancholic tales of displaced country folk battling big city ills continue to resonate.

2013

Bangkok's Wat Prayoon wins the Award of Excellence in UNESCO's Asia-Pacific Awards for Cultural Heritage Conservation, marking the first time Thailand has received the highest honor in this regional competition. The award recognized a six-year project to restore the temple for its technical excellence and how it benefited all stakeholders, from Buddhist monks to community members as well as the public and private sectors.

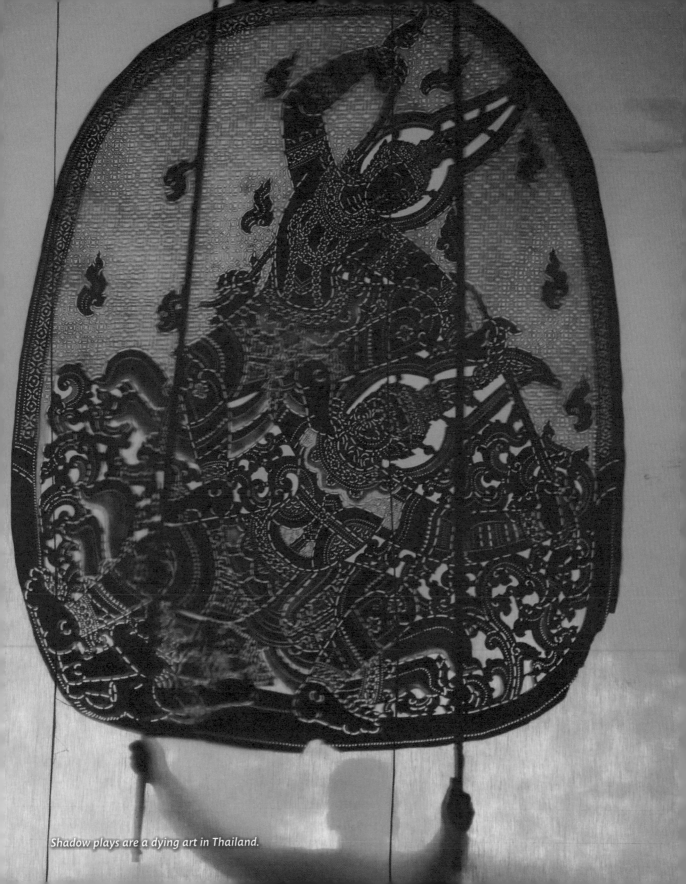

Shadow plays are a dying art in Thailand.

FURTHER READING

- *A History of Thailand,* by Chris Baker and Pasuk Phongpaichit, 2014
- *Thailand: The Worldly Kingdom,* by Maurizio Peleggi, 2007
- *Bangkok: A Cultural History,* by Maryvelma O'Neil, 2008
- *Introducing Cultural Heritage into the Sustainable Development Agenda,* UNESCO report, 2013

REALITY CHECKS

Threats to Thai Heritage

Definitions Given its myriad influences, Thai culture and the intangible construct often referred to as "Thainess" are difficult to define. Regional identities or specific people may be underrepresented, left out or even excluded and the national discourse is often skewed to favor traditional institutions and conservative instincts.

Wrecking ball Long-established communities and buildings in the historic districts of major urban centers are increasingly falling prey to development's relentless onslaught. These initiatives are often shortsighted and fail to recognize that once a city's heritage is lost, it's lost forever. A viable alternative is to develop these areas as historical attractions to benefit the economy, while not disrupting the communities themselves.

Punishing looters Despite being located in a sub-region that is a hub for the illegal trafficking of antiques and artifacts, Thailand has yet to become party to either of the two main international instruments aimed at halting this trade: the 1970 UNESCO Convention and the 1995 UNIDROIT Convention. Endorsing both conventions would be seen by many parties as a major step forward in Thailand's efforts to stop the illicit trafficking of cultural heritage.

Gag orders The arrival of Thailand's first printing press almost immediately brought with it official censorship and this is a situation that has continued to the present day, with film censors notorious for cutting challenging material to the point where the movie

Karen hill tribe member smoking tobacco in Chiang Mai province.

> *"One day the greatness of politicians or important people will disappear. The only thing that will remain is art."*
> **National Artist Thawan Duchanee (1939–2014)**

as a whole no longer makes sense. Such throttling of artistic voices makes headlines; more insidious, however, is the culture of self-censorship and stifling of creativity that these practices foster.

Preserving the past for the future
Cultural preservation in Thailand has been lax over the years. In the field of film, for example, hundreds of historic movies dating back to the fledgling days of Thailand's motion picture industry have been permanently lost. In 2007, the Thai Film Archive (known at the time as the Thai National Film Archives) was recognized for its significant preservation work, but

the massive task of preserving this heritage, and that of other Thai arts, requires a collective effort.

On the cultural fringes Patronage in the Thai arts is usually funneled toward the classical arts such as *khon* (masked theater) and traditional forms of dance, which are also the focus of government efforts to promote Thai culture. Modern artists and art forms that may have a closer connection to the public today, not to mention the richness of voices outside the kingdom's central region, are often passed over for patronage, making it difficult for them to earn a living and maintain their creative careers.

PART 3

IDEAS AND INSPIRATION

Here we showcase the innovative practices, trends, solutions and people across Thailand that are making sustainable development a reality.

The **32 topics** are organized around the four key audiences or stakeholders who can make an impact: the individual, the community, and the private and public sectors.

Numerous case studies, entitled **pioneers**, highlight projects focused on alternative energy, environmental protection, green manufacturing and much more. The interviews and profiles, called **groundbreakers**, sound out the opinions of various experts, which often are just as revealing. We also highlight positive examples of sustainability from **around the world** and the **eye-opening** literature and films inspiring urgency and action.

We were unable to include all the worthy contributions within the limited space of this book. Those featured here are just a sample, selected for their considerable legacies, the influence they have exerted in their fields and their overall creativity.

THE POWER OF THE
INDIVIDUAL

Responsible Consumption

*Personal Participation
& Awareness*

Green Homes

Commuting

So many dilemmas related to the environment and society – such as pollution, climate change, excessive consumption and corruption – seem so vast and insurmountable as to be completely beyond our influence.

But that's not true. Through awareness and moderation of our own personal consumption, the individual can always make a difference.

In fact, consumers today have never had so much power at their disposal. Never has voting with your wallet and conscience been as popular, or as effective with brands both big and small. Every consumer choice we make – from what type of disinfectant we use to what light bulb we purchase – has an effect on both

> *"You must be the change you wish to see in the world."*
>
> *Mahatma Gandhi*

the environment and the bottom line of the firms who produce these products.

Even in our daily lives, whether or not we take public transport to work, can save fossil fuels or add more greenhouse gas emissions to the toxic mix. On the home front we can also make a small difference in the big picture through changes in our habits that cost a little but mean a lot.

In mobilizing bigger efforts there have been some encouraging signs that like-minded individuals are banding together to come up with cutting-edge projects, like developing an app with the best crowd-sourced bike routes to navigate the capital.

Such movements can only grow if they have a mouthpiece to reach other concerned citizens. The new generation has largely forsaken traditional media outlets and forums in favor of social media and online platforms. There is both strength and safety in these kinds of numbers.

In cyberspace, many individuals are pushing the envelope of sustainability issues by fostering the kind of dialogue that questions unsustainable business practices or challenges the country's traditional power structures. For true social inclusion to take place, where every key stakeholder has a voice, public participation – without fear of attacks or reprisals – must become a cornerstone of Thailand's quest for sustainability.

RESPONSIBLE CONSUMPTION

How the individual can make a substantial difference

Today, humanity is living way beyond the Earth's means, using up 1.5 times more resources than the planet can replenish in a year, as compared to only five decades ago when we were consuming only a third of that figure.

In the decades to come, Asia will be at the center of this shopping spree. It is already home to 40 percent of the so-called "consumer class," or around 1.7 billion people with disposable incomes. Thailand has also joined the Western-inspired rush. Malls with brand names, car dealerships with the latest models, and shops laden with high-tech gizmos are abundant throughout the country. Modern, materialistic life revolves around them and a consumerist ethos has become pervasive. In ASEAN, for example, Thailand was second only to Indonesia in the number of new motor vehicles sold per capita in 2014. As this trend continues, your mindset toward consumption and your consumer choices – including choosing to reduce your consumption all together – can play a crucial role in protecting the planet.

Upcycling:
The reuse of discarded objects or materials to create a product of a higher quality or value than the original.

Yes, an individual's choices do make a difference, because every choice you make is a contribution to the big picture created. Consumer preferences and demands can have a direct impact on the dialogue inside boardrooms and government ministries, especially in the new media age, leading to more sustainable products, innovations and policies. That's part of the reason why the UN specifically calls for individuals to be more responsible consumers in Goal 12 of the 2030 Agenda for Sustainable Development.

"Earth has enough for every man's need, but not for every man's greed."
Mahatma Gandhi

Different firms and government agencies are also increasingly forward-thinking, helping

A supermarket aisle in Thailand featuring modern packaging.

people to open their minds, as well as their wallets, to consume in more responsible ways. For the past two decades, the Thai government, for example, has been promoting sustainable consumption and products via various labeling and certification schemes. Green Label Thailand, based on the trailblazing German label Blue Angel, was formally launched in 1994 by the Thailand Environment Institute (TEI) and the Thailand Industrial Standards Institute (TISI) under the Ministry of Industry. Its 123 criteria are designed to distinguish the products that cause the lowest environmental impacts in their sectors. As of February 2017, the list included 591 products in 25 categories from 65 companies.

Meanwhile, other labeling schemes like the energy-efficiency labels for home appliances and Green Leaf for hotels are pushing both consumers and businesses to think beyond price tags and profits. The most advanced scheme is the Green Industry Mark, which focuses on greening the entire value chain step by step, creating networks of eco-savvy manufacturers in Thailand. But these initiatives would be for naught without a growing movement of concerned individuals well aware that their purchasing power can make a huge difference. For the most

part, these are grassroots movements localized to the bigger urban areas. An example are the markets where farmers and other small-scale entrepreneurs can sell their organic fare or green products straight to the consumer.

From there, this revolution is branching out in many different directions. One example is the growing number of eco designers like Deepwear, Rubber Killer and Upcycling Thailand who are **upcycling** used plastic and other kinds of "waste" into creative and fashionable products. Online platforms are also proving popular with the masses. To single out one, the website www.bkkgreenie.com provides reviews of restaurants, farmers' markets and green hotels while dispensing tips on how to dispose of old batteries and other unwanted products.

Much more than just marketing channels, these initiatives are building a vibrant community based on sustainability. They provide workshops on everything from urban gardening to composting and greening your lifestyle. Young and cosmopolitan, these like-minded entrepreneurs, farmers and consumers are catalysts for change who are re-imagining consumption in novel and sustainable ways.

Resources to Green Your Lifestyle

- DISCOVER WHAT HIDES BEHIND YOUR STUFF: Ethical Consumer has been researching products and companies for over 20 years, with a mission to make global business more sustainable through consumer pressure. Check out www.ethicalconsumer.org, which boasts more than 200 ethical product guides. Likewise, PRO-Green has created an online platform to share knowledge about the green economy in Thailand. PRO-Green also has a training center for members of the public who want to know more about this topic.

- FIND SUSTAINABLY PRODUCED GOODS IN BANGKOK AND BEYOND: The website Bangkok Greenie (www.bkkgreenie.com) gives plenty of hard facts and tips on free range organic eggs, restaurants known for organic and healthy food, markets and specialty shops, green travel, recycling guides and other local organizations.

- SHOP ONLINE FOR GREEN GOODS: Looking for eco-friendly options while shopping online? Try Green Shop Café (www.greenshopcafe.com), which provides an array of local and imported environmentally friendly products, from natural skin care lines to organic rice and more.

- PICK THE GREENER OPTION: Choose LED light bulbs over incandescent or even CFLs. When purchasing electrical appliances, select those that have a level 5 energy efficiency certification. Buy products that come in biodegradable packaging or natural packages with banana leaves. Say no to plastic bags, straws and spoons. On weekends, visit one of the many flea markets, both the fashionable ones and the more traditional bazaars to buy second-hand or home-made artisanal goods. Donate your old laptops, phones and TVs. Thai Eco Trade will pick them up, dispose of them safely, and even donate the proceeds to partner NGOs.

- CONSUME LESS: To really downsize your carbon footprint, spend a day at home and avoid all malls, shops and restaurants. Read an old book. Don't eat any meat or fish for a day. Take your family for a bike ride and bring a picnic basket or get inventive with some leftovers at home.

▶PIONEERS

Biodegradable packaging

Producers: Biodegradable Packaging for Environment Public Co Ltd and the KU-Green product line developed by Kasetsart University

Key features: Producing eco-sensitive packages and containers for foodstuffs

Across the country food stalls and restaurants serve up local fare in plastic bags and Styrofoam boxes that fill up garbage dumps and take decades to biodegrade. This push toward disposable products occurred over the last few decades as the country Westernized. Before that, Thai foodstuffs came wrapped in banana and pandanus leaves, or other natural containers that have now become exceptions to the plastic and Styrofoam rule.

However, there is no need to rely on such outdated packaging when bio-plastics have advanced to the point that they can provide a fully scalable and accessible alternative. These eco-products are superior in almost every way. They are safe to use in microwaves and ovens. They biodegrade in less than two months. They can be used as animal feed or mulch and do not bode ill for human health because they're free of toxic substances. They are also produced in an environmentally friendly way and can withstand boiling hot or freezing cold temperatures.

That's a formidable list of benefits compared to oil-based plastic containers, which take decades or centuries to decompose.

The Thai company Biodegradable Packaging for Environment Public Co Ltd (BPE) is a forerunner of eco-friendly disposable tableware. They found an ingenious way of using the pulp that remains after the extraction of sugar cane. Called bagasse, this fibrous waste material is cooked, rolled into sheets and then wet-formed or dry-pressed into the desired shape.

In 2014, BPE was recognized by the Foundation of Science and Technology Council of Thailand and the Science

> *There is no need to rely on such outdated packaging when bio-plastics have advanced to the point that they can provide a fully scalable and accessible alternative.*

and Technology Ministry as one of the leaders in the green industry sector. BPE was also lauded for its production process that minimizes environmental impacts by generating energy from LPG and steam. Shoppers can pick up their products under the Gracz brand in many supermarkets and dining ware supply shops in Thailand.

Another packaging option is the KU-Green product line developed by Kasetsart University. Dubbed "Bio-Dish Food Containers," these foam-like products made from cassava starch feedstock disintegrate when soaked in water for 24 hours. The containers are perfect for any kind of ready-to-eat food.

The Inconvenient Truth About Oh-So-Convenient Plastic Bags

Developed in the 1950s, plastic bags have revolutionized the way we carry our purchases and bundle our garbage. But the inconvenient truth is that they're far too hard on the environment. They pile up in landfills, clog up drains, cause floods and kill marine life (sea turtles often mistake them for their staple food jellyfish). The World Wildlife Fund estimates that every year more than 100,000 turtles, birds, whales and other sea creatures perish by eating this inedible refuse. Now, a global movement is pushing to reduce or eliminate them. One approach is to replace the high-density polyethylene with a biodegradable alternative like cornstarch. However, the more popular approach is to introduce surcharges or outright bans on plastic bags. In 2002 Ireland implemented the first such surcharge for the consumer, a hefty .15 Euro fee. Within five months usage of plastic bags declined by 90 percent.

In early 2016, Indonesia launched a pilot program in 22 cities to test the effectiveness of such a tariff on reducing waste. Thailand could easily do the same. However, the kingdom has been slow to get in sync with such initiatives. Consumers are still given plastic bags, cutlery and/or straws with almost every purchase. Yet, there are signs of progress. Several supermarkets offer biodegradable plastic bags and a number of restaurants are now opting for more eco-friendly options.

▶**PIONEERS**

Socialgiver.com

History: Founded in 2012

Location: Online platform based in Bangkok

Key features: Working with partners like hotels and restaurants to offer consumers discounted rooms and meals, with 70 percent of the money going to charitable causes

Socialgiver revenues benefit hill tribe children in northern Thailand.

Experienced social entrepreneurs Arch Wongchindawest and Aliza Napartivaumnuay discovered an innovative way to empower consumers to make informed choices, fill empty hotel rooms and the coffers of perpetually underfunded NGOs in Thailand all through one business model. Their flagship project, Socialgiver, is an online platform aimed at "creating shared value" for partners by offering consumers "lifestyle with a social twist."

They have also partnered with restaurants and bars and other business to offer consumers a cornucopia of choices to partake in discounted food and beverages. In turn, the businesses donate vacant rooms, unsold tickets, or proffer other services and discounts in return for tax benefits, CSR value, and to attract new customers. That makes

it a triple win for businesses, NGOs and consumers alike.

Consumers purchase these services from Socialgiver knowing that as much as 70 percent of the proceeds from their purchases are donated to support social projects, while 30 percent is used to help Socialgiver grow its list of projects, donor partners and users. Transparency is a lynchpin for building trust, as is working with trusted partners that are thoroughly researched beforehand and followed up with afterwards to make

sure that the donations are used efficiently and reach those in need.

This innovative business has received swift recognition both in Thailand and internationally by winning the Global Social Venture Competition SEA in 2014, the Thai Social Enterprise 'Change Award', The Venture Thailand award, the ASEAN Impact Challenge, and the Singtel-Samsung Regional Mobile App Challenge in 2015. Socialgiver was also featured at the World Economic Forum's 2015 Annual Meeting in Davos.

Co-founder Arch Wongchindawest.

A Socialgiver partner focuses on disabled kids.

Co-founder Aliza Napartivaumnuay.

A MAP TO NAVIGATING A SEA OF LABELS

Labels help companies build a positive brand image and differentiate themselves from their competitors. Most labels, like Fairtrade, are voluntary, meaning that manufacturers willfully seek to fulfill the criteria and receive the recognition. Others are mandatory and used by the government to push for certain minimum standards.

Traditional labeling schemes are based on a set of consensus-determined standards, which are verified by an independent auditing body to help consumers make environmentally informed choices. The criteria can be very specific, like the Forest Stewardship Council (FSC) testifying that the timber and pulp did not come from a virgin tropical rainforest. Others, like the EU Eco Label, cover a wide range of environmental issues such as water and energy use. The world's first eco-label was established in Germany in 1978. Today, the Blue Angel graces the packaging of approximately 12,000 products from 80 categories. Over 1,500 companies have adjusted their production processes and successfully obtained this seal of approval.

Despite strong growth in the past decades, there is still substantial room for improvement. According to the Natural Marketing Institute, less than 20 percent of the commodities on the market today have one or more eco-certificates. Since there are some 458 sustainability labeling schemes used in 197 countries, however, the consumer may now be bewildered by them. As the many labels and their complex technical details can be confusing, below is a guide to help navigate these murky waters.

COMPREHENSIVE ECO LABELS

These are the cream of the green crop. Developed as a joint effort by regulatory agencies, consumer and other associations as well as key private sector actors and companies, they set the standard for sustainable, environmentally conscious products by recognizing those with the lowest impacts. Examples are the Blue Angel, EU Flower, Nordic Swan, Green Label Thailand and others.

ORGANIC

This is the highest assurance of healthy nutrition and low environmental impacts. A third party certifies that these products comply with a set of stringent standards for growing, processing, transport and marketing food. Commonly these include avoiding pesticides, fertilizers and genetically modified crops (GMCs), while using sustainable farming practices. Local examples include Organic Thailand, which is a national standard, and Organic Agricultural Certification Thailand (ACT), which meets typical international criteria.

ISSUE SPECIFIC

There is a wide array of labels and organizations that focus on advancing sustainability in a specific area. Examples include the protection of wildlife (dolphin-friendly tuna or bird-friendly coffee), bio-diversity (Rainforest Alliance), forests (FSC), and eco-hotels (Green Leaf).

COMBINED ENVIRONMENTAL AND SOCIAL IMPACTS

These labels take a more comprehensive look at the entire production and distribution value chain. Along with environmental safeguards they also place a special focus on ensuring economic benefits for producers, farmers and workers. In Thailand, the royal projects and OTOP would fall under this category.

ELECTRICITY EFFICIENCY

As straightforward as the name suggests, these labels guarantee minimum performance standards for electrical appliances or indicate the specific energy use of the item. To sell their products on a particular market, the company's products have to meet or surpass the minimum standards. Energy Star in the US and the Thai Energy Efficiency label are two such paragons.

COMPANY SPECIFIC

Individual corporations develop their own unique standards to distinguish their brands in the cluttered marketplace. Usually these are determined and verified by them, so it's not really possible to easily assess the ecological credentials or compare them with their competitors' claims. Starbucks C.A.F.E. Practices and SCG's green product line are two such cases.

NICE WORDS BUT NO CERTIFICATION SEAL

Some products will be littered with superlatives that sing their praises. Be wary of products featuring such claims as "natural," "cruelty free," "no additives," "non-toxic," "green," "earth-friendly" and other examples of marketing jargon that could be "greenwashing."

The Trickle-Down Effect: Green Public Procurement Power

When it comes to purchasing power, the individual wields considerable clout, but so do those we elect to represent us in government. The procurement orders placed by government departments, authorities and state-owned enterprises make up somewhere between 15 percent and 30 percent of the national GDP in bigger countries and around half of the total GDP in some developing nations.

The public sector is an important client because it places large-volume orders at regular intervals. Moreover, public expenditures incentivize innovations and investments by creating strong demand in a specific sector. Companies see such developments as stable business opportunities and, with adequate financial support, make the required long-term investments for research and development.

By the time economies of scale are achieved, the eco-friendly products (like chlorine-free paper or LED light bulbs) also become available to individual consumers and a national market emerges. Green Public Procurement (GPP) plans also send a clear political message. By spending state funds in line with declared policy goals, governments demonstrate their commitment to promoting sustainable production and consumption.

Idea paper brand from SCG, which uses EcoFiber as a raw material.

LED lighting is one of many products the government promotes.

Thailand's Pollution Control Department (PCD) has been implementing a GPP since 2005. The country's first GPP plan (2008–2011), focusing on central government and administration bodies, is estimated to have delivered reductions up to 25,000 tons of CO_2 with the PCD spending some 62 percent of its budget on eco-friendly products.

The 2nd GPP (2013–2016) was aimed at increasing GPP volume, stimulating more green products, supporting the private sector in green production, encouraging government units to implement GPP, improving monitoring, and promoting sustainable consumption in the public and private sector, and among the general public.

In Thailand, the business sector responded quickly to the introduction of GPP, with the number of certified Green Label products jumping from 191 in 2008 to 591 in 2017. Those figures are sure to increase in the years to come as the government has expanded its list of product categories, including fluorescent lamps, Coolmode fabric and printer toner, as well as services such as office cleaning, oil changes and automobile repairs, with more than 2,000 agencies, both at national and municipal levels, participating.

PERSONAL PARTICIPATION & AWARENESS

How to make informed choices and get involved

A protest against a planned dam in Mae Wong National Park.

Sustainable development cannot be achieved without the active participation of individuals like yourself, not only to act as a check on power and greed and to make business and government interests accountable to citizens, but also to create a constructive sense of engagement and balanced dialogue that creates better long-term solutions for everyone. By nature, sustainable development is an inclusive development program intended to take into account all stakeholder opinions. As the development economist Shavi Kumar observed, "Without people participation, development actions fail."

In a society in which change has historically been enacted in a top-down fashion, that possesses a tradition of non-confrontation and where the questioning of any source of authority or decision-maker may be simply ignored or answered with lawsuits, threats or even violence, simply voicing your opinion can become an act of substantial courage.

Indeed, until recent decades in Thailand, public participation typically took the form of protests, many of which ended in violent suppression. It was only in the latter half of the 20th century that a civil society emerged, with groups to represent farmers, workers, women, students, environmentalists and human rights promoters, among others. Yet mostly these groups have found themselves on the outside looking in, forced to take their cases to the public through street protests, sit-ins and other campaigns to make themselves heard. (Notably, during the anti-government "Black May" protests of 1992, mobile phones were used for the first time to mobilize people, but the peaceful protests ended in military-led bloodshed.)

Thailand is a signatory to the UN's Rio Declaration on Environment and Development (1984), which holds that "environmental issues are best handled with the participation of all concerned citizens." As a signatory, Thailand is committed to disseminating information on environmental issues and actively seeking public input.

Yet for prominent environmental activists, pushing back against unsustainable development has long carried the threat of violence. In a 2014 report by Global Witness, Thailand was ranked the eighth most dangerous country in which to defend land and environmental rights. Some 60 activists who led opposition to coal plants, toxic waste dumping, land grabbing or illegal logging have been killed or have disappeared under dubious circumstances in the past two decades, according to Global Witness. Meanwhile, Freedom House's 2015 country report on Thailand noted that "even in cases where perpetrators are prosecuted, there is a perception of impunity for the ultimate sponsors of the violence."

In practice, the environmental and health impact assessment (EHIA) reports that are conducted and the public hearings that are held are often regarded as mere formalities aimed at appeasing outside observers, while the outcomes are already foregone

THE POWER OF THE INTERNET IN THAILAND

 46 MILLION INTERNET USERS

 67% (46 MILLION) ACTIVE SOCIAL MEDIA accounts as percentage of population

 44.85 MILLION OR 65.74% OF POPULATION Total number of active mobile Internet users.

 46 MILLION FACEBOOK USERS

 42 MILLION ACTIVE MOBILE SOCIAL ACCOUNTS

From Jan 2016-Jan 2017,

 21% there was a 19% jump in the number of active social media accounts

 24% and a 21% jump in active mobile social accounts

Source: We Are Social's Digital in 2017: Southeast Asia Report

FREEDOM UNDER THREAT

FREEDOM ON THE NET SCORE
(0= the best, 100=worst)

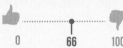

0   66   100

THAILAND'S SCORE IN PUBLIC PARTICIPATION
(examining free and fair association, assembly rights, freedom of expression) **5.3**

Source: Next Generation
Democracy Asia Report 2000-2015

ESTIMATED NUMBER OF URLS BLOCKED SINCE 2007

110,000 URLs

Source: Freedom House's "Freedom on the Net 2016"

ESTIMATED NUMBER OF ENFORCED DISAPPEARANCES AND EXTRAJUDICIAL KILLINGS IN THAILAND FROM 1996-2016 **60**

Source: Global Witness

conclusions. In 2010, for example, the government announced plans to build a dam on the Mae Wong River in Nakhon Sawan province. The EHIA was heavily criticized by protesters who said it ignored the potential impact on ecosystems and how this would in turn damage local communities.

As of 2017, it remains unclear if the project will go ahead. The main catalyst for the delay was Sasin Chalermlarp, head of the Seub Nakhasathien Foundation, who set out on a 388-kilometer trek in 2013 from Mae Wong in Nakhon Sawan to Bangkok to protest the building of the dam. A social media storm kicked up along the way. By the time Sasin reached Bangkok he was met by thousands of supporters, the majority of whom lived in the capital, far away from the dam in question, but nonetheless were inspired to get involved.

Such is the often organic nature of public participation today with social media representing the greatest game-changer. More than half of Thailand's population has a social media account. One person can now communicate with thousands without ever sharing the same physical space, and the instantaneous nature of contemporary telecommunications means that news spreads almost as fast as the events take place. With public opinion amplified like this, corporations, governments and other powerful parties can no longer ignore or avoid it. That said, netizens expressing views must also take care not say something that could get them prosecuted under the broad Computer Crime Act. The law has been widely criticized by free speech advocates who say it is used to silence dissenting voices.

▶**PIONEERS**

Foundation for Consumers

History: Founded in 1994

Founder: Saree Aongsomwang (current director)

Location: Complaints offices in more than 40 provinces across Thailand

Website: www.consumerthai.org

A campaign for 10,000 signatures to push lawmakers to draft a new consumer protection bill.

"Lodging one complaint is worth more than just talking about something 1,000 times," said Saree Aongsomwang, director of the Foundation for Consumers.

Since its establishment, the foundation has received thousands of complaints on all manner of subjects, from beauty clinics offering unsafe services to structural national policy issues in areas such as energy and telecommunications.

The Foundation for Consumers is Thailand's first national consumer advocacy agency that operates based on a participatory process – the foundation and its extensive network in the provinces receive complaints from individuals or groups and work with aggrieved parties to seek redress.

Promoting sustainability and ethical consumerism has long been one of the main tenets of the foundation.

"We work for consumer protection, but we don't solely think about typical consumer issues but of encouraging citizens to be active in society," Saree said. "Our principles for the foundation include ensuring value for money, using money wisely, and encouraging consumers to live sustainable lifestyles."

Over the long term, the foundation's watchdog function helps keep companies in check, encouraging more ethical practices and upgraded production standards by raising awareness among consumers and educating them to have higher expectations, she says.

The foundation has complaint centers in some 40 provinces, but the main office has a dedicated staff of only five, along with a team of volunteer lawyers and a committee of outside experts. The organization's main publicity arm is its *Smart Buyer* magazine, with monthly press releases that focus on the key cases highlighted in each issue.

Saree and her organization have a more-than-two-decade-long history of playing the role of David to the moneyed Goliaths of the kingdom. The Foundation for Consumers' highest-profile victory came in 2006, when the foundation was the lead plaintiff in a successful suit lodged with the Supreme Administrative Court to block the privatization of the Electricity Generating Authority of Thailand (EGAT).

The foundation's more traditional consumer advocacy campaigns have met with similar successes. The key, Saree said, is for the public to refuse to be passive, even in the face of seemingly overwhelming odds. "When we receive a complaint we have to turn it from a passive complaint into an active consumer protection measure," she said. "We tell them: 'If you do nothing, you'll get zero results. If you're an active citizen, I don't believe that you'll always get your rights, but I believe that you can get a better result.'"

Saree has witnessed this ethos take hold in Thailand to a far greater degree than when she started the foundation in 1994. "You can see the increased confidence in the numbers of people who are lodging complaints. The foundation's network is increasing everywhere," she said. "But the mechanism to help them is still behind. People know how to use their rights but they also want to see society improve as a whole, and that's difficult."

> *"If you do nothing, you'll get zero results. If you're an active citizen, I don't believe that you'll always get your rights, but I believe that you can get a better result."*
>
> **Saree Aongsomwang, Director of the Foundation for Consumers (pictured)**

▶PIONEERS

ThaiPublica

History: Founded in 2011

Founders: Boonlarp Poosuwan (still executive editor) and Sarinee Achavanuntakul

Facebook: ThaiPublica (more than 225,300 likes as of May 2017)

Twitter: @ThaiPublica (more than 23,900 followers as of May 2017)

Boonlarp Poosuwan, center, co-founded ThaiPublica in 2011.

Run by an experienced team of professional journalists, ThaiPublica was set up in 2011 as an online platform that publishes investigative journalism focusing on governance in the public sector, transparency in the private sector, and sustainability. Since then, ThaiPublica has become a pioneer in investigative journalism by combining the speed of online technology with the well-rounded perspective of an experienced editorial team that, above all, "dares to speak the truth."

"We believe in the media's role as a watchdog in a vulnerable society," said Boonlarp Poosuwan, editor and co-founder of ThaiPublica.

"Investigative journalism has been an integral part of our work. We believe in the transparent disclosure of information by the government sector and other stakeholders because equal access to information is fundamental to address disparities in society," she said. "If information is disclosed in a more transparent manner, we believe that it will create equal access to opportunities and lower the chance for corruption."

A key function of ThaiPublica is to uncover the kind of hard-to-reach information that mainstream media tends to fall short of acquiring. "One of the weaknesses of Thai journalists is not only in dealing with sensitive issues, but also in access to in-depth information from various agencies. Most of the files are stored in [hardcopy] document format. Therefore, it requires extra time and cost to get hold of the information," Boonlarp said. Even when information

is available for public access in digital format, the data can be too complex for the general public to digest and or understand. The result is that such information is often ignored or overlooked. That's why ThaiPublica strives to provide the general public with greater access to information that is presented in a way that is easily understood and can be easily referenced.

Over the years, ThaiPublica has published numerous investigative journalism pieces that challenge the system by taking a hard look at shady activities in the public and private sectors.

For example, ThaiPublica exposed an embezzlement scam by Klongchan Credit Union Cooperative executives, which affected the assets of 50,000 cooperative members to the tune of 12 billion baht. ThaiPublica's work helped reveal that the money trail led all the way to the Wat Phra Dhammakaya Temple Sect, implicating the abbot, Dhammajayo, who has now been linked to several fraudulent cases and defrocked.

ThaiPublica also helped shed light on abuse in the government's quota lottery system. By using the Official Information Act, ThaiPublica has for the first time revealed the names of companies and persons who benefited from lottery concessions. This revelation drew significant attention to the corrupt nature of benefit sharing among a small group of influential people, despite the original purpose

"We believe in the transparent disclosure of information by the government sector and other stakeholders because equal access to information is fundamental to address disparities in society."

Boonlarp Poosuwan, editor and co-founder of ThaiPublica

of the Government Lottery Bureau, which was to create jobs for people with disabilities. This investigation led to liberal reforms of the lottery trade and an end to the lottery quota.

In addition, ThaiPublica's persistent reporting on a tax dispute between Chevron and the Finance Ministry regarding whether oil from the continental shelf would be categorized as export or domestic trade provided a great deal of clarity on a very complicated issue, and also helped enable the Finance Ministry to seek a tax payment of more than 2.1 billion baht from Chevron.

Understandably, ThaiPublica's notoriety has grown. By March 2017, the website was averaging 600,000 page views per month. Its social media pages also draw the attention of a significant number of Thai users who use the platforms to exchange views and information, and debate key issues.

GROUNDBREAKERS

TUL PINKAEW *founded the Thai chapter of global online petition and campaign advocacy platform Change.org before starting his own consultancy, Sidekick, which advises NGOs and INGOs on how best to engage and mobilize the Thai public.*

Could you describe the challenges you faced with Change.org in a country renowned for valuing a non-confrontational approach?
I realized that this was a new phenomenon in the country and the importance of baby steps. At Change. org, we focused on micro issues such as pressuring city officials to fix potholes in bike lanes, keeping violent movies from being played on public buses. Once people successfully campaign on these small issues, their confidence increases and they feel safer to take on bigger ones.

What are some of the biggest changes you've noted in the evolution of public participation in Thailand in recent history?
I think there were two major events that reshaped public participation in Thailand. The 2004 tsunami saw people from different levels of society come together to volunteer and then that movement grew into people speaking out about other issues relevant to their lives. Then the floods of 2011. The Blue Whale social media campaign, for example, was in response to the confusing information coming from the government about the floods. It conveyed essential information in an attractive way, cartoons. Individuals took the lead, breaking from the long-established 1980s model of activism where NGOs led the charge.

What has been your greatest challenge with Sidekick?
Explaining the benefits of our approach, which is to make the issues relevant to urban audiences – those closest to the people in power, not just solely focused on the areas affected. For real change, urban people need to be informed and passionate about the issues affecting rural people. Many organizations are willing to consider these principles, but to actually act on them would require fundamental organizational shifts that many would be hesitant to take.

AROUND THE WORLD

The Filth and the Fury

The Ugly Indian (TUI) is a group of anonymous Bangladeshis fed up with filth on their streets, and who set a civic example by cleaning up around town – one sidewalk dumping ground or urine-stinking wall at a time.

Volunteers are organized through a central email account and go around town to "spot fix" *paan* (betel) stained walls, open dumps, cigarette butts, unsafe footpaths or public urinal spots.

The next day, TUI members post before and after photos on Facebook (the group has amassed more than 445,000 Facebook likes in only a few years). Oftentimes, this striking transformation alone is enough to inspire locals to take ownership of their neighborhood and, also, to take the lead in maintaining the area. When it's not enough, the TUI reaches out to the community, stressing "respect and dignity" above all and operating on basic guidelines, the first being "no lectures, no moralizing, no activism, no self-righteous anger."

One of the group's adages sums up their philosophy: "Want to change the world? Start with your own street."

Resources

Where else can you go for reliable information or to get more involved with the sustainable development movement? We list some of the many resources available in Thailand.

DATA AND STATISTICS

- **National Statistical Office:** Thailand's statistics and information center is used by the public as well as national and international agencies. Major data and indicators, such as demographic, population, economic, environmental and social statistics, are reported annually.

- **Thailand Development Research Institute:** A public policy research institute that provides technical analyses and reports to public agencies. TDRI's think tanks help formulate policies on socioeconomic development in Thailand.

- **Khon Thai Monitor:** A nationwide survey on Thai views on the country's progress, conducted with an aim to create a collaborative and public platform for all sectors. Its 2014 survey focused on Thai youth and encouraging younger Thais to be active in national development.

EDUCATION AND KNOWLEDGE SHARING

- **Sasin Centre for Sustainability Management (SCSM)** focuses on sustainability management in curricula, consulting, and research. The center offers consulting services and workshops in sustainable development for corporations, nonprofits and government bodies. Some are available to the public free of charge.

- **King Prajadhipok's Institute** provides educational training, seminars, meetings and counseling concerning politics and administration, economic and social development in democratic systems and good governance. The institute also promotes and enhances understanding of civic rights and duties according to the constitution.

- **The National Institute of Development Administration** is among the country's leading graduate schools in the field of national development. Its school of public administration is the first graduate school in Thailand. Its Center for Philanthropy and Civil Society was also established to strengthen the promotion of balanced and sustainable development and the achievement of a civil society. NIDA Poll Center has conducted 459 public opinion surveys (as of May 2015) on political, economic, environmental, social and cultural issues to reveal people's viewpoints.

ONLINE COMMUNITY FORUMS

- **Pantip:** The largest Thai-based community website with over 30 million users per month involved in discussion forums. It has become a platform for locals to exchange opinions on everything from politics to lifestyles. Businesses also use Pantip.com to help drive website rankings, as Thai consumers read reviews of products and services that can impact sales.

- **Facebook:** The social media platform is a powerful mechanism driving public participation in Thailand. The kingdom ranks in the top ten for total number of Facebook users worldwide with an estimated 37 million Thais participating.

INTERNATIONAL SOURCES

- **The Open Society Foundation** implements initiatives and funds civil groups worldwide in a bid to strengthen the rule of law, create a diversity of opinions, promote fundamental rights and encourage a civil society that helps keep government power in check and to shape public policies.

- **International Association for Public Participation (IAP2):** Founded in 1990, the association promotes the values and best practices in public participation. Apart from a semi-annual journal and website, IAP2 also provides comprehensive training and technical assistance to members from 26 countries, including Thailand, in a bid to effectively improve public participation.

GREEN HOMES

At home is where a sustainable lifestyle begins

An example of Thai vernacular architecture in Bangkok.

The home is a perfect platform for applying the adage "Think globally, act locally." Luckily, the incentives to "go green" at home are numerous. Whether living in an apartment, condominium or house, either in an urban jungle or in a remote village, "greening" your household can result in cost savings, increased comfort and health benefits.

Although there is no single definition of a "green home," there are key criteria that make a house environmentally friendly: energy-and-water efficiency; locally sourced, recycled or biodegradable building materials; minimized waste; and indoor environmental quality, such as air quality and temperature.

Most cultures around the world understood these concepts because they had no choice but to plan their abodes to accommodate the environment. By necessity, they also effectively used local resources. In other words, we were all "green" to begin with. "When you are looking to understand how to make a sustainable home, the best place to look is the vernacular house of a country," says Malina Palasathira, a Thai-Italian urban planner who co-runs Design Qua Ltd, a Bangkok-based sustainable building design studio. "The way it used to be done was likely more green than how it is done today."

Later, in the age of globalization, modern technologies such as air-conditioning and prefabricated and manufactured housing materials came into vogue worldwide. "In Thailand, after the advent of air-conditioning, developers were able to build houses in all sorts of unsustainable ways and disregard methods that were previously essential to cool or maintain a house," Malina explains. "People could simply build a glass box or whatever they dreamed up without regard for the heat."

Passive cooling systems:

A way to design a building so that it remains cool without external energy inputs. Passive cooling designs involve using cooling materials, ventilation and orienting the home away from the sun to enhance a house's cooling effects.

"When you are looking to understand how to make a sustainable home, the best place to look is the vernacular house of a country."

Malina Palasathira, Design Qua Ltd

Recently, however, efforts to create more sustainable living spaces are on the rise again worldwide. For example, according to the 2015 Green Building Economic Impact Study released by the US Green Building Council, the US residential green construction market is expected to grow from US$55 million in 2015 to more than $100 million in 2018, representing a year-over-year growth of 24.5 percent. In addition, more people are rejecting the idea that "bigger is better," opting to buy, rent, or build smaller homes to minimize waste and overconsumption. Residents are planting more gardens to improve air quality and increase green space, using energy-efficient LED bulbs, and practicing energy-and-water efficient behavior in their homes. In Thailand, innovators are trying to create homes that can survive floods or other natural disasters.

Energy efficiency is perhaps the biggest environmental hurdle in conventional homes. Cooling and heating account for about 50 percent of an average home's energy consumption. However, thoughtful home designs such as **passive cooling systems** that take a cue from traditional Thai homes can help control temperature and drastically cut down on energy waste. For example, workshops at Maejo Baan Din, (or "Earth Home Village") in northern

THE THAI VERNACULAR HOME

Vernacular architecture reflects practical know-how at its best, showing you the most suitable home designs for local climates before air-conditioning and concrete took over building designs.

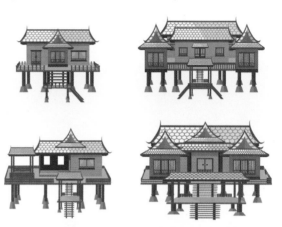

- *Traditional Thai houses are typically made entirely of locally sourced wood. Sometimes thatch or ceramic roof coverings are used.*

- *Their design is brilliantly functional and suitable for the local climate and seasons. For example, they are built with an elevated floor mounted on stilts to prevent flooding during the monsoon season. Raised platforms also protect against dirt and pests.*

- *The ground-level space underneath serves as multi-purpose living, dining, storage or barnyard areas, and is shielded from the fierce midday sun.*

- *Wood, bamboo and other locally sourced lightweight materials cool easily after sunset.*

- *Wide windows and large air vents on all four sides offer better ventilation and natural indoor lighting.*

- *High-pitched, elegantly tapering, outwardly curving roofs channel rainwater away from the house and please the eye with finials and symbolic decorations.*

- *Classic Thai houses are made to expand easily; standardized, pre-assembled walls and panels can be used to turn a single family home into a larger compound.*

Thailand, teach people how dwellings made of mud bricks naturally cool the indoor environment. The earthen walls absorb heat, cooling the structure in the daytime and keeping it warm at night, eliminating electricity expended on temperature control.

Alternatively, simple upgrades to insulation, such as sealing leaks around windows and doors, or installing energy-efficient or dimming windows, can drastically boost energy efficiency and create a more climate-appropriate home. Another easily attainable green-home upgrade gaining popularity is investing in solar panels, which start at about 40,000 baht for an all-inclusive rooftop solar installation. Residential solar systems can result in major long-term cost savings by allowing a household to produce its own free, renewable energy. (For more ideas, see the "Going Green at Home" sidebar on page 244.)

Today, environmental home projects are becoming more financially feasible due to new technologies and increased financial options. Whether retrofitting or building from scratch, greening your home is now incentivized through banks that offer special "green home" loans and "green mortgages" at low interest rates. Now, eco-conscious homeowners can "go green" with less financial burden than ever before.

▶ PIONEERS

Maejo Baan Din (Earth Home Village)

History: Founded in 2007

Location: Chiang Mai

Key features: An organic farm and homestay that also gives workshops in constructing natural buildings

In a hamlet near Chiang Mai the 400 villagers, many of whom are farmers, live in 110 "natural homes" that they have built themselves with the most readily available material at hand: mud.

Maejo village's *baan din* – "earth homes" – have been constructed with sun-dried mud bricks of clay, rice husks and sand. When fortified with natural mortar and built on a cement-based foundation, such clay bricks create sturdy earthen walls.

Once the preserve of the poor, adobe dwellings like these have been embraced by environmentally conscious people worldwide. As a benefit to their residents in tropical climates, the clay buildings'

"thermal mass" absorbs outdoor temperatures and helps insulate interiors against excess heat all day long. "Our adobe cottages are cool in the daytime and warm at night," explained Thai environmentalist Thongbai Leknamnarong, the project's initiator. "Even better, they're built almost entirely of natural materials, which come at a very low cost."

A former Bangkok resident who opted to raise her children in the pastoral surrounds she knew as a girl, Thongbai was inspired by fellow environmentalist Jo Jandai, the founder of the nearby Pun Pun Educational Center and a progenitor of the adjacent Panya Project, both steeped in sustainable living and permaculture. A decade ago Thongbai set about transforming a remote village located beside Sri Lanna Natural Park, which has since become a popular eco-tourism hub, into a model for a more sustainable lifestyle based on traditional practices.

During homestays at Earth Home Village, visitors gain hands-on experience in sustainable living. The Natural Building Workshop is for those aficionados of green architecture eager to design their own space from the ground up. The workshop teaches everything from laying foundations and making adobe

> **"By gently shaping our environment, we are working with nature rather than against it and thinking in the long term."**
>
> **Thongbai Leknamnarong, founder of Maejo Baan Din**

bricks to roofing, woodworking, installing windows, doors and insulation, and even learning how to sculpt with mud for decorative earth art.

What makes Earth Home Village's clay homes a model of sustainability is that the materials are not only ecologically sound, but also culturally sensitive. These natural building practices rely only on what the land provides and require very little cost so that a sturdy, high-quality home is within easy reach of those without the means to build a brand-new Western-style home or condo.

"At Earth Home we strive to live sustainably," Thongbai said. "By gently shaping our environment, we are working with nature rather than against it and thinking long term. We believe that for the sake of our grandchildren our societies need to improve our stewardship of the land and sustain ourselves in an ecologically sustainable way."

Earth house made of sun-dried bricks of clay, rice husks and sand.

▶PIONEERS

YAK01

History: Built in 2013 by architect Ayutt Mahasom of Ayutt and Associates Design

Location: Yen Akat Road, Bangkok

Key features: A naturally ventilated home that uses cooling materials and strategic design to keep cool and cut down on energy use

YAK01, which has been designed with passive cooling systems for energy efficiency.

Traditional Thai homes were built to take advantage of shade and prevailing winds to offer a respite from the relentless tropical heat. However, these passive cooling designs have been largely ignored in modern Thai housing. Instead, dense urban areas like Bangkok have increased ambient heat, while at the same time, clustered condominiums block the wind and a lack of overhangs allows the hot sun to penetrate directly into living areas.

In this hot and humid setting, air conditioners have become major energy wasters, as roughly 90 percent of all urban indoor spaces are air-conditioned. However, a 2001 study found that urban dwellers do not need to be so dependent on air conditioners if their homes are strategically designed for the climate. Naturally ventilated homes in a Bangkok suburb were found to remain comfortably cool for 20 percent of the year. These homes drastically cut down on energy use, saving the homeowner on electricity costs as well as sparing the environment.

"Houses and architecture should be treated as minimal framing devices for the environment, and for the indoor/outdoor lifestyle that comes with this kind of climate."

Ayutt Mahasom, AAD lead architect

Inspired by low-impact and energy-efficient goals, the YAK01 home built by Ayutt and Associates Design (AAD) in Bangkok incorporates design ideas from Thai vernacular housing to create a naturally cool home. "Thailand is a tropical monsoon climate zone, which means average temperatures of above 20°C in every month of the year. In the hot and wet conditions, good designs for Thai houses require naturally ventilated spaces together with a proper canopy and roof to hold off the rain during the heavy rainy season," AAD lead designer Ayutt Mahasom told *The Home Journal,* an online architecture magazine.

In the YAK01 home, cantilevered rooms mimic traditional stilted homes to shade the open terrace and rooms below. The directional orientation of the house also has strategic significance: many of the windows of the house face north, as northern sunlight is the least intense in Thailand. However, AAD was careful to place opposing windows to provide cross-ventilation in every room, much like a traditional Thai home.

In the past, Thai houses were constructed with bamboo, timber and other lightweight materials that cool easily with a drop in temperature after sunset. This style of construction was adapted in the YAK01 design for a modern look

that also has a cooling function. Silver aluminum extrusion strips were installed on exterior bedroom walls to serve as an air buffer, insulating the bedrooms from the outdoor heat.

The landscaping of YAK01's outdoor area also helps control the temperature of the house. A swimming pool placed directly parallel to the house draws cool air into the structure, while increased green space provides shade and helps reduce ambient temperatures. The profuse plants improve air quality and filter out street-level noise, an added benefit for the serenity of the home.

AAD has built four other homes in and around Bangkok adhering to the same ideals. Throughout the design process, AAD focuses on the principle of adapting the house as closely to its natural, regional environment as possible. "Houses and architecture should be treated as minimal framing devices for the environment, and for the indoor/outdoor lifestyle that comes with this kind of climate," Ayutt said.

AAD also has five ongoing projects that seek to advance their approach of sustainable design to the next level. The goal is to minimize the negative environmental impacts of the buildings, as well as to provide more green space and a green façade

Going Green at Home

Respecting the environment begins at home. In addition to having a greener abode, you can enjoy significant cost savings and live a healthier lifestyle if you opt for more eco-friendly practices.

- Indoor plants can serve as natural air filters. Some are also effective absorbers of harmful pollutants emitted by electrical appliances, carpets and furniture. Meanwhile, organic vegetable gardens can further reduce air pollution around the home, while surrounding a house with shady trees and plants will help keep it cooler and reduce the need for air-conditioning.

- LED light bulbs and energy-efficient appliances save electricity and money. To save more electricity, unplug TV sets and computers when they are not in use. When washing or drying clothes use full loads instead of doing laundry piecemeal.

- Eco-friendly, biodegradable cleaners are good alternatives to chemical, often toxic varieties. Baking soda, for instance, is an effective all-purpose cleaner. White vinegar and borax are good cleaning agents while citrus and tea tree oils serve as natural disinfectants. Meanwhile, damp microfiber cloths, which attract dirt, can do wonders even without any cleaning agents.

- Instead of disposing of organic waste, compost it. Fruit and vegetable peelings, tea leaves and eggshells can become excellent soil enhancers. Likewise, if left to decompose on a lawn, grass cuttings make excellent fertilizer for new grass.

- Ditch single-use paper and plastic products in favor of durable goods like ceramics to reduce unnecessary waste. Likewise, choose recycled products when it comes to items like stationery and toilet paper. Instead of dumping old clothes, books and toys, thereby ensuring they end up in a landfill somewhere, donate them to needy children, thereby turning trash into treasure.

- When dealing with e-waste, like old cell phones and computer monitors, you can give them away through online groups such as Freecycle, which has a Thailand branch. Some of the bigger electronics manufacturers have take back programs where you can drop off unwanted components.

- Since the majority of recycling in Thailand is done by the garbage men themselves you can make their job easier by dividing up your glass, paper, metal objects and plastics before putting them out in the garbage can.

- For those with deeper pockets, home automation systems can reduce electricity usage significantly. Smart thermostats and lighting controls, with timers and sensors, can monitor indoor temperatures and natural lighting levels to keep them ambient without the need for extra air-conditioning and lighting during certain periods.

Phi Suea House (right) in Chiang Mai uses a solar-powered hydrogen energy storage system (left) in order to be fully self-sustaining.

GROUNDBREAKERS

SEBASTIAN-JUSTUS SCHMIDT, *a German pioneer in mobile software and digital media, is the man behind the world's first solar-powered hydrogen energy storage, multi-house residential project, which is located in Chiang Mai. He is also the co-founder of the OTT/IPTV provider SPB TV AG in Switzerland and an advisor for several IoT start-up companies.*

What makes Phi Suea House unique? Can you explain the scope of the project?

Phi Suea House is the world's first solar-powered hydrogen energy storage, multi-house residential project. It consists of four family homes and other supporting buildings. All are fully powered by photovoltaic panels harvesting the sun's energy and an innovative hydrogen energy storage system. All power for the site is generated using solar panels with an installed peak capacity of 86kW. During the day, any excess power from the photovoltaic is used to power an electrolyser. This machine uses electric current to separate water into its composite gases: hydrogen and oxygen; and the hydrogen gas is stored in a tank. When there is no sun, the hydrogen gas from the tanks is used in fuel cells to generate electricity again. The energy storage is a hybrid battery-hydrogen system, which maximizes the advantages of both the batteries and fuel cell. The bulk of the nightly demand is covered by the fuel cells using hydrogen, while the batteries help to cover short peaks in demand. About 130kWh of energy can be stored in the hydrogen tank, which is enough to supply the site with 4kW of power from the fuel cells continuously for over 30 hours. The energy system at Phi Suea House is a central solution for community solar power and storage using hydrogen. It is ideally suited for residential or other developments in remote locations or where complete independence from the grid is desired. It is a 100-percent clean, safe process, and is highly flexible. The project is fully operational, supplying power 24/7.

During planning and development of the project, how did you work to ensure that every aspect was sustainability-oriented?

For our family the goal during the planning and development was very clear. We wanted to showcase the highest technology in clean energy production and storage for complete autonomy from the grid, which we accomplished with our unique solar and hydrogen energy system. Furthermore, we wanted to minimize the impact that the Phi Suea House has on the environment and the demand for resources. This principle has guided us throughout the whole development process, no matter if we are talking about the energy system, the heat insulation of the buildings, the garden spaces for organic farming, or the extensive rain and irrigation water collection and reuse system. We also designed and built a water collection system that saves and reuses as much of the rain and irrigation water on the site as possible. From a 1000m3 underground reservoir, water is cleansed by passing through sedimentation and aeration chambers, before it can be reused for irrigation or other uses around the site – all run with energy from the sun. Permaculture – which is the main domain for Erika, my wife – was also designed to help us live a 100-percent self-sufficient lifestyle. What used to be a hole overgrown with weeds was transformed into a fully organic solar-powered fishpond. All sorts of trees are planted throughout the property for environmental benefits, future educational purposes, but also aesthetics.

How are you sharing your learnings and technology?

While Phi Suea House is not a commercial project, we have attracted a lot of interest from various sectors and people through our success with the project. Although we never had the idea to form a business out of it I have started to support businesses which deal in this environment to make it easier for others to adopt similar green technologies in this region. Before the official opening of the project in 2016, we tested our energy system with a reduced load for almost one year, and collected valuable data and experiences that helped us to not only optimize our system in Chiang Mai, but also contribute to the further development of the hydrogen technology in general. We have compiled our experiences during the testing phase in a detailed user report that is available for download on the project website. Doing a project like the Phi Suea House brought us all kinds of visitors. From government officials to green enthusiasts who came from all over the world and really far, far away – from people who are living a life with almost zero spending to some super-wealthy – they all came and shared the same sustainable ideas.

COMMUTING

Taking back the streets from the automobile is an uphill journey

Bangkok's reputation for traffic gridlocks and chaotic streets is well deserved. Indeed, a 2015 report by the British motor oil company Castrol ranked Thailand's capital the eighth worst in the world for traffic, with more than 36 percent of a driver's travel time spent idling.

That makes commuting in this sprawling city as much a science as it is a form of transport. The secret to maximizing the usefulness of the system is to treat its parts like a jigsaw puzzle that you snap together depending on the time of day, whether you're in a hurry or not and what the weather is like. You might take a motorcycle taxi to the Skytrain, or maybe jump on a river or canal boat and hail a taxi when you jump off. Or you may get off the bus and grab a *tuk tuk* to a subway station. Got time but no money? Take the bus. If you have both time and money at your disposal, order a Grab or Uber taxi on your mobile phone. The permutations are endless.

But the most powerful force shaping Bangkok's transportation DNA is that Thais simply love their

Traffic on Yaowarat Road in Bangkok's Chinatown.

cars; some nine million are registered in Bangkok, approximately one per person. They are not only status symbols, but powerful emblems of independence in a culture where many people live at home until well into their twenties. For many who dwell in the suburbs, far from the outer reaches of mass transit lines, automobiles are also the most convenient form of transport.

As the city rushes headlong into the 21st century, one of its biggest challenges is maintaining and expanding its public transportation network. Hobbled by red tape and political uncertainties, projects large and small consistently face delays or hit a dead end and get shelved. Mass transit projects proceed slowly. Bike lanes are used as parking spots or run down the middle of busy sidewalks. But there is hope.

In recent years, a young and increasingly organized population interested in healthy lifestyles or inspired by what they have seen overseas has spearheaded projects that could rejuvenate the capital's infrastructure. For one, Bangkok Sabai Walk is working with officials to tame the city's notoriously unruly sidewalks and get people interested in walking again. The Green World Foundation mapped out bike routes throughout the city and put them in a book

Commuters take the Skytrain in Bangkok.

and smartphone app (called "Punmuang"). On the far side of the Chao Phraya River, King Mongkut's University of Technology Thonburi (KMUTT) has based much of its campus infrastructure on eco-friendly policies like cycling from place to place.

All of this is not to say that the powers-that-be aren't at least trying to play catch up with new mass transit lines in the works. Near Suvarnabhumi International Airport is a cycling park. There is now an eight-kilo-meter bike path around the capital's most historic district of Rattanakosin, while plans are afoot to add another 10 kilometers of routes. Extensions and new routes on mass transit lines are also well underway. Perhaps these initiatives will get Bangkok more on the go.

As the city rushes headlong into the 21st century, one of its biggest challenges is maintaining and expanding its public transportation network. Hobbled by red tape and political uncertainties, projects large and small consistently face delays or hit dead ends and get shelved.

Retaking the Sidewalks

Bangkok Sabai Walk (sabai means "convenient" or "stress-free" in Thai), or BSW, began in 2010 with the mission to make Bangkok's sidewalks safe, accessible, and pleasant for all city residents. Known for their unpredictable and often dramatic bumps, potholes and obstructions, the city's walkways present a challenge to even the most nimble walker.

"Our benchmark is to make the sidewalks usable by those in wheelchairs," said Santi Opaspakornkij, the group's coordinator. "If a sidewalk is usable for those with disabilities, it should be usable for everyone."

Despite working closely with universities, government bodies and community organizations, seeing a campaign through to completion is never easy. "There is no unifying body to coordinate actions by the different utilities, and lax enforcement of regulations means that each utility lays their underground wires, cables and pipes wherever they can," he said.

The red tape involved in digging up and re-paving a sidewalk is a formidable challenge. Some say the challenge is Bangkok itself, where walking is not seen as a viable option for getting around. As Santi explained, "Most Thais have no problem walking when they travel to foreign countries." He attributes the reluctance to walk on sidewalks here to the heat, crowds, sidewalk-blocking vendors and pollution.

To date, the group's biggest achievement is developing a 1.5-kilometer wheelchair pathway on Racha-damri Road between Rama IV and Rama I roads. But they are building momentum. BSW is now collaborating with the Bangkok Metropolitan Administration (BMA), local communities, companies and politicians to improve footpaths in areas with heavy pedestrian traffic. With assistance from Thailand Urban Tree Network, BSW has also begun helping to protect trees and thereby improve shade on urban sidewalks.

"There are many vested interests here, which makes it difficult to get a consensus from everyone. However, we have helped those affected get their voices heard, especially the disabled," he said. "These projects take a long time, and would be easier with more support from the BMA, but I am optimistic. The younger generation, the tourists and the expats who live here realize that quality of life is closely tied to these issues, and are starting to speak up. We'll continue doing everything we can to bring the community together to make Bangkok a better city."

▶**PIONEERS**

Green World Foundation

History: Started in 2011 with funding from the Thai Health Promotion Foundation, an autonomous state agency

Location: Bangkok

Key features:
Crowdsourcing routes from cyclists all over the city and putting them in a booklet and a smartphone app

Saranarat Kanjanavanit oversees the Green World Foundation (GWF), a nonprofit that collaborates with youth, educators and community leaders throughout Thailand to foster the proactive care of local environments. One of the key drivers of this plan is bicycle commuting, which has gained a large number of converts in recent years.

Yet the stigma remains that riding a bicycle in Bangkok is prohibitively dangerous. To counter this, GWF began studying transportation projects in other cities. It soon became clear that the most important factor propelling two-wheeled commuting is an effective and convenient network of routes, which Bangkok was sorely lacking.

Through social media and word of mouth, GWF crowdsourced neighborhood bike routes from riders all over the city, carrying out further surveys to find connecting links between them. The result was the *Bangkok Bike Map*, which is available as a book (pictured above right), or as a sleek app entitled Punmuang that neatly lays out bicycle routes through the city's labyrinthine roads and alleys. But while the argument for bikes as a healthy and non-polluting mode of travel is well established, the infrastructure to accommodate them still has a long way to go.

Despite civic groups banding together to press politicians on their policies for efficient and sustainable transport and an increasingly open-minded political establishment, which is interested in green solutions, the best ideas are usually enacted half-heartedly and the focus remains on automobiles. "Acceptance of bike riding has reached a critical point many times already, but bike-friendly policies are still just a decorative Band-Aid solution that are not fully integrated into the vision of a livable city with an equal transportation landscape," Saranarat said.

Members of GWF look forward to a paradigm shift that will force policymakers to lower the priority given to private cars and come up with more integrated solutions that merge bikes and boats with public transport.

"Acceptance of bike riding has reached a critical point many times already, but bike-friendly policies are still just a decorative Band-Aid solution that are not fully integrated into the vision of a livable city with an equal transportation landscape."

Saranarat Kanjanavanit, Green World Foundation (pictured right)

▶**PIONEERS**

The Green Campus at KMUTT

History: In 2010 the university initiated its Strategic Sustainability Plan, structured around six elements: infrastructure, energy and climate change, waste, water, transportation and education/outreach

Location: Bangkok

Key features: Providing an environment where students take an active role in maintaining the sustainability of their campus to become "change agents" who go on to improve their communities and society as a whole

The development of King Mongkut's University of Technology Thonburi has been guided by six core concepts of sustainability, one of which is bicycle transportation. Not only is the bicycle-centric lifestyle a major part of on-campus transport and student commutes, but it has also played a role in the university's community outreach programs and social activities.

The university's green policies embrace concepts such as energy conservation, water recycling, and reducing carbon emissions, but its most visible element is the Walk and Bike Society. Overseeing construction of bicycle lanes, parking areas, a repair shop and, of course, free bicycles for KMUTT students and staff to use on their commute, is part of an initiative that aims for a 20 percent reduction in car usage by providing covered walking lanes across the campus, clean energy shuttles and half of the student body riding bicycles.

The university also organizes regular bike outings, and offers students bicycle-powered charging stations for phones and tablets, and solar-powered bike locks that alert you if the bike is moved. There are also four self-serving bike share stations, two at KMUTT Bangmod campus and two at the KMUTT Bangkhunthien campus. KMUTT even has bicycle taxis to take you around campus when you can't or don't want to ride yourself. In total, more than 19,000 students and staff took part in KMUTT bike riding programs in 2016.

The effects of the program in surrounding communities are a clear indicator that it is working. Assistant Professor Suchada Chaisawadi, Director of Energy, Environment, Safety and Health, and Director of the Office of Sustainability at KMUTT, said, "Our reputation has led local groups to ask us for help adapting parts of our program into their communities."

More importantly, alumni have lauded the benefits of a bicycle-centric lifestyle, inspiring others to take up this healthy hobby. Indeed, the next step is expanding the Walk and Bike Society to other campuses, creating more change agents and continuing the cycle.

"We plan to expand the Walk and Bike Society to KMUTT's Ratchaburi campus southwest of Bangkok, as well as increasing the functionality of the program to make it an even bigger part of student life," said Suchada. "All of the data from bike usage and the correlating reduction in emissions and power consumption is being recorded."

In 2016, KMUTT's bike riding activities helped reduce carbon emissions by 8.37 million kilograms of CO_2 and reduced benzene fuel usage by 12.84 million liters. Suchada hopes positive results such as these will "inspire others to push for similar policies with an eye to caring for the environment."

Aerial view of KMUTT's green campus.

Students have embraced cycling on campus.

New Cycling Hub in Asia

A few years ago, the city of Taipei in Taiwan needed to make some hard decisions. The capital of 2.7 million people was slowly but surely edging closer to some of its bigger brothers in Asia. The traffic wasn't as grid-locked as Bangkok's, nor the air as soupy-grey as Beijing's, but as the city grew, so did the likelihood that it would soon catch up to megacities for all the wrong reasons.

Then along came YouBike, a bike rental program that, by 2016, had made 7,200 bicycles available from 222 stations around the city. Riders can rent the bikes 24-hours a day, and pay only a small charge.

The program was an immediate success, notching up more than 22 million rides in 2014, almost double the previous year's total.

But the program also highlights

one of the drawbacks of providing an alternative transport system. What if it becomes too successful?

YouBike riders were soon using the bikes in such great numbers that the automated rental stations were often either empty (meaning no bikes for anyone to borrow) or full (meaning that no one could return the bike they'd just finished riding). Clearly, managing the logistics of a popular system like this is not an easy ride.

Another problem was that not all YouBike riders were familiar with road rules for bicycles and were not blending into regular traffic as smoothly or as safely as hoped. This led to riders taking shortcuts along sidewalks and pathways, angering and inconveniencing pedestrians. But they are trying to remedy the situation by educating riders on the rules of the road.

By 2019 the city also plans to spend about US$32 million on more bicycles and kiosks, hoping to increase the total number of rental stations to 400 by 2018.

A Chao Phraya river ferry passes Wat Arun.

Around 40,000 people utilize the Chao Phraya river ferry service each day. Arguably the oldest form of "mass transit" in Bangkok, traveling by river remains a popular choice for local commuters. Today the Chao Phraya ferries serve 34 piers along four routes.

▶PIONEERS

The Pun Pun Bike Share Program

History: Launched in October 2012 by the Bangkok Metropolitan Administration (BMA)

Location: Bangkok

Key features: Allows registered members to rent bicycles at affordable rates for short-term day-time use in the city

Urban bike sharing programs have exploded in popularity around the globe during the last decade. Accessibility, affordability and sustainability have helped to promote the concept of short-term bike rental systems as a win-win for just about anyone who is willing to ditch the use of a car or motorcycle for a bicycle. Commuters can cut costs and avoid traffic jams. Tourists can explore new cities without the hassle of confusing bus routes, costly taxi fares and sore feet. Meanwhile, the environment benefits from a reduction — however small — in greenhouse gas emissions.

Bangkok got on board with bike sharing in 2012. In October of that year, the Pun Pun Bike Share program was launched by the Bangkok Metropolitan Administration (BMA) with just two stations and 16 bicycles. Similar to programs in other major cities around the world, Pun Pun allows individuals who sign up as members to rent bicycles at reasonable rates for short-term use.

The idea behind the program is to encourage the use of eco-friendly transportation rather than cars or motorbikes in some of Bangkok's congested urban areas. To facilitate this aim, the program has bike stations near key destinations such as busy office districts, malls, as well as major Skytrain and MRT stations.

"Sharing" is the key to the concept, and in fact the name Pun Pun is derived from the transliteration of the Thai word for sharing. Today, it has grown to include 50 stations with 330 bikes across Bangkok. Some 10,000 people use the bike share service each month in Bangkok, and Pun Pun has also started programs in the cities of Chiang Mai, Phitsanulok and Udon Thani.

Unfortunately, the program isn't perfect. Users and potential users in Bangkok have complained that there are not enough stations, that stations can be inconveniently located and that on occasion there are regularly no bikes at some stations. Still, it's a good start.

Although the program's website is currently only available in Thai, the process of signing up and renting a bicycle is fairly easy to figure out and there is some English language information about the program on Pun Pun's Facebook page.

After signing up, members are issued a debit card and PIN, which they use to check out and check in rental bikes. Bicycles can be picked up and dropped off at any Pun Pun Bike Share station. Members' cards can also be topped up with credit at any of these stations.

Bicycle usage is actually free if the ride is less than 15 minutes. After that usage rates are tapered, costing 20 baht for 1-3 hours, 40 baht for 3-5 hours, 60 baht for 5-6 hours, 80 baht for 6-8

> *"Some 10,000 people use the bike share service each month in Bangkok, and Pun Pun has also started programs in the cities of Chiang Mai, Phitsanulok and Udon Thani.*

hours, and 100 baht for 8+hours. All bikes have to be returned at the end of the day (closing time for the stations is 8pm), and cannot be kept overnight.

Users who fail to return their bikes on time incur a penalty of 500 baht per day. And don't even think of trying to steal a set of these eco-friendly wheels – all Pun Pun bikes are tracked with GPS.

Of course, riders still have to contend with the fact that Bangkok's streets and sidewalks remain relatively uninviting for cyclists. That's one reason there is an initial registration fee of 320 baht and annual membership dues of 100 baht per year, which cover insurance for the rider.

To register, learn more about cycling events in Bangkok, or see a list of stations, visit www.punpunbikeshare.com.

One of Pun Pun Bike Share Program's "green" bicycle stations in Bangkok.

COMMUNITY
SPIRIT

On the Farm

Area-based Rural Development

Organic Revolution

Integrated Farming/New Theory

By the Forest

Reforestation

Forest Conservation

Wildlife

On the Coast

Saving Marine Habitats

Coastal Resource Management

In the City

Historical Preservation

Urban Development

Green Spaces

Local communities are the driving force behind many noble projects across the country, from reforesting degraded areas to introducing water management systems.

Although Thailand has become an economic powerhouse of Southeast Asia, the spirit of togetherness and the value of communities helping communities – which stem from a more traditional, more agrarian time – still inform much of our daily interactions and relationships. The local skills, knowledge, and traditions of Thai farming communities have also proven sustainable and effective in today's complex world in areas like integrated and organic farming. When such wisdom is combined with the assistance and finances of the royally initiated projects, Thailand has a formidable agent of change to help shepherd it toward sustainability.

Perhaps the best known of these schemes is the Doi Tung Development Project. Initiated by the late Princess Mother, the ongoing project has been hailed by the United Nations for achieving the economic, social, cultural and environmental aspirations of sustainable development. As a result of such projects and other initiatives by NGOs, community groups and academics, Thailand is on the path to repair some of its worst ailments. For example, more than 1.3 million hectares of trees were planted between 1990 and 2010, which has done much to address the issue of deforestation. Over these two decades, the size of planted forests grew from 2.7 million hectares to almost four million. It's an impressive record that attests to the power of community initiative.

In the north and northeast, farming communities have benefited from newer and more sustainable management strategies like integrated farming, which takes a holistic approach to nurturing entire ecosystems and makes farmers more self-sufficient. It is also a move away from the chemically intensive form of monocropping that continues to dominate agribusiness in Thailand.

As more sustainable alternatives grow, the hope persists that agriculture and forestry, and the communities that depend on them, will continue to be rejuvenated by these new developments. Though Thai society has changed to embrace more Westernized habits and consumption patterns in recent decades, the return to a more community-based society with communal values could well be a sea change in sustainability.

AREA-BASED RURAL DEVELOPMENT

Matching the right crop with the right type of land and farming technique

After returning from studying in Switzerland in 1951, King Bhumibol Adulyadej traveled far and wide in Thailand over the decades to meet with villagers and discuss their problems. These discussions and observations formed the backbone of his royally initiated projects that typically promote site-specific measures that are both sensitive to the local environment and also take into consideration the socio-economic conditions and cultural backgrounds of the people. This approach is sometimes referred to in Thailand as "Area-based Rural Development."

In the north, where opium cultivation and slash-and-burn farming were rife, these initiatives promoted the growing of cool-weather crops, like coffee and strawberries, suited to the highlands. In the northeast, where drought kills crops and stunts household incomes, irrigation projects such as the Huay Klai Reservoir slaked the crops' thirst for water.

In promoting this type of rural development, it's important to match the right farming technique with the right terrain. Given the plentiful water and arable soil in the central region, integrated farming is a good choice. In forested parts of the north and south where plots for farming are limited, agroforestry can be put to good use, especially for cash crops like rubber and eucalyptus trees. If done right, agroforestry can also benefit the whole ecosystem because trees, plants and shrubs grow alongside pastures to enrich the earth and improve biodiversity. Farming that benefits the environment is a crucial consideration when looking at alternatives to monocropping, which still accounts for the vast majority of farms in Thailand and relies on harsh chemicals. Natural farming, pioneered by the Japanese farmer Masanobu Fukuoka (see sidebar below), and New Theory farming, the brainchild of King Bhumibol Adulyadej, both eschew a dependence on pesticides.

EYE-OPENERS

BACK TO THE EARTH
ONE-STRAW REVOLUTION: AN INTRODUCTION TO NATURAL FARMING
Author: *Masanobu Fukuoka*
Year: *1978*

The best-selling book *One-Straw Revolution: An Introduction to Natural Farming* by farmer and philosopher Masanobu Fukuoka is an indispensible work in the canon of sustainable agriculture. *One-Straw Revolution* outlines in pragmatic and poetic terms the steps to accomplish Fukuoka's system of natural farming, or "do-nothing farming," which eschews chemicals, tilling, and other farming techniques.

Before World War II, Fukuoka conducted an experiment comparing the yield of crops treated with chemicals to the yield of crops grown without chemicals. He found that although the chemical crops produced a slightly higher yield, the value of the yield did not cover the cost of production. Thus, Fukuoka theorized that farming in a way that simulated natural processes as closely as possible was the best agricultural practice.

In addition to pioneering this system of farming, Fukuoka also led an effort to reclaim desert land through natural farming. He was awarded the prestigious Ramon Magsaysay Award in 1988 for his service to humanity and passed away in 2008 at the age of 95.

Terraced rice paddies in the mountainous area of Chiang Mai's Mae Chaem district in northern Thailand.

Until the invention of the insecticide DDT in 1939, and the gradual spread of industrialized agriculture, all farming had been organic. The harmful effects of such chemicals only came to be widely known thanks to Rachel Carson's landmark book, *Silent Spring*, published in 1962, when the science writer detailed how the use of DDT was killing off songbirds and other creatures. Subsequently, DDT was banned in the United States in 1972, but it's still widely used in Thailand and other Asian nations.

That's how the backlash against chemical-heavy farming began and the resurgence of the organic agriculture movement began afresh, gaining momentum as the "Green Revolution" of the 1960s and 1970s boosted yields but caused environmental side effects from insecticides and pesticides.

In the field of area-based rural development, taking an integrated approach that factors in the land, the people, the crops and culture has proven effective. In this respect, the Doi Tung Development Project is exemplary. Not just a crop substitution program designed to eradicate opium in the northern highlands, the project combines agriculture, education, healthcare and the building of a brand as a panacea for the woes of this once underdeveloped area inhabited by marginalized hill tribes. As with many other royally initiated projects, Doi Tung is also a learning center that sows the seeds of knowledge in visitors who then spread this knowledge.

The Royal Initiative Discovery Foundation's (RIDF) approach to development in Nan province – one of the country's poorest provinces – is also a notable success story. By engaging local stakeholders to take ownership of development projects and better manage their own resources, RIDF has emboldened communities to help themselves. As a rule, RIDF only implements projects in areas where the local community has shown a strong willingness to participate. This grassroots approach has seen communities take the lead on diversifying agriculture practices away from monocropping, creating community funds for microloans, and undertaking the extensive restoration of reservoirs and irrigation systems.

But the real point of these projects is that they have raised spirits and incomes, helped ward off social afflictions like crime and substance abuse, and provided people living in harsh environments with hope for a livable future. The backbone of all such endeavors is the Thai value known as *choomchon khem kaeng* (community spirit).

▶**PIONEERS**

Doi Tung
Development Project

History: Established
in 1988

Location: Doi Tung
Mountain in Chiang
Rai province

Key features: Crop
substitution programs
replaced opium with coffee and
macadamia nuts; the project also
focuses on education, healthcare
and building the Doi Tung brand
for handicrafts, horticulture and
tourism

Doi Tung was once a shady and secluded area in the black heart of the Golden Triangle, which had earned the dubious distinction of being the world's largest opium-producing region. But that was far from the area's only problem. For decades this watershed had been denuded by slash-and-burn cultivation. The down-trodden locals, composed of members from six different ethnic groups, lived in dire poverty with little access to even the basics of running water, electricity, healthcare and schools. To make matters worse, the area was a hotbed of armed militia and cutthroat opium barons, who

made it even more difficult for government officials to provide any assistance to the hill tribes.

That is the backstory of and the catalyst for the Doi Tung Development Project (DTDP), one of the Mae Fah Luang Foundation's four flagship projects. First established in 1988 under the royal patronage of Princess Srinagarindra, the mother of King Bhumibol Adulyadej, on the mountain of Doi Tung in Chiang Rai province, the project area covers approximately 15,000 hectares and benefits some 11,000 people from 29 villages. As a new cash crop for the farmers, Princess Srinagarindra chose Arabica coffee trees. Though not indigenous to Thailand, they flourish under the shade of forests. She also developed a multi-pronged plan that respected the local environs, took into account the local culture and social conditions of the villagers, and aimed to boost their livelihoods. Eliminating crime and corruption were also essential parts.

The long-running endeavor is now the most globally renowned of all royally initiated projects. The crop substitution program, first pioneered by King Bhumibol Adulyadej in 1969, has inspired similar ventures in opium-plagued countries like Myanmar and Afghanistan, although Thailand remains the world's

The crop substitution program has inspired similar ventures in opium-plagued countries, like Myanmar and Afghanistan.

most successful example of eradicating opium production at its roots.

The timeframe for the DTDP is 30 years, which has been broken down into three phases. During Phase I, which ran from 1988 to 1993, the first priority was tackling health issues and providing vocational training. From 1994 to 2002 the project's lynchpin was income generation, as the DTDP introduced the concept of moving up the value chain by building a factory to roast the beans and package them under their own DoiTung brand. The final phase, set to finish in 2017, is about strengthening the business units so that the brand and the community are sustainable, as well as capacity building and education, so that locals can take over the project when it concludes.

To ensure a smooth transition from the old guard to the new upstarts, the DTDP has collaborated with the Ministry of Education on reforming the curriculum in line with international standards to provide opportunities to study and

In addition to coffee, DoiTung's products include homeware, nuts and mulberry paper products.

A hill tribe person harvests coffee cherries.

hands-on vocational training that are both adapted to a local context. The programs are also designed to instill students with cultural pride and provide them with enough skills to find good jobs locally so they can become productive members of the community.

Now that the DoiTung brand has expanded to encompass four different business units (i.e., food, handicrafts, horticulture and tourism), it has become a burgeoning source of career opportunities. Dozens are now employed by DoiTung's two resorts, a high-tech museum in the Golden Triangle Park, and at their restaurants and cafes with outlets in Bangkok, Chiang Mai and Chiang Rai. The DoiTung Lifestyle Shops in those three cities also sell a range of hand-woven clothes, carpets, mulberry paper, ceramics and home décor.

Recognized the world over as a paragon of sustainable alternative livelihood development, all of the project's products sport the seal of the United Nations Office on Drugs and Crime as a hallmark of its success and humble origins. The area is also notable for being a beacon of cultural, social, economic and ecological development that gives equal weight to all of these considerations and blends them into a harmonious whole.

In 2015, the European Union also gave its seal of approval to the coffee brand, granting it the prestigious Geographical Indication (GI) certification, which is only given to products with a uniquely regional flavor that are both produced and processed in that area.

THE INSPIRATION OF PRINCESS SRINAGARINDRA

The Mae Fah Luang Foundation (MFLF) is a private non-profit organization originally founded in 1972 by Princess Srinagarindra as the Thai Hill Crafts Foundation (THCF) under her royal patronage. Following her visits to remote areas of the country, the late Princess Srinagarindra discovered that the ethnic minorities in northern Thailand were disenfranchised and trapped in a downward spiral of sickness, poverty and ignorance. At the same time, she was aware of their many artistic talents and penchant for making handicrafts. She wanted to help them supplement their incomes by marketing their work.

By 1985 the THCF began to incorporate rural development into its programs and activities so the foundation was renamed. Thus, the newly christened Mae Fah Luang Foundation took on a bigger role that culminated in the comprehensive development project in the Doi Tung area.

In honor of Princess Srinagarindra's memory the foundation has gone from strength to strength by focusing on social enterprises with a sustainable development slant. In 2009, Mom Rajawongse Disnadda Diskul, the secretary-general of the DTDP, and her former private secretary, received the Schwab Foundation's "Social Entrepreneur of the Year for the Region of East Asia" award for the organization's efforts to provide people with legitimate livelihoods and a better quality of life, while also keeping environmental concerns at the forefront of their agenda.

Insect Farms are Studies in Sustainability

For travelers visiting Thailand, insect vendors on the streets of Bangkok and Chiang Mai make for fantastic photo ops and drunken dares. But now, insects, long an alternative source of protein in rural parts of Asia, have finally shed some of their stigma as a repugnant snack among Western consumers.

Indeed, research has shown that caterpillars have more protein than red meats or chicken and they come with much less saturated fat. They are also packed with vitamins and minerals. "Eating a few insects is like taking a multivitamin," Patrick B. Durst, a senior FAO official who co-authored a study on Thailand's edible insect industry, told *The New York Daily News* in 2014.

Insect farms are easy on the environment too. The water and food used to nurture them is but a drop in the bucket when compared to cattle farms, which are one of the biggest causes of deforestation. To produce a pound of beef takes 25 pounds of feed, 2,900 gallons of water and a lot of room for bovines to roam. To produce a pound of crickets, con-versely, takes a mere two pounds of feed, one gallon of water and a cubicle. None of the most contentious supplements like growth hormones or antibiotics are needed. Nor do insects produce bursts of methane gases from either end like cows do.

For farmers in the northeast used to reaping only one rice harvest per year, crickets can be harvested every two months. They are also more resistant to periods of drought (a common occurrence in the northeast) and with around 200 different species for sale in Thailand offer plenty of diversity.

Besides crickets and palm weevils, which are farmed, other common insects are taken from the wild. These include bamboo caterpillars, weaver ants, giant water bugs, silkworm pupae and grasshoppers.

With some 20,000 small-scale farms operating in Thailand, according to the FAO, the kingdom produces around 7,500 tons of edible insects per year, making it the world leader. In 2013 the UN agency released a book called *Six-legged Livestock: Edible Insect Farming* that chronicles Thailand's success story in opening a new chapter in health food, as what was once unpalatable to Western palates has turned into something

Vendor sells fried insects at a market in Phuket.

of a delicacy, with energy bars that consist of ground-up crickets turning up on the shelves of health stores in the US, and the first American cricket farm beginning operations in 2014.

Concluding their report on Thailand, the authors wrote: "The collection of edible insects in Thailand is an historic practice, but their farming is relatively new. Incomplete information nationwide indicates a growing and healthy market. However, knowledge gaps regarding sustainable wild collection and best management practices for farmed insects are a major risk for the industry. The current lack of government involvement in the promotion of the industry is seen as a major weakness. Edible insects have huge potential as a protein source with significance both domestically and internationally in helping to feed the burgeoning global population."

Bamboo caterpillars are a popular snack in Chiang Mai province.

▶**PIONEERS**

The Nan Model

History: Started in 2014 by the Royal Initiative Discovery Foundation

Location: Nan province

Key features: Takes an inclusive, collaborative approach to repairing reservoirs and irrigation infrastructure

Given how much time and effort have been put into the issue over the decades, the management of water resources in Thailand's rural areas should not be such a daunting task. But in practical terms, Thailand still struggles in this area, with bureaucratic "red tape" serving as the primary obstacle to making water management more efficient.

Unlike most countries, Thailand does not have laws governing the management of water resources and thus cooperation in this area is fragmented between many bodies. Currently, 30 different state organizations under seven different ministries apply more than 50 different laws to the multiple issues of water management. Bureaucratic obstacles, from overlapping authority to disunity within the various departments responsible for water management, end up delaying the decision-making process and impeding budgetary disbursements. It can also result in flabbergasting gaps in execution that sometimes leave beneficiaries high and dry.

For example, in some cases reservoirs are built in villages but without the installation of accompanying pipework to siphon water into agricultural areas (often, this is due either to budgetary constraints or a lack of foresight). Meanwhile, the maintenance of reservoirs and distribution systems is even more complicated due in part to the misuse of budgets. According to the law, local villagers cannot carry out repair work by themselves because the reservoirs are government-owned,

and while some local organizations may have the budgetary resources to carry out repairs, they more often than not lack the technical skills to do so. It should come as no surprise then that thousands of Thailand's reservoirs and irrigation systems in agricultural areas are in a ruinous state.

In Nan province, where the mountainous terrain renders large-scale irrigation systems almost impossible to install, local people struggled for years to make do with inadequate small-scale reservoirs. The concerned authorities had also never developed appropriate flood management systems, meaning that during droughts, farmers had insufficient water for crops, while almost every rainy season locals faced flash floods that inundated villages. Lack of foresight also led to failures like the Nam Lieb reservoir in Chiang Klang district. While the reservoir was completed in the 1980s, effective irrigation systems were never built to supply water to nearby villages, resulting in chronic water shortages and low agricultural yields.

In 2014, the Royal Initiative Discovery Foundation sought to encourage a better model of development – one that cuts through unnecessary red tape and involves all relevant stakeholders. To that end, the foundation began collaborating with the Nan provincial governor's office, local communities, civil society groups and other organizations to repair and service small-scale reservoirs and irri-

gation systems across the province in an effort to replenish water resources. This became known as the "Nan Model."

As part of this Nan Model initiative, the foundation and relevant authorities conducted repair work on 560 reservoirs and other related structures, some of which had been inoperable for decades. The Nan Model encompassed construction of pipework that enables locals to manage water resources on their own, development of water management plans to boost productivity, and raising awareness about the benefits of reforestation and the adverse effects of slash-and-burn agriculture.

Overall, more than 41,700 households benefit from the Nan Model project, with newly restored infrastructure supplying an adequate amount of water to more than 16,000 hectares of cropland.

Bureaucracy remains a huge impediment to improving water resource management. But as the Nan Model demonstrates, through collaboration that involves all relevant stakeholders, progress is possible. To help cut through the red tape that binds up the decision-making process and final implementation, the government could also streamline and consolidate bureaucratic processes and legal apparatuses, decentralize authority in terms of which bodies have final approval over local projects, and draw up provisions that allow for the fast-tracking of approval and implementation of urgent projects.

The Nam Ngim reservoir, in Song Khwae district, Nan province.

ORGANIC REVOLUTION

The movement continues growing and sprouting different offshoots

Farmers at work in the rice fields of the central plains.

"Organic" is one of the most widely used buzzwords of recent years, but it's the harbinger of much more than just chemical-free produce and toiletries. Emerging over the past few decades as a reaction against large-scale agribusiness, the organic movement seeks to support and propagate crops grown without pesticides or herbicides to improve land management, promote responsible resource use, reduce pollution and preserve biodiversity.

But what does it mean to be "organic"? The International Federation of Organic Agriculture Movements (IFOAM) defines this form of agriculture as "a production system that sustains the health of soils, ecosystems and people. It relies on ecological processes, biodiversity and cycles adapted to local conditions, rather than the use of inputs with adverse effects." In general, these inputs refer to synthetic chemicals (pesticides, herbicides and fertilizers), as well as biogenetic processes (such as genetically modified food). But the organic movement also influences and is influenced by many other concepts that impact economy and society, from serving as a lifeline for farmers mired in debt to giving forward thrust to the economic engines of rural communities.

In Thailand, the certification infrastructure and government support for organic farming already exist.

Although the majority of Thailand's modern agriculture is monoculture, its detrimental effects on the environment have long been known here. That backlash resulted in the formation of the Alternative Agriculture Network in the early 1980s. It was the first step to establishing a certification body, the Organic Agriculture Certification of Thailand (ACT), in the 1990s. From tiny beginnings, organic farming has been on the upswing, increasing from roughly 1,000 hectares of land under cultivation in 1998 to 45,587 hectares in 2015. Most of Thailand's organic crops are produced for export. The main crop is rice, while other cash crops, such as soybeans, peanuts, tropical fruits, asparagus, tea, coffee, herbs and rubber, bulk out the total.

The movement in Thailand faces many challenges. According to Green Net, an organic agriculture research and advocacy agency, less than 0.2 percent of all arable land in Thailand is under organic cultivation today, and only 0.2 percent of all farming households were certified organic. Notably, almost 60 percent of the organic food sold at retail outlets in Thailand is imported. Overall, Thai awareness of and interest in such crops and products is relatively low as well. Certification by a globally recognized body such as the International Federation of Organic Agriculture Movements (IFOAM) can also eat up

ORGANIC FARMING IN THAILAND BY THE NUMBERS

THE PREDOMINANT ORGANIC PRODUCTS IN THAILAND ARE

RICE   **VEGETABLES**   **FRUIT**

which represents about **0.2%** of Thailand's total farmlands.

THREE CHANNELS WHERE ORGANIC PRODUCTS ARE SOLD:

CONVENTIONAL SUPERMARKETS   **SPECIALIZED SHOPS**   **DIRECT SUPPLY**
such as farmers' markets or cooperative memberships.

LAND USE FOR ORGANIC FARMING BETWEEN 2000 AND 2015

2014

2000

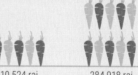

10,524 rai
(1,684 hectares)   284,918 rai
(45,587 hectares)

**Estimated value
of organic produce
in Thailand in 2015**
2 BILLION BAHT

Sources: Thai Organic Trade Association, Green Net, Earth Net, World of Organic Agriculture 2017

20 percent of a farm's profits per year. Yet there are encouraging examples from inside the kingdom, such as the Palang Punya project, which is aimed at solving the issue of acidic soil in the northeast through organic farming practices, and also helping the farmers make a living by finding markets for their organic products.

That can be a challenge. According to sustainable development consultant Jeff Rutherford, the organic movement here is currently dependent on individual purchases, which makes for small sales and slow progress for the movement as a whole. But a look at Europe, the US and Japan shows a growing institutional demand for organic foods and products.

In Thailand, this untapped market of institutions, such as government departments, municipal bodies, schools and corporations, has the potential to raise the demand in a major way by contracting farmers or procuring organic products to supply entire institutional populations. The certification infrastructure and government support for organic farming already exist. But making more inroads into marketing and distribution, as well as forging more partnerships between farmers and consumers, will play decisive roles in the movement's future growth or stagnation here.

▶ PIONEERS

Chaiporn Phrompan

History: Began organic farming in 1990

Location: Suphanburi province

Key features: His success, high-profile awards and innovative techniques have been major inspirations to his peers

Chaiporn Phrompan, a rags-to-riches organic rice farmer.

While most Thai rice farmers struggle with heavy household debt and poverty, one rice farmer who transitioned to organic production is now known nationwide as the "millionaire farmer." Chaiporn Phrompan has become a pioneer of organic rice farming in Thailand, serving as an inspirational example of the economic opportunities that await the maverick farmer.

Like many struggling rice farmers, Chaiporn Phrompan was no stranger to hard work for few rewards. He grew up in a farming family from Suphanburi, which never saw much profit despite year-round toil and heavy investment in much-touted "miracle chemicals."

Now Chaiporn has succeeded beyond his wildest dreams. He owns 108 rai of paddy, he's bagged the Ministry of Agriculture's Outstanding Farmer award multiple times, and he's been hailed as one of Thailand's most successful farmers. What happened?

In the late 1980s, Chaiporn and his father learned about organic farming under the tutelage of the organic rice scientist Dr Decha Siripat, renowned for melding cutting-edge technology such as soil testing and microorganism cultivation with the time-honored practices of integrated farming found in Thailand. "Dr Decha asked my father to experiment with organic farming on five rai of land. At the time, I was not interested and thought it was useless," said Chaiporn.

By that point, he had already taken courses on synthetic aids and chemical pesticides in the hopes of increasing his family's yield. During the years 1983 to 1989, he invested heavily in chemical methods and produced an average of 12 tons of rice per year from 25 rai of paddy. However, because he had to go into debt to pay for those chemicals, he could barely turn a profit.

Decha, who founded the Khao Kwan Foundation, an influential learning and research center, told Chaiporn in 1996 to experiment with organic manure and an organic pesticide made from neem, a tropical tree. Chaiporn noticed that not only were the organic crops doing well, but the land also retained its fertility more than areas where the chemical crops grew, and required less expenditure on fertilizers. That was the impetus for him to expand the family's organic operations.

Within a mere three or four years Chaiporn turned his whole plot of land into an organic rice paddy. As the profits rolled in, he bought more land. Chaiporn said he has not been in debt since 1990: a poignant fact given that farmers are counted among the nation's highest debt holders.

Chaiporn believes that the organic trend is here to stay. "People care more about their health and environment these days. Farmers have to find ways to guarantee that their rice is safe from chemicals," he said. "Organic farming is the answer."

Rice farmers never saw much profit from harvesting due to heavy investment in chemicals.

Decha Siripat, founder of Khao Kwan Foundation, which teaches the virtues of organic farming.

▶PIONEERS

Raitong Organics Farm

History: Founded in 2007 by Bryan Hugill and Lalana Srikram

Location: Farm and office in Sisaket

Key features: Accredited by IFOAM and certified organic by the EU and Canada, Raitong distributes organic rice internationally and uses innovative marketing strategies domestically for 100 percent natural ginger beer, organic baby food and rice-based pasta

Raitong Organics Farm, which boasts a 51-rai organic rice farm in the northeastern province of Sisaket, is also capitalizing on the new and growing demand for organic produce. Owned and operated by Bryan Hugill, from South Africa, and his wife Lalana Srikram, from Thailand, Raitong Organics Farm not only grows organic rice, but also mills, packages and distributes it directly to consumers in Thailand and through wholesale partners overseas.

Of Thailand-based companies, Raitong has taken an international approach to marketing their main commodity. They have invested in accreditation by the International Federation of Organic Agriculture Movements (IFOAM) and have been organic certified by the EU and Canada for several years. Without these expensive yet essential labels, Raitong could not distribute their rice as an organic product in Europe and Singapore.

While most Thai rice farmers sell their harvests to millers and middlemen, remaining anonymous to both retailers and consumers, Raitong has reached out directly to consumers by branding and packaging their own rice and selling their products at a network of farmer's markets. In these face-to-face transactions, Raitong can add personal touches to its branding campaigns by directly engaging with consumers.

In another visionary move, Raitong Organics Farm has taken advantage of Thailand's trend-conscious social scene and forged relationships with upscale organic restaurants, such as the much-lauded Bo.lan. For both of them it's a win-win situation. While Raitong is able to move large amounts of product to Bo.lan, the restaurant can also bolster its reputation as a hub of healthy, conscientious eating.

Overall, the Raitong brand is continuing to grow with the addition of other products, such as "Ginger Fizz," a naturally brewed ginger beer. In the recent past, they also hosted the food// hack@BKK campaign, a series of public workshops on urban gardening and sustainable living. They are also currently working on the development of an innovative peer-to-peer organic auditing process that takes advantage of GIS mapping and smartphone technology to bring consumers and farmers closer together. And they have a line of organic baby food and rice-based pasta in the pipeline too.

By committing to these initiatives, Raitong Organics Farm is establishing itself as a creative brand that is also devoted to fostering a sense of community among those enamored of natural products and healthy living.

Organic products that go straight from Raitong Organics Farm to consumers.

Organic Jerusalem artichokes

▶ **PIONEERS**

Farmers' Friends Rice Community

History: Founded in 2014

Location: Yasothon province

Key features: Fair trade subscriptions for rice benefit both community and consumers

Members of Farmers' Friends Rice Community paid a visit to farmers in Yasothon province and helped them plant organic rice.

As the kingdom's top crop, rice is critical to the country's economy. Changing the way it is cultivated has the potential to improve both output and farming standards throughout Thailand.

The Bangkok-based Farmers' Friends Rice Community cooperative supports rice-farming families in Yasothon province who have committed to replacing chemical farming practices with organic methods. Members of the cooperative pay for a share of the rice harvest upfront, ensuring that the farmers do not have to go into debt to acquire seeds and equipment for that year's crop.

Farmers' Friends individual subscribers purchase 50-100 kilograms of organic rice per year, while corporate members are asked to commit to an annual amount of 200-500 kilograms. The membership cost of 100 baht per kilogram goes directly to farmers and the cooperative, allowing participants to earn 2,000 baht more per kilo than the market price. Meanwhile, members enjoy the choice of brown, jasmine, red jasmine or riceberry rice, in addition to the health benefits of organically grown produce, and free delivery to their doorsteps.

While this fair trade system cuts out the kind of middleman that often saps farmers' profits, it also contributes to a Farmers' Security Fund that is used to support farmers whose crops are affected by forces beyond their control, such as drought, floods, blight or unstable market prices.

The environment also benefits from the cleansing properties of this natural school of farming. On average, after three organic planting seasons, the land becomes cleansed of chemical traces and the soil quality is able to meet international organic standards. Proving how well rounded the program is, society's most needy also stand to benefit from Farmers' Friends, as cooperative members are encouraged to donate excess rice to charities and orphanages.

Co-founder Oraya Sutabutr said that so far, Farmers' Friends has met or surpassed many of its goals. To date, some 200 farming families and 100 individual members have joined the program. Farmers' Friends has also been able to attract a number of corporate members who want to feature organic rice on their menus including Cafe Now by Propaganda, Mint Cafe by Peppermintfield, Bo.lan, Sri panwa Hotel, and Kids' Academy International Pre-school. Each year Farmers' Friends now produces between 20 and 50 tons of milled rice.

However, the main challenge remains in raising consumer awareness and changing consumer habits to create a larger market for organic rice. "We need to do more work to increase members

Rice seedlings waiting to be planted.

on the consumers' end and we plan to do more activities such as farm visits and organic fairs, held jointly with other organic rice groups," Oraya said.

In the coming years, Oraya says Farmers' Friends is planning to innovate by creating rice-based products such as medicine capsules, and to diversify the cooperative's products by growing other crops such as banana and *plai*, a popular ingredient in massage oils and balms.

▶**PIONEERS**

Pun Pun Center for Self-Reliance

History: Started in 2003 by Jon Jandai

Location: Mae Taeng district in Chiang Mai province

Key features: Multi-purpose enclave for the organic community that is also a seed bank and school

Tucked away down a country road in this northern province, the Pun Pun Center for Self-Reliance is a seed bank, educational center, natural housing community and organic farm that shares its resources and knowledge with both local and international communities. The guiding philosophy at Pun Pun asserts that the four "necessities" of life – shelter, food, materials and medicine – are easily attainable through self-reliance rather than through mass consumption.

Life at Pun Pun attempts to emulate nature as closely as a community can: homes are built of plentiful materials such as dirt, straw and bamboo; integrated farming methods mimic the natural states of jungles and forests; no chemicals or synthetic products are used on the land; and community members produce almost all of their own food and daily sundries such as soaps and shampoos at little or no cost.

Meaning "a thousand varieties" in Thai, Pun Pun is committed to biodiversity and working to propagate and reintroduce indigenous seed varieties to Thailand and beyond. In the process, it's become one of the most comprehensive seed banks in the region.

Today, the use of such organic growing practices has not only reintroduced a diversity of plant species to Pun Pun's previously barren 20-rai piece of land, but it's improved the soil quality too. The farm now produces roughly 400 vegetable varieties that feed the permanent community of 20 at almost zero cost. Because the farm produces a surplus of food and other products, it has launched two sister restaurants and a market in Chiang Mai where organic vegetables and hand-made soaps, nut butters and jams are sold. These venues have contributed to the creation of a larger market for other local, organic farmers to sell their wares.

> *"We can learn to do things ourselves, then tell our friends about it. My hope is to change things from the bottom up."*
>
> **Jon Jandai, Founder of Pun Pun Center for Self-Reliance**

Pun Pun is also a well-respected learning center, hosting three-day to two-week immersive courses in self-reliance and sustainable living, such as natural building, organic farming and making soap, all year round. Although initial interest in the learning center came mostly from the international community, today Thais make up roughly 80 percent of trainees. Many of them are employees from companies with CSR programs.

The overriding philosophy of Pun Pun is that change must start on a grassroots level with the individual. As Pun Pun's founder, Jon Jandai, explained, "We can only start with ourselves. We can't think about things on a policy level very much. We can learn to do things ourselves, then tell our friends about it. My hope is to change things from the bottom up."

Seeds are dried and stored for the seed bank.

The homes at Pun Pun are built by hand from mud bricks and natural materials.

▶**PIONEERS**

Panya Project

History: Started in 2004

Location: Mae Taeng district

Key features: Organic community and learning center

The term "permaculture" initially stood for "permanent agriculture" and focused on sustainable farming methods. It has since evolved to mean "permanent culture."

In 1978, the founders of permaculture, Australians Bill Mollison and David Holmgren, outlined their agricultural, economic and interpersonal principles. To make their ideas and methods of implementation accessible to more people, they devised a permaculture curriculum and certification program.

The three major tenets of permaculture are "care for the earth," "care for the people," and "return of surplus." The first two are clear enough while the third, also known as "fair share," encourages people to live within their means and not become overly greedy. A fairly new concept in Thailand, permaculture is taking root in the country through the efforts of a handful of small-scale endeavors and institutions. Its most famous and pure application is found in Chiang Mai's Panya Project. Located on a 10-acre mango grove in the Mae Taeng district, Panya's origins illustrate a major principle of permaculture: collaboration. In 2004, Panya was built next to Pun Pun, an established self-reliant community in Mae Taeng. The two communities, though serving different purposes, are committed to pooling their knowledge, manpower and resources to support each other. This cooperative spirit embodies "care for the people," one of the three core tenets of permaculture.

Since Panya Project's inception, the venture has grown from a private community to an internationally known education center. These days, Panya offers accredited courses year-round on "Introduction to Permaculture" and "Permaculture Design," which provides credit for permaculture accreditation. A certificate in permaculture can be earned after doing 72 hours of theory. From there, students can go on to earn a diploma through Gaia University in Colorado and other accredited schools. In 2017, Panya Project also began offering 6- and 12-month 'Permaculture Internships' to qualified candidates.

The environmental design principles taught in the "Permaculture Design" course abound throughout the Panya site. The residents produce their own methane to fuel their stoves, collect and filter their own water, cultivate microorganisms for use as liquid fertilizer, and create compost out of organic waste and human manure. The garden adheres to principles of integrated farming and has been contoured into terraces to promote even water distribution. Surrounding the community, the fruit groves have been carefully planned to create a "food forest." Moreover, the diversity and placement of fruit and nut trees is like a real forest, complete with a canopy, understory, brush and groundcover. Not surprisingly, chemicals and pesticides are never used on the property.

The community's goals are to attract more locals to participate in its program, and to ultimately expand its work, bringing permaculture solutions to Thai communities outside the property's food forest perimeter.

A fairly new concept in Thailand, permaculture is taking root in the country through the efforts of a handful of small-scale endeavors and institutions.

Panya Project volunteers return from the field.

Residents are essentially self-sufficient, producing their own fertilizer, fuel and more.

PRECISION FARMING AND THE SCIENCE OF SOIL

To get a good grounding in agriculture, you need to understand soil; it holds the potential to revolutionize farming.

By analyzing the nutrient content of soil all over Thailand, Prateep Verapattananirund believes that farmers can tailor their use of fertilizers to the soil's needs and implement more organic practices, cutting costs and boosting yields in the process.

This practice of fine-tuning and customizing agricultural equipment and chemicals to cut waste is called "precision farming," which is a move toward more natural ways of working the earth. Rather than over-fertilizing with unnecessary chemicals, farmers can analyze the nutrients inherent in their local soil to make the most of what nature offers.

However, the two biggest problems facing Thai farmers right now, Prateep said, are a lack of information about precision farming and a lack of access

The Department of Agricultural Extension promotes its tailor-made fertilizer project.

Prateep (standing) trains farmers on how to measure soil nutrients.

to proper technology to measure the soil's nutrients. That's why, in 2012, with support from the Kyuma Fund, he launched his soil clinic initiative in partnership with the Department of Agricultural Extension.

From the beginning, the goal was to establish a soil clinic station in every district across the country. The purpose of the clinics is to analyze and monitor the amount of essential nutrients in farmland, drastically cutting the amount of chemicals used and offering farmers strategies for the precision management of their crops. As of January 2017, there were more than 880 clinics nationwide, with some 17,640 members.

The actual process of operating a soil clinic is simple. Participating farmers simply dig soil samples from 15 different locations across their land to create a composite of the earth's quality. These samples are then analyzed with a test kit that measures the soil pH, and the amount of nitrogen, phosphorous and potassium (a com-

bination of essential nutrients known as "NPK") is measured. Based on the chemical and nutritional makeup of the soil, recommendations can be made on how much and what kind of fertilizer should be used.

According to Dr Prateep, on average, precision farming has lowered the use of chemicals by 30 to 50 percent and increased farmers' yields by 10 to 20 percent. Farmers have also seen their profit margins rise due to lowered production costs.

He hopes that farmers who participate in the clinics will train others how to assess their farmland and eventually launch their own training initiatives. This will facilitate the kind of knowledge transfers needed to foment an agrarian revolution on a grassroots level.

In the future, more advanced forms of technology, such as drones and soil sensors, may be deployed to show how such high-tech aids can produce even more high grade soil, he said.

INTEGRATED FARMING

Combining different kinds of farms into one is a winning prospect

Integrated farming is a holistic farm management system that incorporates multiple species of produce and livestock to ensure sustainability for both the land and the people who work it. This type of farm is designed as a complex ecosystem in which each plant, animal and type of land is both interdependent and functional.

It's a style of farming that comes naturally to Thais as, in one form or another, it has been practiced for centuries. While integrated farms make up just a fraction of all farms today, according to the National Economic and Social Development Board, integrated farming is still the most commonly used method of sustainable agriculture practiced in the kingdom.

However, despite the country's history in such practices, virtually all farms in Thailand became **monoculture** during the Green Revolution of the 1960s and 1970s. Although this increased productivity and made Thailand a major agricultural exporter, the landscape suffered severe side effects: soil erosion, deforestation, dwindling biodiversity, increased dependence on chemical pesticides and fertilizers, deepening debt, poverty and bad health for farmers, who suffered from the constant exposure to a barrage of chemicals.

Seeing these tragic effects, King Bhumibol Adulyadej outlined his New Theory of agriculture, a specific model of integrated farming. He advocated the return to sustainable, holistic farms that encouraged food security and self-sufficiency, as well as water and nutrient retention. Some of these ideas overlap with the more traditional practices of integrated farming and organic agriculture.

By cultivating a mixture of products, such as vegetables, fruits, rice, livestock and fish, integrated farming methods offer farmers greater opportunities for self-reliance. A variety of seeds and varietals ensures a good yield despite the unpredictability of weather

Monoculture:

Also known as "monocropping," monoculture is an agricultural practice that devotes vast tracts of land to the cultivation of just one crop. Monoculture often requires the use of pesticides and chemical fertilizers to keep pests at bay.

Farmers walk with their cattle inside Phu Pan Royal Development Cente

and market forces. Diverse crops keep nutrients in the soil and encourage a diversity of insects, which naturally help control pest populations and decrease crop vulnerability. Having a variety of products to sell also helps farmers earn a decent living.

On an earthier level, the coexistence of livestock with water sources, crops and trees returns both moisture and essential nutrients to the soil while enhancing its fecundity, which in turn helps to conserve plant and animal species. A potent example is the combination of agriculture and aquaculture (or rice and fish farming in colloquial terms), the most common type of integrated farming in Thailand. It's especially prevalent in the northeast, where this integrated system not only creates food security and increased profits for the household, but also contributes to better water management and soil use for the entire region, which is wracked by droughts and all sorts of other ecological woes.

Just the same, this kind of approach is not without its challenges. One drawback is that it tends to be more labor intensive than monocropping. Integrated

INTEGRATED FARMING IN THAILAND BY THE NUMBERS

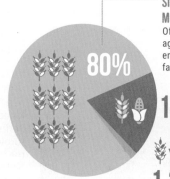

SINGLE-PRODUCT FARMING MONOCULTURE

Of all Thailand's 5.9 million agricultural holdings, 80% engage in single-product farming monoculture.

80%

18%

1.3%

TWO & THREE COMMODITIES

Of the other 5th, around **18%** have two commodities while a mere **1.3%** boast three or more different products.

46.7%

Nearly half **(46.7%)** of all argricultural holdings are in **THE NORTHEAST REGION** of Thailand, including the greatest number of multiple-product farms, with about **one quarter offering two products** and **just over 1% producing three or more commodities.**

33%

33% of all argricultural holdings producing freshwater fish are located in **THE NORTHEAST**, the highest total in the country.

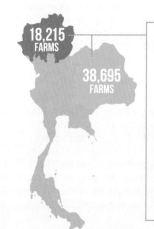

18,215 FARMS

38,695 FARMS

In total, 78,903 farms simultaneously cultivate crops, rear livestock and culture fish in fresh water. The majority of them (56,910) are located in the country's northeast (38,695) and in the north (18,215).

practices also require a broad range of skills, including raising pigs and poultry, crop farming, growing grass and aquatic plants, as well as farming fish. Given these requirements, calling on the skills, muscle and help of other farmers, or a local cooperative, may be necessary, especially during harvest time.

However, despite extensive studies and numerous ideas on the benefits of integrated agricultural practices, there is no single formula that will lead to a successful integrated farm. The differing locations and growing conditions of each farm require careful planning, and the selection of produce, livestock and their appropriate ratios often require some trial and error to get the balance right.

Integrated farming may be particularly relevant for underdeveloped areas, arid regions and small-scale and household farmers. With the potential to alleviate poverty and help rehabilitate the environment, these multi-faceted farms have a range of exciting implications and possibilities for a sustainable future.

Source: Thailand's 2013 Agricultural Census

A "NEW THEORY" TO ADDRESS OLD WOES

The king receives papayas at Pikun Thong Royal Development Center.

New Theory Stage One - Family-oriented self-sufficiency: During the initial stage the goal is for the family to have enough to live on, with good shelter and security. Thus the land is divided into four parcels in a ratio of 30:30:30:10. One 30-percent parcel is made into a pond to store rainwater for use during the dry season. Aquatic animals and plants such as fish, shrimp and morning glory may be raised in the pond for food. Another 30-percent parcel is for rice farming for consumption. The other 30 percent is for growing fruit, herbs and vegetables for consumption and, in case of a surplus, for selling at market. The remaining 10 percent of the land is for a residence, a corral for livestock, and other buildings. Significantly, this initial stage, which calls for ponds to store rainwater, can help solve chronic drought problems.

New Theory Stage Two - Community-based Agriculture: Farmers can then join hands and set up a cooperative to prepare soil, distribute seedlings and make an irrigation system among other activities. The community-based cooperative may market its farm products, find a rice-drying space, build a rice-storing barn or a rice mill. Moreover, the cooperative may provide loans for member farmers as well as scholarships for their children.

New Theory Stage Three - Lending Facilities: Farmers or cooperatives may find financial sources such as banks and private firms to provide funding to distribute or export goods outside the local community. Cooperation between farmers and lenders may potentially result in the sales of farm products, such as rice, at a fair price as opposed to an unjustly lowered price. Likewise, the cooperatives can also benefit by buying consumer products on a wholesale basis at lower prices from private traders.

In keeping with the site-specific tenets that formed the building blocks of many royally innitiated projects, King Bhumibol pointed out that this basic model could easily be modified to suit large or small holdings in regions where soil, water and crop conditions varied. However, he also noted that this model "was not easy to implement because the one who uses it must have perseverance and endurance."

The New Theory never really caught on with farmers in a major way, but its influence can still be felt in the many royal initiatives that continue to thrive. Many communities have recognized that monoculture is no longer an economically viable option. As a result, they have decided on their own to engage in a form of diversified farming in line with the New Theory method. Several such communities now serve as learning centers for others who want to acquire knowledge about integrated farming.

Concerned about the security and struggles of Thai farmers, especially in remote and drought-stricken areas, King Bhumibol Adulyadej formulated his "New Theory" farming system in the mid-1990s primarily to alleviate poverty. As such, the theory addressed the pressing issues of land and water management, crop diversification, and promoted self-sufficient farming methods. However, seen in a more contemporary context of food security during an era of climate change, the New Theory framework also offers a uniquely Thai form of sustainable agriculture with a local, cooperative and integrated approach. With its prudent, knowledge-based approach to development and sensitive, sensible allocation and use of resources, the New Theory bears many of the hallmarks of SEP thinking.

The idea came about almost serendipitously when King Bhumibol's Chaipattana Foundation began to develop land in Saraburi province. The initial plan was to build a community center where people could learn new farming techniques. King Bhumibol suggested a pond be dug to store water for the dry season. When the king visited the Wat Mongkol Chaipattana Area Development Project, he approved of the way the land had been divided: 30 percent was allocated to the pond, 30 percent to rice paddy, another 30 percent to mixed crops and 10 percent to other household uses. This formula became the basis of his New Theory.

For practical implementation of the New Theory model, King Bhumibol proposed three stages of development:

▶PIONEERS

Maha Yu Sunthornchai

History: Began developing his integrated farm in 1947

Location: Surin province

Key features: A very Thailand-specific model of integrated farming that has helped to alleviate poverty and improve soil fertility throughout the northeast

Maha Yu Sunthornchai was a humble farmer who hailed from Surin province in the northeast, or Isan, the largest and poorest part of the country. Agriculture is the mainstay of the economy and the main livelihood in an area afflicted with drought and poor soil.

Through his own observations and patient practice, Maha Yu developed a form of integrated farming specific to this region that restores soil fertility, provides plenty of foodstuffs and other items for personal use, and alleviates poverty. His farming model is now widely used in Surin and beyond.

In 1947, Maha Yu inherited seven hectares from his parents. At the time, Thai agriculture was transitioning from subsistence and small-scale farming to agribusiness by employing new technologies. It also developed an increased dependence on the use of chemicals. Maha Yu sought a way to make his farm more in line with Thailand's new goals of productivity and profit without stripping the region's environment of its scarce resources or sacrificing his financial independence as a farmer.

Until roughly 1960, Maha Yu practiced a form of mixed agriculture that he had inherited from his parents. The practice involved selecting the best seeds and varietals for the farm's conditions, ensuring good yields despite unpredictable weather, labor shortages and fluctuations in commodity prices. Yet he still cultivated these crops separately.

It was not until 1970 when he visited an integrated farm where fish, rice and pigs were raised together that he discovered that diversity for its own sake does not necessarily yield ideal results; rather, it's how these elements feed and play off each other that does the trick. While fish swam in the rice paddies and fertilized the soil, rice husks fed the pigs, and the pigs' manure provided the fish with food. Meanwhile, fishponds collected rainwater and maintained the moisture that was badly needed in this arid land. The end result was a self-reliant, high-nutrient, low-cost farm that recycled resources and did not rely on expensive chemicals.

Upon returning home, Maha Yu dug fishponds in his rice fields and began experimenting with the twinning of agriculture and aquaculture. According to

Geese share the balanced, integrated farmland.

him, a well-balanced integrated farm provides the farmer with every need if a one-rai plot of land incorporates eight essential commodities: rice, fish, pigs, poultry, vegetables, fruits, herbs and medical plants. This formula is beneficial in several ways. First, producing all of these commodities frees the farmer from having to purchase many consumer goods. Second, the farmer sells only surpluses, whatever he and his family cannot consume, giving the household greater autonomy over what is consumed and what is sold at the market. Third, the byproducts from the agricultural commodities are used to foster the farm's productivity, as well as to regenerate nutrients and resources. Finally, the farmers' dependence on human and animal labor and their independence from chemicals, machines and other external inputs may help them move out of the red and into the black.

Built on long-standing local and traditional knowledge, Maha Yu's model of integrated farming also embodies some of the tenets of King Bhumibol Adulyadej's New Theory and Sufficiency Economy Philosophy and his farm now acts as a learning center for other farmers interested in adopting these principles. In 2002 he founded a farmer's group called Local Wisdom of Isan. While Maha Yu died in 2007, judging by the fact that the region now has the highest concentration of integrated farms in the country, he has remained an inspiration for a whole new generation of farmers.

Thanks to the Maha Yu model, integrated farming is being put into practice in the northeast.

▶**PIONEERS**

Baan Huay Hin

History: Started in the early 1980s

Location: A village in Chachoengsao province

Key features: Innovative agro-forestry project encourages farmers to become self-sufficient by diversifying the crops grown on their land, giving them new sources of income

In the early 1980s, Wiboon Khemchalerm pioneered an innovative agro-forestry scheme in the central province of Chachoengsao, when he switched from growing a single crop, cassava, by diversifying into growing many different plants and trees on his land organically.

As the headman of Baan Huay Hin village, Wiboon encouraged the other villagers to join him. Diversifying in this way means farmers enjoy new sources of income and are less reliant on a single crop. Most of the villagers have adapted to the scheme and use it themselves.

Wiboon (pictured at right) also became an expert on medicinal herbs and advises others who would like to follow his lead, while his son Kanchit carries on the fieldwork. The family has around 10 rai of land and today grows around 700 to 800 different crops. These include vegetables for eating, flowers for selling, and herbs for use in products like shampoos, mosquito sprays and toothpaste.

The locals are not rich but they are no longer burdened by debt, Kanchit said, adding that they are not focused on money and care more about health and happiness. They also warmly welcome anyone who wants to visit the village and learn about the project.

Among these visitors are many young graduates who want to take the skills they learn back to their home villages, although they sometimes face resistance from their parents, who expect them to take the conventional route and get a job in an office rather than return to the farm. In the future, Kanchit hopes to expand the scheme and show more people the rewards of diversifying their crops and becoming more self-sufficient.

TIPS FOR INVESTING IN AN INTEGRATED FARM

As the Food and Agriculture Organization (FAO) of the UN points out, "It is useful to remember that most farmers in the world have little margin for taking risks." Thus, careful planning is required to successfully transition or invest in an integrated farm. Here are some tips from the FAO:

1. Invest in social relationships: Friends, neighbors and kinfolk can act as insurance against risk, helping in times of crisis. A social network can also provide shared resources, such as money, equipment and labor.

2. Assess available resources: Before purchasing anything, ask what resources are already available on the farm or in the community. Which resources are abundant, scarce, over- or underutilized? Having a firm grasp on the resources at hand on a particular farm will help your planning.

3. Create a farm budget: Add up all the prospective costs, subtract the potential income, then figure out the difference. This will give you a better idea of how the farm operates financially, and in which areas you can cut back or maximize your profits with other revenue streams.

4. Learn by example: Talk to other farmers and study successful farms as role models.

THE ABC'S OF CSA

Community-supported agriculture (CSA), a growing trend in sustainable agriculture and organic movements worldwide, is vastly under-utilized in Thailand. In CSA, consumers in a distribution area pay in advance for weekly deliveries of produce from local farmers. Each week, the farmers harvest their produce, pack it into boxes and deliver them to a pick-up point, or direct to the customer's door. Each box contains a week's supply worth of fresh, locally-sourced and often chemical-free produce.

In particular, integrated farmers can reap rewards from this system because it dovetails perfectly with the model of surplus sales. While integrated farmers may have trouble catering to an industrial system only interested in bulk sales to supermarkets and other big chains, CSAs sidestep mono-cropping requirements by serving a market and demand for seasonal, local produce. The farming family also continues to consume what they need at home, selling only what they cannot use.

Because CSA subscribers pay upfront, the risk for farmers, barring seasonal catastrophes and infrastructure breakdowns, is low. If a farmer were to sell produce exclusively at a farmers' market, there would be no guarantee that all his or her products would sell. But farmers who produce this kind of produce often sell at both markets and CSAs for added security.

With subscribers paying a premium for organic fruits and vegetables that are fresh and locally sourced, CSAs also offer farmers higher profits. According to a CSA operating in Chiang Mai province, a weekly box of produce costs 200 baht. With 50 participating consumer families purchasing from a network of five farming families, a month of CSA subscriptions earns each farming family roughly 8,000 baht. Assuming that they work 20 days per month, this figure exceeds the average monthly salary of a person earning the national minimum wage by 2,000 baht.

Meanwhile, the consumer is guaranteed a direct relationship with the producers. Everything they purchase is bound to be ethically sourced and environmentally sound. That makes it a win-win situation for both farmers and consumers.

CSAs in Thailand:

ADAMS ORGANIC BANGKOK CSA
Contact: 5/26-29 Saladeang Rd., Silom Bangkok
Telephone: 086-655-3078 and 087-905-3521
Website: http://www.adams-organic.com/adams-organic-csa
Facebook: https://www.facebook.com/adamsorganic/

CSA THAI
Facebook: https://www.facebook.com/people/Csa-Thai/100008197182739

▶**PIONEERS**

Mab Ueang Agri-nature Center

History: Founded in 2001

Location: Chonburi

Key features: To teach students natural farming techniques in the name of sustainability and self-reliance

Somsak Kruawan's Natural Farm

History: Changed over to natural farming in 2003

Location: Song Salueng village in Rayong province

Key features: Used natural farming techniques that inspired him to develop a form of fermented manure, which he then marketed to other farmers

Inspired by King Bhumibol Adulyadej's New Theory farming system and the pillars of the Sufficiency Economy Philosophy, Wiwat Salyakamtorn set up the Mab Ueang Agri-nature Center in the Ban Bueng district of Chonburi province to showcase natural farming techniques.

Wiwat was working for the Office of the Royal Development Projects Board when he closely followed the king on his visits to rural areas, where the monarch dispensed wisdom and schooled farmers in these ideas and techniques. After resigning from the civil service in 1997, Wiwat put these theories into practice in the little-known village of Mab Ueang.

Before opening the Mab Ueang Agri-nature Center in 2001, he experimented with natural farming to test it for himself. On an acre of land that had dry, alkaline soil, he planted a paddy field without the use of chemicals. The first crop was blighted by disease and plagued with pests. Undaunted by this

failure, Wiwat worked the bugs out of the system by using organic manure fermented from grass, herbs, molasses and animal excrement, which yielded pest-free, greenish husks in the second crop, and some two tons of rice.

At the Agri-nature Center, Wiwat and volunteer colleagues still give orientation classes of four to five days for those interested in learning about the ins and outs of this farming system. The students are given free food and lodging during the rigorous program that starts at 5am each morning and finishes around 10pm at night.

One of his students was Somsak Kruawan, the headman of Rayong province's Song Salueng village, who was desperate to dig his way out of debt. Over seven years, he had racked up debts of around US$30,000 dollars largely due to the price of insecticides, herbicides and fertilizers he bought to cultivate several acres of orchards.

During his first few years of chemical-dependent farming, Somsak grew durians that developed severe infestations of fungi and insects. To eradicate these pests, he doused the trees with insecticides and herbicides. Yet the profits from the sale of those fruits did not make up for the cost of the chemicals.

After that, on the same parcel of land, the farmer tried to grow rubber trees, but the saplings became riddled with fungi. In desperation he turned to tangerines only to find the orchard choked with weeds. The herbicides he used on them not only killed the weeds but the tangerine trees too.

After studying with Wiwat in 2003, he decided to ditch chemicals in favor of naturally fermented manure made from animal feces, fallen leaves, branches and moldering plants easy to find on his farm. As a way to increase his supply of this natural fertilizer, he raised pigs and chickens to collect their droppings. The fermented manure not only protected the plants from fungi but also nourished the soil which made the trees bloom.

As the years went by, his neighbors looked on with envy at his fruitful farm. Naturally they wanted to know his trade secrets. By then he had stockpiled 400 tons of manure. Somsak put the manure up for sale in and around Rayong at US$55 per ton. Eventually, he found himself making more money from selling manure than fruit. The combined profits allowed him to wriggle out of debt and he now makes up to US$2,000 dollars per month just from his natural fertilizer.

Somsak is but one of the Mab Ueang Agri-nature Center's successful trainees. On average some 3,000 trainees come every year to the center, which has expanded to include 70 other such branches across the country with many more planned to open in the future.

Wiwat Salyakamtorn.

Maitree Sakuna, who follows the king's New Theory
farming model in Nakhon Si Thammarat province.

REFORESTATION

The battle to reforest the country is taking place on many fronts

Monks walking through the rainforest in Doi Inthanon National Park.

As Oxford University researchers put it, forests are one of the "most promising technologies" we have for combating climate change and promoting sustainable development. For one thing, forests are carbon sinks. Using energy from the photons of the sun's light, and combining it with carbon dioxide – the primary greenhouse gas emitted through human activities – forests create carbon and store it in their wood, leaves, roots and the soil surrounding them. That's how carbon dioxide "sinks" into forests, ensuring that it doesn't escape into the atmosphere where it becomes the main agent of global warming. The more degraded forest areas that are restored the more carbon dioxide is removed from the atmosphere and the less the planet warms.

According to Oxford University, forests are one of the "most promising technologies" we have for combating climate change.

But these jungles and woodlands also provide humankind with a host of other benefits, from providing different foods and medicines to acting as safeguards against soil erosion and natural disasters such as floods. In northern Thailand, for example, where most of the kingdom's remaining forests are situated, they act as key watersheds, playing an elemental role in supplying water to the intensively cultivated and densely populated lowlands.

In this region, the Huai Hong Khrai Royal Development Study Center has been a paragon among reforestation projects. The mountainous areas located upstream were reforested with trees to help the land retain moisture and prevent water runoff and soil erosion. Then more trees were planted in the degraded forest. Downstream, eight large reservoirs that hold approximately 3.3 million cubic meters of rainwater were constructed. In addition, smaller reservoirs, check dams and smaller channels were constructed throughout the area to distribute moisture and to rejuvenate the forests.

Through this and other royally initiated projects, King Bhumibol was a catalyst for reforestation efforts. One of his strategies, dubbed "three forests, four benefits," recommends planting diverse forests over monoculture plantations. In a royal address to villagers living near Khao Yai National Park in 1977, he requested that they "grow fruit trees, trees that yield firewood and bamboo in the compound of the reservoir, and grow trees especially along the mountain crest near the watershed areas, to secure the soil and conserve moisture."

The emphasis that he placed on such diversity flew in the face of conventional (and often misguided) reforestation initiatives in Thailand, which had begun earlier on in the 1970s. For years, government officials had focused on single cash crops, like eucalyptus, oil palms or rubber trees. Although some experts argue that such reforestation methods are positive as they provide locals with a livelihood and stop primary forests from being converted into fields for agriculture, other ecologists believe that such plans are shortsighted because they do not focus on restoring forests to their true biodiversity. In fact, single-crop plantations actually have negative impacts on ecosystems and biodiversity.

A GROWING MOVEMENT

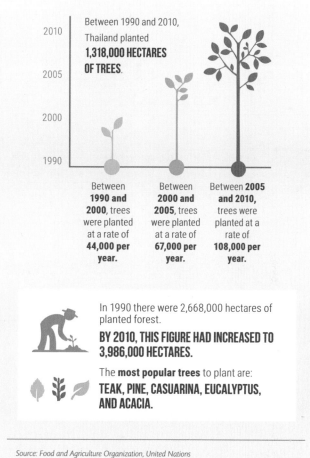

Between 1990 and 2010, Thailand planted **1,318,000 HECTARES OF TREES.**

Between **1990 and 2000**, trees were planted at a rate of **44,000 per year.**	Between **2000 and 2005**, trees were planted at a rate of **67,000 per year.**	Between **2005 and 2010,** trees were planted at a rate of **108,000 per year.**

In 1990 there were 2,668,000 hectares of planted forest. **BY 2010, THIS FIGURE HAD INCREASED TO 3,986,000 HECTARES.**

The **most popular trees** to plant are: **TEAK, PINE, CASUARINA, EUCALYPTUS, AND ACACIA.**

Source: Food and Agriculture Organization, United Nations

To celebrate King Bhumibol Adulyadej's Golden Jubilee in 1996, a plan was implemented to reforest more than 8,000 square kilometers of denuded land nationwide with diverse forests. Since then, many such efforts have favored this approach over monoculture plantations, which are neither able to function as effective carbon sinks nor provide the full spectrum of benefits that diverse forests can. In the long run, local communities also benefit from the creation of a more sustainable source of water and foodstuffs, improved irrigation and soil, and thus better human security overall.

Revealing their foresight, King Bhumibol and Queen Sirikit's efforts became global concerns over the last few decades. At the UN's summit on climate change in New York in 2014, governments and multinational companies pledged to restore hundreds of millions of hectares of formerly forested land. All in all, the declaration promised to restore 150 million hectares of degraded landscapes and forest areas by 2020, with an additional 200 million to be reforested by 2030. If that promise is realized, the world would regain an area of forests greater than the size of India. Pulling off such a feat is a monumental undertaking with many positive ramifications.

In Thailand, however, for the most part the reforestation movement is fragmented. Much of it is done for public relations purposes: small-scale tree planting programs that businesses hype as corporate social responsibility (CSR) programs.

In moving forward, three questions have to be considered. Who will pay for the kind of large-scale reforestation programs needed? How will such programs be realized in remote areas with primitive infrastructure in place? Thirdly, how will local communities be affected by these plans? Whether the government treads on human rights by removing the rural poor from their land, or instead includes the local people in managing and implementing these programs will define the government's reforestation record in the years to come.

▶ **PIONEERS**

Khao Paeng Ma Reforestation Project

History: Began in 1996 to honor King Bhumibol Adulyadej's Golden Jubilee

Location: The Khao Paeng Ma protected forest reserve in Nakhon Ratchasima province, near Khao Yai National Park

Key features: Some 5,000 rai of forest have been restored, allowing the return of diverse animal species, including more than 200 increasingly rare gaur

Gaur family on a hill at Khao Phaeng Ma.

Chokedee Poralokanon.

Following the Golden Jubilee, when the Royal Forestry Department (RFD) mobilized reforestation efforts across the country, it earmarked a 5,000-rai plot of degraded land in a protected forest reserve at Khao Paeng Ma in Nakhon Ratchasima, near Khao Yai National Park, for regeneration.

The RFD handed over the reins to Chokedee Poralokanon, more commonly and affectionately referred to as Uncle Choke, who, along with Oratai Jotklang and Nikom Putta, spent 13 years transforming a denuded mountain into an oasis of greenery.

When discussing the project, Choke took great pains to explain all the hard work the locals put in to ensure its success. For six years prior to the project's inception, he worked with villagers, "changing their minds," as he put it, about not encroaching on forest reserves to instead reforest their own land to provide them with an array of benefits. Though it took some convincing at first, eventually they bestowed on him all their "local wisdom" about which local trees and plants would be integral to the project's success.

Choke first planted a diverse forest on his own land, which in turn attracted various birds and beasts, and allowed him to harvest many different foods and herbal medicines. His emphasis was on local species of plants and trees. "We have to believe in nature," he said, explaining why he planted 30 local tree species and more than 300 local plant species on the 5,000 rai of degraded land. "If you plant only one tree species, there will be nothing else to grow in its place if it gets attacked by a disease. In nature, there are many different species working together. If one of the tree species I planted had died out, there would still have been 29 others. The local trees were here already, so we didn't need to bring other species here. What we already had was enough."

The initial phase of the project, doing the research and winning over the villagers, took six years. The actual reforestation work took a little more than twice that. The project was finally completed in 2007.

Although the forest does not belong to the villagers, Choke believes they have gained valuable wisdom that they can plow back into their own land. Perhaps the most incredible part of the project is that it has brought gaur, also known as Indian bison, back to the land. The largest and mightiest of all bovines, more than 200 of these brawny creatures roam the area. In addition to providing ecological benefits such as seed dispersal and fertilizing the land with their droppings, the gaur have also put the area on the map for eco-tourists.

But these positive effects have not just stayed in the community. Community leaders from other villages have come to learn about the project, as well as other organizations working on reforestation, PhD students from Mahidol University, and more than 1,000 young learners from a network of 40 primary and secondary schools in Thailand's northeast.

These days, Choke focuses on educating the visitors and enjoying the natural splendor. "What I got from bringing back a healthy ecosystem to this land is happiness," he said. "I enjoy a better life here every day."

▶ **PIONEERS**

Plant Banana Trees to Save the World

History: Started by Nikom Putta in 2005

Location: Doi Luang Chiang Dao in the Chiang Dao district of Chiang Mai province

Key features: Dozens of forest-dependent communities near Doi Luang Chiang Dao are committed to growing banana trees, which help reforestation efforts and protect watershed forests among other benefits

Nikom Putta began studying forests while working as a volunteer in Khao Yai National Park. Although he was a political science major at Ramkhamhaeng University then, he had always been wild at heart and always yearned for greener pastures.

When he was helping out Uncle Choke on the Khao Paeng Ma Reforestation Project, he noticed that banana trees grew naturally around the base of the area's mountains, where their thick canopy and ability to retain moisture served as a barrier against forest fires.

"Because the banana trees are heavy with moisture, the fire only burns the grasses around them. Then it fizzles out before it can reach the forest," he said.

Around this time, Nikom discovered that for reforestation efforts to work, moisture must be returned to these dry and barren areas. That made him think of banana trees again. Because he did not study forestry, he said, it enabled him to think outside the box and "learn from nature," as his mentor Uncle Choke said.

He also gleaned a lot of vital information from traditional Thai beliefs. "In Thailand we believe that if there are bananas, we can survive," said Nikom. "Bananas can be used for cooking. The leaves can be used as plates or to wrap up different foods. Even the flower inside can be used for food or given to pigs."

Upon moving back to his childhood home of Chiang Dao 10 years ago, he founded the project Plant Banana Trees to Save the World. After kicking off the project with an event for some 300 young people, he started spreading the word about all the beneficial aspects of these trees that can also protect rivers from soil erosion.

Part of the project entailed Nikom planting banana tree plots on his own property as demonstration sites, which became the centerpiece of his meetings and discussions with villagers from Doi Luang Chiang Dao. Most of the local hill tribes were planting corn, which tends to dry out the land. Nikom advised them to plant banana trees to keep the soil moist and fertile.

His work has helped to maintain freshwater sources in the area and as far away as Bangkok. By maintaining the forest watersheds of the northern highlands, the project has been beneficial for the Ping River, a tributary of the Chao Phraya River that connects Chiang Mai with Bangkok. Without those watersheds in the north, Bangkok would not have access to so much clean water.

Nowadays, dozens of villages in and around Doi Luang Chiang Dao have planted plots of banana trees. The seeds

> *"Bananas can be used for cooking. The leaves can be used as plates or to wrap up different foods."*
>
> **Nikom Putta, founder of Plant Banana Trees to Save the World**

of these trees have now been dispersed all over the area by wild boars, bears, squirrels, birds and bats, so the project has borne even more fruit.

Nikom continues to spread his knowledge across the kingdom, sometimes in unusual ways, like spending three months walking the length of the Ping River, stopping along the way to teach people about the benefits of planting his favorite multi-purpose tree.

Doi Luang Chiang Dao, Chiang Mai.

Nikom Putta's project up north helps Bangkok have enough clean drinking water.

▶ **PIONEERS**

Chiang Mai University's Forest Restoration Research Unit (FORRU-CMU)

History: In 1997, FORRU-CMU partnered with the Hmong hill tribe villagers of Ban Mae Sa Mai to restore their degraded plots of forest land

Location: Ban Mae Sa Mai village in Doi Suthep-Pui National Park in Chiang Mai province

Key features: With the organization's help the villagers have restored their degraded land and continue to maintain a nursery for reforestation efforts

FORRU-CMU's work in Ban Mae Sa Mai in Chiang Mai led to this incredible transformation.

When early attempts to reforest Thailand began in earnest in the 1970s and 1980s, they floundered for several reasons. One of the most significant was that proper research had not been done on the thousands of tree species that make up the country's varying ecosystems. In 1994, Chiang Mai University sought to remedy the problem by establishing the Forest Restoration Research Unit (FORRU-CMU) to study native tree species and develop techniques that could be used to make degraded land more pristine.

Since 1997 it has worked with the villagers of Ban Mae Sa Mai in Doi Suthep-Pui National Park, the largest Hmong hill tribe village in northern Thailand, to develop an experimental nursery and trial plots. These villagers had worked hard to show the authorities that they could be capable environmental stewards, so they were allowed to remain living in the now protected forest area.

In partnering with these villagers, FORRU-CMU learned a lot about local ways of living off the land and in harmony with nature. The villagers helped identify which trees were able to colonize the abandoned areas and how to attract seed-dispersing animals. In return, the villagers gained plenty of technical expertise.

At the villagers' request, FORRU-CMU helped them build a tree nursery and trained them on how to manage it. The nursery, currently sponsored by the Rajapruek Institute Foundation, now produces more than 20,000 trees per year. In the end, the trial plots were successful. After planting 20-30 different tree species, the trees survived and grew well, creating a multi-layered canopy within six and a half years. The development of structural complexity attracted seed-dispersing animals, particularly when the planted trees began to fruit. This resulted in the seedlings of 73 other (non-planted) tree species recolonizing the planted sites mostly from seeds brought in by birds and mammals.

When such recruit species are added to those planted, the total comes to more than 100 tree species growing in the restored sites above Ban Mae Sa Mai, creating lush forest and effectively repairing a degraded watershed, which now supplies the villagers with irrigation water vital for their crop production lower down in the valley. Meanwhile, the once degraded forest has become far more bountiful and is now rich in bamboo, flowers, vegetables, fruits, mushrooms, and more.

FORRU-CMU credits the partnership with the villagers, and the implementation of a model relying on the diversity of tree and plant species. Using the site as a template, the research unit has expanded to Krabi, as well as China, Cambodia and Indonesia.

GROUNDBREAKERS

*One of Thailand's leading forest ecologists, **STEPHEN ELLIOTT**, co-founder and research director of Chiang Mai University's Forest Restoration Unit (FORRU), talks about the present and the future of forest restoration in Thailand, which he defines as "re-establishing the original forest ecosystem that was present before deforestation occurred."*

Forest restoration, by definition, promotes more diverse forests than conventional reforestation, such as monoculture plantations. Why is promoting diversity in forests important? The beauty of the economics of restoring forest biodiversity is that it creates highly adaptable forest ecosystems. The objectives of a diverse forest are more encompassing: recreating original wildlife habitat, preserving biodiversity, mitigating climate change, eco-tourism, producing a vast range of forest products, watershed conservation and so on. The idea is to create a self-sustainable ecosystem because it doesn't have any management costs after the establishment costs have been paid. You can't get more sustainable than that.

Is species diversity also important in terms of mitigating climate change? A diverse forest is always going to be much better at sucking up carbon than a monoculture forest. If you've got a monoculture, the trees are all the same age, which means they're all the same size, which means they've all got exactly the same habitat requirements. They're all fighting with each other for the same resources at the same levels. But if you've got a forest ecosystem where every tree is sitting next to one of a different species, and you've got trees with their crowns and roots at all different levels, then the light and CO_2 not taken up by one tree, will probably be grabbed by its neighbor. If you have a monoculture of trees all of the same size, then if light or CO_2 gets through the upper canopy, there's nothing to catch the leftovers.

Do you see forest restoration efforts moving forward in Thailand? There's been no let up in the number of organizations that contact us each year. The problem is that people, when they want to plant trees, want to do so instantly. A sponsor comes along and wants an instant restoration project: they'll just grab any old trees they can get their hands on, and often those trees die within a year, because the wrong species were selected and wrong techniques used. But the mindset is there. Attitudes have completely changed. Twenty years ago, it was marching up the mountain with eucalyptus seedlings and planting them in rows.

Given the UN plan, how will this play out on a global level? I think the problem now is the sheer scale on which forest restoration must be done, if we are to mitigate climate change or the biodiversity crisis. The UN is calling for 350 million hectares to be restored by 2030. But when you think about it, all the flat ground next to roads is already used for agriculture. So most of the sites available for restoration are not going to be sitting comfortably next to a car park. They're going to be miles and miles away. So it all hinges on the availability of labor and the sheer physical difficulty required of actually getting enough people to walk up steep slopes, across rough terrain, usually miles away from proper road access.

▶**PIONEERS**

Huai Hong Khrai Royal Development Study Center

History: Founded in 1982

Location: Doi Saket in Chiang Mai province

Key features: Water management, reforestation, soil rehabilitation and wildlife conservation

Chiang Mai province is situated in a rain-fed watershed in a resource rich part of the country that has suffered rampant deforestation due to illegal logging, monocropping and forest fires. As a result, large tracts of once-arable land have become barren. Recognizing the urgent need for reforestation, water management, soil rehabilitation, wildlife conservation, and agricultural development, King Bhumibol Adulyadej founded the Huai Hong Khrai Royal Development Study Center on an 8,500-rai plot of denuded land in the Khun Mae Kuang National Forest Reserve.

To restore fertility and moisture to the land and to help villages in the vicinity to enjoy bumper harvests again, the project leaders adopted an integrated watershed management model of "upstream forestry, downstream fishery, and in-between agriculture." The mountainous areas located upstream were reforested with trees to help the land retain moisture and prevent water runoff and soil erosion. Then more trees were planted in the degraded forest. The trees served three different purposes, as timber, as fruit and firewood, while also propagating biodiversity and providing local communities with valuable resources.

Downstream, eight large reservoirs that hold approximately 3.3 million cubic meters of rainwater were constructed. In addition, smaller reservoirs, check dams, and smaller channels were constructed throughout the area to distribute moisture and to put a damper on forest fires. These reservoirs and streams not only contributed to the water management system, but also provided a space to conduct studies on fish cultivation and harvesting that continue today. Meanwhile, the space at the foot of the mountain where groundwater collects before it meets the

reservoirs and fisheries became fertile agricultural spaces. In these areas, the center continues to carry out studies in integrated farming, combining livestock rearing, crop farming, and aquaculture such as frog and fish cultivation.

Since the center's founding in 1982, studies have shown the merits of these programs. Mixed deciduous forest, which once only covered 16 percent of the study area, now covers 54 percent. The number of plant species has increased from 35 to 104. The density of trees has gone up from 100 trees per rai to anywhere between 242 to 663 trees per rai. All the trees have provided more crown cover that has increased humidity and rainfall in the area while lowering the rate of evaporation. Since 1995 there have been no forest fires in the area.

Today, the Huai Hong Khrai Royal Development Study Center continues its research but also focuses on spreading these positive effects to neighboring communities. By doing so, the center serves as a living model and learning space for the public, providing agricultural and land management services free of charge. It has spawned 112 "model farmers" who serve as volunteer teachers, and four community rice centers. It also now receives as many as 70,000 visitors per year.

The center serves as an active showcase of how integrated land and water management, which is sensitive to its environs, can benefit an entire region environmentally, agriculturally and educationally.

Reservoir built downstream to conduct studies on fish cultivation at Huai Hong Khrai.

Today, the Huai Hong Khrai Royal Development Study Center continues its research but also focuses on spreading these positive effects to neighboring communities. By doing so, the center serves as a living model and learning space for the public.

Ideas and Inspirations for Tree Planting

- Planting in June, in the early part of the rainy season, is essential for all the country's regions.

- Plant 20–30 tree species. Diverse forests produce more benefits and are more sustainable.

- Plant local trees rather than imported trees. There's a reason they grew there and not here.

- Choose a good location for tree nurseries. It must be flat with good drainage, sheltered and partially shaded, close to a supply of clean water with no risk of flooding, and near clean soil.

- Plant species that will attract animals. They will help disperse seeds and reduce costs.

- Have a team monitor possible forest fires in the hot and dry season (February to April).

- Livestock can hinder reforestation by grazing on young trees. Make sure they remain in agricultural areas.

- Use fertilizer; it increases survival rates and accelerates the growth of trees.

- To build a tree nursery, make a shaded area for germination and seedling growth, a work area for seed preparation, a lockable store for materials and tools, and a fence to keep out stray animals.

- Sow only the highest-quality seeds available. Sowing low-quality seeds could encourage the spread of diseases and is a waste of resources.

- Weed control is essential. Don't simply slash them as they will re-sprout. Dig out the roots. Do not use fire to clear your plots – this kills off soil nutrients.

EYE-OPENERS

MAN OF THE FOREST AND THE PEOPLE
FOREST MAN

Director: *William Douglas McMaster*
Release Date: *2013*
Available on: *YouTube, National Geographic Website*

Forest Man is a short film from 2013 that documents the huge impact one man has had on an island in India.

Just north of the town of Jorhat lies one of the largest rivers in India: the Brahmaputra. Millions of people live along its shores. Every year during the monsoon, the river floods everything, destroying homes and farms. In the process, it erodes hundreds of square kilometers of land. One of the worst affected areas is Majuli Island: the world's largest river island is home to over 150,000 people. Since 1917, the island has lost over half of its landmass to erosion – and the rate of erosion has accelerated. Scientists believe that in 15 to 20 years the island could be completely gone.

But since 1979, one Majuli islander, Jadav Payeng, has been planting trees to save the island. In almost four decades, it is estimated that he has planted 550 hectares of forest, transforming what was once a barren wasteland into dense forest. Elephants, deer, rhinos and tigers have since returned to the land.

As Dr Arup Kumar Sarma of the Indian Institute of Technology Guwahati said, "Payeng has shown the example that if one person can, through his own effort, do this kind of plantation, then why not others? I hope that through reforestation, we can solve the problems of flood and erosion to a great extent." Payeng certainly sets an inspiring example.

FOREST CONSERVATION

Alliances between communities and authorities preserve nature

In Thailand, over one million people live in and depend on the country's forests. These people also play a major role in preserving them. Not only do forest-dependent villagers know wooded terrains the best, they also depend on forests for their livelihoods, making their protection a top priority.

While the Thai government recognizes the role villagers can play in forest preservation, it has not gone far enough in empowering them to do so. Neither the **National Park Act** nor the National Forest Reserve Act of the 1960s, which set aside protected areas and suddenly turned those who had lived on the land for generations into trespassers, recognized the rights of locals to participate in the decision-making process about how to manage these areas. In a speech to lawyers in 1973, King Bhumibol Adulyadej acknowledged the problem and stated: "They [the villagers] have human rights. It's a case of the government violating the people, not the people violating the law."

It is no wonder, then, that this state-managed approach did nothing to stop rampant deforestation, as the amount of forested area decreased from 60

Koompassia Excelsa Taub, *a giant tree in a community forest in the northe*

percent in 1961 to 25 percent in 1998. Since then, when the government realized the importance of involving local communities, the latter figure rose to around 28 to 30 percent.

By no means is this solely a local phenomenon. In 2014, the World Resources Institute released a report providing evidence that deforestation has decreased considerably in places where local peoples' community forestry rights are given strong legal recognition as compared to those where they did not enjoy such rights. In the Bolivian part of the Amazon, deforestation rates were found to be six times lower; in the Brazilian Amazon, they were 11 times lower; and in Guatemala's Petén Basin, 20 times lower.

Across Thailand there are now more than 9,000 officially-registered "community forests," where local residents have been empowered to manage these important resources, sometimes collaborating with government agencies, civil society groups and even Buddhist monks. Experts believe that many of these projects have made a significant contribution to the maintenance of healthy forests, although the effect on local incomes and poverty reduction has not been as great as hoped thus far.

National Park Act:

A law that enabled the government to designate public land as national parks for the benefit of public education and pleasure.

A grasshopper at Kanchanaburi's Pu Toey-Pu Lad community forest.

Positive examples abound. One such project is the Joint Management of Protected Areas (JoMPA) based in one of the country's most biologically rich areas, the Western Forest Complex, where the Seub Nakhasathien Foundation has served as a go-between for villagers and government officials to join forces. By equipping both sides with GPS systems to survey the land, the project has proven that combining modern technology with local wisdom is a win-win situation.

Meanwhile, the "Tree Bank," launched in 2006, has encouraged farmers to plant trees that they can use as collateral for low-interest loans. And from small beginnings in northern Phayao province in the late 1980s, the concept of "ordaining" trees in a Buddhist ceremony that entails wrapping them with sacred sashes as a way of protecting them from loggers, has spread across Thailand and to neighboring countries. The many royal projects to build check dams across the country have also become watersheds in forest conservation.

All in all, the battle to preserve Thailand's forests for future generations will not be easy, but as these projects show, it's far from a lost cause.

Keeping it in the Community

In recent decades, numerous community forest projects aimed at empowering villagers to protect the environment and at reducing poverty have been launched. There are now some 9,177 community forests across the country registered with the Royal Forestry Department (RFD).

"Community forestry is an alternative approach to getting people involved in sustainable forest management for supporting local livelihoods," said Ronnakorn Triraganon of the Regional Community Forestry Training Center for Asia and the Pacific (RECOFTC), an NGO headquartered in Bangkok with other offices in Southeast Asia and Nepal. "If they have the benefits, that provides the motivation for them to take care of the forest."

Giving local communities access to forest products, such as bamboo, wild fruits, vegetables, and honey, which allows them to move from subsistence-level farming to small-scale enterprises, can reap substantial rewards. Some communities earn up to two million baht a year from selling such products, said Ronnakorn.

But in general, community forests tend to have a greater effect on conservation than on local economies. That's why RFD aims to register around 20,000 community forests, though there is still debate about how to include people living in protected forest areas. Currently, there is no recognition for community forests that overlap with these zones, affecting between one and two million people who depend on forest resources. By law it is illegal to harvest forest products from protected areas, said Ronnakorn, though local agreements are often made between park officials and local representatives through advisory bodies set up in each national park.

In this growing field there have been some noteworthy success stories, like the village of Ban Pred Nai, which is working hard to restore a mangrove forest in Trat province where many villagers harvest crabs. Another example is Ban Mae Kampong in Chiang Mai province, which has become a top ecotourism destination and campsite. Such grassroots movements are quickly proving how small, community-based conservation efforts can have a big impact.

▶PIONEERS

The Joint Management of Protected Areas (JoMPA)

History: Began in 2004

Location: The Western Forest Complex (WEF-COM), which refers to the largest forest reserve in Thailand, covers 18,000 square kilometers in six provinces: Kanchanaburi, Suphanburi, Kamphaeng Phet, Uthai Thani, Tak and Nakhon Sawan.

Key features: Reduces ongoing conflicts between local residents of protected forest areas and government officials through mediation

In 1999, the Department of National Parks, Wildlife and Plant Conservation (DNP) initiated a project called Joint Management of Protected Areas (JoMPA) to help reduce conflicts between forest-dependent people and government officials in western Thailand. To carry out the mission and lay the groundwork they approached the Seub Nakhasathien Foundation.

Before the endeavor got off the ground in 2004, tensions were running high between the local residents of forest reserves and the government officials managing the protected areas, said Sasin Chalermlarp, the secretary-general of the Seub Nakhasathien Foundation. "They didn't even speak to one another. The government would come to arrest the villagers who were using land for agriculture, and the villagers would fight back by blocking them from leaving the area. Sometimes they even threw rocks at the forest rangers and authorities when driving past their offices."

In some ways, it's hard to blame the villagers for retaliating in the face of what they perceive as grave injustices. Many of the estimated one million people who depend on forests for their livelihoods in Thailand live in jungles and woodlands where their forebears had free rein until these areas became forest reserves under the National Forest Reserve and National Park Acts of the early 1960s. Land that had been theirs for hundreds of years suddenly did not belong to them anymore. And so they lived hand to mouth, forever dodging the authorities, with no security and little peace of mind.

The Seub Nakhasathien Foundation's project was centered on bringing the two warring sides together. To develop relationships and build trust between them, Sasin and his team spent many hours working with the villagers and the government officials separately, eating with them, spending time with them, and just getting to know them.

What the foundation did after winning the trust of both sides was to give them access to GPS systems that enabled the local residents and the government to work together on surveying the land. Together they helped each other determine which lands could be used for agricultural purposes and which lands would have to remain protected forest. This was revolutionary, for rarely had the authorities and locals teamed up to protect the forests; the approach was quite unlike the usual top-down decision-making process.

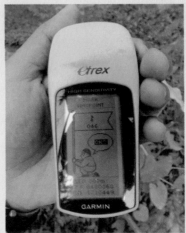
GPS equipment is crucial for local residents and forest officials to work together on surveys.

The cooperation between these former rivals led to the development of a color-coded map showing the areas of reserved forest, agricultural land and other zones. White dots depicted the communities in protected areas while green dots showed uninhabited forest. The collaboration also resulted in the creation of a so-called "Handbook for Local Participation and Conflict Management of the Western Forest Complex," which laid down guidelines for conflict resolution in protected forest areas. It now serves as a blueprint for other forests.

Local residents and officials survey the forest along the buffer zone of the Western Forest Complex.

As Sasin explained, the project has engendered a new sense of trust and amity. "In the past seven years, there hasn't been a single arrest in the area," he said. "Now the villagers and government officials are friends."

Such amity has allowed the Seub Nakhasathien Foundation to focus on its post-JoMPA project to enhance occupational skills among local residents. The goal is to enable villagers, particularly breadwinners, to earn income for their families without having to encroach on protected land to grow commercial crops. One of the crops that has been successful thus far is cumin production, which grows naturally in the local forests.

Households send the cumin grown in the communities to Chaophraya Abhaibhubejhr Hospital, Thailand's first hospital introducing herbal products in modern forms to both domestic and foreign markets. Cumin from the area accounts for up to one-third of the total resources used in the herbal products branded under the hospital's name and manufactured according to global organic standards. "The goal is to have a few households be really successful in what they're doing," Sasin says. "Then others will naturally follow."

"The government would come to arrest the villagers using land for agriculture, and the villagers would fight back by blocking them from leaving the area. Sometimes they even threw rocks at the forest rangers..."

Sasin Chalermlarp, secretary-general of the Seub Nakhasathien Foundation

Sasin Chalermlarp.

Growing cumin is an additional source of income for local residents living in the forest.

The Ecological Martyr

Referred to by many as Thailand's greatest conservationist, Seub Nakhasathien was a champion of both the environment and wildlife. He was also a scholar and activist, who protested against the building of dams, campaigned for the rights of local people to live in forests, and added his voice to the rallying cries to put a stop to poaching and to protect the environment better.

While working as the superintendent of the Huai Kha Khaeng Wildlife Sanctuary in Uthai Thani province (a position that he took rather than accepting an opportunity to pursue a PhD in England because he believed this would better position himself to protect the environment in his homeland), Seub grew disenchanted with the rampant destruction of nature and the ceaseless slaughter of wildlife.

On September 1, 1990, at age 40, Seub took his own life in a stand against the plundering of nature, famously writing in his suicide note, "If any animal will be killed in Huai Kha Kaeng, it must be me." Only 17 days after his death, Seub's friends and colleagues received permission from the Ministry of the Interior to establish an organization in his name to carry out his work – the influential Seub Nakhasathien Foundation – that carries on to this day.

▶PIONEERS

The Mae Wong Network

History: Founded in 1996

Location: Lat Yao district of Nakhon Sawan province

Key features: Reforesting 5,000 rai of degraded forest, boosting livelihoods, stopping forest fires and providing a lair for tigers.

Narong Raengkasikorn, the chairman of the Mae Wong Network, has no background in environmental science. Instead, like his good friend, Nikom Putta, instigator of the Banana Trees to Save the World Project, he said, "I just learned from nature."

Narong first became interested in forests as a teacher at a local school in the Lat Yao district of Nakhon Sawan province, where he taught for 30 years. The schoolgrounds were situated on barren land, so he began growing trees there with the help of his students. As the years went by, he thought about the community at large. "Many villagers would cut down trees. They wanted to expand their farms and crops and were not educated about the benefits of the forest," he said.

When the government came to Narong in 1996 with a plan to plant trees in the Mae Wong forest and to keep the nearby national parks of Mae Wong and Khlong Lan as green as possible, Narong was a natural choice to head the project and function as an intermediary between the government and the local people. That year, he formed the Mae Wong Network, which consisted of representatives from 18 villages who would help get the project up and running, as well as a team of 22 that formed a think tank. The latter group also served as educators for locals and consultants for the government on decisions affecting the project.

The Mae Wong Network began with the planting of 5,000 rai of new trees on degraded land in the Mae Wong Forest, where the 18 villages were located. Three years later, the network expanded its work to the nearby Mae Wong and Khlong Lan National Parks, where they worked with teams of locals representing 22 different communities on the edges of the parks from districts in Nakhon Sawan and Kamphaeng Phet. They collaborated with government officials to preserve the forests in those parks.

One of the network's main functions in the national parks is preventing forest fires. As Narong said, "We haven't had a forest fire in 10 years," he said, attributing this spotless record to three reasons. The first is that the Mae Wong Network has built many small dams in the area to retain moisture that staves off fires while not disrupting the natural flow of streams and brooks. The second is that the network sends out teams of 15 people as fire-spotters once a month during the wet season and every day during the dry season from December to May, which Narong called the "crisis time." And the third reason is that villagers have been educated about the need to preserve forests.

Since the inception of the Mae Wong Network, there are no longer forest fires; villagers have diversified their livelihoods through selling non-timber forest products such as rattan, mushrooms, medicines, silks and dyes; and wildlife has returned in significant numbers, even some 10 tigers that feed on the wild boars and deer that now roam the area. As Narong said, "When the tiger returns, that means the forest is abundant and healthy."

Narong Raengkasikorn.

Asian golden cat in Mae Wong National Park.

Forests in Thailand have fallen victim to land cultivation practices and fires set by poachers.

▶**PIONEERS**

Tree Bank

History: Launched in Chumphon province in 2006

Location: Nationwide

Key features: Encourages farmers to grow trees, which can then be used as assets by the farmers to secure loans

Planting trees is good for the environment, but also for farmers. That is the main idea behind the pioneering Tree Bank project. Launched by conservation officials and the Bank for Agriculture and Agricultural Cooperatives (BAAC) in 2006, the initiative encourages farmers to grow trees on their land alongside cash crops. Members can use the trees as economic assets, either as security for loans or by depositing them with the bank, which pays interest on them as they grow. When the mature trees are eventually cut down and sold, the farmers pay back the loans with interest. At least 300,000 farmers nationwide now take part in the scheme.

Forest conservation official Phongsa Chunaem was one of the founders of the project. The inspiration came from seeing how debt-laden farmers in the southern province of Chumphon were losing land they had deposited as security with banks. So he envisioned the tree bank as a way of allowing them to clear their debts while encouraging tree planting on a large scale to help the government with its reforestation efforts. The project also fits in well with the Sufficiency Economy ideas proposed by the late king. So much so that the government has contributed its own financial resources to developing the project over the last decade.

From its humble beginnings in Chumphon, Tree Bank has spread its branches across the country. The program is now active in every province as well as in Bangkok. The staff travel around the country explaining the benefits to farmers, endeavoring to change a common mindset that typically

Now, due to Tree Bank's efforts, trees are beginning to be seen as assets in their own right.

sees timber as only having value when it's been cut down.

Now, due to Tree Bank's efforts, trees are beginning to be seen as assets in their own right. When the mature trees are eventually cut down and sold, farmers are encouraged to plant new ones to replace them. Indigenous species are preferred, and the bank keeps a database of what trees are grown and where.

Meanwhile, other new benefits are being reaped – from enjoying the shade to picking mushrooms, farmers are enjoying a greener life.

Phongsa Chunaem, who spearheads the Tree Bank project.

Tree Worshippers

"A tree is a wonderful living organism that gives shelter, food, warmth and protection to all living things. It even gives shade to those who wield an axe to cut it down." - Lord Buddha

The Lord Buddha was born, achieved enlightenment and passed away in a forest. Scattered throughout his teachings are references to the importance of nature. Shrines are often placed at the entrance of forests and parks, reminding visitors to respect the sacred surroundings and warning loggers and poachers that cutting trees and killing animals are sins.

So it should come as no surprise that monks are sometimes at the forefront of conservation efforts in Thailand. Among them is Phrakhru Manas Natheepitak, abbot of Wat Bodharam in Phayao province. In 1988 he conducted what is believed to be the world's first "tree ordination" in a bid to stop indiscriminate logging.

The idea is simple. By "ordaining" trees and wrapping them in saffron robes, they become holy. Locals believe that felling them is a sin. The practice echoes the Thai tradition of young men being ordained as monks for a short period. It also has roots in animism, which venerates nature.

Since the first tree was given monastic vows, the practice has spread throughout Thailand and to other predominantly Buddhist countries. In fact, the abbot has been awarded a copyright license for the practice.

More than 100,000 trees have been ordained in the forests around Wat Bodharam, and now only a small minority of locals chop them down

to sell the wood or to use in construction. In 1996, in honor of King Bhumibol's Golden Jubilee, Thais attempted to ordain 50 million trees, and villages across the country took part in the project. The inclusiveness of the campaign, and its endorsement by King Bhumibol, also helped ensure its continuing popularity.

Amazonian Biodiversity

One of the world's most vital frontiers is the Amazon, which contains the largest tropical forest on the planet. According to Greenpeace International, the Amazon Basin is home to an estimated one-quarter of all known land species, and according to the Nature Conservancy, it also shelters almost one-third of all the known species on the planet.

It is also home to more than 30 million people who depend on this vast wilderness for their survival. As the forests dwindle, the inhabitants, deprived of many vital resources, struggle to survive in what is an increasingly hostile environment.

The Amazon contains between 80 to 140 billion tons of carbon, which must remain stored if we are to have a chance of fighting off the effects of climate change for a more sustainable future. If these rainforests continue to be logged and cut down for cattle pastures that feed the world's

fast-food industry, the side effects of global warming could be much more punishing.

A 2015 report by the Overseas Development Institute (ODI) found that the Amazon's ability to soak up carbon dioxide is worsening. In Brazil, which contains the largest tracts of this ecosystem, the government has spent 100 times more on subsidies given to industries that cause deforestation than the country normally receives in international conservation aid to prevent it.

That news underlines the urgency of action as this global treasure reaches a tipping point.

Inspiring Respect for the Forest

King Bhumibol Adulyadej and Queen Sirikit long encouraged reforestation practices and conservation principles, believing that these would help lead Thais toward more sustainable livelihoods.

King Bhumibol, for example, helped change attitudes in Thailand through his idea commonly referred to as "three forests, four benefits," which recommends planting diverse forests over monoculture plantations. By conducting reforestation in this way, local villagers can meet their own needs for timber, fruit and firewood, while also receiving an additional fourth benefit: the conservation of soil and watersheds.

Meanwhile, Queen Sirikit, when visiting remote border villages in the north and the northeast during the 1980s, observed poverty, a lack of security owing to the harvesting and trade of opium, as well as deforestation caused by shifting cultivation. Based on her belief that the people could peacefully coexist with the forest, the Pa Rak Nam watershed conservation program was initiated

in December 1982 in a northeastern province of Sakon Nakhon's Song Dao district. The queen assigned a team to prepare a rai plot of deteriorated land for her to replant about 100 saplings as a sample for local villagers to follow. Local residents then helped look after the trees. As part of this project, the king's "three forests, four benefits" idea was also put into practice, and villagers bred fish in a man-made pond to create a source of protein for their diets.

Soon local villagers presented more deteriorated land plots and the project was expanded to cover thousands of rai in Sakon Nakhon and nearby Udon Thani provinces. With full participation of the villagers and support from local authorities and the military, Her Majesty personally supervised the initial stage of the project to serve the purpose of watershed conservation. The queen often stressed, "Let residents live their normal lives. We can promote their quality of life, helping them earn more income, and be in better health."

Inspired by the happy scenes from Laura Ingalls Wilder's *Little House in the Big Woods*, the queen also initiated the first Ban Lek Nai Paa Yai (Little House in the Big Forest) project in Ban Huai Lo Duk, a village in Om Koi forest in Chiang Mai province in 1991. At that time, the overall area was already lush, fertile and a host to wild animals. The aim was not to restore the forest but to improve interdependence between the forest and the people. Poultry, swine and cattle breeds suitable for highland agriculture were supplied and a rice bank was set up for consumption during the dry season. Villagers were supported in terms of their agricultural practices as well as local craftsmanship to ensure they could support themselves. In 1995, the project was expanded to a nearby village Ban Huay Pu Ling and the forest conservation area was announced. A large number of local hill tribes also volunteered to be forest rangers so they could help protect their own homes from forest fires, illegal logging and drug trafficking.

These two projects initiated by the queen have been duplicated throughout the country in various watershed areas.

Check dam at a watershed forest in the north.

WILDLIFE

The campaign to save wildlife is going high-tech

Many wildlife conservation efforts focus on the endangered tiger.

As threats to Thailand's endangered species and biodiversity continue to grow, one of the key long-term solutions is to encourage society to play a greater role in wildlife protection. From promoting community awareness about the value of wild animals in the ecosystem to assisting former poachers to develop alternative sources of income, there are plenty of ways that the government and NGOs can help. But the most important thing is for locals to gain an understanding of why they should protect wild animals. They must build a sense of real partnership with organizations and feel the satisfaction of helping to build a sustainable future.

Wang Mee district is situated a short distance from Khao Yai National Park. A little over 15 years ago, the area was a hotbed of poachers and illegal loggers.

Today, thanks in part to the support and technical advice of the Freeland Foundation, the inhabitants have found alternative livelihoods like growing organic mushrooms or mulberry trees. Although some have been lured back to the illegal trade by sky-high prices for endangered species, a combination of poverty alleviation programs and strict enforcement has led to a reduction in poaching levels of up to 75 percent in some areas. "We take a holistic approach towards biodiversity conservation and protected area management," said Tim Redford, who oversees Freeland's Surviving Together program. "Local communities just need a bit of help to get on their feet."

PRICES FOR ILLEGAL WILDLIFE PRODUCTS

ONE KG OF TIGER BONE
US$1,000-1,500

TIGER PELT
US$2,000

A LIVE BABY ELEPHANT
US$30,000

WORKED IVORY
US$5,000 PER KG

PANGOLIN
US$320 PER KG
OR AROUND US$800
PER WHOLE ANIMAL

SLOW LORIS
US$650 FOR A PET

Source: Freeland Foundation

Successful wildlife conservation requires strong collaboration between government agencies, NGOs, community-based organizations and the private sector: all of them working toward the sustainable management of biological resources. Defining the role of each partner is crucial. By law, the Department of National Parks, Wildlife and Plant Conservation (DNP) is responsible for managing Thailand's protected areas. But the support of foreign NGOs who have the technical expertise and the ability to fund and monitor long-term conservation projects can help the DNP and other government agencies to function even more effectively. And both NGOs and wildlife officials in turn need to engage with local communities.

Of course, there are plenty of other approaches that could help. Visionary leadership at the highest levels of government is vital if the country is to follow a path that puts conservation before short-term profit and ensures strict enforcement of wildlife laws. Encouraging villagers to live in harmony with nature could also go a long way toward conserving the country's rich natural heritage. "You have to teach people that if they want clean air and clean water for the next generation, then they must protect wildlife

and the watersheds," observed Wayuphong Jitvijak, a project manager at World Wide Fund for Nature (WWF Thailand).

The good news is that endangered species do have the potential to recover, so long as the conditions are right. Take the Western Forest Complex on Thailand's border with Myanmar. This region, covering 18,000 square kilometers, contains what is probably the best tiger habitat in Southeast Asia. While current estimates put the tiger population there at only 100, the region could easily support far more. And because the tiger is the top predator in the natural forest ecosystem, its existence would indicate the prevalence of wild ungulates like banteng, sambar, gaur and muntjac.

"The key is to have good protection," said Anak Pattanavibool, Thailand program director of the Wildlife Conservation Society (WCS) and a lecturer at Kasetsart University. "Education is important for the long term. But first you have to stop the poaching." Conservation work is immensely challenging, but it's the only chance we have to save our rapidly disappearing wildlife and to keep these dynamic ecosystems intact.

Winning the Conservation Battle

Thailand may still have some distance to go before it can solve the problems of the illegal wildlife trade. But the country can take inspiration from the example of Nepal, which, against many odds, has successfully turned the tide against poachers.

Ten years ago, this Himalayan nation was a major center for animal trafficking. Poachers regularly killed more rhinos here than in the rest of Asia put together. The horns were then smuggled mainly to China where they can be sold for tens of thousands of dollars for use in traditional medicine. Tigers, elephants and leopards were also killed for their valuable body parts and skins.

Not any more. In 2010, recognizing the pressure these animal populations were under, the government, backed by the army, committed to protect its biodiversity and launched an unprecedented crackdown on poaching through better intelligence and technology, increased cooperation with local communities and a policy of "zero tolerance." More than 300 poachers have been arrested and tough new penalties on wildlife crimes have been imposed.

The results have been outstanding. In 2014, the government announced that there was "zero poaching of rhinos, tigers, and elephants" during the previous year. Tighter enforcement has also resulted in the dismantling of one of the major rhinoceros smuggling syndicates.

Of course, rhinos and tigers continue to face plenty of uncertainty in Nepal. But what makes this success story all the more extraordinary is the fact that it occurred against a backdrop of prolonged civil unrest, political deadlock and widespread poverty. Nepal's example should give hope to countries like Thailand that, in spite of other hurdles, they can also fulfill the hopes of many across the kingdom and safeguard their natural biodiversity and ensure its future survival.

▶ **PIONEERS**

SMART Patrol

History: Launched in 2005

Location: Huai Kha Khaeng-Thung Yai Naresuan Wildlife Sanctuaries, western Thailand

Key features: SMART features a GPS navigation system with electronic maps to plot the routes of poachers and up-to-date information on the distribution of wildlife; 1,500 rangers have been trained to use it

Huai Kha Khaeng-Thung Yai Naresuan Wildlife Sanctuaries are among the last strongholds of tigers and elephants in Southeast Asia. Located in the remote Dawna Range along Thailand's border with Myanmar, these adjoining sanctuaries support an extraordinary diversity of species, including 120 types of mammals, 400 types of birds and 96 different species of reptiles. It is little wonder that UNESCO recognized Huai Kha Khaeng-Thung Yai Naresuan as a natural World Heritage Site due to "its outstanding natural beauty and great scientific value."

Like other pristine areas in the country, however, this vast wilderness has found itself on the front lines of the war against poachers, with tigers, elephants, clouded leopards and banteng being mercilessly trapped or gunned down and their parts sold for enormous profits.

In 2005, the parks were chosen to pilot an innovative monitoring and reporting system called SMART Patrol. Backed by the WCS in collaboration with Thailand's Department of National Parks, Wildlife and Plant Conservation, the aim of SMART is simple: to harness information technology for enforcement efforts, enabling park rangers to more effectively protect endangered wildlife.

Since its launch, incidents of poaching have fallen sharply, and the population of adult breeding tigers in Huai Kha Khaeng has begun to recover, rising from around 38 in 2007 to 90 today, according to data from camera traps. That figure is still critically low, but at least it's heading in the right direction. According to WCS, Huai Kha Khaeng and the surrounding Western Forest Complex of protected areas is the single most important site for recovering wild tigers in Indochina. This vast habitat could potentially support 2,000 tigers.

The gradual increase in tigers is not the only positive development. Anak Pattanavibool, director of WCS Thailand's program, says that the SMART system has enabled a force of around 500 rangers to effectively patrol an area of 6,400 square kilometers, and thanks to information on the

Incidents of poaching have fallen, and the population of adult "breeding tigers" in Huai Kha Khaeng has begun to recover...

whereabouts of poachers, they have been quick to identify and arrest them.

The effectiveness of SMART in Huai Kha Khaeng-Thung Yai Naresuan has led the system to be adopted by 35 protected areas across the country, with 1,500 rangers trained to use it. Now upgraded and simplified, it's being modified for possible use in marine national parks, potentially opening up a new frontier.

But technology alone does not guarantee success. Thailand's rangers need increased funding and better weapons if they are to match the firepower of well-armed poaching gangs. They also need to be trained how to fight in shootouts, as dozens of them have been killed in recent years.

SMART is just one of a new generation of technology-based solutions designed to combat wildlife crime. The hope is that data-based systems like this could throw a lifeline to some of Thailand's most endangered species, helping to protect the country's national parks and wildlife sanctuaries for generations to come.

▶**PIONEERS**

Kui Buri National Park

History: Established in 1999

Location: The Tenasserim Hills in Prachuap Khiri Khan province

Key features: Kui Buri occupies approximately 1,200 square kilometers, consisting of dry and moist evergreen forests that support around 250 wild elephants and 150 gaur

Kui Buri National Park is where you are most likely to see elephants in the wild.

Just a four-hour drive south of Bangkok, Kui Buri National Park is famous for its magnificent scenery: its steep mountains cloaked in luxuriant foliage, its tropical evergreen forest and open grasslands. More than anything, it is known as the place where you are most likely to see elephants in the wild, thanks to the park's relatively small land area of more than 1,200 square kilometers and easily accessible open areas.

But the park's elephants have also found themselves in conflict with local farmers. Between 1999, when the park was established, and 2005, herds of elephants, numbering up to 80 at a time, would frequently leave the park in search of food. They would raid the nearby pineapple plantations, destroying land and damaging crops. The farmers would fight back, putting up electric fences, building ditches and sometimes killing them.

These days, elephant incursions have been greatly reduced thanks in large part to a royally initiated project to restore Kui Buri's forests and wildlife to their former glory. As part of an agreement between park authorities, local communities and the government, land that had previously been encroached upon has been set aside for open grassland to be used by the elephants. Even better, the authorities have planted native trees, established artificial mineral licks and increased water storage capacity to ensure that water is available for the elephants well into the dry season.

"Cooperation between the local community, government agencies, NGOs and the private sector is something you normally only read about in textbooks," said Wayuphong Jitvijak, the WWF manager who has directed the project for almost a decade. "But here in Kui Buri we have been able to achieve a real partnership."

> *"Cooperation between the local community, government agencies, NGOs and the private sector is something you normally only read about in textbooks, but here we have been able to achieve a partnership."*
>
> **Wayuphong Jitvijak, WWF manager**

Today, the population of elephants in the park has risen to about 250 from 84 in 1999. Rare animals like the tapir, clouded leopard, sun bear and even tigers have been spotted, demonstrating the park's improved biodiversity. Locals have been among the first to benefit through a community-based wildlife tourism program. There are now 25 village guides, seven homestays and 35 trucks servicing the growing number of tourists who come here.

As such, Kui Buri is a remarkable example of what can be done to mitigate human-elephant conflicts, which affect at least 12 other provinces around the country. But its success cannot hide the bigger underlying problems in Thailand. Elephants require vast amounts of land in which to forage. And to ensure viable breeding populations, they require a habitat that is linked by migratory land corridors.

The real question for the authorities, then, is to figure out who takes priority – the elephants, the locals or the tourists?

Ensuring a future for Thailand's emblematic species is unlikely to be easy. But as Kui Buri shows, if done right, it is certainly possible. And the benefits are likely to be felt as much by the local communities as by the wild creatures.

▶PIONEERS

Sarus Crane Breeding Program

History: Efforts to reintroduce sarus cranes to Thailand began in 1983. In 2016, the first chick was born in the wild

Location: Buriram and Nakhon Ratchasima provinces

Key features: Breeding sarus cranes at Korat Zoo in Nakhon Ratchasima province and re-introducing adult birds into the wild in Buriram, while also encouraging organic farming

Standing at a height of up to 1.8 meters, an adult sarus crane is a formidable creature. So when the last of the "tallest flying birds in the world" disappeared from Thailand's northeastern wetlands in the early 1980s, their absence was rather conspicuous. Fortunately, they were sorely missed.

As a result, in 1983, Bubphar Amget of the Royal Forest Department (RFD) and Pilai Poonswad from Mahidol University traveled to India for a conference on cranes hosted by the government of India and the International Crane Foundation. There, they told the story of the extirpation of sarus cranes in Thailand. George Archibald, a co-founder of ICF, was moved by their story and in early 1984 he visited Thailand to discuss crane reintroduction with RFD. In November, he returned with six juvenile Australian sarus cranes, three male and three female.

Over the years, the Zoological Park Association of Thailand and the crane breeding program at Korat Zoo, which took over the Sarus Reintroduction Project, have had great success in breeding sarus cranes in captivity. As of 2016, some 70 sarus cranes have been released into the wild in Buriram and more than 40 have survived and

are regularly spotted near the release site, according to ICF.

In other ways, 2016 had been a notable year. "This year, for the first time, eight pairs attempted to breed in the wild. One pair has been successful thus far," said Archibald, adding that the immaturity and inexperience of many of the pairs, and the flooding of two nests, "attributed to their lack of success in breeding."

Cranes are sensitive to early-experiences, a phenomena know as imprinting. "If they are raised by humans without an association with cranes, they will want to pair with humans when they are sexually mature. To avoid incorrect imprinting, humans wear crane costumes and use hand puppets that look like the head and neck of a crane," explained Archibald.

Reintroduced cranes remain vulnerable to threats including habitat destruction, degradation and disturbance, excessive pesticide use, hunting and accidental injuries. These threats have already had an impact on the reintroduction program, resulting in a 60 to 70 percent survival rate. A changing climate also presents a threat to the national population of Eastern Sarus cranes. Increased periods of drought

result in higher pressure on key water reservoir areas, causing water levels to drop and wetland habitats to degrade.

Still, Archibald is optimistic. "The cranes appear to be thriving in the agricultural landscapes of Buriram province," he said.

"It is hoped that local communities will embrace their return and will actively participate in the wildlife conservation," said Dr. Boripat Siriaroonrat, Assistant director, Bureau of Conservation Research and Education, Zoological Park Organization. In fact, many local farmers have already ceased using harmful chemicals, opting instead for organic practices, a move that has paved the way for the cranes to flourish. An added benefit and impetus to continue and expand organic farming is that these farmers have seen growing profit margins as their rice now sells at a premium in urban supermarkets. "It is hoped that organic farming practices will eventually replace the use of chemicals for the welfare of both humans and wildlife," Archibald said. However, he said more funding is needed to properly monitor the cranes and to educate the public so that the cranes and humans can "live in harmony."

Until 2016, a sarus crane hadn't been hatched in the wild in Thailand for over three decades.

▶PIONEERS

The Gibbon Rehabilitation Project

History: Founded in 1992

Location: Phuket

Key features: Rescuing, rehabilitating, breeding, and releasing white-handed gibbons into their native habitat

Mother with baby gibbon born in the wild.

A tout with a baby gibbon in Phuket .

The white-handed gibbon is native to Phuket, but was poached to the point of extinction in the area in the 1970s and early 1980s. That alone is not so surprising. Many of Asia's animals are hunted to extinction for a range of unethical reasons, such as demand for their organs for use in traditional medicines, or for their pelts or ivory. But the reason behind the plight of Phuket's gibbons might just be the most despicable yet: for years, the mothers of baby gibbons have been killed off so that their cute, furry little offspring can be used by touts as a photo prop with tourists.

As it turns out, baby gibbons are big earners. A tout arrested on Phi Phi Don with a baby gibbon in 2015 revealed to authorities that he was earning as much as 5,000 baht per day in the low season and more than 20,000 baht a day during high season. Luckily, a number of organizations are collaborating to combat this problem and are also working to repopulate the local area with gibbons.

One such organization is the Gibbon Rehabilitation Project (GRP), which is the longest operating gibbon re-introduction project in the world. GRP, which became a research division of the Wild Animal Rescue Foundation of Thailand (WARF) in 1994, was originally established in 1992 by Noppadol Preuksawan, the chief of the Royal Forestry Department in Phuket, with support and assistance from the Asian Wildlife Fund, American zoologist Terrance Dillon Morin and WARF.

The project's goal has always been to save gibbons and their rainforest habitat through rehabilitation and reintroduction of the animals, and to end the demand for the illegal use of gibbons as tourist attractions and pets. Reintroduction remains a relatively new division of the gibbon conservation movement as well as somewhat uncharted terrain for research. In the 1990s, GRP's earliest reintroduction attempts failed to yield positive results. But the organization has continued to test new methods. In 2002, GRP embarked on a groundbreaking gibbon breeding program in the Khao Phra Thaew forest on Phuket. The forest was selected because it is the largest remaining evergreen rainforest on the mainland of Phuket, covering an area of 2,228 hectares.

Starting in 2002, GRP began to successfully reintroduce pairs of white-handed gibbons into the forest. In a sense, it could be said that the forest needs gibbons as much as they need the forest. Due to their largely fruit-based diet, a single gibbon can essentially help to plant around 10,000 trees per year by "dropping" seeds in one manner or another.

The reintroduced gibbons have already begun to reproduce, according to WARF secretary-general Thanaphat Payakkaporn. "We released three gibbon couples into the wild several years ago," Thanaphat told the *Phuket Gazette*. During a follow-up monitoring trip, all three pairs were found and it was discovered that one couple had given birth. Thanaphat said that WARF records show that to date at least 17 gibbons have been born in the wild around Phuket, including one that is "second generation," proving that the program is working and sustainable. Meanwhile, others continue to be born at the GRP's rehabilitation site, where they are paired up and later released.

An educational campaign directed at tourists has been crucial in helping to stop the use of gibbons as photo props. However, Thanaphat said that some touts have simply moved onto using other species like the Slow Loris because it is smaller and can easily be hidden when authorities approach.

"We're proud that we are a worldwide model for successful species rehabilitation," he said. "Their numbers were sparse in Phuket just a few years ago, but now they're breeding again and their population is growing. We're glad to see them back on the island after being hunted for tourism activities."

WOMEN AND WILDLIFE CONSERVATION

Soraida Salwala established the world's first elephant hospital.

They are celebrated elephant conservationists, tireless activists, outspoken wildlife crusaders and, most surprising of all, in the male-dominated conservation world in Thailand, Soraida Salwala and Sangduen "Lek" Chailert are women.

In some ways that makes their achievements all the more admirable. Yet what's important is that Soraida and Lek show us how compassion, perseverance and incredible courage can change things for the better.

For Soraida, it began as a child, when she saw a badly injured elephant with two long tusks lying by the side of the road. The elephant had been hit by a truck and was eventually shot to put it out of its agony. Why couldn't the elephant be taken to a doctor? asked the uncomprehending eight year old. Soraida never forgot those words. In 1993, she established the world's first elephant hospital in Lampang. Since then Soraida and her team have treated close to 4,000 elephants for everything from abuse by humans to traffic accidents and stepping on landmines. She has also valiantly campaigned for the rights of these animals, who have often been reduced to begging for food in the streets of Bangkok.

Lek's passion for elephants also began as a child. Today, she runs the Elephant Nature Park, where elephants saved from tourist and illegal logging operations live out their days in peace. But it's going to take a lot more than animal welfare to change the way we think. And so on a Friday afternoon in March, there is Lek in a Bangkok shopping center lecturing well-dressed city folk on why it is wrong to use elephants in circuses or to teach them to play football. "We have to speak out about the cruelty that goes on in Thailand," she explained. "There are so many ways humans and elephants can live together rather than teaching them unnatural tricks."

Despite considerable local and international support, both women have been repeatedly threatened by elephant traders, ivory traffickers and "influential" businessmen who fear a backlash from such high-profile public campaigns. Yet intimidation has done nothing to stop their quest for justice. "I used to be frightened, but now I don't fear. I fight," said Lek.

Whether these formidable conservationists can save Thailand's dwindling elephant population, estimated at less than 3,000 in the wild, is a thornier question. One fully grown elephant consumes 200 kilograms of food a day and requires at least 20 gallons of water. Yet vegetation and water resources are rapidly shrinking thanks to human encroachment and climate change. "Fifty years from now, I fear that the elephant will be gone from Thailand," said Soraida with a note of melancholy in her voice.

Lek is more upbeat. She passionately believes in the power of education to bring about genuine change. As a result, she works with school children, celebrities and occasionally even politicians to spread her message. She also assists village communities to live sustainably with nature. "So many people care about animals," she explained. "But they do not have a leader. I try to make an example."

Fortunately, both Soraida and Lek have every intention of continuing their fight to raise awareness about the plight of Asian elephants and their habitat, which means there is still a chance that these lumbering giants may have a future in Thailand.

Sangduen Chailert runs the Elephant Nature Park.

GROUNDBREAKERS

Renowned wildlife investigator and Freeland Foundation director **STEVE GALSTER** *talks about wildlife conservation.*

Can campaigns to stop the wild-life trade succeed in Thailand given the enormous profits for rare animals? Yes, they can. A lot of Thais, in particular the younger generation, love animals. They know what is behind the trade and they think it is wrong. In many ways the wildlife trade in this country has become a welfare issue. However, we must not be complacent. Visible reminders of this ugly trade continuing here are Chatuchak Market's animal section, as well as a plethora of online markets.

In 2013, the governing body of CITES identified Thailand as a major transit center in the illegal ivory trade. Has enough been done to crack down on it? The Thai government has definitely stepped up its awareness campaigns to stop people from buying and trafficking ivory. They get high marks for this. However, we still have not seen any major ivory trafficker caught and brought to justice in Thailand. That remains a blemish.

What has been the single biggest success in Thai enforcement in the past decade? Thailand led the effort to form the ASEAN Wildlife Enforcement Network (ASEAN-WEN). We now have 10 countries and 10 task forces exchanging information on the wildlife trade. Since 2006, seizures have increased 11-fold around the region. However, ASEAN-WEN is not getting the same support as it was before from the Thai government. I hope this will change.

Should wildlife protection and community support programs go hand in hand? Community and outreach programs are very important. One needs to minimize friction between rangers and local communities and make them allies. But you can't step back and expect the government to do it all themselves because decision-makers and political leaders change so fast. This kind of work should be led by civil society. NGO staff stick around for longer and must help carry the torch.

Name three things that need to be done to ensure the survival of endangered species in Thailand. Firstly, we would like a total ban on trade in all globally listed protected wildlife, meaning zero tolerance for the sale of any endangered or rare species. At present it's just too easy for crooks to manipulate the system and say they bred some rare species here in Thailand, when we know they smuggled it in from some other country. Secondly, we need to see major wildlife criminals get arrested and have their enormous assets seized. Third, we need to see increased support for rangers and their bosses through higher salaries, life insurance and better training.

EYE-OPENERS

YEAR OF THE WOLF
WOLF TOTEM
Written by: *Jiang Rong (2004)*

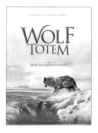

Not since Jack London's *Call of the Wild* and Farley Mowat's *Never Cry Wolf* has an author written with such passion and conviction about wildlife as Jiang Rong in *Wolf Totem*, an autobiographical novel detailing the author's experiences during the Cultural Revolution on the Mongolian grasslands, where wolves attack and kill dozens of army horses during the "white-hair blizzards." Through the Mongolian herdsmen, descendants of Genghis Khan's armies who conquered the world (partly by imitating wolves' hunting strategies), Jiang Rong gives us a written record of a dying culture, a genocide, and a predator that is a spiritual totem. In 2015, a feature film was released. What the book and film so impressively document is a harsh yet fragile environment where hills, pastures and lakes are threatened by the removal of a top-tier predator as grasslands are turned into farmland: a chilling lesson in biodiversity.

SAVING MARINE HABITATS

Innovative projects are coming to the rescue

Until the 1980s marine preservation was not on the radar for Thailand's politicos and bureaucrats. However, a growing, if diffuse, awareness of the myriad benefits (fiscal, environmental and social) that such spawning grounds provide for a range of development sectors was acknowledged with the designation of marine and coastal protected areas.

Today, it's much more widely understood that these areas, such as marine national parks and restricted areas for projects like turtle breeding, are essential for maintaining the capacity of critical ecosystems to support sustainable development. That said, gains made by protecting some areas have been offset by losses created by ongoing threats such as illegal fishing, the encroachment of man-made infrastructure, the destruction of coral reefs by commercial fishing and tourism and the building of shrimp farms in mangrove forests.

What impresses about some of these pioneering projects are their sheer simplicity and innovation, often spurred by limited financing...

On the positive side, combating these threats has united an array of talents, from economic planners and policymakers to members of the private sector, who have teamed up with academics, activists and local residents. Such collaborations have made it possible for a number of projects to proceed. In Trat province, the Ban Pred Nai Community Forestry Group has banded together to preserve the area's mangrove forests. Renowned for their abilities to stop soil erosion and protect villages from storms, mangroves are also known to protect juvenile marine creatures that hide in their tangled roots. By protecting the forest, the villagers have also protected their own livelihoods as fishermen and preserved their pastoral cultures.

Many seaside communities like Ban Pred Nai are dependent on fishing for both cash and calories. That's why several have adopted Ban Pred Nai Community Forestry Group's pragmatism as reflected in their saying: "Stop catching a hundred – wait for a million." This could well be a rallying cry for sustainable development among such rural communities in Thailand, but it would fall on deaf ears if it were not for financial incentives. Ban Pred Nai Community Forestry Group's projects have used a system of financial bonuses and penalties to make sure villagers cooperate. They have also set up community savings funds to help farmers in need, but only if they play by the rules.

As far back as the 1980s, King Bhumibol Adulyadej recognized the need for better marine resource management. Concerned about the deterioration of mangroves and the correspondingly negative impact on livelihoods, in 1981 the king established the Kung Krabaen Bay Royal Development Study Center, with the center and related projects covering 13,090 hectares in Chanthaburi province. Focusing on aquaculture, mangrove conservation and improving the occupational skills of local people, the center has had notable success in carrying out mangrove restoration initiatives, promoting eco-tourism and increasing the output of local fisheries and shrimp farms.

Meanwhile, in 2013 the Reef Biology Research Group within Chulalongkorn University's Faculty of Science announced that it had successfully completed its tests to breed and release warm-water corals – the first organization to do so in Thailand. The experiment, backed by Princess Sirindhorn's Plant Genetic Conservation Project and the Royal Thai Navy, was hailed by marine scientists in the kingdom as a major breakthrough. According to the research team, the survival rate of corals transplanted with their method was around 40 to 50 percent (within the first two years), whereas the survival rate in nature is only about 0.01 percent.

Coral colony on a reef top at Koh Chang.

Certainly it was a timely discovery. An estimated 75 percent of the world's coral reefs are now threatened. Mangrove forests, currently being depleted at a higher rate than any other type of forest, are not doing much better, though they are making something of a comeback in certain areas of the kingdom where conservation projects and protective measures hold sway.

A 2014 report from Kasetsart University said that there are some 244,000 hectares of mangroves in the country. In parts of Phetchaburi province, where the royally initiated Laem Phak Bia Environmental Research and Development Project began in 1990, they are making a slow comeback. The report pointed out that in a cluster of four villages mangroves are growing at a rate of 3.7 hectares per year. Considering the benefits that these ecosystems provide, from purifying the air and storing carbon, to serving as nurseries for juvenile fish and havens for bird-watchers and eco-tourists, they are extremely valuable. Each hectare, the report estimates, is worth around 424,000 baht (about US$12,200) per year.

The fact that many such efforts have spread to other communities and inspired many younger people is proof positive that these disparate movements, which are united by a unanimity of purpose – to preserve the coasts and seas for their descendants – will have staying power.

THE OCEAN OF LIFE
THE OCEAN OF LIFE: THE FATE OF MAN AND THE SEA
Author: *Callum Roberts*
Year: *2012*

The *New York Times* called Callum Roberts' book, "a *Silent Spring* for oceans," referencing Rachel Carson's seminal work on the detrimental impacts of pesticides on the environment, which led to a ban on DDT in the US. In the award-winning *Ocean of Life*, Roberts, a professor of marine conversation at York University, presents a searing portrait of the damage we are doing to our often-overlooked oceans. Topics covered include overfishing, deep-sea mining, pollution, climate change, the destruction of wetlands, soil erosion, dams and more. Not only a wake-up call to take these issues more seriously, the book also presents background on the ocean's beginnings and fascinating details on the life of the sea and its underwater creatures. In its review of the book, *The Economist* wrote: "The enormity of the sea's troubles, and their implications for mankind, are mind-boggling. Yet it is equally remarkable how little this is recognized by policymakers – let alone the general public....There is also a dearth of good and comprehensive books on a subject that can seem too complicated and depressing for any single tome. Callum Roberts, a conservation biologist, has now provided one."

▶PIONEERS

Ban Pred Nai Community Forestry Group

History: Founded in 1986 by locals

Location: Ban Pred Nai in Trat province

Key features: Reviving mangrove forests and bringing back species that disappeared after disastrous experiments with shrimp aquaculture

For over a century, this bucolic community in Trat province relied on rice farming and aquatic animals from the mangrove forest as sources of food and income. The money earned from fish, crabs and shellfish was viable revenue that helped tide the community over when rice crops failed or the price of such commodities plummeted.

That all changed in the 1980s with the arrival of a private company that was granted a state concession to convert some of the mangrove forests into shrimp farms. As part of a larger aquaculture trend that took coastal communities by storm in the 1980s, this was not unusual. The government estimates that between 1961 and 1996, the nation lost 50 to 60 percent of its mangroves, mainly due to shrimp farming.

In Trat, the company built dykes that blocked the seasonal flow of saltwater from the sea into the land. As the number of aquatic species began to decline, the villagers learned of the harmful side of this business. The realization was a call to action for a number of villagers. In 1986 they joined forces to protest against the concession and lobby the government to demand a change of policy, said Supakij Huangnam, one of the original members.

Even though the movement did not trigger a national policy change, it did make villagers aware that there is strength in solidarity. And so the Ban

Pred Nai Community Forestry Group was formed. Following the nationwide collapse of shrimp aquaculture in the late 1990s, the company left the scene. For the villagers it was a time of renewal. They spent the next decade replanting parts of the decimated forest while allowing other parts of it to naturally regenerate. As they tackled a range of problems, the villagers reached out to a number of stakeholders, from government officials to religious leaders, teachers and a technical body, the Regional Community Forestry Training Center for Asia and the Pacific (RECOFTC), to assist them. "Without these collaborations, we could have ended up just living to make ends meet," said Supakij.

In addition to stopping the loss of biodiversity, their diligence and foresight have brought back many displaced species, from wetland birds to bees, monkeys, shellfish and fish.

Involving RECOFTC was a turning point. Local leaders gleaned much from the scientific experts. "There are three layers of plants in our forests, each of which caters to different types of plants," said Supakij, adding that putting the right plants in the right places was crucial to the project's success as it enabled them to lure back species that had died or fled, such as crabs.

In addition to stopping the loss of biodiversity, their diligence and foresight have brought back many displaced species, from wetland birds to bees, monkeys, shellfish and fish, according to a report by RECOFTC. Consequently, the waning populations of crabs and other aquatic animals have rebounded, and in some cases even doubled.

Learning from the costly lessons taught by shrimp farming, the community realized that it was high time to put some controls in place. As of

2002, they agreed on some safeguards, such as refraining from catching crabs for six days during the mating season in October every year. "The motto we used to encourage other villagers to follow was simple. 'Stop catching a hundred — wait for a million,'" said Supakij.

They also convinced other villagers to stop netting crabs in the special conservation areas and to not target juveniles. The group used a financial incentive to help cement their cooperation. Those who got with the program were entitled to borrow money from the community's savings fund. In another innovative move, they set up a community watch program to make sure nobody was logging any of the mangroves. As a result of these practices, the local economy and environment are thriving in tandem, said RECOFTC, and their prospects for the future look excellent.

Working on such a far-reaching project with so many different repercussions on the ecosystem made the villagers aware that other communities in the mangrove swamps face similar issues. The Ban Pred Nai Community Forestry Group reached out to them. Their goal was admirable: to share the knowledge they had accrued with other villagers in the name of protecting these increasingly rare ecosystems that are being lost at a faster rate around the world than any other type of forest.

As a result of reaching out to nearby communities, the group's network now extends all over Trat and other eastern provinces. The extended group meets once a month. In the Pred Nai community itself, the villagers have built a learning center that provides common ground for discussions about all manner of forestry issues and conservation.

If projects like these are to succeed in the long run, however, they have to engage and inspire the young. This is why members of the group now offer impromptu study sessions in the village's primary school about conserving mangrove forests for the future and all the precious creatures that find refuge in them.

Young students participate in a mangrove reforestation project.

Mudskippers abound in mangrove forests.

Black winged stilts.

▶**PIONEERS**

Transplanting Coral Reefs

History: Started in 1995 by Prasan Sangpaitoon

Location: All over the eastern seaboard

Key features:
Academics combined an everyday item with a technique used in dentistry to arrive at a watershed in marine conservation

To be an innovator does not always require inventing some fabulous new high-tech device or pioneering a new school of philosophy. Sometimes all it takes is putting an ingenious spin on an already existing product, like the PVC pipe, ubiquitous in Thailand, combined with a well-known technique used in dentistry to fill root canals.

This was the double-edged thrust and modus operandi of Prasan Sangpaitoon's work to revitalize the coral reef colonies along the east coast of Thailand, which had been battered by fishing, tourism and storms that bleached or ruined the reefs. Prasan, a lecturer at the Rambhai Barni Rajabhat University in Chanthaburi province, used PVC pipes as breeding grounds for reef fragments to grow in the seabed that is their natural home. "The PVC pipes we chose are the same as those used for supplying drinking water supplies. To me, if they're safe for humans they should not poison the reefs," he said.

The process is simple. Prasan attaches PVC tubes to a small window-sized frame, then puts a coral fragment into each hole in the frame and attaches it with a screw. After that he puts the frame in the seabed. Unlike metal bars, which could turn rusty and kill the coral, PVC pipes are durable and recyclable, he said. They are also cheap. Each PVC frame costs only about 500 baht (US$14.25).

Still, it's a slow process: Corals only grow by around 1.5 to 2 inches a year. That's a large part of the reason why they

are so grievously endangered. A World Resources Institute Report from 2011 claimed that 75 percent of the world's reefs are under threat: a worrying rise of 30 percent over the previous decade.

Knowing the importance of his task supplied much of the motivation for the professor. For his first forays into coral transplantation he chose the Samaesarn beach area in Chonburi province. After enjoying excellent results there, he expanded his venture to other nearby islands, such as Koh Kham, Koh Samet, Koh Wai and Koh Talu. Choosing the right coral for these underwater terrains (in this case the *Acropora* species) was decisive.

After rainforests, coral reefs are the second-most productive ecosystem on the planet. They serve as shelters and sources of food for many different marine creatures. Once these degraded areas were restored to a semblance of their former glory, the fish and other creatures returned en masse: a resurgence that also helped to shore up local fisheries.

Prasan's project is one of several partnerships between academics, private organizations and local communities to restore degraded coral reefs using different methods, such as sexual repro-

duction in an artificial environment, or reattaching coral fragments to natural substrates. Even though some fellow academics have questioned his approach and the way he has added unnatural elements like PVC pipes to nature, they have had to concede that this method has brought about a much higher survival rate of coral.

To continue and expand his work, Prasan has established a small foundation to run this successful undertaking that has now transplanted some 40,000 corals over 20 years.

▶**PIONEERS**

Crown Property Bureau Foundation

History: Coral reef replantation initiative commenced in 2012

Location: Coastal waters off of Chanthaburi, Chumphon, Krabi, Prachuap Khiri Khan, Phang Nga and Songkhla provinces

Key features: Planting artificial coral reef structures to help restore marine resources and improve the lives of local communities

Since 2012, the Crown Property Bureau Foundation has put 30 million baht toward planting 5,464 artificial reef structures off the coasts of Chanthaburi, Chumphon, Krabi, Prachuap Khiri Khan, Phang Nga and Songkhla provinces. The foundation's approach focuses on inclusive engagement of all stakeholders. For example, the Department of Fisheries, local administrative organizations and local fishermen have worked together to survey and identify appropriate locations for coral plantation.

Most of the conservation work has been near other royally initiated project sites. In addition to increasing habitats for marine life, these initiatives have had the knock on effect of increasing the income of local fishermen, improving their quality of life, and prompting them to adopt more sustainable practices. The initiative also contributes to efforts to achieve Goal 14 of the United Nations' 2030 Agenda for Sustainable Development.

"For each artificial-reef zone, many benefits are generated for the economy, society, environment and sustainable natural-resource conservation. The monitoring has shown that after the artificial reefs are installed, sea life in the area increased," said Dr. Wimol Jantrarotai, Director General of the Department of Fisheries. "By installing artificial reefs, we have also provided food sources for local small-scale fishermen."

This latter point, in turn, has helped encourage locals to think more about conservation. These communities are now working together with the government to help maintain the artificial reefs. "Committees have been set up at the local level to lay down guidelines on how to manage fishery-related resources. These committees have also mobilized the local population to watch over their marine zones and prevent illegal fishing from occurring. They now realize the need to limit the number of fish taken from the sea to ensure that the fish population continues to grow," said Dr. Wimol.

Dr. Voranop Viyakarn of the Reef Biology Research Group at Chulalongkorn University, said it is important to consider the long-term impacts of artificial reefs. "We have to find the right balance in setting up artificial reefs. People should erect artificial reefs in areas where resources are adequate but scattered."

"We must take care of natural reefs," he said. "It is better to devote resources to protecting them instead of creating artificial ones."

Laem Phak Bia Environmental Research and Development Project

History: Founded in 1990 as a royal project

Location: Laem Phak Bia sub-district of Phetchaburi province

Key features: Cleaning up the polluted Phetchaburi River through natural means and improving residents' standard of living

Known today for its popular palm sugar sweets, such as Thai custard, and its famous beaches, such as Had Cha-am, Phetchaburi province used to be far more notorious for its heavily polluted rivers. The usual culprits of agriculture and shrimp farms, urbanization and industrial waste were to blame.

Its negative reputation has disappeared since the royally initiated Laem Phak Bia Environmental Research and Development Project, dedicated to cleaning up the Phetchaburi River and its tributaries, was launched in 1990. Named after the sub-district where it's situated, the project followed King Bhumibol Adulyadej's suggestion that these messy problems could be cleaned up through cheap, easy and practical means.

As with so many of the monarch's projects, natural solutions are deployed to right man-made wrongs. The project channeled wastewater from Phetchaburi town through an 18.5-kilometer pipe to Laem Phak Bia where three innovative treatment systems treat the polluted water. The first system uses the natural self-purification function of lagoons to clean the wastewater before it is discharged or reused. The second treatment is a plant-and-grass filtration system that uses soil to filter wastes. Plants then consume the waste, making the water safe to discharge into rivers and streams.

The third is an artificially constructed wetland system, which uses the natural functions of vegetation, soil and organisms to treat wastewater, mimicking the purifying powers of real wetlands. Again, the system does not require any elaborate or expensive technology.

After some 25 years, the project has improved the water quality in the Phetchaburi River and brought other benefits. Treated wastewater is used for agriculture or discharged into the sea. Farmers use the lagoons for fish farming, and the natural fertilizer can be used on crops. It's also an important learning center, disseminating knowledge and best practices to neighboring communities through the Chaipattana Foundation.

COASTAL RESOURCE MANAGEMENT

Team efforts pay dividends

Thailand's coasts are home to 23 sea-straddling provinces, where the people's wallets and stomachs are inextricably linked to marine fisheries, coastal aquaculture, agriculture and eco-tourism. Owing to perennial trends like industrial and tourism development, pollution, illegal fishing and the resulting depletion of marine resources, these coastal communities have had difficulty keeping their heads above water at times. Faced with shrinking opportunities and mounting debt, many locals have had to make do with living hand to mouth or moving to bigger urban areas in search of better opportunities.

Aquaculture plays an important role in many of these provinces, as well as in the food security and economy of Thailand as a whole. Freshwater aquaculture, especially on a small scale, is mainly aimed at domestic consumption and is a crucial source of protein for the Thai diet.

In contrast, brackish water aquaculture, particularly shrimp farming, is export-oriented. According to the Food and Agriculture Organization, aquaculture is one of the fastest-growing food producing sectors globally, with a projected increase in market share of up to 62 percent by 2030, mainly due to decreasing marine resources and increasing demand from the global middle class.

Tambon Administration Organization:

The governmental body that administers a tambon, the local unit that forms the third administrative level in Thailand under provinces and districts. Tambon are usually made up of about ten villages.

If responsibly developed and practiced, aquaculture, both small and large-scale, can generate lasting benefits for global food security and economic growth. Thailand's shrimp exports for the first six months of 2016 alone totaled some 94,000 tons, ranking the kingdom third globally behind Ecuador and India, according to the Food and Agriculture Organization. However, aquaculture here is increasingly facing environmental and human rights challenges. Much of the current controversy is centered around irresponsible practices such as land rec-

Koh Panyee fishing village in Phang-nga province.

lamation, coastal construction, drainage and the discharge of wastewater and sewage from intensive shrimp farms into the sea, rivers and irrigation canals. Shrimp aquaculture has also resulted in the physical degradation of coastal habitats through the conversion of mangrove forests and destruction of wetlands, salinization of water for agricultural and drinking water purposes, and land subsidence due to groundwater contamination. Heavily criticized for relying on illegal migrant labor, the industry is also being challenged to use resources more efficiently by adopting responsible aquaculture farming management and practices.

Prior to the decentralization of governing bodies and agencies in the late 1990s, communities traditionally came together under an umbrella of local groups whenever they wanted to improve these situations and had a say in the management of their resources, which was usually decided by the powers-that-be in Bangkok anyway. But the gradual process of decentralization has provided more social inclusion and public participation, just as it's inspired the formation of local administrative bodies, such as **Tambon Administration Organizations**. These organizations have allowed the coastal communities to mobilize

and network on a scale that would have been unimaginable just a little over a decade ago.

Today, an increasing number of coastal community residents have decided to fight for more sustainable development practices, namely by mobilizing their peers to call for changes to government policies that would allow them to have a say in the state's management of the marine and coastal resources. There is strength in numbers, but also in diversity – an astounding array of local leaders from monks to village heads, teachers to activists, marine biologists to lawyers, have leading roles in this mobilization.

The involvement of local and national academics and research institutions, which have lent their support through in-depth research, analyses and practical solutions, has helped strengthen the community agendas. Finally, local, national and international non-governmental organizations have also facilitated the networking and mobilization efforts and helped link the grassroots efforts to larger calls for policy changes. These combined efforts have helped fill gaps in coastal communities, which otherwise would lack science-based knowledge and modern management skills.

GROUNDBREAKERS

For some three decades, the marine conservationist and activist **BANJONG NASAE** *has been at the forefront of current affairs affecting the country's coastal communities.*

As governments come and go in Thailand, policies fluctuate with them, which makes advocacy a difficult proposition. Who suffers from these frequent changes of the political guard and their self-serving policies? The small-scale fishermen that do not have the money or connections to compete with the clout of big commercial fishing outfits, tourism operators or other business interests.

A native of Songkhla province with a law degree, Banjong Nasae decided to pursue a career with NGOs because of his personal connection to the people living in coastal communities and his attachment to the sea. "After graduation, I saw local people bearing the brunt of considerable hardship due to the depletion of natural resources. So I decided to use my degree in law to help them," he said.

To give them a voice and a platform, Banjong has often turned to the media. "The media is a channel that has echoed the voices of local people and played a role in helping us to achieve our missions," he said. "Social media has provided an additional dimension for people to reflect the problems in their areas and deliver their messages to the public."

As the director of the NGO known as the Thai Sea Watch Association, Banjong and his cohorts are locked in a long-running battle over the Southern Seaboard Development Scheme that includes building petrochemical plants and other potentially polluting industries, which can have massive repercussions on marine ecosystems and fish stocks. "We'll continue fighting against it. If we lose the fight, we'll also lose our sources of food and income," he said.

Even after three decades of triumphs and setbacks, Banjong is still ready to rock the boat and his aspirations remain the same: bringing prosperity and peace of mind to fishing communities that have enjoyed little of either.

▶**PIONEERS**

Mae Klong Community Network

History: Formed in the 1990s

Location: Samut Songkhram province

Key features: Bringing disparate communities together to protect their waterways by creating a water management plan to show the authorities, as well as branding and promoting their own products

Before the building of dams, dykes, highways and large-scale fisheries in the 1980s, Samut Songkhram, the hometown of the famous "Siamese Twins" and Thailand's smallest province, enjoyed the distinction of being the "town of three waters" (those being the Mae Klong, Tha Chin and Nakhon Chaisi rivers). As a coastal province some 72 kilometers southwest of Bangkok, Samut Songkhram is at the confluence of where freshwater from the Mae Klong River and the many canals branching off it combines with the saltwater from the

Pla tu, *a well-known fishing product of Mae Klong.*

Since they had a smaller number of resources to work with, locals focused on adding value to the area's signature products.

Gulf of Thailand to form a truly original ecosystem.

That diversity makes for a fertile and fruitful province, home to orchards, salt farms and marine fisheries. Because local operators and farmers were so well off and financially independent, they tended to live quite isolated lives. Community spirit was lacking. As Surajit Chirawet, a former senator from the province, said, "Each of us was like a solo artist. We didn't rely on the others nor live in the close vicinity of villages or communities."

The "three waters" that were the lifeblood of the community began to lose their currency in the 1980s, when the construction of four large dams upstream from the Mae Klong River, and numerous small dykes built along the canals, disrupted the flow and polluted them. In addition to those developments, shrimp aquaculture and the extensive use of trawling nets reduced the number of other marine species such as crabs and mackerel.

After watching their revenue streams dry up for over a decade, Surajit, as the chairperson of the provincial chamber of commerce, saw the need to bring people together to have a say in the state's policies that affected their hometown. Bringing all the different groups together was an uphill battle. So he settled for uniting smaller enclaves with similar interests. "We formed networks of people who grow this type of fruit or catch this kind of fish or trade in this sort of good," he said.

From there it was essential to build up databases of local research and statistics so the groups could deal with the state agencies and their bureaucratic processes. By 2004 the Mae Klong Community Network was ready to submit a proposal on water management and the maintenance of the "three waters" to the provincial authorities.

The network also had to create a new marketing plan for their products. Since they had a smaller number of resources to work with, the locals focused on adding value to the area's

signature products: palm sugar, palm oil, shrimp paste and mackerel. To give their seafood another promotional push the network masterminded an annual Mackerel Eating Festival every December. It has now been running for nearly 20 years. The food fair features musical performances, cooking contests and a bustling bazaar. As Surajit said, "The festival has added value to our fish products and expanded our distribution channels."

Even though many of their products are now famous around the country and the province's economy and "three waters" are in better shape, the network still meets regularly to make sure they can sustain that success in tandem with the waterways that make it possible.

A vendor transports coconuts, which are abundant in the Mae Klong–area floating market.

Finding cockles on the muddy beach is a traditional Mae Klong way of living.

▶PIONEERS

Pak Phanang River Basin Royal Development Project

History: Started in the 1990s

Location: Pak Phanang River Basin in the provinces of Nakhon Si Thammarat, Phatthalung, and Songkhla

Key features: Building a watergate to keep out seawater and store freshwater; zoning the land for different kinds of farming and water usage

Situated in a shallow bay below sea level, the Pak Phanang River Basin covers a total area of 300,000 hectares and is home to around 600,000 people. The basin, located in parts of four different provinces, is fed by freshwater, brackish water and saltwater. Once a fertile plain known as the rice bowl of southern Thailand, the basin began to deteriorate in the 1980s due to urban encroachment and the destruction of the watershed, which triggered heavy floods during the monsoon season and disastrous droughts during the dry season.

Many rice farmers converted their rice fields into shrimp aquaculture, an export-oriented business that has boomed in Thailand in recent decades as demand for shrimp soared around the world. However, intensive shrimp aquaculture typically relies on antibiotics and chemicals to increase yields and leads to saltwater seepage, polluting the groundwater and ecosystem, rendering it unsustainable for farming in the future. Wastewater from shrimp ponds was not treated properly in the Pak Phanang River Basin and was released into the river, damaging agricultural land, especially paddy fields, and forcing some rice farmers to leave their land. Diseases among shrimp populations, which are common due to the monocropping nature, can cause the spread of pathogens in wetlands.

All these problems ultimately boiled over into a prolonged local conflict, pitting the freshwater farmers against the shrimp farmers. Into the fray stepped King Bhumibol Adulyadej. Following his advice, the Uthokawiphatprasit Watergate was constructed in 1999 about three kilometers from Pak Phanang district. This watergate would help to address the problems of drought, flooding, salinity and a lack of freshwater.

Its main functions were twofold: preventing saltwater from seeping into the river and contaminating farmland, as well as storing freshwater to use for agricultural purposes and household consumption. The project also included

Shrimp farmer in a saltwater zoning area at Pak Phanang River Basin.

the construction of a long embankment to separate freshwater and saltwater, and an irrigation system to tackle flooding and water pollution by channeling water to a special holding tank in the basin.

Zoning was another integral part of the plan. This meant that farmers residing in the different zoning areas had to consider changing their crops based on the type of water available in their localities. For example, shrimp farmers converted their saltwater farms into freshwater agricultural plots. The zoning has also helped end prolonged local conflicts. In addition, the watergate has mitigated flooding in the area, while the saltwater irrigation has helped increase the productivity of shrimp aquaculture.

The project has not only restored the livelihoods of local people but also boosted their revenues. Since 2004 the average income of the locals has increased by 28 percent. Overall, the Pak Phanang River Basin Royal Development Project's greatest strength lies in its multifaceted approach that has boosted local standards of living just as it's been a boon for the environment.

The Pak Phanang River Basin development project began with the construction of Uthokawiphatprasit Watergate in 1999.

HISTORIC PRESERVATION

Preserving the past for posterity's sake through community participation

"A nation's culture resides in the hearts and the souls of its people." This quote by the spiritual leader, Mahatma Gandhi, delineates the essence of the cultural heritage of a nation-state. Historical monuments, archaeological sites and the handing down of traditional practices and disciplines, such as crafting, performing and writing, are traces of the existence of previous generations. They are integral to the identity of a people, and they represent the story of a nation. Thus, preservation work is not simply about protecting one's heritage from physical wear and tear, but it is also about safeguarding and promoting the communities connected to it.

In Thailand, balancing the value of the past with constant striving for the new is a tremendous challenge. Developers, mass tourism, cultural commodification and a general disconnect from past traditions – in short, modernization – conspire to make heritage management a field fraught with tensions. But over the past several decades Thailand has seen some notable success stories that have effectively balanced economic, social and environmental factors in the name of cultural preservation.

Some of these, such as the ancient cities, cover large-scale historic areas, while others, such as the floating markets, cover everyday activities from bygone days. Many of these sites and activities are drivers of the country's tourism industry, which generates as much as 7 percent of GDP. Community involvement in such preservation work varies in form depending on the project or activity.

State-sponsored renovation projects at ancient cities such as Ayudhya, Sukhothai and Udon Thani's Ban Chiang, all of which are UNESCO World Heritage Sites, create only limited, direct community involvement because the renovation work of these protected areas relies mostly on experts and technical and theo-

Historic preservation: Traditionally, the term "historic preservation" includes four different activities: preservation, rehabilitation, restoration and reconstruction.

"The value of such heritage lies in the way of life, the traditions and activities of the people in that community."

Yongtanit Pimonsathean, lecturer in the Faculty of Architecture, Thammasat University

A floral installation by Wat Chaiwattanaram in Ayudhya.

retical methodologies. Villagers are hired for their labor and can benefit from selling goods and offering tourism-related services. In addition, the sites and their ability to attract global travelers create a sense of pride, history and identity for the locals.

Thailand also boasts some excellent examples of community-based heritage conservation, such as the Shadow Puppet Troupe of Wat Khanon in Ratchaburi province, which won the Better Practices in Communities' Intangible Cultural Heritage Revitalization by the Asia-Pacific Cultural Center for UNESCO in 2007.

Another award-winning community is situated around Lampang's Wat Pongsanuk, where the young and the old have worked with the monks to restore the temple to its original grandeur. After saving a venerable form of northern Thai (or "Lanna") architecture, the project received international kudos, including a Heritage Award of Merit from UNESCO in 2008, as an example of community-led conservation.

Yongtanit Pimonsathean, a lecturer in the Faculty of Architecture at Thammasat University, said that at

Crowning Achievements

Since 1997 the Crown Property Bureau (CPB) has worked to preserve hundreds of the kingdom's most cherished architectural sites, from temples and palaces to the colonial flourishes of the Oriental Hotel's Author's Wing. The CPB's renovation projects have four main cornerstones: extensive research on the history of the premises, examining all architectural details, fine-tuning the renovations, and writing detailed plans for the care and maintenance of the property.

These conservation projects are also intended to provide learning experiences for the general public and concerned parties who may want to restore their own properties. With the proper permission, anyone can visit these sites to consult with the relevant specialist on the finer points of architectural restoration.

One remarkable aspect of these prize-winning projects is that these are not museum pieces inhabited only by ghosts of the past. No, many of them are fully functioning buildings with tenants. According to Oranuch Im-Arrom, head of CPB's Conservation Management Department, many are proud to reside in such celebrated abodes, which is also the key to preserving them. "Bringing back historic buildings to their former state of glory and to serve people as they used to is the key to achieving sustainable preservation, while sharing this knowledge among the general public will provide a better understanding of architectural values and what these buildings really mean," said Oranuch.

historic sites situated within an active community, the locals serve an integral role in preserving the spirit and authenticity of the site. "The value of such heritage lies in the way of life, the traditions and activities of the people in that community," he said.

Yongtanit pointed out that preserving the past is not only a source of pride for these communities, but also presents a number of financial benefits, such as earning money from tourism. That's one reason why the residents of the rustic Sam Chuk district of Suphanburi province agreed to use some of their own funds to renovate the old shophouses in their area. Collaborating with experts in architectural conservation, the community rebuilt and reinvented itself as an award-winning enclave of tourism.

The success of any such preservations efforts, whether it's the tangible heritage of timeworn monuments or the intangible heritage of shadow plays, can be achieved through a joint venture among locals, academics, state agencies, artists and even monks to determine how to manage heritage in a way that respects not only its significance in the past, but also creates value and meaning for the future.

▶**PIONEERS**

Sukhothai Historical Park

Visitors: 1 million in 2015

History: The first Thai kingdom lasted from 1249 to 1378

Location: Sukhothai

Key features: Zoning management, protection of the historical sites, arts and crafts promotion. This UNESCO World Heritage Site features famous temples, monuments and Sukhothai-style Buddhas; the designation also includes the ancient sister cities of Si Satchanalai and Kamphaeng Phet

Ancient Buddha statue in Sukhothai Historical Park.

As the birthplace of the Thai nation, no city has more historic or cultural significance in the country than Sukhothai. Declared a UNESCO World Heritage Site in 1991, the ancient kingdom dates back to the 13th century and is still revered for its royal palaces, Buddhist temples and masterworks of iconography that set the standard for many Thai arts and crafts.

These relics were almost destroyed by tomb raiders pilfering them until the government stepped in to give the area official protection in 1962. Since then it has fallen under the purview of the Fine Arts Department, which has a five-pronged plan to protect it by maintaining the historic town as a cultural center; creating connectivity with adjacent tourist destinations; generating income for locals; supporting community involvement; and conserving the natural environment.

With dozens of ruins spread over a sprawling park, maintaining it is a Herculean task that must strike a balance between preserving the site's authenticity and maintaining it as a tourism attraction. The former implies keeping it real so that the cracked mortar of pagodas exposes the patchwork of bricks beneath and Buddha images

are realistically weatherworn. The latter infers that the park's infrastructure can cope with mass tourism.

In this respect, Sukhothai is exemplary in providing an array of amenities for visitors without clumping the crowds all together. The park is easy to navigate. Restaurants and restrooms are plentiful but not obtrusive. Cars are forbidden, bicycles encouraged. The lack of vehicular traffic is crucial to preserving the park's natural integrity and atmosphere. Big trees provide shade, flowerbeds provide color, and ponds dappled with sunlight and water lilies provide a serene and photogenic setting.

Nongkran Sooksom, the director of Sukhothai Historical Park, said the park and its sister sites, Si Satchanalai and Kamphaeng Phet, attracted more than one million visitors in 2015. Those visitors filled local coffers, with restaurants, bars, transport companies, travel agencies and hotels enjoying the benefits.

In heritage-related tourism, historic attractions are but the tip of the pagoda. Many visitors are also keen to shop for antiques and curios with an exotic and local edge. In the case of Sukhothai, which has a grand artistic legacy to uphold, that's a challenging proposition, but locals have managed to revive

the age-old techniques of Sukhothai handicrafts. One brainchild of the present villagers is Sukhothai celadon pottery that mimics the ancient style. This pottery, which was a major export in the golden age of Sukhothai, is once again a revenue stream. Home factories equipped with old-looking kilns are now spotted around the town.

Tourism has also revived the goldsmith industry, which also aims to copy ancient Sukhothai designs. The traditional weaving of the Tai Puan ethnic people in Ban Had Siew village has made a comeback too. As Nongkran said, "Historical sites must be able to put money in the pockets of villagers through such means because cultural preservation is intimately linked to sustainable development and economics."

For Sukhothai, the big challenge ahead is maintaining its local flavor in the face of rising mass tourism, as an interconnected road network among Asian countries is set to bring more traffic. According to Nongkran, in recent years land owners surrounding the ancient sites have been moving away, selling their plots to opportunists from outside the community who are thus unable to convey or represent the identity of Sukhothai people.

▶ **PIONEERS**

Wat Khanon Shadow Puppet Troupe

History: Over 300 puppets have been in the temple's possession for over a century and were used in shadow plays by generations of performers until the 1960s when modern entertainment arrived.

Location: Ratchaburi province

Key features: An inspired abbot sought to revive the art to benefit the local community; now regular performances of shadow plays draw crowds and instill cultural pride and artistic acumen in local youth

When the haunting sounds of a traditional *pipat* ensemble strike out at Wat Khanon in Ratchaburi province, dozens of young boys dressed in red outfits gather in front of the stage inside an open-air theater. One boy holds up a large puppet made of cowhide to begin the shadow puppet performance known as *nang yai*.

"*Nang yai* is the pride of Wat Khanon and the locals because the troupe was established by our ancestors," said Phrakhru Pitak Silpakom, who supervises the troupe and is the abbot of the temple.

In Asia, shadow plays are well over a thousand years old. According to UNESCO's "Performing Arts in Asia" publication from 1971, shadow puppets were brought to Java in the ninth century by Jayavarman II after he established Angkor. From there the art spread to Indonesia, India, China and Thailand, where the first references to it date back to 1458.

As televisions became more affordable in the 1960s, the shadow play began to wane. In 1990, Phrakhru Pitak and local puppeteers decided to stage a comeback by training local youth to enact the old performances. The current troupe of puppeteers/dancers at Wat Khanon is made up of 35 boys from nine years old and 15 teenage girls, mostly students from the temple's nearby school.

The performances are based on popular episodes from the *Ramakien*, a Thai version of the Indian epic poem *Ramayana*, featuring choreographed battles and dances, live music and lyrical narration. The free, 30-minute shows are held every weekend at the temple.

> ## "Nang yai *is the pride of Wat Khanon and the locals…*"
>
> **Phrakhru Pitak Silpakom, abbot of Wat Khanon**

Phrakhru Pitak said the key to sustaining this project depends on the support of three factions: the young performers who spend time rehearsing; the parents who encourage their children; and the spectators who come to support them. However, the abbot believes a bigger network is vital to the survival of such time-honored arts. "We will not be able to prolong the Thai traditions by ourselves. We need to make connections with other preservationists and traditional performers so that we have a network," he said.

After receiving the 2007 award for Better Practices in Communities' Intangible Cultural Heritage Revitalization by UNESCO, the abbot has future plans, such as building a Nang Yai Exhibition Hall for their collection of 313 vintage puppets. "We're constantly learning more about how to continue and thrive. But our future projects will still focus on creating more awareness about cultural heritage among our young people," he said.

AROUND THE WORLD

Hoi An: A Model Town

Inscribed on the UNESCO World Heritage List in 1999, the Vietnamese town of Hoi An is an exceptionally well-preserved example of a Southeast Asian trading port dating back to the 15th to 19th centuries that melds the influence of Chinese, Japanese and European settlers.

As Vietnam opened up in the 1990s and tourism began to boom, the economy of the town was given a dramatic injection of foreign capital. All sorts of tourism-related services,

accommodation and restaurants sprouted up, with vendors, touts and tailors to staff them.

Such growth has put untold pressure on the site and raised concerns about how to sustain this momentum while preserving the old town's myriad charms. So far, those fears have been unfounded. Hoi An has proven to be a model of heritage management that is also self-sustaining, partly because revenue raised from entrance fees is reinvested into tourism promotion, heritage conservation and management that ensures its ongoing viability.

URBAN DEVELOPMENT

Building a proper platform for low-income communities

Owing to its oversized role in the country's politics and economy, Thailand's largest urban center, Bangkok, has exercised a gravitational pull over the rural poor who flock to the capital in search of better jobs, brighter lights and higher living standards. They form the backbone of the city's underclass and constantly increase the ranks of urban poor living in some 1,000 slums across the city.

In general, slum areas have typically been passed over by governments for improvements, further exacerbating Thailand's growing divide between rich and poor and creating a society of haves and have-nots. Instead of hope and opportunity, the slums breed crime, violence and substance abuse. Most are overcrowded firetraps with hovels and tin shanties for rooms, which offer little in the way of safety or security for slum dwellers. Many of the residents are squatters with few rights who can be evicted at any time on short notice.

Low-income housing communities are common along Bangkok canals.

Proper housing is crucial for sustainable development.

"Our main objective is to empower poor people to deal with their own problems."

Thipparat Noppaladarom, former director of CODI

"Creating new urban environments must involve developing communities in a humane and sustainable manner," said Somsook Bunyabancha, a preeminent housing rights advocate with decades of experience in urban planning. "Proper housing is extremely important for the sustainable development of any urban area."

Somsook is one of the masterminds behind the government-run Community Organization Development Institute's (CODI) Baan Mankong ("Secure Housing") Collective Housing program.

The program assists communities of the capital's urban poor in upgrading their living environments within slums through infrastructure subsidies and housing loans. Working hand in hand with municipal authorities, experts, urban planners and non-governmental organizations, enterprising slum dwellers can take the initiative in collaborative efforts to make their communities more livable through the creation of better housing and public spaces. They can also receive legal and technical support.

Set up in 2003, the agency has helped locals build almost 100,000 new homes in more than 1,800 communities as part of some 933 collaborative grassroots projects. These projects are active not only in Bangkok but also nationwide. For this bottom-up rather than top-down form of development to really work, the program demands that participants take a proactive approach.

"Our main objective is to empower poor people to deal with their own problems," explained Thipparat Noppaladarom, CODI's former director. "People have to be the principal actors behind sustainable development projects by taking charge and ownership of their communities. In the past, many poor people

In 1988, **SOMSOOK BOONYA-BANCHA** *helped found the Asian Coalition for Housing Rights, a non-governmental, regional organization that promotes the rights of slum-dwellers and seeks to create better housing conditions for them. Until recently she was also the director of Thailand's Community Organization Development Institute, an agency advocating for housing rights for the urban poor.*

were content to be passive recipients of help from the government in a top-down approach. Now more and more of them are seeking to help themselves through bottom-up initiatives."

This theme of self-empowerment runs through many other initiatives in the capital that aim to help the disadvantaged by allowing them to help themselves. The Human Development Foundation, situated in the city's biggest slum of Klong Toey, is one such organization. Providing proper schooling for the offspring of slum dwellers to give them a shot at a decent future outside the slum may be its most renowned feature, but the foundation is really a microcosm of urban woes for the less fortunate, as it also serves as an orphanage for homeless children, a hospice for terminally ill patients, a center for legal aid and a credit union for impoverished women who can receive start-up loans for small business ventures.

As its name suggests, the Human Development Foundation puts people first, a trait that is also prominent in CODI's housing projects, and addresses the social inclusion aspect that is so integral to all forms of sustainable development.

How can more socially inclusive and livable environments be created in Bangkok for the disadvantaged? Urban planning should be a process through which experts engage people in collective decision-making. For plans to work, they cannot be done in a top-down approach. They need to be based on a bottom-up approach that helps create a practical vision for communities according to their own needs. We sorely need land reform in this country and because of the complex land ownership situation, planning must often involve a fine balancing act between the needs of landowners and informal communities that squat on their lands. According to a land-sharing arrangement that we have devised and that has worked well over the years, landowners get to develop the more commercially valuable street-front portion of a plot while squatters retain the back portion for their own more modest housing arrangements. It's a matter of compromise but can be a win-win situation.

What do you see as the greatest challenge in creating more comfortable urban environments in Bangkok? Because of the breakneck speed of economic development over the past decades, much of the city's growth has been helter-skelter with little foresight and proper planning. The city has grown too fast for its own good. The infrastructure of well-to-do commercial hubs in the city has been well developed but less prosperous areas have often been neglected. The city's economic development has become uneven and unequal.

What is the way forward for the city and its residents? We need to capture the energy of new human potential. All sustainable development models need to be based on our greatest asset: people.

▶PIONEERS

Mercy Centre, a school and second home to Klong Toey kids.

In a school flanked by the shacks of squatters, scores of children are learning the alphabet. They're the offspring of destitute city dwellers and economic migrants from the countryside – manual laborers, construction workers, street vendors and scavengers.

The school inside a Bangkok slum is just one of three dozen kindergartens and preschools that have been built over the years by the Human Development Foundation, a nonprofit with a decades-long track record of looking after the poorest of poor children in Thailand's capital, while providing more services for the slum's indigent inhabitants.

> **The Mercy Centre's staff and volunteers, in collaboration with the local community, have built or rebuilt some 10,000 homes in Bangkok's slums.**

"Over the past 40 years more than 50,000 children have learned to read and write in our kindergartens," explained John Padorr, a long-time adviser to the foundation. "Many of them have continued their education past primary and secondary schools and now make their own way in a world far beyond their parents' and grandparents' dreams."

The foundation's flagship humanitarian enterprise is its Mercy Centre, a well-equipped, multi-story establishment that is home to a large orphanage for homeless street kids, a kindergarten for 300 students, a school for special-needs youngsters of various ages, and a hospice for terminally ill patients. Since its inception in 1973, the Mercy Centre has been run by its founders, Joseph Maier, a Roman Catholic priest from the US popularly known as "Father Joe," and Sister Maria Chantavaradom, a Catholic nun.

The building has long served as a welcoming haven for neglected and abused children in the hard-up district of Klong Toey. Yet Father Joe refuses to let them see themselves as victims by urging them to take control of their lives. To do that, education is key.

Wannee Kitswad, the center's general manager, espouses the same faith in learning. "Education in deprived communities is paramount. These children's parents are often uneducated themselves but they want their children to learn and have a better life. They're very supportive. Every day we have garbage collectors and manual laborers bringing their children to school."

In a testament to the transformative power of education, several of the Mercy Centre's teachers are graduates of its schools. Beyond teaching essential skills to children, the nonprofit's teachers also instill basic virtues in them: diligence, honesty, prudence and generosity.

The Mercy Centre's staff and volunteers, in collaboration with the local community, have helped build or rebuild some 10,000 homes in Bangkok's slums following frequent flash fires that can devastate entire blocks of shanties within minutes. It also runs a center for legal aid and a thriving credit union for some 2,000 impoverished women.

As Father Joe once mused, "A hero in Klong Toey is a soul beaten up but never beaten. It's someone who keeps going no matter what."

With chronic problems related to housing, health, crime and access to education, Bangkok's Klong Toey district has posed numerous community development challenges over recent decades.

▶PIONEERS

Four Regions Slum Network

History: Formed in 1998, the network is comprised of eight community rights groups: Community Development Center, Under the Bridge Community Development Group, South-West Railway Community Network, Rama III Slum Railway Network, Homeless People Network, Community Network for Development, Creative People Network, and Southern Rights Development Network.

Location: Nationwide

Key features: Advocates for community and housing rights, and works to improve the lives of the urban poor.

Understanding and upholding community rights can still be a challenge in Thailand where forced evictions remain a common occurrence. According to the Four Regions Slum Network (FRSN), thousands of families across the kingdom face the possibility of eviction over land development and infrastructure projects including the construction of highways, high-rises and high-speed railways.

Often, FRSN is the first organization that would-be evictees turn to for help. Acting as a buffer between communities targeted for eviction and the authorities

and property developers who would seek their removal, members of FRSN insert themselves into these precarious situations with the goal of finding a long-term housing solution for the evictees. Ideally, this means that if a community has to be relocated, the relocation site will be in an area that is not too remote for the residents' work or school commute.

"The state must ensure that relocation will benefit the evictees. Even though they live on land that belongs to others, their houses are their everything," said FRSN Chairperson Nuchanart Tanthong, adding that access to schools and jobs are of utmost importance. "If there is no concern over this, they will definitely move back to the city."

By encouraging all stakeholders to think about long-term housing security, FRSN has helped parties in numerous disputes to reach viable solutions.

For example, in the year 2000 when a community that had lived on private land in Bangkok's Asoke area for more than 50 years was suddenly faced with impending relocation, FRSN stepped in and implored the landowner not to utilize the authorities to carry out an eviction. Instead, FRSN helped the parties reach an agreement whereby the community could temporarily rent the land while negotiations took place, which ended up taking around 15 years. Eventually, the landowner agreed to purchase a 15-rai plot in Minburi to be

"The state must ensure that relocation will benefit the evictees. Even though they live on land that belongs to others, their houses are their everything."

Nuchanart Tanthong, chairperson of Four Regions Slum Network

jointly owned by the benefactor and the 213 evictee households.

Another notable success in 2000 occurred when FRSN secured a Memorandum of Understanding with the State Railway of Thailand to ensure that 61 communities in the Klongton area of Bangkok were able to rent land at a fixed price of 20 baht per square meter – dramatically cheaper than the typical market price.

FRSN is also well known for staging protests. However, since 2015 the group has seen its more visible acts of advocacy blocked by the Public Assembly Act. Passed by the junta-installed National Legislative Assembly in July 2015, the act requires anyone who wants to stage a public rally to seek permission from authorities at least 24 hours in advance via a request that states the purpose, date, duration and venue of the assembly.

Some members of FRSN were even taken for "re-education" after they demonstrated publicly against the curtailing of their housing rights protests. Since then, the group has had to ask for permission every time they have wanted to stage a protest.

A housing rights demonstration by Four Regions Slum Network.

▶**PIONEERS**

Bang Bua Canal Community Upgrading, part of the Baan Mankong ("Secure Housing") Collective Housing Program

History: Baan Mankong ("Secure Housing") Collective Housing was founded in 2003 by the Community Organization Development Institute (CODI)

Location: Nationwide

Key features: Collaborating with indigent urbanites in hardscrabble neighborhoods to make permanent homes for them

Until a few years ago, the residents of this model community on the Bang Bua Canal, which has been lauded internationally as an example of do-it-yourself urban regeneration in low-income settlements, languished in squalor. "This used to be a seedy slum," recalled Sanit Supaka, the community's vice chairwoman who is also a housewife. Crime, drug abuse and violence ran rampant in a grubby and overcrowded settlement of squatters living on public land where they were facing the looming prospect of eviction.

"We came together to work out a way to improve our living conditions," she added. "It started small with a few people, then the projects grew and grew."

Denizens organized clean-ups of the canal, stopped dumping rubbish and pumping untreated sewage into it, and set about tearing down their old shacks so they could build permanent new homes with help from the Community Organization Development Institute (CODI). Through its Baan Mankong ("Secure Housing") program, CODI provides enterprising slum-dwellers and villagers around Thailand with financial loans, legal counseling and architectural guidance for upgrading their dwellings.

CODI's Revolving Fund provides microcredit for those without means in slum communities so they can muster up the capital to undertake these home improvements. Now about 3,400 families live in the 12 informal settlements that line the 13-kilometer stretch of Bangkok's Bang Bua Canal, many of them vendors, laborers and daily-wage workers.

As a guarantee of their dedication to those projects, which can take several years from conception to completion, participating communities need to set up "saving groups" among residents and put up 10 percent of the collective loan they take out. "We say to them, 'We trust you. We know you can do it,'" explained Thipparat Noppaladarom, CODI's former director, noting that the repayment rate on CODI's loans has been a robust 97 percent. "Families help each other out with their savings. For many low-income people it's a dream come true: secure and much better housing in what used to be slums."

> *"We came together to work out a way to improve our living conditions. It started small with a few people, then the projects grew and grew."*
>
> **Sanit Supaka, vice chairwoman of CODI**

FILM: URBANIZED
URBANIZED
Director: *Gary Hustwit*
Release Date: *2011*

"Cities are always the physical manifestations of the big forces at play – economic forces, social forces, environmental forces," says an expert in *Urbanized*, a feature-length documentary by the New York-based independent filmmaker Gary Hustwit. Through interviews with city planners, architects, developers, mayors and thinkers, this must-see documentary explores the ins and outs of successful urban designs and ingeniously repurposed urban habitats, from Beijing to Bogota, in an increasingly urbanized world where half the world's population of seven billion already lives in cities. That figure is projected to rise to 75 percent by 2050, which will place enormous pressure on already limited resources. City spaces come in all shapes and sizes, the film demonstrates. Yet whatever the local conditions, improvements are often just a matter of initiative, whether in economically depressed Detroit, or gridlocked Bogota.

GREEN SPACES

Bringing nature back into the urban jungle

Few laments are as commonly voiced in Bangkok as "We need more parks!" This is often uttered soon before or soon after "We have too many malls" and "We have too much traffic." The underlying sentiment of all three protests is that Bangkokians long for the outdoors – and the fresh air, peace and open space it offers.

Space, proper shading and comfortable connectivity are important to a city's sustainability. They not only impact the physical and mental health of urban residents, but also promote economic productivity as well as a sense of community.

Unfortunately, few city residents in the world enjoy less public green space than those in Bangkok. The City of Angels has only 5.46 meters of green space per person, making it one of the worst endowed metropolises in the region. The WHO's recommended standard of urban green space per capita is 16 square meters. By comparison, Kuala Lumpur has 12 square meters of green space per capita, while Singapore residents enjoy 66 square meters of green space per person, thanks to decades of proper city planning, which has been largely absent in freewheeling Bangkok.

Poor planning and zoning as well as endless construction projects have gradually denuded the city of its greenery and led much of its once world-famous network of canals to be filled in, exacerbating the city's "**urban heat island**" effect. Air, noise and nocturnal light pollution have also taken their toll on the capital's livability.

Urban heat island:

An urban heat island (UHI) is when a city is significantly warmer than the surrounding rural areas, owing to the heat generated by its buildings, infrastructure, vehicles and human activity.

Having recognized the need for more green space, the Bangkok Metropolitan Association has gradually been turning back the clock, with 980 rai of green space added in 2013, an additional 1,459 rai in 2014, and another 1,350 in 2016, including nine public parks. Fortunately, Bangkokians are also innovative with the space they have been given. With little room left for ground-level development across much of Bangkok, there is nowhere

Lumphini Park – Bangkok's first green lung – in the heart of the city.

to go but up. The Skytrain's shaded, open concourses known as skywalks, for example, are now favored over cluttered and damaged sidewalks. As new high-rise projects sprout up in areas already densely populated, many developers seek to compensate for the absence of green space at ground level by going green vertically. Rooftop gardens and landscaped terraces are becoming increasingly common. A case in point is the relatively new EmQuartier mall, located in the heart of Sukhumvit Road. The high-end shopping center has a 3,000-square-meter indoor tropical garden with dripping ferns, stand-alone trees, cultivated orchids, artificial lotus ponds and a cascading waterfall.

Several new high-end apartment complexes, which dot Bangkok's skyline in towering clusters at premium locations around town, offer alternative green spaces in the form of sky courts, sky gardens and landscaped terraces, which are often integrated into the core designs of new structures. They include such luxury private residence projects as Tropicana Eco Green Condo in Bangna and Circle Sukhumvit 11 at the eponymous location. The elevated architectural platforms of such buildings may not compen-

The WHO's recommended standard of urban green space per capita is 16 square meters.

sate for the loss of natural ecosystems and habitats at ground level, but they do help add to the sum of available urban greenery. In addition to providing some much-needed greenery and alternative social spaces, landscaped terraces and rooftop gardens can also help reduce ambient temperatures and filter out the perpetual din from the street-level bustle for the residents of hybrid buildings.

Another intesresting project is Metro Forest in the northeast district of Prawet. Developed by PTT on company-owned land, the initiative has transformed most of the available 12 rai into a self-sustaining native forest with its own indigenous ecosystem. The project seeks to reconnect locals to the environment in a sustainable manner while also serving as an example for a national reforestation campaign. Making Bangkok greener will entail many more similar forward-looking initiatives and far more responsible urban planning.

Bangkok's Green Lungs

Bangkok is home to more than 6,000 sites classified as parks, green spaces or public sites, but for all that, its major public parks make up less than half a percent of the total land area of the metropolis. However, that doesn't mean there are no escapes.

The city's first and most famous major park is the sprawling yet well-manicured Lumpini Park. Created in 1920 by Rama VI, the park boasts an artificial lake, 2.5 kilometers of paths for jogging and walking, and numerous playgrounds.

One of the city's most gorgeous and popular islands of greenery is the Chatuchak Discovery Garden. This oasis of green space and fresh air, made up of three interlinking public parks (Chatuchak, Queen Sirikit and Rot Fai), is situated on a total of 112 hectares. It boasts an inviting patchwork of well-tended lawns, scenic fields, lotus and lily ponds, shady woods, botanical gardens, nature trails, bicycle lanes and exercise areas. The garden supports its own mini-ecosystem that comes complete with squirrels, monitor lizards and exotic birds.

Across town, Rama IX Garden features an 80-hectare expanse of greenery with an artificial lake, picturesque walkways, Zen gardens, pavilions and playgrounds. In between, a series of smaller parks and other green spaces – Sanam Luang, Suan Somdet Ya, Santi Phap, Bang Kachao, Benchasiri – are playgrounds for nature lovers, picnickers and joggers alike.

▶PIONEERS

Big Trees

History: Started in 2010

Location: Bangkok

Key features: Network of concerned citizens and civic groups that lobbies to protect the city's trees from the developer's axe

Once known as the "Venice of the East" for its abundance of leafy trees and network of canals, Bangkok has become a concrete desert with a few oases of greenery. The statistics speak for themselves. In 2011, the Asia Green City Index revealed the glaring disparity in public green spaces between Bangkok, which only had three square meters per person, and Singapore, which provided 66 square meters per person. The average of all 22 cities on the index was 39 square meters.

While the powers-that-be have failed in altering this landscape despite numerous promises from a string of Bangkok governors, a group of concerned citizens have banded together to protect the city's big trees by working from the grassroots up.

Oraya Sutabutr, the founder of BIG TREES, spoke for many of the city's denizens when she said, "No one can deny that the weather has become hotter, the air more polluted and the traffic even worse despite mass transit systems like the Skytrain and subway. People can no longer put up with traffic jams, but they start walking and they realize that the streets are too polluted and too hot, because of the lack of trees and green areas."

One of the most popular and active civic groups in the city, BIG TREES started in 2010 when they launched a campaign to save a huge rain tree in Sukhumvit Soi 35. The tree, located on private property, was going to be cut down as the landowner wanted to sell the plot to a nearby shopping complex. Using social media sites like Facebook, the group united nature lovers at loggerheads with Bangkok's mall-heavy development schemes.

The campaign could not save that tree, but it did succeed in sowing the seeds of awareness about the value of such trees, which give off oxygen, suck up the CO_2 produced by burning fossil fuels and provide a natural form of air-conditioning for houses surrounded by them. Parks and trees tend to boost property prices, the group noted. In Bangkok, big real estate developers such as Sansiri have begun to preserve the original trees around their condos and create green spaces as these attract high-end buyers and add value to units.

Since that first headline-grabbing campaign, the group's social media sites have become popular places to congregate for nature-loving people to share their campaigns to protect large trees in their community, or save green spaces in the public domain from being bulldozed for commercial ventures. Indeed, the BIG TREES Facebook page now has some 140,000 followers.

A few years ago, BIG TREES spread its roots. They joined forces with an association of architects, architecture students and other civic groups to ask the State Railway of Thailand to convert a large tract of greenery in the middle of the city into a park instead of selling

MALL SPACE VERSUS GREEN SPACE

25 BIGGEST

SHOPPING MALLS

IN BANGKOK (1,157 ACRES) **2,926** RAI

VS.

25 LARGEST

PUBLIC PARKS

IN BANGKOK (1,011 ACRES) **2,556** RAI

Source: Thai Climate Justice Working Group

A large rain tree in Nan province.

it to developers. That is an ongoing campaign. More recently, the group championed a campaign to prevent the authorities from revising a town plan that would turn the city's "green lung" in Bang Krachao into another industrialized suburb of Bangkok. Through big campaigns like these, the group has garnered more and more advocates, including some high-profile supporters such as Anand Panyarachun, a former prime minister.

More recently, BIG TREES helped established a network of 70 organizations called Thailand Urban Tree Network, which is working to promote arboriculture and collaboration with the private and government sectors in order to preserve and improve urban green spaces in Bangkok and elsewhere.

BIG TREES has joined hands with the Bangkok Municipal Administration's professional arborists to conduct work-shops that train people how to take care of trees. The group is also collaborating with the Royal Forestry Department and Fine Arts Department

to preserve trees in historical parks and trees planted via royal initiatives.

Its other efforts have also yielded positive results. In 2014, the group lobbied CentralWorld to spare the trees that blocked access to a store exit. In the end, the mall agreed to design a new exit to save the trees. Since then BIG TREES has begun working on urban greening projects with a number of other corporate partners such as Ben & Jerry's, SCG, PTT, SCB, and the Stock Exchange of Thailand.

Oraya believes that Bangkok, despite its many environmental challenges, is far from a lost cause. When she started

doing volunteer work two decades ago, her friends told her she was out of her mind because the situation was hopeless. "Now I see more civic groups and individuals come to help. Bangkok residents do not sit idle any longer, waiting for the authorities to do things for them. They realize that they cannot wait for them and they start to help themselves," said Oraya.

The success and popularity of BIG TREES is proof positive that urbanization is not just an issue for town planners and commercial developers, and that power is also vested in the little people who make up these big cities.

> *"Real urbanization must create active citizens who are not afraid to speak up, who spend time monitoring the authorities and making sure officials do their jobs. It is not real urbanization if the city is fully populated by passive citizens who just complain about problems, but sit idle and let authorities and developers do as they wish."*
>
> **Srisuwan Janya, activist and founder of the Stop Global Warming Association**

▶PIONEERS

EnerGaia

History: Founded in 2009

Location: Bangkok rooftops

Key features: Cultivates spirulina, or blue-green algae, which reduces greenhouse gases and acts as a food source

EnerGaia's rooftop gardens may not be all that visually pleasing. They consist of plastic cylinders, each occupying one square meter of space, with viscous green goo in them. But these gardens make up for their lack of aesthetic appeal with plenty of environmental benefits. Part of the burgeoning Bangkok-based urban farm trend, EnerGaia's project is aimed at growing edible algae, which are high in nutritional value and easy to cultivate in bioreactors situated on otherwise unused space on the roofs of highrises across the city.

Spirulina is a cyanobacterium, or blue-green algae, increasingly cultivated worldwide both as a dietary supplement and as a staple of vegetarian diets. It is rich in a superior form of plant protein as well as in a variety of vitamins, amino acids and essential minerals. As part of EnerGaia's product line, the algae come in several edible forms, from flavored energy drinks to pasta. As an added bonus, the algae obtain their energy through photosynthesis, just like plants, thereby absorbing carbon dioxide and releasing much-needed oxygen into Bangkok's atmosphere.

"We take greenhouse gases from the atmosphere and convert them into some of the most nutritious products on the planet," explained EnerGaia's Ingo Puhl, who is a German engineer and agroindustrial entrepreneur. "In other words, it's a sustainable source of protein and has a negative carbon balance by absorbing more carbon than it emits. Everyone who wants to do so can have their own clean and odorless algae-cultivating system in their backyards."

EnerGaia's algae-cultivating skyline project has grown to include three sites in Bangkok, and the company is planning to launch a new project in Phuket with the backing of a European partner to grow spirulina on a commercial scale. EnerGaia has also established partnerships with 13 restaurants, hotels, and health food retailers in Bangkok.

Puhl, who is a long-time resident, sees plenty more potential in spirulina cultivation. A single bioreactor can

grow around one kilogram of algae in a month, meaning a few bioreactors can supply a family with plenty of protein all year round.

Additionally, some types of algae are increasingly being touted as one of the cheapest, most productive sources of biomass for alternative energy. Algae can reduce greenhouse gases not only through photosynthesis, but also by replacing fossil fuels as an energy source. The oxygen-creating potential of algae is significant, too: scientists at the Hydrology Institute in China have found that 1.5 cubic meters of algae can produce enough oxygen for a man weighing 70 kilograms to live for one day.

"I believe in creative, technology-driven solutions to common environmental problems," Puhl says. "The best solutions need to be bottom-up. With these bioreactors you can grow your own food with no fuss and in the process you can reduce your carbon footprint and make Bangkok greener."

The initiative is now gaining some recognition. In 2016, EnerGaia's spirulina project was one of ten to win the Blue Economy Challenge sponsored by AusAID and the Australian Department of Foreign Affairs and Trade.

One of EnerGaia's rooftop gardens, which grows spirulina, a rich source of protein and vitamins.

▶ **PIONEERS**

The Metro-Forest Project

History: Opened in 2015

Location: Khet Prawes, Bangkok

Key features: A previously derelict, 12-rai plot of land belonging to PTT Group was repurposed to become a forest in the city, with 75 percent dedicated to forest area, 10 percent to water and another 15 percent mixed use; also features an on-site learning center

If Bangkok residents have neither the time nor the means to reconnect with nature, then why not bring nature to them? This was the thinking of PTT Group and its then CEO Pailin Chuchottaworn when they developed the idea for the Metro-Forest Project. Taking a 12-rai plot of land on the outskirts of Bangkok, Thailand's richest company decided to pull off the unlikely: regenerate a native forest on a completely derelict piece of land. Fast forward three years and now there is a canopy of foliage rising over a once empty urban landscape.

How did it happen so fast? Well, the PTT company, ecological forest consultant Dr. Sirin Kaewlaierd and Landscape Architects of Bangkok (LAB) used a natural ecosystem model pioneered by Dr. Akira Miyawaki of Japan. This method restores the forest by using only indigenous species and random planting patterns to imitate nature, leading to a natural forest. The Metro Forest Project contains rare and wild plants and traditional tree species native to Bangkok as well as lowland evergreen species found in the central region of Thailand along rivers. On this plot, once earmarked for a petrol station, around 300 plant species are now spreading their roots.

Water systems within the project area create a stream that has different depths at different levels in order to circulate the water and hydrate the forest and prevent flooding. Aquatic plants and vetiver grass are grown around the banks to prevent soil erosion and promote water conservation. The entire park requires no future landscape management once implemented. On site, a skywalk, viewing tower and learning center make the forest and its lessons accessible to the public.

The Metro-Forest Project is part of PTT's long-standing, national refor-

estation work through the "1 million rai reforestation" project. Across the country, the company has been active in analyzing and improving soil quality and reforesting different regions with appropriate tree species. No one typically associates Bangkok with images of green jungle, but there once were thriving ecosystems where the urban jungle stands today. By turning back the clock, PTT hopes to plant the feet of condo-dwelling city people back on firm ground.

In 2016, the project received awards from both the American Society of Landscape Architects and the Thai Association of Landscape Architects.

PTT's Metro-Forest Project in outer Bangkok.

THE ART OF DESIGNING PUBLIC PARKS

FREDERICK LAW OLMSTED: DESIGNING AMERICA

Directors: *Lawrence Hott and Diane Garey*

Release Date: *2014*

Frederick Law Olmsted's legacy is all around us. An iconoclastic, 19th-century landscape architect, Olmsted designed New York's Central Park – arguably the world's most famous city park – before designing a slew of other landmark green spaces across North America. Olmsted regarded public parks not only as essential mainstays of healthy urban living but also as works of art in their own right.

In the PBS documentary *Frederick Law Olmsted: Designing America*, modern experts pay homage to the master who continues to influence the designs of city parks and green spaces around the world.

Both a pragmatist and a visionary, Olmsted set out to create self-contained, man-made natural spaces within often crowded and unwholesome modern cities. In the process, with his architect partner Calvert Vaux, he invented the parkway, created the first integrated park system and helped launch the profession of landscape architecture.

He helped preserve nature as part of urban landscapes by incorporating it seamlessly into man-made designs. Most importantly, Olmsted helped mainstream the idea that well-designed public parks are integral parts of the modern cityscape.

PRIVATE SECTOR
ENTERPRISE

Sustainable Business

Green Buildings

Renewable Energy

Green Manufacturing

Ethical Sourcing

Waste Management

Sustainable Tourism

Social Enterprise

Restaurants

Green Finance and Banking

Indices

Countering Corruption

A sustainable world is only a utopian fantasy without the cooperation of the private sector. As the driver of development, the incubator of ideas, the maker of products, the investor in innovation, an employer and trainer of human resources and a provider of services, the private sector touches all of our lives in direct and indirect ways.

Thus the private sector holds both the potential and the responsibility to drive sustainable development. A single company can have the impact of a thousand households, through greener manufacturing practices, investment in energy efficiency, sensitive labor practices, better waste management or procurement policies that reward companies who also believe in sustainable development. Toshiba, one of the leading electronic manufacturers in the world, has built a green factory in Thailand that has cut down emissions by 75 percent. Mitr Phol, one of the largest sugar producers in the world, has revolutionized waste-to-energy technology and implementation, while amply demonstrating the profitability of such innovations.

In addition, by integrating the principles of the Sufficiency Economy Philosophy into their business practices after the 1997 Asian Financial Crisis, Thai firms – and the private sector in general – has been able to weather several global economic downturns.

Of course, money talks and competition and shareholders demand that firms focus on profits. But studies increasingly reveal that those companies that consider the "triple bottom line" of environmental impact, social wellbeing and good governance, commonly known as ESG, perform better over the long term.

Indeed, new indices that measure this success are increasingly taken into account by investors, and capital is flowing into sustainable development initiatives such as alternative energy projects. In Thailand, banks are allocating profits to provide seed money for green ventures. Such small businesses, from independently owned restaurants in Bangkok to coffee farms in Mae Hong Son, are also influencing the way the public views consumption, inspiring them to recycle, to buy local and to be responsible consumers. In the travel business, some boutique tour operators and eco-lodges take environmental protection seriously, employ locals and promote traditional cultures to travelers.

Business is finally realizing that it cannot be a bystander, that it must take a leadership role in improving society as a whole, that it is indeed not separate from society or the environment at all but dependent on their sustainable development.

SUSTAINABLE BUSINESS

The advantages of moving from CSR to CSV

In the past, companies either thought of sustainability as a luxury or as a temporary answer to accusations of poor social and environmental performance. However, as stakeholders push and pull enterprises in different directions and unforeseen events temporarily disrupt operations, some businesses have gradually come to see sustainability as a core tool for their survival and growth in today's complex economy. In Thailand, for example, the great flood of 2011 cost domestic businesses US$45.7 billion and created shortages in the global supply chains of several industries. Thailand's fishing industry has lost business from developed countries due to human trafficking uncovered in its labor pool. Rising social and environmental movements around the globe underscore both the urgency and magnitude of these problems. Business leaders are recognizing that their own unsustainable business practices directly affect the environment, place burdens on society, and ultimately add to the cost of business itself.

Sustainable business should not be confused with corporate social responsibility (CSR). CSR initiatives assess and take responsibility for a company's impact on the environment and social welfare. The concept began with companies merely striving to be "responsible corporate citizens" by paying taxes and complying with laws. Today, CSR has evolved into companies acting as "good corporate citizens" through philanthropy. In contrast, a business that operates on the principle of sustainable development incorporates long-term, strategic planning that pairs business growth with positive environmental and social continuity based on fairness and equity. In other words, they make sustainability part of the core of their operational mission.

Since the beginning of the 21st century, in addition to integrating the concepts of sustainable development throughout their supply chains, their labor practices and the very products they create, some companies have increasingly tapped into Michael E. Porter and

Effective logistics management is part of sustainable business.

Mark R. Kramer's idea of "creating shared value" (CSV) as a way to enhance competitiveness while advancing economic and social value. Now, CSV is seen as another tool to make business more sustainable.

Business can create shared value in three distinct ways: First, it can reevaluate its products and markets. Targeting the unsolved social needs of the billions of customers in developing countries at the Bottom of the Pyramid, or "BOP," can provide both substantial profits and societal benefits.

The second way to create shared value is to redefine productivity. Issues within long and complex supply chains often increase costs to a business, as well as to society and the environment. Areas of opportunity include energy efficiency in processes and logistics, recycling and reusing resources, improving procurement by increasing supplier access to inputs, updating technology and financing, revamping distribution channels, and engaging employees to increase their productivity.

Finally, building supportive clusters around the company creates shared value. The deficiencies of any party in the cluster – such as partner firms, suppliers, public assets and logistical infrastructure – add to the cost of business. Companies should enable fair and open markets, give suppliers incen-

Integrating SEP in the Private Sector

Sustainable development actions in the private sector are sometimes inspired by Sufficiency Economy Philosophy (SEP) principles. Both government and non-government organizations have also sought to support SEP's integration into corporate policy, value chains and governance through contests that reward companies for best practices or through the creation of new industry standards firms can achieve.

The Office of the Royal Development Projects Board (RDPB) organizes a national contest to identify SEP businesses. The winners are appointed as learning centers on how to run an "SEP business." The first-time winners among large, medium and small businesses were SCG, Chumporn Cabana Resort and Nithi Foods, respectively. The second-time winners were Bangchak Petroleum, Bathroom Design and Porntip Phuket, respectively.

The RDPB, Thailand Research Fund, the Thai Chamber of Commerce and the Board of Trade of Thailand have also supported Mahidol University in launching the "Sufficiency Economy Business Standard," which is now being used by agencies such as the RDPB, Thailand Sustainable Development Foundation and Thai Credit Guarantee Corporation as a tool to promote the adoption of SEP among business people.

The Ministry of Industry has introduced Industrial Standard No. 9999 which was developed based on SEP for Thai companies. Meanwhile, the Office of the National Economic and Social Development Board has established a network of large Thai companies such as SCG, PTT Group, Bangchak Petroleum and Toshiba to implement SEP with their trade partners throughout their value chains. All of these actions are helping to promote not only the late king's ideas but also the principles of sustainable development.

tives for quality and efficiency, improve the sourcing practices and labor standards of local communities, and address gaps or issues surrounding the cluster.

A good example of a CSV in Thailand is Bangchak Petroleum Plc, the country's second-largest gas station operator. With co-investment between BCP and co-ops, its gas stations allow all co-op members to jointly own a business and an opportunity to receive an annual dividend. In addition, BCP supports community members to become business owners and facilitates their education on how to operate a business. Members also receive a discount for oil products. By sharing profits and dividends with the gas station, communities enjoy improved income distribution. Moreover, the co-op gas stations serve as centers to educate people about renewable products such as gasohol and biodiesel.

The idea that companies should seek the "Social License" of communities affected by their projects is also gaining traction. Social License has been defined as existing when a project has the ongoing approval of the local community and other relevant stakeholders. As Social License is rooted in the beliefs, perceptions and opinions of the local population and because such views are subject to change if and when new information is acquired, a company must earn the Social License, and then work to retain it.

More and more, companies are starting to see the "sound business" argument for CSV, sustainability, and operating with the Social License of local stakeholders. While increasing eco-efficiency offers both financial and marketing benefits, corporate risk is reduced by better management of sustainability challenges. Products that draw on sustainability concerns are also becoming more profitable and popular with the public. Sustainability is not a luxury, but a way to overcome challenges and create winning business opportunities.

▶ **PIONEERS**

Chiva-Som

History: Wellness resort and spa founded in 1993, located on the Gulf of Thailand, with offices in Bangkok and Hua Hin

Location: Hua Hin district, Prachuabkhirikhan province

Key features: Uses a "quadruple bottom line" to cut costs and drive its initiatives in energy efficiency, waste management, community and staff programs and continued sustainability

For almost 20 years, Chiva-Som has remained one of the world's top wellness resorts. Its holistic approach to balancing and rejuvenating mind, body and spirit has attracted high-end international customers. Nevertheless, not many people are aware that it is a sustainable business leader in Thailand.

Several firms use the concept of a triple bottom line as their sustainability backbone, but Krip Rojanastien, chairman and CEO of Chiva-Som, created a "quadruple bottom line" system to uphold his company's values.

The first bottom line is environmental wellness. Boonchu Rojanastien, the former finance minister of Thailand and Krip's father, turned his private beach house into this health resort. He installed a wastewater treatment system that stored wastewater in the resort's lake before it was used again to water the gardens and lawn, and to feed the air cooling system. Conserving water saves Chiva-Som 800,000 baht per year.

Body-temperature water is at the heart of most of the resorts' spa-treatment programs, but it requires a lot of energy to heat. In 2006, Chiva-Som installed a 300 square meter solar water heater to supply 100 percent of its hot water, saving 200,000 baht in electricity costs annually.

Chiva-Som also engaged with the Clinton Climate Initiative's Energy Efficiency Building Retrofit Program to analyze its energy efficiency. As a result of the findings, the resort installed the latest water-cooling technology in its air-cooling system and also installed LED lighting throughout the resort. The retrofit resulted in a 26-percent reduction in electricity costs and a 20-percent reduction in its carbon footprint.

As for other environmental footprints, in 2009 Chiva-Som reduced its plastic bottled water use by 97 percent by giving the guests a premium-grade stainless steel bottle. The resort also produces 28 percent of the vegetables, fruits and flowers it uses to reduce long-distance transportation. In addition, Chiva-Som hired three sustainability engineers to focus on how to continually make its 55 guest rooms and 72 treatment rooms more sustainable.

The second bottom line is community wellness. Back in 2004, Boonchu founded the "Preserve Hua Hin" group to promote environmental awareness and preservation in the local community through various events and advocacy. In 2015, PHHG launched "Krailart Niwate," a mangrove preservation project and ecotourism center to protect the last remaining urban mangrove forest in Hua Hin. Chiva-Som sponsored the construction of the 1,000-meter boardwalk throughout the mangrove.

Personal wellness is the third bottom line. Chiva-Som has approximately 200 innovative treatment programs. While these treatments set Chiva-Som apart from its competitors, Chiva-Som's unique care and excellent customer service are key to improving customer health and wellness. For staff wellness, Chiva-Som offers competitive salaries, medical services and higher education, resulting in a 90 percent employee retention rate.

The first three bottom lines have consequently generated the fourth bottom line: business sustainability. Over the past 10 years, Chiva-Som's average occupancy rate has been 70 percent, while its yearly profit growth before taxes has been nearly 10 percent. Krip has been approached by several global hotel operators to expand Chiva-Som to other destinations, but he insists on applying his expansion rule: all partners would need to adhere to Chiva-Som's full sustainability policy and budget at least US$2,500 per room per year for sustainability.

Krip strongly believes that the only way to reach sustainability is to create the quadruple bottom line, focusing on so much more than just financial profit. The inherent wisdom of this commitment is reflected in the fact that Chiva-Som has received several international accolades for sustainable development and corporate responsibility over the years.

▶ PIONEERS

Dairy Home

History: Established in 1999; Thailand's first organic dairy producer

Location: Pak Chong district, Nakhon Ratchasima province

Key features:
Successfully markets organic dairy as premium products; customers willingly pay a higher price for Dairy Home's sustainable practices that also ensure higher quality products

Dairy Home, Thailand's first certified organic milk producer, has proven that becoming a sustainable business represented not only a new business opportunity but also a competitive advantage.

In 2004, when the Thai government signed the Free Trade Agreement (FTA) with Australia and New Zealand, Pruitti Kerdchoochuen, the CEO of Dairy Home, predicted the demise of Thai dairy production because Australia and New Zealand dairy products were much cheaper per liter than Thailand's. To survive as a business, Pruitti knew he could not compete by cost. As a result, Pruitti shifted the positioning of Dairy Home to be a premium product and became the first Thai organic milk producer.

Ten years ago, there was no such thing as "organic dairy" in Thailand. The Department of Livestock had not yet developed a standard for organic milk. Instead, Pruitti applied international organic dairy standards to his supplying farmers. Given the unsustainable farming practices common at the time, Pruitti had to essentially start from scratch, starting with cattle feed, such as corn, grass and cassava. These had to be produced from organic-certified crops, which were not available at the time. Instead of purchasing processed animal feed, Pruitti promoted his supplying farmers to grow the feed themselves to better control the chemical-free environment. Each cow was also required to have a corral of five square meters, as well as green space to roam freely, instead of being locked up in a crowded pen. Synthetic supplementary feed, chemical insecticide and antibiotics were entirely banned. Only herbal insecticide sprays and medicines were permitted.

Pruitti worked closely with farmers to motivate them to make these changes. He also offered a premium price of six baht more per kilogram if the milk met all the other organic criteria.

Pruitti's efforts to go organic eventually lowered his suppliers' costs, especially in cattle feed, which accounts for 55 percent of overall cattle farming costs. Additionally, when cows are not locked down and forced to milk, they are more satisfied and don't easily get ill, dropping the cost of medicine. Thus, Pruitti's suppliers ran their farms at a lower cost, received the same amount of milk per cow, and reaped a higher income.

The move proved profitable for Pruitti as well. Dairy Home products continued to sell for their higher quality and nutrients, despite prices that were 40 percent higher than competitors. According to many customers, Dairy Home products were also tastier. Based mostly on word of mouth, Dairy Home is rapidly penetrating into modern trade channels countrywide, resulting in annual revenue growth of 15–20 percent.

In addition to organic dairy, Pruitti has also considered the environmental

impact of the business's other operations. Dairy Home uses solar thermal to produce hot water and saves 30 percent on electricity costs. Dairy Home is currently obtaining ISO 50001 to reach the highest "green factory" standard of Thailand.

They also use biodegradable PLA plastic packaging made from sugar. The Thailand Greenhouse Gas Management Organization found Dairy Home's carbon footprint from its production processes to be smaller than any of its competitors. With continual profit growth, Dairy Home has been a champion in demonstrating that a small sustainable business does not have to give up profit to become green.

▶PIONEERS

SCG

History: Founded by royal decree in 1913

Location: Headquartered in Bangkok, operating throughout Southeast Asia

Key features: Environmentally friendly products; eco-friendly mining and production processes; "Zero Waste to Landfill" initiative; compliance with sustainable forest standards

SCG demonstrates that even a huge conglomerate employing thousands of people and operating complex production processes can exercise sustainability and become a leader in the field.

Founded in 1913 following a royal decree by Rama VI to manufacture

cement during a vital time in the development of the country's infrastructure, SCG produced cement locally in order to cut the costs of importing it and to reduce the influence of Western countries during the colonial period in Asia.

As one of only a handful of Thai companies in operation for over a century, SCG has proved its economic sustainability and has survived crises as wide-ranging as World War II, domestic political unrest, and the 1997 financial crisis. In 2016, the company's revenue was 423,442 million baht.

After restructuring several times, SCG now has three units: SCG Chemicals, SCG Packaging and SCG Cement-Building Materials. The company has more than 200 subsidiaries across 10 Southeast Asian countries and employs more than 51,000 people.

Not only has SCG proven its commitment to environmentally sound business practices, it has also balanced its triple bottom line in its production process, from acquiring raw materials to delivering end products. Even though cement, building materials, packaging and chemicals are perceived as environmentally harmful products, SCG has pioneered eco-friendly technology to create sustainable business practices in the industry. For example, SCG's Lampang cement plant adopted the Semi Open Cut mining process to reduce noise and dust pollution, and to minimize damage to the mountain ecosystem. SCG implemented a "Zero Waste to Landfill" policy to minimize its industrial waste. In 2014, SCG only sent 4.6 percent of its non-hazardous waste and no hazardous waste at all to a landfill. Even though the pulp industry consumes a tremendous number of trees, SCG Packaging was the first Thai company to comply with the Forest Stewardship Council's (FSC) sustainable forest standards.

In 2009, SCG launched its environmentally friendly product label called "SCG Eco Value." The company's more than 80 SCG Eco Value products range from marine cement, paper, and bathroom tiles to biodegradable plastic pellets. SCG's product life-cycle assessment ensures their products have minimal environmental impact.

SCG developed another product line called High Value Added Products and Services (HVA) to meet the needs of both today and tomorrow. These products include SCG's HEIM modular house, which features an earthquake-safe steel structure, is simple to construct, and comes with 20 years of after-sales check-ups. In 2016, its HVA products accounted for 38 percent of its revenue.

These are only a few examples of SCG's sustainability practices. The company is also considered a leader in green building, green procurement, and employee engagement. SCG's sustainability initiatives have been awarded by world-class organizations, including the Dow Jones Sustainability Indices (DJSI).

The HEIM modular house is designed for earthquake safety.

▶PIONEERS

Sal Forest

History: Founded in 2013

Founders: Sarinee Achavanuntakul and Pattraporn Yamla-or

Key features: Conducts research and social impact assessments on key sustainability issues in Thailand, and fosters public discourse about sustainable business through print publications, online media and events.

"Sustainable Business Accelerator"

Pattraporn Yamla-or conducts a training session in sustainable business for Sal Forest.

Sarinee Achavanuntakul knows that it takes more than a sound bite to explain sustainable development and the complexities of the obstacles in its path. As the co-founder of sustainable business research firm Sal Forest, she has spent much of the past four years trying to unravel these issues to build a base of research that can be tapped to raise public awareness.

"If you look at sustainability issues, whether it's smog in the north or overfishing in the south, first of all, there is no one clear culprit – it's the concerted outcome of many different players' practices and incentives," she said. Sarinee's goal is to map out these main players and the incentives that drive them.

"I was interested in this issue of sustainable development for many years and I also had been trying to push the

Sarinee Achavanuntakul.

media to do more investigative work on this," said Sarinee. But she realized that a greater accumulation of information and data was needed so that journalists could better do their jobs.

"There needs to be much more research done first before the reporters can come in," she said. "In the sustainable development arena, even basic information like what the problem is, who is involved, who is acting, who isn't acting – all of this isn't really known. That gap made me realize that in order to really bring out the data and information on a lot of these problems and situations in Thailand, we would need a proper research company."

Thus, together with Pattraporn Yamla-or, Sarinee founded Sal Forest in 2013. The pair have since joined forces with five other like-minded individuals focused on helping Thailand's business community to transition to legitimate sustainable development practices.

The firm now does in-depth, specialist research, receiving grants to look deeply into critical sustainable development issues. In general, the firm's research falls into two categories: sustainable banking, and sustainability in the supply chains of various sectors.

For example, the UNDP funded Sal Forest's research to delineate the producers and supply chain responsi

"If you look at sustainability issues, whether it's smog in the north or overfishing in the south, first of all, there is no one clear culprit – it's the concerted outcome of many different players' practices and incentives."

Sarinee Achavanuntakul, co-founder of Sal Forest

ble for the deforestation for corn planting and the resulting smog in Nan province.

In 2014, Sal Forest mapped the shrimp feed supply chain in Songkhla province for Oxfam. Then in 2015, with a grant from the Rockefeller Foundation, Sal Forest produced a report on sustainable banking in Thailand.

Sarinee recognizes that bringing this database to a generation notorious for short attention spans won't be easy, but she says that new channels, such as social media, can be vital in raising public awareness. "To me it's not really a hopeless situation, media just have to adapt," she said. "Sustainable development is really about the future of your country, the future of the planet, future generations – the issue itself is important, we just need to find ways to communicate."

GREEN BUILDINGS

The cost savings and other benefits of eco-friendly offices

Eco-buildings are gradually gaining ground in Thailand. Since 2007, when Thailand earned its first **LEED** certification for a manufacturing plant of the carpet maker InterfaceFLOR, green buildings have been cropping up. The greening of buildings has not been limited to factories. Eco-friendly structures have been commissioned by myriad institutions, from schools (International School Bangkok's Cultural Center) and foreign governments (USAID's Regional Development Mission for Asia) to blue-chip companies (Unilever's Rama 9 House, PTT's Energy Complex and SCG's 100th Year Building).

These buildings are designed (inside and out) to operate on key sustainability principles, including energy efficiency, reduction of waste and greenhouse gas emissions, and water conservation. They are built with careful consideration of materials and the local environment. Utility, practicality and comfort are also important factors because a green office must also be a healthy office for its workaday inhabitants.

Today, about 140 certifiably green structures dot the country, from a Toyota showroom in Nakhon Ratchasima province to a 7-Eleven outlet in Bangkok, to a six-story KASIKORNBANK Learning Center in Chachoengsao province, as well as the Thai Health Promotion Foundation Office. Partly driving the trend are companies branding themselves as eco-friendly. When a conglomerate like SCG, whose core businesses have environmental impacts, sells "green products," their headquarters becomes a part of their brand and corporate image.

Meanwhile, just as the internationally recognized LEED certification program by the United States' Green Building Council has been gathering momentum in Thailand, the kingdom's own Thailand Green Building Institute (TGBI) is offering a domestic equivalent for eco-friendly certification: the TREES system. Since the TGBI was founded in 2009, 76 buildings have been certified by TREES.

LEED, or Leadership in Energy and Environmental Design:

The world's leading certification program for green buildings. To be LEED certified, buildings must meet strict criteria laid out by the US Green Building Council.

Green building practices are increasingly critical when you consider that shopping malls, commercial buildings and apartments are among the three biggest energy users in the business sector, according to the Thai Climate Justice Working Group. When you combine all of these in Thailand, they consume more energy per year than the entire nations of Cambodia and Laos put together. Working at optimum efficiency, green buildings use up to 50 percent less energy and 60 percent less potable water. They produce 70 percent less solid wastes and emit 35 percent less CO_2. During construction, companies also save money by cutting waste by up to 80 percent. Whether starting small by refitting an office with energy-saving light bulbs or designing a whole new building, the money saved on water and electricity can be significant over the long term.

In Asian countries, McGraw Hill Construction estimates that companies with green buildings can save around 21 percent on operating costs over five years with new buildings, and 13 percent on buildings renovated to be more eco-friendly. This means that the return on investment for such structures is approximately seven years, faster than in many Western countries where construction costs are significantly higher. Over time, these savings will defray the exorbitant costs of constructing such monuments to sustainability. SCG's 100th Year Building, for example, set the company back 3.5 billion baht, much more than a normal building of that size.

Owners of green buildings enjoy an additional incentive: they can charge higher rents. The Park Ventures Ecoplex, a landmark building in Bangkok, charges 1,300 baht per square meter per month while normal buildings charge around 800 baht. This trend would surely wither, though, if people did not enjoy working in green buildings. The specially coated and insulated glass panels help reduce sound, light and heat absorption at the Park Ventures Ecoplex, creating an enjoyable place to work.

Unilever House, a custom-built sustainable and energy-efficient office building, is the Thailand and Asia-Pacific regional headquarters of Unilever.

Towering Examples of Sustainability

1. PTT's Energy Complex

is the first building in Southeast Asia to be awarded LEED Platinum status. In 2011 it took first prize in the New & Existing Building Category at the ASEAN Energy Awards.

2. KASIKORNBANK's Learning Center

in Chachoengsao province earned a score of 10 out of 10 in Water Efficiency, 24 out of 26 in Sustainable Sites, and 4 out of 6 in Innovation, among other LEED categories.

3. Park Ventures Ecoplex,

another LEED Platinum green building in Bangkok, boasts a design that simulates two hands pressed together in the traditional wai greeting.

Shopping malls, commercial buildings and apartments in Thailand consume more energy per year than the entire nations of Cambodia and Laos put together.

▶**PIONEERS**

SCG 100th Year Building

Designed by: Design 103 International

Location: Bang Sue district of Bangkok

History: Opened to celebrate the Siam Cement Group's centennial in 2014. SCG's core businesses are petrochemicals, cement, building supplies, packaging and logistics

Key features: The 23-story building houses 36,600 square meters of office space for 1,700 employees with solar-reflective tiles that reduce the need for air-conditioning, balconies that serve as sunshades and solar panels

Accolades: Certified LEED Platinum, the highest category for green buildings in this global body

Lights turn off automatically unless motion sensors detect the presence of occupants. They are powered partly by solar panels located on the roof of an adjacent 10-floor parking building.

The large conglomerate SCG takes sustainable development seriously. The company's most powerful testament to this is its headquarters: the 100th Year Building, which opened in January 2014 to celebrate the centenary of the company's founding in December 1913.

The 23-floor structure is an architectural marvel, boasting Streamline Moderne design elements that give it a distinctly aquatic look. Light-blue balconies resembling rippling waves cascade down glass-fronted sides, creating a shifting mirage effect when viewed from different angles. The effect is further accentuated by its sprawling, leafy, holiday resort-style surroundings with trim lawns, landscaped gardens and tall *kapok* trees. "It feels less like a traditional workplace than like a village or a university campus," observed Thanasak Phakdee, who works at the building's Corporate Communications Office.

But there is far more to the building than greenery and appealing aesthetics – the myriad eco-friendly features start right at ground level. The 36,600-square-meter edifice stands on a 1.2-meter artificial foundation paved with porous turf blocks, which are manufactured by SCG itself. They support the growth of natural grass and, just as importantly, let rainwater seep into the soil unhindered, helping it drain downhill into an adjacent pond. The pond serves as a reservoir for the irrigation of plants on the premises and for flushed water from the buildings' eco-friendly toilets. The porous tiles are made largely from recyclables to lower the need for "virgin" materials and help reduce storm water runoff during the monsoon season.

Also underfoot are solar-reflective tiles that bounce back sunrays, keeping the office tower naturally cooler in the year-round tropical heat. At the entrances sprawl dust-absorbing micro-fiber carpets that reduce airborne dust from outside.

That's just for starters. The structure itself has been constructed with auto-claved aerated concrete, a lightweight and thermally insulating material, and the windows in the entire building are made from reflective and insulating laminated glass. Both serve to minimize heat absorption. In another energy-saving design, the eye-pleasingly undulating balconies act as sunshades on each floor for the offices inside while also aiding the natural ventilation of offices, which are equipped with air-quality sensors to ensure a healthy working environment for the office tower's employees.

Inside the main building's restrooms, lights turn off automatically unless motion sensors detect the presence of occupants. They are powered partly by solar panels located on the roof of an adjacent 10-floor parking building, which can house 1,000 vehicles and features a 270-meter open-air running track for workers. The panels generate 84 kilowatts of energy per hour and a total of 99,000 kilowatts per year.

There's more. An automatic temperature control system ensures an ambient climate inside offices all day long, while daylight sensors monitor the levels of natural light from outside and adjust indoor artificial lighting accordingly. All these designs and devices help cut the building's energy consumption by 2.3 million kilowatts per year, amounting to a 30 percent saving in electricity use.

Such energy-saving measures are especially useful in Bangkok, where year-long high temperatures and levels of humidity adversely impact buildings' natural ventilation, often requiring the nonstop use of air conditioners that add greatly to energy usage. Thanks to all these eco-friendly features, SCG's 100th Year Building has been certified LEED Platinum.

But there is a downside: construction costs. This skyscraper came with a sky-high price tag of 3.5 billion baht. That's several hundred million baht more than a similar building without many of the green features.

"The way companies have to look at it is that while they may have to spend more on construction costs, they will be able to recoup their additional investments over time by saving on operating costs," Panupant Phapant, manager of SCG's Green Portfolio Management unit, explained.

It's also true that building an ethical, environmentally sensitive brand is good for business and a way to act as a trailblazer for all of the companies and suppliers under this top conglomerate's wing. In that light, SCG is also looking at logistics and how to minimize its carbon footprint by cutting back on business trips and meetings through means like video conferencing. "We have green factories and we want to encourage the whole supply chain to care more about the environment so we need to lead by example," said Kanapapha Akapha, the cement company's green solutions manager. "More and more stakeholders we work with are embracing the green business concept."

Making Old Buildings Greener

Although new buildings with the latest green technologies may offer the best eco-solutions, older structures can also be retrofitted with energy-saving fixtures and apparatuses. There are also some simple behavioral changes you can make in your office to help. Here are some ideas:

- Newly installed water-saving appliances can help water efficiency. While older flush toilets often use up to around five gallons of water with every flush, modern eco-friendly designs do the job with only 1.28 gallons.

- Electricity-saving light bulbs such as compact fluorescent lamps (CFLs) and light emitting diodes (LEDs) help save electricity.

- Lights, computer monitors and air-conditioners can be turned off during breaks and lunch hours.

- Ceramic mugs and glasses should replace disposable paper and plastic cups.

- Cutting down on printed office materials in favor of digital versions saves paper and trees.

- Promoting the three R's ("Reduce, Reuse, Recycle") sets a positive example both at home and in the office.

- Employees can be encouraged to use bicycles and public transportation to commute to work, and monthly no-car days can be organized. The owners of hybrid cars and smaller, energy-efficient vehicles could be allocated special "eco-car" parking spaces.

- Urging staff members to use video-conferencing tools for long-distance communication rather than traveling in person to meetings cuts costs and downsizes carbon footprints.

- Office supplies that are recyclable and sourced locally help the bottom line and reduce the need for long-distance haulage.

GROUNDBREAKERS

A pioneer of ecological design in Thailand, **SINGH INTRACHOOTO** *is head of the Creative Center for Eco-design at Bangkok's Kasetsart University and serves as design principal at OSISU, the country's leading eco-design production house. Singh received Elle Décor's Designer of the Year award in 2007 and won the Top Environmentalist award from Thailand's Department of the Environment in 2008.*

What features and functions make a building truly eco-friendly? If we focus on being responsible to the environment, we must be sensible about the materials we use. Sadly, many architects know little about the materials they use. They do not know where materials come from, how much energy we need to manufacture them, what toxic chemicals may be used in their production, etc. Basically, many people are clueless about the materials' lifecycle impacts, yet the design and building industry consume tremendous amounts of materials every day.

What considerations should architects and developers bear in mind while designing and building high-end green structures? Architects and developers should bear in mind that we may be destroying the planet even as we create new structures. The big question is: should we build more and more, or should we focus instead on renovating sub-standard buildings that are in need of upgrading? We can turn older buildings into high-end developments with our creativity instead of taking up yet more space by constructing more new buildings.

How popular do you think the idea of green buildings (both homes and offices) is in the country? The idea of green buildings is becoming very popular, but real concerted action about them remains sporadic. Most Thais know green buildings are good and they are quite supportive of them, yet this widespread awareness has yielded few tangible results so far.

How would you rank Thailand in terms of its green building development within ASEAN and in the larger world? Advocates of green development in Thailand are mostly academics. Thailand has green building voluntary standards (the TREES certification program), Green Hotel Standards and Green Office Standards through the Department of Environmental Quality Promotion, but their implementation systems are not sustainable. I would say that Thailand ranks high within ASEAN in terms of green awareness but low in terms of real action.

What needs to be done to encourage homeowners and corporate stakeholders alike to invest in greener buildings? The government needs to provide further incentives to homeowners and corporate stakeholders, both through laws and regulations and through subsidies and rebates. There also needs to be government pilot projects that showcase green designs that are smart, elegant, efficient and ecologically friendly. We need to make sure that people equate being green not just with high performance and efficiency but also with beauty, flair and style.

Solar panels attached to a roof to save energy.

Office wall decoration made of recycled paper.

Certification Programs at Home and Abroad

Eco-buildings range across a spectrum of shades of green and are classified accordingly. Their rankings are based on their degrees of compliance. The premier global green building certification program is the US Green Building Council's LEED (Leadership in Energy and Environmental Design) system, which evaluates structures according to numerous criteria. They include design and construction; sustainable site location; public transport accessibility; interior environment quality; sourcing of materials; maintenance; energy-saving strategies; and water resource management.

Buildings are classified into four categories on a scale of points: Certified (40–49 points), Silver (50–59), Gold (60–79) and Platinum (80+).

Thailand's Green Building Institute also has its own rating system, TREES (Thai's Rating of Energy and Environmental Sustainability), which evaluates newly constructed eco-friendly buildings around the country. It is based on the international LEED system and likewise considers a wide range of similar criteria, such as building management, use of renewable energy sources and water conservation practices. In total, more than 130 buildings in Thailand have been LEED-certified and 76 have been certified by TREES.

LEED Platinum-certified K-Bank Learning Center in Chachoengsao province.

BUILDING A BETTER FUTURE
THE FUTURE OF ARCHITECTURE IN 100 BUILDINGS
Author: *Marc Kushner*
Publisher: *Simon & Schuster/TED*
Year: *2015*

In *The Future of Architecture in 100 Buildings* from Simon & Shuster and TED – the organization behind the inspiring TED Talks – Mark Kushner looks at some of the most original edifices in the world that are pushing the barriers of building design. Drawing on his background as an architect and as the social media mogul behind Architizer.com, he is especially adept at seeing how the public can shape architecture and how architects themselves must respond to these challenges by designing new buildings with them in mind.

These cutting-edge structures will redefine how most of us look at buildings, whether it's an inflatable concert hall, or a research lab that can move by itself through snow, or a building that could endure the harshest ice storms on earth or provide shelter on the moon for a future colony of humans.

For those interested in the future of green buildings, Kushner's 2015 book provides much fodder for thought and debate in the form of structures built of repurposed materials or offices that can eat smog. Paired with handsome photos of the corresponding designs or 3D models, this book is full of architectural wonders that are both down-to-earth and fantastical.

RENEWABLE ENERGY

Priming the private sector for an energy revolution

While fossil fuels grow scarcer, Thailand is making efforts to wean the country off of its fuel dependency. While some forward-thinking businesses are taking advantage of this trend toward alternative energy to cut down on their own energy costs, some investors are cashing in on highly incentivized opportunities to produce clean energy.

Thailand was one of the first Asian countries to implement a feed-in tariff, or "adder" program, incentivizing renewable energy development. Thailand's adder program offers renewable energy producers long-term contracts to sell electricity at attractive rates. Companies that generate power through biomass, biogas, hydro, solar, wind and waste energy are eligible for the adder program. Incentives like these to switch to renewables remain vital for Thailand's energy sector, but alternative energies may soon be able to stand on their own in the marketplace.

A wind power plant in Thailand.

Jatropha curcas is used to produce ethanol.

Thailand's adder program offers renewable energy producers long-term contracts to sell electricity at attractive rates.

In Thailand, solar is emerging as the lead alternative energy industry as photovoltaic technology continues both to improve and become more cost-effective. Although the costs associated with photovoltaic technology were once astronomical, the price of solar power has tumbled from nearly $75 (2,640 baht) per watt in 1972 to less than 45 cents (16 baht) in 2016. Today the use of solar is growing faster than any other power source worldwide, and soon it will be the cheapest form of energy on the planet.

A decade ago solar amounted to less than one-tenth of one percent of Thailand's power. By 2014, it produced about four percent of the nation's electricity or about 1,288 megawatts (MW), a figure that jumped to 2,149 MW in 2016. Though still modest compared to global solar leaders like Germany and China, Thailand maintains its position as the region's solar powerhouse, producing more than the rest of ASEAN combined. By 2036, Thailand aims to increase solar capacity to 6,000MW in order to meet the needs of three million households.

Several other renewable technologies are showing great promise as well. Thai Solar Energy Public Company Limited built Southeast Asia's first concentrated solar thermal power plant in Kanchanaburi, and windmill farms are being tested around the country, as are small hydro applications. Biogas, biomass and biofuel also present huge potential. The concept of using agricultural waste as a resource offers multiple avenues for profit: not only is it useful for power generation, but it also provides waste management solutions and produces other "clean" materials such as fertilizers, chemicals and plastics (see "Biofuels: A Natural Fit for Thailand" sidebar).

One of Thailand's leading green energy producers, Bangchak Petroleum, is in the midst of a six-year, 90 billion baht business plan with a significant focus on environmentally sustainable power plants and

renewable energy. Bangchak already operates a solar farm with a sales and distribution capacity of 118 MW, in addition to a biodiesel business that averages 365 thousand liters per day.

In the short-term, costs are the biggest challenge facing alternative energies as the initial investment is higher in comparison to fossil energy solutions. Alternative energy production systems, technologies and infrastructure to connect to the national power grid all require large upfront investments and can take decades to pay for themselves. Every form of alternative energy also comes with its own set of logistical challenges: solar and wind power require vast tracts of land, while gasohol and ethanol require gas stations capable of distributing the fuel.

Further steps can, however, be taken to help promote the proliferation of alternative energies. Technological improvements that increase productivity and lower costs are certain to drive private sector investment. Banks offering green loans can help fill the financing gap for alternative energy projects, especially for small- to medium-sized enterprises (SMEs). Finally, raising public awareness about the cost, health and environmental benefits of clean energy can potentially create a groundswell of demand that shifts the goals and operations on the supply side.

The Energy Conservation Fund

Thailand's Energy Conservation Fund (ENCON Fund) is an innovative way to fund alternative energy projects. Instituted in 1992, the ENCON Fund is sourced from a levy against petroleum products, the rate of which is established by the prime minister. As of 2015, the rate stood at $0.002 per liter. Annually, it generates around 7 billion baht, making it capable of significant achievements in sustainable energy development.

Through the levy on petroleum products, the ENCON Fund operates under the "polluter pays" principle. The government can effectively influence fossil fuel pricing (and usage) depending on the levy rate imposed, and guarantee a source of funding for energy efficiency programs. The Energy Conservation Promotion Act requires the ENCON Fund Committee to administer the funds, keeping it separate from the annual government budget allocation system. The Ministry of Energy presides over administrative matters relating to money and fund disbursement. Every five years, the committee establishes a conservation program to serve as a guideline for utilization of the ENCON Fund. The fund has also helped develop programs like Thailand's Energy Service Company venture capital scheme, the renewable energy feed-in tariff scheme, the provision of tax incentives for energy efficiency projects, and various grant programs. Given its large budget, how the ENCON Fund is managed will likely play a large role in whether the government can achieve its energy efficiency targets under the Energy Efficiency Development Plan (EEDP).

BUDGET ALLOCATION BY ECONOMIC SECTOR

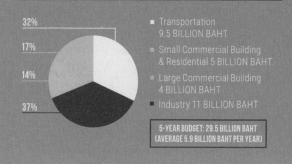

- 32%
- 17%
- 14%
- 37%

- Transportation 9.5 BILLION BAHT
- Small Commercial Building & Residential 5 BILLION BAHT
- Large Commercial Building 4 BILLION BAHT
- Industry 11 BILLION BAHT

5-YEAR BUDGET: 29.5 BILLION BAHT (AVERAGE 5.9 BILLION BAHT PER YEAR)

▶PIONEERS

Solar Power Company Group (SPCG)

History: Founded in 1993

Location: Corporate headquarters in Bangkok, with major solar farms operating in northeastern Thailand

Key features: The largest solar power generation company in Thailand, SPCG has found a way to partner with foreign companies and private investors to develop solar power in Thailand. Headed by CEO Wandee Khunchornyakong, SPCG is poised to continue to grow not only in Thailand but also in ASEAN.

A decade ago, solar power in Thailand produced less than one-tenth of one percent of the country's power, but by 2014, it produced four percent of the nation's electricity. That figure is set to increase yet again this year.

This growth is due to government subsidies, the rising cost of conventional power, the falling cost of alternative energy and to forward-thinking investors. One important person who fostered that growth is self-made businesswoman Wandee Khunchornyakong, the pioneering CEO of Solar Power Company Group, or SPCG Public Co Ltd, which she founded in 1993. The company is now Thailand's largest solar power firm.

SPCG has set up 36 solar farms across 10 provinces in sunny northeastern Thailand, which are thus far generating a total of 260 megawatts. For perspective, in 2008 Thailand's total solar capacity was less than 2 megawatts.

Wandee got into solar development early, installing small arrays in remote corners of Thailand in the 1980s. However, Wandee's first chance at creating utility-scale solar farms came in 2007 when Thailand's Ministry of Energy offered an 8 baht "adder" per unit of solar power produced for the following 10 years. This adder fee, offered to private solar power producers, was part of a 15-year renewable energy development plan. Other alternative energies were offered smaller adders.

Wandee made large-scale solar development in Thailand possible by tapping into private funding from the World Bank Group's International Finance Corporation. She also approached multinational electronics producer Kyocera to partner with SPCG on building 35 utility-scale solar farms. Kyocera agreed to the joint venture, supplying over a million solar panels for the project. Construction began in 2010, and by July 2014, the final solar farm was completed and connected to the utility grid.

Today, Wandee envisions a future where SPCG will mass-produce solar tiles as a roofing product that generates electricity, so every household in Thailand can produce energy for its own consumption. For her efforts, Wandee received recognition with a UN Momentum for Change award for "transforming Thailand's renewable energy capacity with utility-scale solar farms," increasing clean energy and stimulating economic growth in Thailand's impoverished northeast where the solar farms were built.

Private investors have also gained confidence in solar farms because of Wandee's success. Thailand is now striving for alternative energies to fulfill a minimum of 20 percent of its power needs by 2022. Thailand's solar farms are also models for neighboring countries, Wandee has said, adding that SPCG is currently expanding into ASEAN countries and Japan by investing both in solar farms and solar roofs, while adapting to match the opportunities and challenges presented by each country.

Wandee's prophetic investment move into solar power has paid off handsomely for personal gain as well. In 2015, *Forbes Magazine* ranked Wandee among the 50 most powerful businesswomen in Asia.

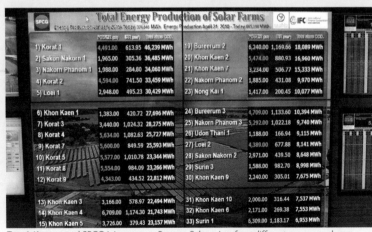

Top: A Kyocera and SPCG joint venture. Bottom: Solar prices from different energy producers.

A solar rooftop panel in one of Thailand's upcountry regions.

GROUNDBREAKERS

Architect, professor, solar-home owner and environmentally concerned citizen **SOONTORN BOONYATIKARN** *talks about Thailand's need to conserve energy.*

What have you learned from the solar house you built 14 years ago? You have to think of the house as a whole. It is important to conserve energy first. You have to think of the home, not just solar panels. People need to consider their consumption and decrease what they use, not just install solar panels.

Do you think solar offers the most alternative energy potential in Thailand? For now, I think solar cells are the most viable. Thailand has a lot of sun. Solar lasts a long time with no maintenance. You can install solar panels on your roof and then you can forget about them and 15 years later they are still producing power.

What other alternative energies hold potential for Thailand? Biogas, biomass and wind all have potential. But they all require maintenance. Thailand needs to manufacture windmills locally and save a lot on costs, and employ Thais.

What other environmental concerns standout in Thailand? You fly over Thailand and you see the biggest problem, deforestation. The area without trees marks the Thai border. Water management is number two. We need to learn to conserve. Media is third. The media need to inform the people, not advertise for business.

What is your favorite sustainable building in Bangkok? The government building at Chang Wattana. It is a very big building and it uses so little power. (Editorial note: Dr Soontorn designed the building.)

What is your favorite environmentally friendly city? Singapore is better than Bangkok, but in Europe there are others that are better.

Do you consider yourself an environmentalist? I would say I am one who is very concerned about the environment.

EYE-OPENERS

THE TIMES THEY ARE A CHANGIN'
THE GREAT TRANSITION: SHIFTING FROM FOSSIL FUELS TO SOLAR AND WIND ENERGY
Authors: *Lester R. Brown, Janet Larsen, J. Matthew Roney, Emily E. Adams*
Year: *2015*

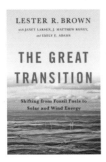

This important and timely book looks at the changing global economy and how it is forcing a shift from coal and fossil fuels to clean power.

The research comes from environmental analyst Lester R. Brown and his colleagues at the Earth Policy Institute, which he founded. The book argues that the shift to renewables is being forced by three things: economics, government policy and a growing acceptance that climate change is real and accelerating.

"As fossil fuel resources shrink, as air pollution worsens, and as concerns about climate instability cast a shadow over the future of coal, oil, and natural gas, a new world energy economy is emerging," Brown says. On the horizon is a long-lasting change for how we will power the economy of a shrinking world. It "amounts to a massive restructuring of the global economy," the authors write.

Government incentives started the shift towards clean energy, but market forces "favoring both solar and wind energy" have become even more influential. Now this shift is happening much faster than expected by even the most optimistic environmentalists.

Chapter five, "The Solar Revolution," stands out with an insightful history of solar energy from when Bell Labs introduced photovoltaic technology in 1954 to the falling price of solar in 2014. This is an important book about global economics, not just alternative energy.

Biofuels: A Natural Fit for Thailand

Biofuels are emerging as an alternative energy solution for Thailand. Often associated with biomass and biogas, biofuel can be a confusing concept. Here's a quick breakdown of the key elements:

BIOFUEL: The fuel derived from any source of natural product other than fossil fuels. This can include biomass or biogas.

BIOGAS: A specific form of biofuel made from the decay of organic matter to produce a gaseous fuel high in methane and carbon dioxide.

BIOMASS: The organic matter – such as animal waste, rice husks, cassava, palm oil, algae, etc. – that can be converted to energy. Biomass can be converted indirectly into energy by distillation into a biofuel, such as ethanol and biodiesel. Biomass can also be turned directly into energy by incineration or combustion.

Using waste to generate power is a logical strategy for sustainable development in Thailand, where a massive agricultural sector creates millions of tons of waste each year. Farms and other waste producers can now sell their refuse as feedstock and generate more profit.

However, even with abundant waste resources, biofuel plants in Thailand face several challenges: as higher demand drives the price of feedstock up, biofuel becomes less profitable, the variability in the quality of feedstock poses operational and cost risks, and the variability in the quality of technology can lower production or increase maintenance costs.

Alternative fuels to power vehicles, on the other hand, face even bigger challenges. Alternative fuels are often mixed with gasoline, making them vulnerable to oil price instability. However, the Thai government is supporting the development of alternative fuels. The Ministry of Energy has set targets to increase Thailand's alternative fuel production. It has also created new feed-in tariffs for energy-crop power production, which are good for 20 years and adjustable with inflation.

Finally, biomass offers opportunities for Thailand beyond the energy sector. With current technological capacities, biomass energy uses only 7 to

A Thai worker fills a container with used vegetable oil before it is processed to be used as biodiesel.

10 percent of the total value of biomass. The use of integrated technologies for turning biomass into other materials, such as bioplastics, fertilizers, animal feed and chemicals, can increase this to 70 to 80 percent. Rubber wood refuse, for example, was once used to create particleboard. Today, it can be ground into a powder and used to form a biodegradable, plastic-like material. Such bioplastics are already being used to create consumer products such as plastic cutlery and straws. Biomass can be composted and enhanced to create fertilizer, while the sugars from decaying plant refuse can be rendered into building-block chemicals used in manufacturing, agriculture and more.

In addition to biofuels, the creation of bioplastics and other raw materials presents a massive economic and labor opportunity for Thailand. A growing, labor-intensive industry of turning waste into resources can provide added value to the economy, as well as provide more employment. In the bigger picture, these products may go far in displacing the fossil economy entirely. Thus, biomass is not only about power; it may also prove a key strategy for Thailand's sustainable future.

GREEN MANUFACTURING

Opportunities knock at factory doors

Making products with a low environmental impact such as energy-efficient light bulbs, additive-free foods, or toys made from wood rather than plastic is often referred to as "green manufacturing." However, in the broader context of private sector industry, "green manufacturing" means the application of "green processes" to all types of manufacturing – processes that reduce energy and waste on a production line, for example, or industrial designs that allow the use of sustainable raw materials.

Green manufacturing differs slightly from the more general term of "green industry" because not all industries are concerned with manufacturing. Yet in Thailand, which has nationally adopted green manufacturing principles under the Green Industry (GI) Program, manufacturing generates around a third of the kingdom's gross domestic product and around three-quarters of its exports.

Green manufacturing is especially concerned with reducing the materials, energy and waste consumed by manufacturing processes to minimize environmental impacts while maximizing the efficient use of resources. A comprehensive approach to green manufacturing takes into account the entire lifecycle of a product: from the design and procurement stages, the manufacturing, packaging, distribution, and customer use of the product, all the way up to the "remanufacturing" stage, or the recovery, reuse and recycling of products.

In the face of increasing consumer demand and regulatory pressure for sustainably manufactured products – especially in Thailand's key export markets of the EU and the US, which have stringent environmental standards – Thai industry leaders and government authorities see the "greening" of manufacturing here as a means to safeguard the existing sector and to ensure its sustainable growth.

Several Thailand-based manufacturers have already

ISO 14001:
The key standard within the ISO 14000 series, a set of environmental standards released by the International Organization for Standardization. ISO 140001 defines the requirements for an adequate environmental management system (EMS) implemented by any company.

SCG packaging paper is certified by the Forest Stewardship Council.

taken the lead in green manufacturing. SCG and Amphol Food Processing are among the growing number of Thai companies that have implemented programs under the **ISO 14001** standard to ensure that any environmental impacts are monitored and improved. The related idea of "lean" manufacturing – a system of waste reduction – is used by companies such as Toyota and Toshiba in their Thai factories.

Green manufacturing has been promoted in the private sector in Thailand as national policy since 2011 as part of the 11th National Economic and Social Development Plan. As of 2016, more than 13,000 factories in 30 key industries had joined the Ministry of Industry's Green Industry Program, which aims to establish Thailand as a green manufacturing hub for the ASEAN region. The program has now been integrated into the International Organization for Standardization (ISO) program as ISO 26000. So far, Thailand and more than 80 countries have adopted the program as an internationally accepted standard for corporate social responsibility.

On a practical level, the implementation of green manufacturing can pose significant initial costs, but analysts say early adoption can help reduce the long-term costs to more than cover the initial outlay. A recent study of green manufacturing by

Quality control is crucial for green manufacturing.

proof" their operations against the potentially more expensive costs that could arise from future environmental regulations, such as penalties for high energy consumption. Green manufacturers also stand to cash in on rewards for low emissions.

In Thailand, focusing on the carbon footprint of a product or product chain has already presented forward-thinking Thai companies with an opportunity to develop and market low-carbon products. Measures to reduce that footprint on the manufacturing side can include redesigning products to make low-carbon versions with similar functions, or reconfiguring a supply chain to reduce carbon emissions at key stages.

Brahmanand Mohanty of Bangkok's Asia Institute of Technology highlighted how the reduction of a manufacturer's carbon footprint can reduce costs and waste. The study also pointed out that adopting green manufacturing methods can help companies "future-

---

THE JOURNEY FROM CRADLE TO CRADLE
WASTE EQUALS FOOD
Director: *Rob van Hattum*
Release Date: *2007*

The American documentary *Waste = Food* looks at the development of a new manufacturing philosophy of using non-toxic and recyclable materials for all man-made products.

The 2007 film focuses on the work of American architect William McDonough and German ecological chemist Michael Braungart, who have co-authored books on their famous "Cradle to Cradle" principles, which take into account the full life cycle of manufactured products and the wastes that result. Their ideas extend to groundbreaking environmental architecture, including McDonough's

designs for Ford's new River Rouge plant, a GreenHouse factory for the Herman Miller company, and a model village in rural China.

The film also explores how some of the world's leading manufacturers are experimenting with clean and sustainable production methods, going on-site at a Swiss textile factory, a German clothing manufacturer, the Nike shoes headquarters, a US furniture manufacturer, the Ford Motor Company and a government housing project in China to illustrate examples of on-the-ground green manufacturing initiatives.

▶**PIONEERS**

Toshiba Semiconductor (Thailand) Company Limited

History: Toshiba Corporation was founded in Japan in 1938; its first joint venture opened in Thailand in 1969

Location: Its flagship "green manufacturing" plant is in Prachinburi province

Key features: Toshiba Corporation operates comprehensive environmental management controls across its operations worldwide

Toshiba Semiconductor Company Limited's flagship green manufacturing plant in Thailand is its semiconductor factory in Prachinburi, which fabricates small signal devices and photocouplers for use in tablets, smart phones and digital sensors – important component markets with strong forecasts for growth. When its semiconductor plant at Pathum Thani near Bangkok was inundated by floodwaters in 2011, the company took the opportunity to design and build a new plant optimized for the latest green manufacturing processes.

The newer Prachinburi factory employs about 500 people and is situated at an industrial park about 140 kilometers east of Bangkok, well outside the Chao Phraya River's flood plain. It is almost one and a half times the size of the Pathum Thani plant it replaced, giving the factory's operations room to expand in response to the expected growth in the market for its components. Advanced process control and energy-saving technologies allow the plant to optimize its operating efficiency and productivity.

The silicon chip production process, for example, has gotten an efficiency upgrade at the Prachinburi plant. Because silicon chips are mass-produced at high temperatures on silicon-ceramic wafers, being able to use larger wafers means that more chips can be made at one time. Carefully controlling heat levels during the process also helps to reduce waste and energy usage while maximizing production. By supporting these changes with other adaptations in the distribution chain, Toshiba has also become more responsive to the fluctuating demands of the marketplace.

The new plant is outfitted with high-efficiency LED lighting, and the factory buildings are built on a large, landscaped site that helps offset carbon emissions. High-efficiency chiller and air-conditioning systems help reduce the use of energy and chemicals and greenhouse gas emissions, while an electronic deionization (EDI) water treatment system allows the reduced use of hazardous chemicals such as hydrochloric acid during the chip production process.

The results are both measurable and remarkable. Compared with the peak output of the Pathum Thani plant in 2010, the Prachinburi plant uses around one-fifth of the energy to produce the same number of semiconductor components, while emitting around a quarter of the greenhouse gas emissions and around one percent of the waste in total, with equally

Electronic deionization water treatment system.

dramatic falls in the use of water for processes and the emission of wastewater.

Like all Toshiba semiconductor plants around the world – in Japan, China, Singapore, the Philippines, Korea and the United States – the Prachinburi plant is certified to the ISO 14001 international standard for environmental management. In 2013 the Prachinburi plant also met the Thai government's Level 3 milestone for the Green Industry mark for "green systems," which stipulates detailed policies on the reduction of waste and energy use to prevent environmental harm and greenhouse gas emissions.

▶**PIONEERS**

Plan Toys

History: Founded in 1981

Location: Trang province

Key features: An innovative Thai company that makes toys from safe and sustainable materials

Until just a few decades ago, the extensive rubber plantations of Thailand, Malaysia and Indonesia – which produce about 70 percent of the worlds' latex rubber – were simply burned off at the end of their productive life so new rubber trees could be planted. But companies like Thailand's Plan Toys have helped change all that by developing an innovative approach to green manufacturing and exploiting the real value of rubber wood as a sustainable raw material.

Based in the rubber-producing region of Trang in southern Thailand, Plan Toys has earned a worldwide reputation for creating imaginative and child-safe toys made from sustainable materials, principally mature, recycled rubber wood. The factory started production in 1981 and now consumes about 1,000 tons of rubber wood a year. It produces more than five million sets of toys each year, mainly for export to the United States and Europe, as well as more than 60 other countries, and boasts sales of more than US$16 million per year. Europe, in particular, has emerged as a key new market. "[Europeans] share the same values as us. They intrinsically understand the concept of sustainability and child development, making it easy for us to reach them," said Kosin Virapornsawan, Managing Director of Plan Creations Company.

At the heart of the company's operations are principles of environmental and social sustainability, as well as a highly developed approach to green manufacturing. Plan Toys uses a chemical-free kiln process to treat rubber wood, thereby protecting it from fungal and insect damage. This key innovation has helped turn what was once disposable rubbish into a sustainable resource and valuable raw material. Once treated, the dense rubber wood is tough, and does not split or splinter. Its natural colors do not fade and its contours do not bend. What's more, the workers only use non-toxic glues and dyes. For packaging and promotional purposes, the company only uses recycled or recyclable materials.

Recent developments have led Plan Toys even further down the road of sustainability: the company now creates valuable composite materials from the waste products of their existing processes. Waste wood, bark and small branches unsuitable for manufacturing are used as thermal fuels to drive the factory's kilns and other processes. The scrap wood is burned in a five-megawatt biomass gasification waste-to-energy plant that the company built in 2010 next to its toy factory at a cost of more than 350 million baht. All in all, the biomass plant and solar power panels installed at the site produce enough electricity to power factory operations, which creates a saving of more than 15 million baht a year on energy consumption. The factory is also able to supply several nearby villages with power.

In 2012 the company launched a new line of toys made of "PlanWood," a non-toxic composite wood made of the sawdust that piled up in Plan's factories. The new material saved the company 32 percent on manufacturing costs, and in 2016, the sale of PlanWood toys accounted for 30 percent of their overall profits. More recently, in 2017, Plan Toys launched 38 new products including their new PlanMini series for children age 3+ which aims to enhance learning development through play.

Kosin said it is important for children to be introduced to the concept of sustainability at a young age. "We believe that with a strong foundation, those children will grow up with sustainability all the way into their adulthood and hand down this spirit from generation to generation," he said.

For this forward-looking company, sustainability principles and green manufacturing ideas inform every step of the manufacturing chain, from the early design stages to production, packaging and marketing. Plan Toys has also successfully tapped into a growing market for environmentally sound toys and built a global brand. Of course, the path to building a successful sustainable business is not an easy one.

"The most important challenge would be that the cost of green manufacturing is higher than the cost of regular manufacturing. This is due to the fact that we need to make sure our production processes lessen impacts on the environment and society, including all of our employees," said Kosin.

A non-toxic sawdust used in child-safe toys.

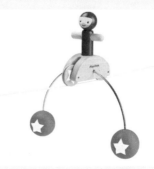

An acrobat from Plan Toys' new series.

ETHICAL SOURCING

Ensuring sustainability through the supply chain

Companies that follow the principle of ethical sourcing ensure that their products are sourced, manufactured and supplied without exploiting people or the environment. In the private sector, ethical sourcing policies are developed by some companies to verify transparency in each link of their supply chain, assuring customers that ethical standards are being upheld. This is important not only for public image and customer satisfaction, but also because the behavior of suppliers and partners can pose risks to a business's operations and viability. Ethical issues can arise from almost any aspect at any stage of a supply chain – from the sourcing of food, raw materials or components, to factory pay rates and working conditions; from industrial waste disposal, to transport, packaging and delivery.

Ideally, an effective ethical sourcing policy ensures that all the materials and services that go into making a company's finished products are created in safe facilities by workers who are treated well and are fairly paid, and with regard to environmental sustainability. Working to enhance transparency throughout the supply chain not only mitigates risk, but also lifts communities out of poverty.

Some of the modern concepts of ethical sourcing have origins in the **"fair trade"** movement, which

Eucalyptus trees from agroforestry in Kanchanaburi are used for paper.

gained global influence in the 1970s with a focus on fair crop prices for farmers in developing countries. Fair trade concepts helped inspire many global ethical sourcing initiatives, including the Fairtrade International certification program, one of the best known in the field. Non-governmental organizations such as Oxfam have campaigned for many decades to highlight ethical sourcing issues and to develop sustainable trade-based aid programs.

A growing number of consumers worldwide expect the companies they buy from to behave ethically, spawning an "ethical consumerism" movement. These kinds of consumers prefer brands with agreeable ethical standards, known as "positive buying," or they may adopt a "moral boycott" by refusing to buy products from companies they see as ethically harmful.

In recent years, ethical consumer movements have focused on issues such as the safety of globally sourced food products – highlighted by China's contaminated baby milk scandal in 2008 – and on wages and workplace conditions in developing countries, including the Asian electronics factories where Apple's digital products are built. Thailand's fishing

Fair trade:

An international trading practice that aims to lift suppliers (mostly farmers in developing countries) out of poverty through giving a fair price for commodities, cutting out the middle man, and verifying fair labor conditions and environmental sustainability.

Thanks to ethical sourcing, a paper manufacturer guarantees tree farmers a fair price for timber.

Mulberry leaves for feeding silkworms.

A growing number of consumers worldwide expect the companies they buy from to behave ethically.

industry has recently come under international scrutiny for sourcing ingredients for its products from trawlers using what some observers say is tantamount to slave labor.

Increasingly complex, global chains of supply and manufacture mean that raw materials and components are often sourced from different parts of the world with widely differing standards for the protection of workers and the environment. With so much potential for risk to be introduced at any stage, one of the most common mistakes of companies with multi-tier supply chains is failing to look below the first tier.

Although transparency can be hard to manage at deeper levels of a multi-tier supply chain, it is imperative for a truly ethical business. While many companies make efforts to eliminate risk on tier one (the direct suppliers) of the supply chain, audits show that a greater number of environmental and social risks are found deeper down the supply chain on tier two or three. What's more, the risks found on deeper tiers are more critical than the risks found on the first tier. Sedex, a nonprofit organization dedicated to improving ethical business practices in global supply chains, found in 2013 that tier two on the supply

chain posed 18 percent more risk issues than the first tier; however, only a third of companies globally seek transparency below the first level.

While managing transparency through a far-reaching supply chain can be tricky, there are various management systems to ensure ethical sourcing on all levels, even sub-tiers. Companies are encouraged to use and monitor data from deeper in the supply chain to mitigate risk, including site visits and audits to gather information. Hiring third-party supply chain consultants can identify ethical chinks in a supply chain, and pooling information with direct suppliers can help assess problems further down the chain.

Although ethically sourced materials and workplaces usually pose additional costs to a company, a progressive ethical sourcing policy has been shown to have a strong positive influence on consumers who are willing to pay higher prices for ethical products. Thailand already boasts several successful examples of flagship "green" and ethical companies, such as Plan Toys in southern Thailand and the Jim Thompson Thai Silk Company, both of which have put their sustainable and ethical sourcing policies at the heart of their commercial branding to powerful effect.

▶PIONEERS

Jim Thompson Silk

History: Founded in 1948 as the Thai Silk Company

Location: Nakhon Ratchasima province

Key features: A park-like farm and estate at the edge of the Khao Yai mountain range is the center of the company's sericulture program, a sustainable textile venture

The Jim Thompson Silk Farm in Nakhon Ratchasima province is a noted cultural attraction and the hub of a silk production – or sericulture – program that embodies the pioneering vision of an ethical and sustainable silk industry, decades before many of the modern concepts of fair trade were developed.

"Without Isaan, there would be no Jim Thompson Silk," said Eric Booth, the company's assistant managing director who doubles as a director of the trust that cares for the Jim Thompson legacy.

Booth said Jim Thompson fell in love with what he called "the lumps and bumps" of Thai silk. He was committed to retaining the traditions of hand-spinning and hand-weaving that many farming households in rural villages depended on for additional income. The premium Jim Thompson placed on these traditional products helped ensure a local employment opportunity very much in touch with native aesthetics and values.

More than 50 years later, aside from a few adaptations, hand-spun and hand-woven silk from Isaan remains the company's signature product. Although the modern technologies employed in hand-weaving mean that little of the work is still carried out in homes, weavers at the company's silk factory in Nakhon Ratchasima are able to choose their own hours and work rate to match their household or farming commitments.

Unlike the model of large silk farms used by overseas manufacturers, the company adopted a sericulture program to distribute a unique breed of silkworm eggs to around 500 farming families in Isaan, who raise the delicate silkworms on trays of mulberry leaves and then sell the raw silk cocoons back to the company.

For the team at Jim Thompson Silk Farm, the Isaan connection is a vital and living part of the company's identity and community links. Communications director Chutima Dumsuwan said around 80 percent of the company's staff are from Isaan families, even those employed in the Bangkok head office, like herself.

A Jim Thompson employee weaves silk in Isaan.

Jim Thompson Furnishing Fabric.

EYE-OPENERS

FAIR TRADE GOES NUTS
THE LUCKIEST NUT IN THE WORLD
Director: *Emily James*
Release Date: *2002*

The Luckiest Nut in the World mixes animations and documentary film footage to tell the story of the ascendency of the American peanut as an international trade commodity during a period when the prices for cashew nuts, brazil and ground nuts – mainly grown in poor countries in South America and Africa – crashed from effects of liberalized world trade.

The same liberalization rules don't apply to the American peanut – the "luckiest nut" of the film's title – because the American peanut industry is heavily subsidized and protected by tariffs.

Meanwhile, subsistence farmers in some poor countries have endured decades of collapsing prices for their nut crops. The film explains how pressure on debt-ridden countries to embrace "free market" globalized economics has actually driven some poor countries and poor communities even further into poverty.

Written and directed by Emily James, the film was screened at numerous festivals and won Best Short Documentary at Full Frame. The UK-based NGO Christian Aid used the film and its characters in their "Trade Rules Are Nuts, Let's Crack'em!" campaign.

▶**PIONEERS**

Ethnic Lanna

History: Started in 2011 by Carlos Mantilla

Location: Chiang Mai

Key features: A fair trade enterprise committed to women's empowerment, livelihood development, and preservation of cultural practices

Hill tribe women sell Ethnic Lanna products at a market in northern Thailand.

An unassuming showroom shop-front in the center of Chiang Mai's old city, Ethnic Lanna can easily be missed. It is, however, an impressive thriving business set up by Mexican-born Carlos Mantilla. Inspired and excited by the colorful and detailed hand embroidery presented by Hmong sellers at Chiang Mai's historic Wararot market, Carlos started a partnership with the sellers in 2011 and established Ethnic Lanna Ltd Co. The business works directly with Hmong women to design, craft and distribute textiles for the international market.

Ethnic Lanna has a strong commitment to women's empowerment, environmental preservation, and fair trade practices. The business works closely with 30 artisan groups in three Hmong villages in the northern province of Phayao.

Exquisite patchwork and cross-stitched garments are "up-cycled" into vintage textiles (handbags, apparel and accessories) for export. The villagers also produce unique products using age-old indigenous methods of batik and indigo dying on organic cotton and hemp fabric. Repurposing vintage garments, including some lavish outfits crafted for one-time-use for ceremonial purposes, also extends the life of the fabric, ensuring both cultural preservation and environmental sustainability.

Ethnic Lanna's core focus on "Fair Fashion" extends to Chiang Mai city, where (not far from the showroom) Thai and Burmese sewers machine-embroider custom-ordered textiles in the nearby workshop. Disillusioned by

the number of machine embroidered fabrics imported to Thailand from China, Mantilla decided to set up this side of the business to provide local employment opportunities to some of the city's most disadvantaged. Conscious of keeping ahead of the curve, Ethnic Lanna designs and produces contemporary textiles inspired by everything from Indian to Guatemalan indigenous influences.

In 2015, following a period of rapid expansion, Ethnic Lanna also down-scaled the number of artisans it was partnering with in order to make the project more sustainable and manageable. Mantilla reflects: "Fair trade is very important to our business; it is something very important to me personally. I was born in Mexico and grew up around small villages. I've always been fascinated by the making of their special crafts, and handmade products. At Ethnic Lanna we make sure that whoever is touching the product from the very beginning is part of fair trade, and adheres to the guidelines of fair trade".

Deep in the mountains of Northern Thailand, Ethnic Lanna also regularly organises capacity building workshops for Hmong artisans to enhance their small business skills, with the view to diversifying their products and increasing their independent revenue.

Finally, the business reaches out to its customers through the Ethnic Lanna Helps Foundation to raise funds for projects focused on environmental conservation, and on improving the quality of life of communities that don't have access to government funding. Previous initiatives include supporting a Karen school and an Elephant Sanctuary. Ethnic Lanna is currently expanding this side of the business by working closely with non-profit organizations in Chiang Mai to identify projects for 2017.

"At Ethnic Lanna we make sure that whoever is touching the product from the very beginning is part of Fair Trade, and adheres to the guidelines of fair trade."

Carlos Mantilla, founder of Ethnic Lanna

▶PIONEERS

Doi Chaang Coffee Original Co

History: Founded in 2003 by local farmers who partnered with a Canadian businessman in 2006 to represent the company on the international market

Location: Chiang Rai province; and Vancouver, Canada

Key features: A sustainable model of coffee production and an example of ethical sourcing that goes "beyond fair trade"

As in other highland areas of Northern Thailand, the idea to grow Arabica coffee was first introduced to Doi Chang hill-tribe villagers as part of the Royal Projects program in 1962. Situated in the highlands of Chiang Rai province, the area around Doi Chang is particularly well suited to grow Arabica coffee trees. The endeavor was not to be an overnight success, but in time the hill tribe villagers including Akha and Lisu eventually established themselves as independent, successful coffee producers.

They built their own processing plants, drying facilities and storage warehouses. They cultivated their coffee in small family gardens with everyone committed to sustainable agriculture and inflicting a minimal impact on the natural habitat. By the late 1990s, the farmers knew they had a high quality product, but they also recognized that they needed outside help to market their single-origin coffee and to open up business opportunities that at the time remained out or reach.

In 2002, Wicha Phromyong and the family of Panachai Pisailert began putting their heads together to find a viable solution. In 2003, together with Pitsanuchai Kaewpichai, they founded Doi Chaang Coffee Original Co Ltd. The goal was to assist Doi Chaang coffee growers in getting a better price for their coffee, to develop the Doi Chaang Coffee brand both in Thailand and internationally, and continue to refine and improve the quality of their coffee.

Wicha, who passed away in 2014, knew that in order for villagers to begin earning more, the company needed outside assistance. So in order to reach other Asian markets and markets in both Europe and North America, in 2006 the company formed a strategic partnership with John M Darch, a Canadian

"If you are able to give the farmers dignity, and the ability to earn more money through an ownership in an entity (not just selling beans) you break the cycle of poverty and create a cycle of prosperity."

John M Darch, co-founder of Doi Chaang Coffee Co (Canada)

businessman. In recognition of their contributions, the growers and the Canadian group established an equal partnership for the international distribution of Doi Chaang Coffee. This means the farmers have 50 percent ownership of the Canadian Doi Chaang Coffee Co, which is fully funded by the Canadian side. In essence, it is a relationship that goes "beyond fair trade."

Darch explained: "We believe the farmers need more than a Fair Trade price (essentially minimum wages) to elevate them above survival. The farmers must have a way to improve their lives and be self-sufficient, not just poorly paid farmers at the bottom of the ladder, or they and their children are condemned to poverty with little prospect of a better life."

"It is our belief that if you are able to give the farmers dignity, and the ability to earn more money through an ownership in an entity (not just selling beans) you break the cycle of poverty and create a cycle of prosperity. Additional money means an ability to have an education [for farmers and their children] leading to greater production and more prosperity."

Many youth from Doi Chang families are now acquiring higher levels of education and then returning home to apply what they have learned to the benefit of the community. "In this way, the farmer, their families, [and] their community are elevated and Doi Chaang Coffee has greater production. A truly win-win situation," Darch said.

Doi Chaang Coffee's 'beyond fair trade' model has helped to empower indigenous hill tribe people.

▶PIONEERS

The Sampran Model Project

History: Founded in 2010 by the owners of Sampran Riverside hotel

Location: Nakhon Pathom province

Key features: A model for converting conventional agriculture practices to organic, and linking organic farmers directly to buyers using fair trade principles

Sampran Riverside, a four-star resort and traditional cultural activities center, has developed a model for converting conventional agriculture to organic, and linking the farmers directly to buyers via an organic value chain based on fair trade principles. Founded in 1962, the resort is located on 28 hectares of ISO organic-certified land an hour west of Bangkok in Nakhom Pathom province.

Conventional farmers in Thailand rely heavily on middlemen who dictate prices and specifications for their produce. As a result, farmers have a difficult time negotiating a fair price for their goods. The vision of the "Sampran Model" project, launched in 2010, is to restore balance to the local food system by finding new market channels – bulk and retail – for farmers who Sampran Riverside helps convert from conventional to organic agriculture.

By circumventing middlemen and offering a premium product, farmers are able to get far better prices for their produce. Another important aspect of the initiative is the relatively large amount of guaranteed sales each month for these smallholder farmers. Sampran Riverside itself sources approximately eight tons of organic vegetables, fruits and herbs, and three tons of organic rice per month from local farmers.

"The intention from a business perspective was to have a selling point to differentiate Sampran Riverside from other hotels by having organic fruits and vegetables for our guests," said Arrut Navaraj, managing director of Sampran Riverside.

In 2010, the Sampran Model opened a weekend farmer's market called Talad Sookjai for organic farmers to sell their produce and delicacies, with monthly workshops about organic farming and holistic health offered to visitors. Other market channels include Sampran Riverside itself, the Sookjai Market Roadshow and the Sookjai Organics Website.

In early 2016, Sampran founded a "Farm to Functions" initiative with nine business partners and the Thailand Conventions & Exhibitions Bureau (TCEB) to expand the impact of the Sampran Model. The alliance of prominent hotels and convention centers purchase organic products directly from farmers and, along with TCEB, promote sustainable agriculture

and healthy eating. The first phase of the initiative links farmers in Thailand's least-developed region, the northeast, to large businesses in Bangkok, so about 300 farmers benefit as part of the Amnatcharoen Satjatham Rice Community Enterprise Network. The project aims to order a minimum of 600 tons of rice in two years. All of the jasmine, riceberry and aromatic black (hom-nin) rice is IFOAM-certified organic in-season rice.

In late 2016 the initiative added an agreement between 13 hotel operators in Bangkok and the Ban Thap Thai Organic Agricultural Cooperative in Surin province. The operators committed to purchasing seven tons of jasmine rice per month. Besides helping farmers and the environment, "Farm to Functions" also helps boost the hotels' eco-friendly image and their competitiveness. In addition to the hotels, the restaurant chain Sizzler and the Seefah Group directly purchase jasmine rice through the Sampran Model.

As of 2017, Sampran Riverside has been hosting workshops with various government agencies to further expand the model. "Before the government just thought about selling," says Arrut. "They didn't understand how the farmer thinks and they didn't understand the value chain. Now they are beginning to grasp things from a more systematic approach and perspective. The goal in the future would be for the model to be expanded to a provincial level."

A farm inspection (left) helps to verify that organic standards are being met. Sampran model farmers sell produce at a weekend farmers' market (right).

WASTE MANAGEMENT

Refuse is a precious resource with many uses

Modern ideas of sustainable waste management involve a reevaluation of the concept of "waste" itself. Even a generation ago, in Thailand as in the rest of the world, most waste was destined only to be dumped or burned – a legacy that persists today as hazardous dump sites and polluted waterways dot the countryside, and residents still raise a stink about the sporadic fires in garbage dumps.

But for this generation and those to come, the concepts of waste and waste management have a whole new meaning. Waste is now seen as a potential source of valuable resources, whether as processed materials that can be reused or recycled, raw materials for factory processes, or fuel for power generation, which can offset the demand for fossil fuels. The concept of "waste as resources" has already been put to good use in the private sector, as businesses across Thailand develop effective and even profitable ways to manage their waste.

The disposal of waste from industrial processes in Thailand is governed by relatively strict environmen-

Wat Larn Kuad, meaning "temple built of a million bottles," in Si Saket province.

Waste is now seen as a potential source of valuable resources, whether as processed materials that can be reused or recycled, raw materials for factory processes, or fuel for power generation.

A man sells used machine parts, mostly from engines, in Bangkok.

tal laws and overseen by several government departments. Hazardous wastes that can't be recycled or reused, including "biohazard" wastes from hospitals and clinics, are either incinerated or stored in secure waste facilities designed to stop pollutants leaking into the environment. But municipal waste – mainly from households, shops and markets – is not subject to such strict controls or costs, and the false classification of industrial waste as municipal waste is a recurrent problem.

The hundreds, perhaps thousands, of illegal dumping sites dotted around Thailand are subject to similar dangers. However, authorities have vowed better enforcement of existing laws on industrial waste and tighter controls on municipal waste, coupled with efforts to build awareness of the dangers of illegal dumping and the benefits of safe waste management. But such measures will take time.

In recent years there has been a sharp increase in "e-waste" – digital electronic detritus such as old cellphones, computers, laptops, batteries, televisions, cameras and printers. E-waste can contain small amounts of valuable minerals and components, but

also toxic materials. A local industry has sprung up in some "e-waste villages" in northeast Thailand to process e-waste imported from overseas in shipping containers. While valuable materials and components are recovered and sold, the unprofitable waste from these village ventures has been piling up at local dumps, where subsequent tests have found hazardous levels of lead and arsenic.

Modern waste-to-energy technologies offer the promise of near "zero waste" industrial processes and a partial solution to Thailand's problems. Waste-to-energy power plants are already well established in Thailand, producing power for factory processes and electricity primarily from renewable biomass sources such as rice husks and bagasse waste from milling sugar cane. Biopower plants that use methane gas from waste and bioethanol fuel are also being implemented.

The outlook for sustainable waste management in Thailand is hopeful, offering potential solutions for a number of environmental issues, including the reduction of pollution and landfill sites, while generating more renewable energy.

Recycling Outside the Box

Tetra Pak packaging – the coated paper cartons used extensively as aseptic packaging for drinks and other foodstuffs – is one of the most successful packaging innovations of our time, with more than 50 billion Tetra Pak–packaged products made each year in more than 170 countries.

But Tetra Pak technology is not perfect: the plastic- or aluminum-coated paper cartons can't be recycled through the usual processes, and only a quarter of the packaging is recycled worldwide. The rest mainly ends up in landfills and dumps.

Since the 1980s, the company has worked hard to develop its recycling efforts and it now operates more than 100 specialized recycling factories around the world, including a plant near Bangkok. The first in Southeast Asia, it was built and continues to operate in partnership with the Bangkok Metropolitan authority. In 2011, Tetra Pak Thailand launched a recycling initiative with the establishment of 10 collection centers around the country. So far, the project has collected more than 55 million discarded cartons and used them to make sheets of roofing material, which are distributed through the Princess Pa Foundation and the Thai Red Cross for use in building emergency shelters for disaster victims.

▶ PIONEERS

Wongpanit Recycling

History: 1976

Location: Nationwide, headquartered in Phitsanulok

Key features: Somthai Wongcharoen built a nationwide recycling business from scratch and is now one of the country's most respected experts in waste recycling; he also offers business management guidance to Wangponit franchisees around Thailand

Somthai Wongcharoen began building his recycling empire as a young man. Today, his Phitsanulok-based company, Wongpanit Recycling, employees more than 250 people directly and has trained more than 6,000 others on recycling as a business model.

"There is no such thing as waste – merely misplaced resources," said Somthai. As a testament to this, the Phitsanulok headquarters of Wongpanit Recycling is filled with towering stacks of plastic, paper, glass, scrap metal and more – the waste streams that the company transforms into recycled raw materials on an elaborate production line of Somthai's own devising.

Somthai has achieved a measure of fame through his unstinting efforts to change the way Thailand thinks about waste. As a young man in Phichit province, he was inspired by one of his aunts who earned money by buying and selling recyclable waste. Following her example, he abandoned his small business selling garlic and went into the waste business with just 1,000 baht and an old pickup truck, buying household waste and junk for cash and finding new ways to sell what he found. Some 40 years later, Somthai still marvels at the results.

"I never thought it would all get so big. It was destiny," he said. Today the Wongpanit headquarters is fed by a network of 1,250 Wongpanit recycling centers around Thailand, each an independently owned business guided by Somthai's expertise. The headquarters can process up to 500 tons of waste per day. The processing lines consume the stacks of plastic, metal, glass, office paper, construction debris, and industrial and electronic waste purchased through the buyer network, turning them into recycled raw materials for sale to commercial customers, including bales of scrap metal and aluminum cans, bags of recycled plastic chips for use as raw plastic for new products, drums of color-sorted glass, and bins filled with recycled e-waste, such as copper connectors and the titanium bearings used in hard disk drives. Even organic waste, such as overripe market vegetables, is composted and sold.

Somthai has personally trained thousands of people in waste recycling and business techniques through the Wongpanit recycling network, which supplies its well-known branding to help recycling businesses start up in new areas. It's a concept that has even reached the shores of the United States: a Thai expatriate has set up America's first Wongpanit recycling center in Maryland. Franchisees pay a small

"There is no such thing as waste — merely misplaced resources."

Somthai Wongcharoen, founder of Wongpanit Recycling

deposit for the Wangponit branding, but remain free to sell their recycled materials to any buyer.

The Wongpanit network also benefits from Somthai's "Waste Bank" scheme, which he pioneered at a school in Phitsanulok. Because small items of recyclable waste don't earn much money, Somthai devised a system of "waste accounts" to record the value of the items each student brought to the recycling program. As they brought in more recyclables, they banked more money until they accrued a decent amount. The "Waste Bank" program proved so popular at the pilot school that it is now used at more than 5,000 schools, factories, hospitals and even military bases across Thailand.

Wongpanit's International Coordinator, Wimonrat Santadvatana, explained that the company only buys waste that it has established a market for, but Wongpanit has developed processes that can recycle almost everything. A key recent development is a production line to process a mix of plastics and dry organic wastes into bags of Refuse Derived Fuel (RDF). The calculated mix of different types of wastes in the bags determines the thermal value of RDF fuel and the prices of the bags.

Wongpanit currently supplies its RDF as fuel for a large cement kiln, but it's the growing market for waste-to-energy electricity that has sparked Somthai's imagination. Wangponit headquarters now has a prototype for a prefabricated waste-sorting line that will produce bags of RDF fuel for small waste-to-energy plants. Somthai designed the assembly himself with local conditions, materials and budgets in mind. The goal is to bring down the cost of an RDF plant to prices that local governments or single factories can afford. The addition of a small generator to Somthai's "green RDF machine" would supply enough energy to fulfill several villages' electricity demands, a development that offers the hope of low-emission, waste-to-energy generation at the local level that can help ease demand for fossil fuels.

▶PIONEERS

Scholars of Sustenance

History: Incorporated in the US in 2014, and launched in Thailand in 2015

Location: Bangkok

Key features: Acquires food surplus donations from hotels, restaurants and supermarkets and distributes surplus food to people in need. SOS also has a composting program and an urban gardens initiative.

What is the true value of food? And how can food surplus and food waste be better utilized in Thailand? These are the questions the people behind Scholars of Sustenance (SOS) want individuals and businesses to consider more closely. In doing so, SOS hopes to raise awareness about the importance of "the food cycle", promote food security and assist communities in need.

After numerous visits to Thailand, Danish software developer Bo Holmgreen decided to establish the SOS foundation in Bangkok in 2015 along with his business partner Turid Kaehny.

"I was traveling through the region multiple times a year, seeing the incredible food waste from large hotels. I started to interview hotel staff and other entities – thinking about what a non-profit could do to re-distribute all that excess food."

Holmgreen, whose home country is at the forefront of food waste minimization, had been reading for years about problems associated with global food waste. "I knew Turid's and my early retirement was around the corner as we were selling the software company we ran, so we designed SOS around our desire to reduce landfill gases, help hotels/retailers be greener, feed those in need, and above all to use the food rescue 'business' as a vehicle to lift bright children out of poverty."

Thailand would certainly prove to be a bountiful location for SOS's flagship project, which is dubbed "Happy Spoons." For starters, the vast amounts of food surplus and waste generated by the food service industry and by the more than 30 million tourists who visit the country each year offers an invaluable resource when repurposed.

The logistics of sourcing, collecting and handling that food is the hard part. This starts with SOS approaching hotels, restaurants and supermarkets to establish a systematic relationship of food surplus donation. To encourage participation, the program is designed to be as convenient as possible for donors. SOS covers the cost of all logistics; provides containers, kitchen facilities and transport; ensures donor anonymity as well as minimal donor staff impact; and is responsible for inspecting the condition of food before sending donations on to people in need. SOS also assumes responsibility for any resulting issues.

After food surplus is collected and inspected by a food hygienist, it is then redistributed to communities in need, either uncooked or as pre-cooked meals. As of 2017, SOS is distributing food in seven communities through partnerships with Halfway Home, Mercy Center and Asylum Access. Altogether, the program benefits around 1,500 people.

"People are already on the right track [in Thailand]. We only need to help push it further," said Abigail Miguel, who was brought on as director of SOS in 2016.

In addition to Happy Spoons, SOS also runs the "Little Sprout" urban gardening project and the "Sunny Soil" composting program. Little Sprout focuses on teaching students how to start their own vegetable gardens, while Sunny Soil supports local communities in turning food waste into fertilizers.

Tesco-Lotus is a regular donor to the Happy Spoons project.

Scholars of Sustenance director Abigail Miguel (right) and food quality and safety project assistant Tanaporn Oi-isaranukul inspect and wash donated produce.

▶**PIONEERS**

Mitr Phol Bio-Power

History: Mitr Phol was founded in 1946; Phu Khieo power plant was commissioned in 2002

Location: Phu Khieo Bio-Energy power plant and nationwide

Key features: Thai pioneer in the field of waste-to-energy and biomass power systems

Staff closely monitor every step of the waste-to-energy process.

The Mitr Phol Group, Thailand's largest sugar producer and the fifth largest in the world, is a leader in the waste-to-energy field. Its efforts were inspired by a historic problem in the sugar industry: What to do with "bagasse," the fine fibrous remains of sugar cane that has been milled for sugar juice.

Sugar is an important crop in Thailand, especially in highland areas that are unsuited to rice farming. But a sugar mill produces around three tons of wet bagasse waste for every 10 tons of crushed sugar cane, and the annual sugar harvest at any one of the six Mitr Phol sugar mills in northeast Thailand creates mountains of dry, pulp-like bagasse fiber.

The massive furnaces at Mitr Phol's Phu Khieo biopower plant in the northeastern province of Chaiyaphum are the burning heart of a technology that offers a sustainable solution to some of Thailand's waste and energy problems. This 102 MW sugar-waste-fueled plant not only powers the entire operations of the sugar milling and refining process, but also offers around half its power output as electricity to consumers. At the same time, the use of modern incinerator technology and waste management techniques are helping to reduce the need for landfills.

The total electricity generating capacity from Mitr Phol's six waste-to-energy plants in Thailand is currently at 466 MW, or around 17 percent of Thailand's total biomass power generating capacity. As many of the power plants upgrade to high-pressure steam systems, even more power will be generated from the same amount of biomass.

Mitr Phol is also Thailand's largest biofuel producer, using molasses waste from the sugar refining process as a feedstock for bioethanol at four produc-

tion plants with a combined capacity of more than a million liters a day.

Behind hydropower and solar, waste-to-energy and other biomass power plants are the country's third leading renewable energy resource, with a total capacity of around 2,718 MW—more than 5.5 times that of Thailand's installed wind power capacity. Mitr Phol is also investing in solar rooftops and solar farms, and expanding its biomass power operations that utilize rubber wood chips, rice husks, and cane leaves.

In an effort to scale up research and development, the company opened a second R&D center in 2014 to augment the work being conducted at its first R&D center, established in 1997. Some of the work being done includes the development of bio fertilizers, the production of lactic acid, and the development of Thai yeast strains to help enhance the efficiency of ethanol production and reduce the costs of yeast imports.

Incinerator technologies reduce landfills.

The total output of 76 megawatts from Mitr Phol's sugar-waste-fuel plant in Chaiyaphum powers operations. Extra electricity is sold to consumers.

Turning Waste Heat into Alternative Energy

Waste heat recovery (WHR) is the process of extracting energy from the heat vented by a factory, often by using heat emissions to drive an electrical power generator. Modern technologies and new applications mean that more waste heat from industrial processes can now be captured effectively, and the efficiency is improving all the time.

WHR is often implemented as heat exchangers that trap heat from industrial kilns, boilers, smokestacks and cooling towers. The heat exchangers typically drive generators that feed electricity back into the factory, where it can be used to drive manufacturing processes. This increases the overall efficiency of the manufacturing processes by utilizing a waste resource that otherwise would be emitted into the atmosphere. WHR systems can also help reduce air pollution in the atmosphere by offsetting a factory's demand for electricity from the national grid, which in Thailand is largely generated by burning gas and oil.

The use of WHR systems is limited by their relatively high cost and the technical limits of the equipment. In the past, low-temperature waste heat was difficult to use without large, expensive heat exchangers. But new technologies are reducing the cost of WHR systems while increasing their efficiency. Also, new applications have been developed specifically to utilize sources of low-temperature waste heat, including the production of biofuels. For example, waste heat under 100 degrees Celsius has been found effective in growing algae farms used for biofuel.

In Thailand, SCG has been developing waste heat recovery power generation at three of its cement plants since 2007, at a projected cost of 3.4 billion baht over 20 years to build and maintain a total of 70 megawatts of generating capacity. The projects include the WHR system at SCG's largest plant, located at Lampang in Saraburi province. It is the largest cement factory in Southeast Asia. High kiln temperatures of up to 1,450 degrees Celsius are used to turn limestone and clay into "clinker": nodules of silicate minerals that make the binding material in cement. The waste heat from the main boiler and kiln processes at the Lampang plant is estimated to produce around 152.6 gigajoules per kilogram of cement clinker that it produces.

Since the implementation of a WHR system at SCG's Lampang plant in 2009, heated air from the main boilers and the clinker cooling process is used to generate steam to drive an 8.46-megawatt electricity generator that supplies electricity for SCG's cement production line. The WHR system also reduces the levels of heat vented into the local environment, and less water is now used to cool waste gases before venting.

In concert with other energy-saving efforts, according to SCG, the WHR system at Lampang produces 51,590 megawatt hours of electricity in a year, saving about one billion baht in electricity from the national grid – roughly a quarter of the plant's consumption – while reducing greenhouse gas emissions by the equivalent of 29,301 tons of carbon dioxide per year.

SCG designed its WHR project at Lampang in part to set an example to other industries in Thailand by promoting WHR as a best practice in the field of waste management. Two more WHR power-generating plants are being built by SCG Thailand, while several non-generating WHR projects, which do not produce electricity but recover heat for other purposes, have been implemented in different SCG factories. The group's expertise with WHR is also spreading to its subsidiary companies outside the country. In 2013 the group built a waste heat power generator at its main cement factory in Cambodia, and later implemented WHR technology at its plants in Vietnam and Indonesia.

SUSTAINABLE TOURISM

Low-impact, community-based travel is gaining ground and winning converts

When the term "eco-tourism" was coined in 1983 by Héctor Ceballos-Lascuràin, a Mexican architect, it came to encompass all sorts of nature-based trips where the onus was on the tour operator to provide an environmentally friendly excursion. As the term morphed into "responsible tourism," the onus was more and more on the traveler to behave responsibly by respecting the local environment and customs. But as the market continued to grow, it evolved into "sustainable tourism," the term most commonly used today, where the onus is on all stakeholders, from the hotels to the agents to the traveler.

Aerial view of green rice paddies in Nakhon Sawan province.

All of these terms represent branches of the same tree. They are guided by similar principles, such as low volume/low impact, which is only common sense: the fewer travelers on any given outing, the less impact they will have on the environs. The necessity of sourcing local foods and reducing transport is also easy to grasp; both contribute to a lower carbon footprint for each visitor and the company.

Equally important is that the members of the communities visited also benefit from the influx of travelers. For tour operators, that means employing local guides and drivers. For resort owners that means hiring local cooks, maids, handymen and performers to ensure that the tourism industry becomes part of the community instead of an entity that stands apart from it.

An even more integrated version of this approach is sometimes called "community-based tourism." In this niche market of sustainable travel, the hottest buzzword of the last five years is **"authentic experiences."** That could mean a trip to the oldest tea plantation in Thailand for a few lessons in plucking and making tea, or it might be a visit to

Authentic experiences: A specific experience in a specific place that may involve things like eating a traditional dinner in a traditional manner, or taking part in an ancient ritual like giving alms to Buddhist monks in the morning.

a local shaman with a translator in tow for a consultation, or it may entail a performance of songs and dances from members of a hill tribe.

Sometimes called "learning experiences" or "local encounters," these are not your run-of-the-mill side trips that coachloads of other tourists will join for the sake of posting a few photos on social media. These are boutique experiences meant to sincerely involve and benefit local communities and to provide the tourist experiences to be savored and remembered, as the elders pass down their hard-won, time-honored nuggets of wisdom or experience to help preserve these customs and traditions for future generations. When the villagers see visitors taking interest in different aspects of their culture, it's a powerful reminder that such heritage has an intrinsic value that is only partly monetary. All over Thailand and other parts of the world, tour operators are helping to preserve indigenous arts, crafts and folklore through exactly this kind of exposure.

Part of the reason why there has been so much "greenwashing" in the travel industry is that hotels, tour operators – and even countries – have been able to make the most outrageous claims about

PATA: Partners in Sustainability

When the Pacific Asia Travel Association (PATA) began in 1951, the total number of international arrivals to the region stood at around 100,000 per year. Now it's close to 550 million. As a true originator in the region, PATA has played a pivotal part in that phenomenal increase.

Initially, PATA served as an advisor to national tourism organizations in the Asia Pacific region as tourism plans were developed, marketing campaigns organized, and infrastructure built from the ground up. In the 1970s the nonprofit association was instrumental in promoting then up-and-coming destinations like Chiang Mai but, more significantly, it became one of the first such international tourism bodies to push environmental concerns to the forefront of its agenda.

That dedication has continued to resonate through many parts of the organization and its various events, like the PATA Travel Mart, started in 1978 as a hub of networking, a showcase of destinations and a trove of best practices in sustainability. More recently, the association started the PATA Foundation, a charitable arm that lends a helping hand to rural communities engaged in tourism, and inaugurated the PATA Adventure Travel and Responsible Tourism Conference and Mart with high-profile events in Nepal and Bhutan.

their tours and destinations without having to back them up. The tour operators and hotels decide on policies and implement them at their own discretion. There are no official government policies to follow except when it comes to complying with certain rules governing national parks or wildlife sanctuaries. That has changed to some extent with increased regulatory oversight and pressure from governments and the rise of certification bodies like Earth Check, Green Globe, the Thai Ecotourism and Adventure Travel Association and Travelife.

All of these sincere efforts can seem like a hopeless battle in the face of the ever-expanding tourism sector. Putting the impacts of tourism in a rather frightening context, one study estimated that local activities account for less than 5 percent of the CO_2 emissions of the tourism industry, while transportation, especially air travel, may be responsible for around 70 percent, with the remaining 25 percent created by accommodations. These numbers support the adage that the most sustainable form of tourism is to simply travel closer to home.

But that's not a winning sales pitch or proposition for any country as reliant on global tourism and business

as Thailand is. The expansion of air travel across the world, region and Thailand is certain to translate into more tourists, as well as power-guzzling trips, hotels and shopping complexes to cater to them. Having long ago embraced mass tourism and its economic benefits and with a budget airline industry essentially in its infancy, integrating the practices of sustainable development throughout the tourism industry to limit negative impacts may become a critical issue for the country in the years ahead.

▶**PIONEERS**

Lisu Lodge

History: Opened in 1992 by Asian Oasis

Location: Mae Taeng, 40 minutes outside of Chiang Mai city

Key features: 24 rooms and 16 staff members; all building materials sourced locally except for floor tiles from Bangkok and water heaters from the United States; organic rice and vegetables; no TV, wi-fi or air-con in rooms

Accolades: Winner of Responsible Tourism Award 2013 from Wild Asia, among many others

In the hospitality trade, one of the biggest offshoots of sustainable tourism is the eco-lodge. These structures are built to be as environmentally sound as possible, using local materials like bamboo, and are designed so they blend into their surroundings as naturally as possible. An exemplar in the field is the Lisu Lodge in northern Thailand.

When the Thai Ecotourism and Adventure Travel Association came to audit the Lisu Lodge as part of its certification process in 2012, the lodge was awarded 97 points out of 100. Lisu Lodge would have had a perfect score, but in the "recycling food" category there was

To provide employment opportunities, all staff at Lisu Lodge come from nearby hill tribe villages.

no box to tick for feeding leftovers to the villagers' pigs.

That's a good illustration of how comprehensive the lodge's approach to sustainability is. The building, and by extension the entire brand, has been constructed around such key pillars as preserving the environs and keeping the local communities intact by respecting their cultures and providing them with employment opportunities. All of the staff come from the nearby Lisu hill tribe village who supplied the inspiration for the name, and who work as guides, drivers, cooks, receptionists, maids and performers, and for every visitor that comes to stay in the lodge the owners put 60 baht into a village fund.

The lodge itself has been constructed in a Lisu style, using only locally sourced materials save for some tiles from Bangkok and a super-efficient gas water heater from the US that produces no emissions.

In an effort to bolster local incomes and provide another activity for guests, Lisu Lodge pioneered white-water rafting in the area. Now that it has become wildly popular, spawning dozens of independent tour operators and attracting hundreds of visitors, the lodge is spearheading efforts to minimize the environmental impacts. The waterborne excursions offered by Lisu Lodge are only one of a raft of outdoor activities that embody all the most basic elements of eco-tourism, like trekking, cycling and visiting Thailand's oldest tea plantation to learn how to pick leaves and brew tea.

Preserving and appreciating the local culture is another specialty. That means trips to the Lisu village for guests, who can also take in performances of songs and dances by hill tribe members. They can even request consultations with the local shaman on everything from health to mystical matters. Interactions like these are one reason that the owner of the company, Chananya "Ann" Phataraprasit, does not sell room nights, like many other eco-lodges do. Instead, her company, Asian Oasis, only sells packages ranging from one to five nights or more. "If I sell room nights,

> *"We have to adapt to the growing demands of the clients. People today want to experience something totally different from 20 years ago, or even 10 years ago."*
>
> **Chananya "Ann" Phataraprasit, owner of Lisu Lodge**

people will skip the community. People will skip learning. People will skip the experiences. It's part of what we have a commitment to do," she said.

Ann notes that this is not a field with huge revenue potential. Simply put, these kinds of lodges do not have the large number of rooms or guests to ever become massive moneymakers. They also tend to be one-off experiences; repeat visitors are rare. Eco-lodges are, in fact, a niche market, but there is potential for steady growth.

To ensure long-term growth, government bodies like the Tourism Authority of Thailand (TAT) would have to begin promoting the kingdom as a sustainable tourism destination, similar to how the TAT promoted Thailand's spa industry. This would take some of the marketing pressure off individual resorts and tour operators. Ann welcomes more competition in the field and the chance to band together with other entrepreneurs to make the local industry more dynamic. "You need more people so you can market it as a destination of sustainable travel," she said. "We have to adapt to the growing demands of the clients. People today want to experience something totally different from 20 years ago, or even 10 years ago. People are looking for bigger rooms, more comfort and different tours. We have a huge rice field because people are concerned about organic living and we grow organic vegetables too. So we are adapting all the time to the changing market. That's what makes your business sustainable."

▶ PIONEERS

Chai Lai Sisters Trek and Tours

History: Founded in 2016 by five ethnic Karen women

Location: Mae Wang village in Chiang Mai province

Key features: A tour company run entirely by indigenous women and focusing on culturally and ecologically sustainable tourism

Soe Aubomo, one of the Thai-Karen guides of Chai Lai Sisters, rides an elephant at dawn.

Nukul Chorlopo was nervous when she journeyed to Chiang Mai to register her tourism company with the Tourism Authority of Thailand (TAT). She was a young Thai-Karen woman who grew up in a hill tribe village in Mae Wang, about two hours southwest of Chiang Mai city. She had just earned her tour guide license, but she'd never dealt with city officials before.

In all her 26 years, she'd never heard of a woman from her village starting a company. But in September 2016, Nukul's company was successfully registered and Chai Lai Sisters Trek and Tours was born.

Nukil and four other Thai-Karen women from the hill tribe villages of Mae Wang became the founders of a sustainable tourism company that focuses on ecologically sound homestays, cultural exchange, and community empowerment. Chai Lai Sisters Trek and Tours is likely the first tour company in Thailand to be run entirely by indigenous women, although TAT does not keep records on the ethnicity of company founders.

Regardless, not only is this a major achievement for the young founders, but also for the Karen community, who are an ethnic minority group who has seen decades of civil war in their home country of Myanmar. Many of them now live in Thailand as refugees, migrant workers, or illegal immigrants. Some, like Nukul, have been born or naturalized in Thailand and are allowed to move and work freely in the country.

The Chai Lai Sisters offer tourists immersive cultural experiences with homestays, nature walks, and eco-minded activities in the founder's home villages. Many of the activities are gender-sensitive, encouraging an understanding and appreciation of the skills, knowledge, and customs of Karen and Thai-Karen women.

"I don't want to waste my culture," said Golo Wararom, another co-founder. "With Chai Lai Sisters, I can show our life."

One of the company's missions is to hire only Karen women, providing employment opportunities for a long-neglected and disadvantaged demographic. This breaks new ground in Thailand's hill tribe tourism industry, which does offer homestays but does not put a premium on gender sensitivity or women's employment.

Golo said that village women are becoming more motivated to learn English after interacting with tourists. They are also taking greater pride in their culture and developing a stronger sense of self-worth. On the surface, these may seem like small impacts, but they are big

Nukul Chorlopo co-founded Chai Lai Sisters Trek and Tours in 2016.

steps in an ethnic minority community that is under-educated and given few resources.

The venture is supported in part by Daughters Rising, a Mae Wang–based nonprofit that provided the founders of Chai Lai Sisters with the necessary language skills, vocational training, and seed funds to become tour guides.

▶PIONEERS

Khiri Travel

History: Founded by Willem Niemeijer in Bangkok in 1993

Location: The company operates in seven countries in Southeast Asia, including Thailand, Laos and Indonesia, and also has a sales office in the US

Key features: Specializes in sustainable tours with cultural, natural or educational angles

Accolades: "Best Asia Pacific Responsible Tourism Website" award at the 2014 TravelMole APAC Web Awards

Thi Lor Su waterfall in Tak province is a famous eco-tourism destination.

Founder and CEO of Khiri Travel, Willem Niemeijer, says that his company adopted sustainable tourism early on because the company was not interested in mainstream travel. Khiri's niche tours revolve around cultural, natural and educational excursions. It's a model that makes long-term economic sense. "To manage a company well, 360 degrees well, you have to look not only at profit, but giving back to the local community and training and retaining staff, as well as providing a better working environment for the staff. Because if it's only about money, your staff will only be there for the money and leave for another company that offers more," he said.

Destination management companies (DMCs) like Khiri have a special responsibility and power because they control so

much of the travel supply chain. DMCs recommend hotels and restaurants, hire drivers and guides, and provide other logistics. These choices can make a big difference. "If we have a choice between two hotels, the one with the better sustainability practices wins," said Niemeijer.

But with the growth of eco-tourism came a wave of "greenwashing" as tour operators and hotels began billing their operations as environmentally sound. At the time, there were no certification bodies in place to challenge these claims. That changed with the emergence of organizations like Earth Check for hotels and Travelife for tour operators.

Khiri Travel spent a year preparing for the Travelife audit in Thailand, Vietnam and Myanmar. The process included audits of its offices to measure everything from energy use to waste management, and audits of their work in the field to ensure minimal negative social and environmental impacts, as well as positive fiscal or cultural contributions to rural communities. Finally, they received the coveted certificate in 2015.

Obtaining the certification, which is rapidly becoming an industry standard in Europe, taught the company a lot about its operations. "We learned that it's possible to make your business 100 percent sustainable by buying carbon offsets and managing your water usage and garbage output. You can measure what you do and try to improve every year," Niemeijer said.

Khiri Group's Sustainability and Responsible Tourism Manager Graham Read said that the company saves money by reducing water and energy use, and certification comes with other benefits: "Consumer demand for sustainable products is growing. Companies that take sustainability seriously will score higher on customer satisfaction, staff motivation and business efficiency with positive effects for their competitive advantage. As a B2B company, with many agents from Western Europe, certification would give us a competitive advantage over other DMCs in the region."

The company is also turning its CSR wing, Khiri Reach, into a foundation. One hundred percent of the money Khiri fundraises for development projects goes to affected communities. Khiri looks for NGOs and charities to manage the projects, but the company also encourages its own employees to serve as ambassadors. Every year, each staffer devotes about five or six work days of company time to a particular project.

Despite Khiri's progress, Willem is realistic about the nature of his business. "As an incoming long-haul business our main customers are people coming from Western countries, anywhere from a 5- to 24-hour flight away. That in itself is not very sustainable. But if you want sustainable travel and you're a European, go on a cycling trip in your own country. We cannot turn back the clock. International travel is the norm for many people now."

GROUNDBREAKERS

SOMSAK "PAI" BOONKAM *earned a bachelor's degree in engineering and an MBA in sustainability before he began to work on development projects with hill tribe communities in Thailand. After working for the Mae Fah Luang Foundation at Doi Tung in northern Thailand, he co-founded his community-based tourism company, Local Alike, with Noon Pakavaleetorn in 2012. Local Alike offers sustainable, socially responsible and immersive travel experiences that preserve culture and generate income for local communities. The outfit has won many competitions in Thailand and Singapore for social enterprises, attracting much media attention in the process.*

How does community-based tourism benefit these villages?

They get 70 percent of the net price of the tour and another 5 percent of the profit gets put into their community fund. Having visitors and making money also helps to keep these families together and makes them proud of their cultures. So the communities benefit in many different ways.

At first you could not get traditional investors, so you chose to enter social enterprise competitions. Have you been able to secure investors now?

At first, we tried to get a few investors interested but we did not succeed. So I came up with the competition plan. We won a few competitions for social enterprises in Thailand from the British Council and AIS. We have also won The Venture award. In Singapore we won the Young Entrepreneur Award from the Singapore International Foundation. The awards helped us raise money and get publicity. After that, the investors came to us. Now we have investors for the main company and are now looking for investors for LocalAlike.com which is a marketplace of sustainable tourism experiences for South East Asia.

What were the origins of the company?

After I worked in Germany I saved up enough money to travel in rural parts of India, Myanmar and Laos where I came to realize that tourism could be an important force for not only employment and income but also a source of hope for communities. Coming from a poor rice-farming village in northeastern Thailand and growing up without electricity, I could relate to the people and their struggles.

What sort of rural communities are you working with now and what sort of tours are you offering?

We work with 58 different communities and have 60 different tour packages. The communities are all over Thailand, in Chiang Rai up north, in Trat province in the east, in some southern communities like Nakhon Si Thammarat and the mostly Muslim village of Koh Yao Noi. We have a few tours around Bangkok, too, like "A Day as a Fisherman," where you visit the district of Bang Khuntian to learn about fishing and tie-dying with natural dyes, riding kayaks on the canals and trying local food. We also now offer customized packages.

Local Alike now offers around 60 different socially responsible tour packages.

The "Phi Phi Model" of Development

Koh Phi Phi has been one of Thailand's top travel destinations for over three decades. It now draws more than 1.4 million tourists a year, according to the Tourism Council of Thailand. Consisting of six islands that form part of Hat Nopharat Thara-Mu Ko Phi Phi National Park, the area was made famous as the setting of *The Beach*, starring Leonardo DiCaprio. But the years of high tourist traffic and poorly enforced regulations have taken a heavy toll. Coral reefs in the Andaman Sea are bleaching due to rising sea temperatures, tourists are producing around 10 tonnes of trash per day, and wastewater discharge from hotels and resorts is despoiling once pristine waters.

In 2015 Assistant Professor Thon Thamrongnawasawat, deputy dean of the Fisheries Department at Kasetsart University and an adviser to the Ministry of Natural Resources and Environment on marine resource management, decided something should be done to better protect Koh Phi Phi. In 2015, he began posting photographs of the environmental degradation on and around the islands. Those first posts sparked a significant level of public outcry, and since then, Thon has used social media to gather momentum for a conservation plan dubbed the "Phi Phi Model," which is now being implemented with support from local communities and businesses.

Under the Phi Phi Model there are now quota limits on the number of visitors allowed, tour operators have been compelled to register tourist boats, there are strict bans on shark and parrot fish meat, hotels face the risk of losing their licenses if they do not properly dispose of wastewater, and entry fee collection is more tightly monitored to help root out endemic corruption.

Before implementation of the Phi Phi Model regulations, only about 90 tourist boats had been registered and the collected fees were less than 70 million baht per year. Since the regulations came into force in 2015, more than 1,500 boats have been registered and fee collection has risen more than 500 percent, according to park officials.

"We are trying to turn the hopeless island into a hopeful place," said Thon, who has a knack for witty turns of phrase. "Phi Phi island is an extreme case of how unsustainable tourism has destroyed nature. We chose this place because if the Phi Phi Model works, it means we can apply it to other marine national parks across the country."

The Department of National Parks, Wildlife and Plant Conservation is considering taking things further and is mulling over a plan to close Koh Phi Phi Lay annually during the low season (from June to September) in an effort to allow the environment and ecology to recover. The off-season moratorium, if implemented, would mean the loss of at least 40 million baht per month in revenue during that period. That said, the policy is said to be backed by local communities, local businesses, conservationists and park authorities.

▶**PIONEERS**

Wonderfruit Festival

History: Founded by Pranitan "Pete" Phornprapha in 2014

Location: About 20 minutes outside of Pattaya city

Key features: An emphasis is on sustainability, featuring zero plastic, an array of eco-themed talks and workshops, event structures made from recyclable and natural materials, and active support of conservation projects elsewhere to offset the festival's carbon footprint.

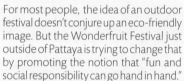

Revelers take part in the Wonderfruit Festival outside of Pattaya.

For most people, the idea of an outdoor festival doesn't conjure up an eco-friendly image. But the Wonderfruit Festival just outside of Pattaya is trying to change that by promoting the notion that "fun and social responsibility can go hand in hand."

Founded in 2014 by Pranitan "Pete" Phornprapha, the festival has an underlying emphasis on sustainability. In addition to live music, wellness and art, it features a wide variety of eco-themed talks and workshops. Organizers have a firm commitment to "zero plastic," instead opting to use recycled or natural materials for event structures, seating, water taps, containers and more.

"Our ethos is that we need to demonstrate to people that sustainability can be fun and easily integrated into our lifestyles. Sustainability needs to be given a fresh makeover, to be seen as sexy, aspirational and even luxurious," Pete said, adding that Wonderfruit's program content strives to reinforce these ideas.

The festival's "Scratch Talks" series features speakers from Thailand and beyond, with topics ranging from rainforest preservation and natural capital, to the shared economy and more. "We have the founder of unicorn app GO-JEK, Michaelangelo Moran talking

about social empowerment through technology and movie director Craig Leeson talking about his award-winning documentary, *A Plastic Ocean*. In addition to the talks, lots of our interactive workshops follow a sustainability theme, including wood carving and traditional farming techniques," Pete said.

Wellness activities at the festival include a cornucopia of yoga and meditation classes, as well as martial arts, belly dancing, obstacle courses, games and microlight flights. Food is also a big part of Wonderfruit, and much of the produce is grown organically at an on-site farm. This in turn is served up in recyclable containers and reusable bottles. Last but not least, the festival hosts some 40 different musical acts on its eco-friendly stage.

Wonderfruit's mission statement declares that those behind the festival "aspire to be a catalyst for positive change

and seek to find creative ways to live sustainably and responsibly, and to have a net positive influence on our planet."

With that in mind, Wonderfruit's orgnizers attempt to offset the festival's carbon footprint by investing in the Rimba Raya Biodiversity Reserve, a protected forest in Indonesian Borneo that, in addition to a wide array of biodiversity, hosts one of the few remaining populations of wild orangutan.

"Twenty-five years ago, my dad developed an environmental campaign called 'Think Earth' and that was really inspirational for me. Back then, sustainability was a very new concept and the importance of it has stuck with me ever since," Pete said.

"I think it's important for everyone to think about protecting something that is greater than ourselves. That sense of selflessness is liberating," he added.

"Our ethos is that we need to demonstrate to people that sustainability can be fun and easily integrated into our lifestyles. Sustainability needs to be given a fresh makeover, to be seen as sexy, aspirational and even luxurious."

Pranitan "Pete" Phornprapha, founder of the Wonderfruit Festival

SOCIAL ENTERPRISE
Combining wellbeing with profit

Young entrepreneurs who started an eco-tourism business to benefit loca...

One strength of Thai culture is a willingness to take care of each other. Thais of all walks of life regularly give time and money to support communities, temples, schools and other causes. Leading companies and patrons believe it is their responsibility to support village development projects and education. Social enterprises in Thailand promise to take this desire to do good even further.

Bridging the divide between the public and private sectors, social enterprises apply business strategies and the power of the marketplace to solve environmental and social problems. The Social Enterprise Alliance, a US-based organization, defines social enterprises by three distinguishing characteristics: a successful social enterprise directly addresses a social need, drives revenue through commercial activity, and holds the common good as the enterprise's primary purpose.

Although the term "social enterprise" was coined in the UK as recently as the 1970s, the concept has spawned a rapid proliferation of social enterprises worldwide and a young, hopeful generation of entrepreneurs. Numerous organizations are now dedicated to incubating and funding social enterprises and connecting entrepreneurs with impact investors. Social entrepreneurs give sophisticated consumers more opportunities to channel their buying power into community-conscious, sustainable products and services. Meanwhile, impact investors fund social enterprises to create meaningful change.

The social impact of a social enterprise is measured by the non-financial benefit that a venture contrib-

utes to society. According to the Social Enterprise Alliance, social enterprises often create social impact more efficiently than government, more sustainably and creatively than the nonprofit sector, and more generously than business. Whether organized as for-profit or nonprofit, social enterprises fill the gaps between the public, private, and nonprofit sectors.

In Indonesia, for example, Dr. Gamal Albinsaid's Garbage Premium Insurance Clinic Program allows people who cannot afford health insurance to bring an equivalent of $0.85 in recyclable waste to the clinic to receive healthcare in return. The clinic then trades the recyclables for cash. This program not only provides healthcare for those in need, but it also addresses the issue of waste management plaguing many Indonesian municipalities. But as a financially sustainable venture, it also ensures that the clinic and its staff are properly paid for their services.

While charities and nonprofits also strive for the common good, they have limits. Things offered freely are not always needed but always accepted, leading to wasted resources. Often, donations must be replenished to maintain a desired outcome. Also, those giving generously of their time or money do

> **Bridging the divide between the public and private sectors, social enterprises apply business strategies and the power of the marketplace to solve environmental and social problems.**

not always understand the need to evaluate whether they are achieving the value they intend to create.

Social entrepreneurs can have a significant impact on business in Thailand. If these social entrepreneurs succeed, they can leverage that success for even higher impact by attracting impact investors. If they are solving important challenges and willing to run their ventures as well-managed businesses, they can attract part of the estimated US$700 billion in impact investments available around the world, resulting in further capital inflows into Thailand.

Accordingly, the Thai government is recognizing this sector's value and is ramping up structural support for social enterprise. In 2009 the government formed the National Social Enterprise Committee to increase funding and awareness for social entrepreneurs. In 2011, it launched the Thai Social Entrepreneurship Office (TSEO), backed by US$3.2 million in funding.

Finally, social entrepreneurs also benefit themselves. They give themselves meaningful work and allow themselves to gain an appropriate reward for that work. As they succeed, they become role models and pave the way for a more sustainable society overall.

d.light: Solving a Global Problem through Social Enterprise

A d.light distributor in Kenya.

According to San Francisco-based d.light, roughly two billion people in the developing world do not have consistent access to electricity. Many rural households remain unconnected to national power grids, but light is still needed after dark for myriad tasks: cooking, cleaning, studying, and making handicrafts, food or other products that can be sold for extra income. Founded in 2008, d.light is addressing this need by providing affordable, reliable solar-powered lights to households and small businesses in over 62 countries.

With field offices in 10 countries, including India, China, and Kenya, the company has sold over nine million solar lanterns and lights, touching the lives of 51 million people. The demand is still strong, and supplying these families in need are local dealers who live in the same developing communities. In this way, d.light not only offers a product with social impact, but also offers employment for local entrepreneurs who lack many opportunities.

d.light products come with a two-year warranty and dealer support through field-based staff. Products range from single lanterns to mobile power grid systems that light entire homes and businesses. Backed by easily accessible customer support, these products are empowering those who would otherwise be living in the dark, changing the way people access, pay for, and use energy. It is also a move away from dependency on national power distribution, often generated through coal and natural gas, and a move toward clean, efficient, renewable energy.

▶**PIONEERS**

Klongdinsor

History: Founded in January 2013

Location: Bangkok

Key features: Designs, manufactures, and sells educational tools for children with disabilities; raises social awareness about special needs to create a more inclusive society

Bangkok-based Klongdinsor, which means "pencil box" in English, was founded with the aim of creating a more inclusive society for those with special needs. The company is achieving this by creating and distributing educational tools for children with disabilities.

As a regular volunteer at the Bangkok School for the Blind, Klongdinsor's founder and CEO Chatchai Aphibanpoonpon identified a need in the educational community. He noticed that visually impaired children often lacked the educational tools they need to synthesize information from their lessons. This, he discovered, was a common plight throughout Thai schools. Because many schools don't have the funds to import proper educational materials from other countries, teachers tried their best to improvise. Chatchai, too, often made tools at home and gave them to students. Although these efforts were appreciated, the handmade materials often broke. This lack of resources both lowered the quality of education and made teachers' jobs harder. Ultimately, the lower quality of education also meant fewer opportunities for students with special needs once they reached adulthood.

A graduate of Thammasat Business School, Chatchai founded Klongdinsor with the triple-pronged goal of improving education, providing employment opportunities and raising awareness of special needs. To do this, the "not-only-for-profit" company

designs, manufactures, and sells the Lensen tactile drawing kit for visually impaired children. The kit uses a special pen that dispenses yarn onto a Velcro board so that visually impaired children can draw or write, then touch their creations. Art becomes a multisensory experience, allowing students to use tactile means to tell stories and express thoughts. The boards have also proven useful for helping students understand lessons in geometry, physics and English. The Pythagorean Theorem or the alphabet, for example, can be taught effectively through tactile information.

Klongdinsor only uses affordable materials to manufacture Lensen products, making them accessible to poorly funded schools and even to parents at home. Today, every school for the blind in Thailand has a Lensen drawing kit, and the products are also being exported to Australia, Germany, Hong Kong, Italy and the United States. The growing reach of Klongdinsor is not only raising awareness but improving the education level of children both at home and abroad.

To fulfill a goal of providing employment opportunities, Klongdinsor hires people with mental disabilities to package Lensen products. Working with a partner organization that provides support for people with mental disabilities, "We hope to expand their service into more of our supply chain," Chatchai

told UNESCO Bangkok. In addition, Klongdinsor is partnering with an art university "to create an art curriculum on Lensen so teachers and parents can use it more effectively," Chatchai said.

A social enterprise with growing social impact, Klongdinsor's effective branding, marketing and distribution are ensuring the fulfillment of its social goals as a fully solvent, financially sustainable business.

Klongdinsor only uses affordable materials to manufacture Lensen products, making them accessible to poorly funded schools.

Blind children use Lensen to develop their learning capacity.

▶**PIONEERS**

Wanita Social Enterprise

History: Established in 2016

Location: Yala, Narathiwat, Pattani and Songkhla provinces

Key features: A social enterprise that empowers women and helps them find markets for their products

Muslim women who are members of the Wanita Social Enterprise weave hand-made products.

Women are often disproportionately affected by war, even if they are not directly impacted by hardships such as personal injury or the loss of a family member. The situation in Thailand's conflict-ravaged Deep South is no different. Since 2004, the conflict has claimed over 6,400 lives, resulted in over 11,000 causalities and made widows of more than 2,850 women in Yala, Narathiwat, Pattani and Songkhla provinces.

One repercussion of the conflict is that many women have been pressured to take on breadwinning responsibilities to help supplement their family's income. Over the years these women have formed collectives as a way to spend their extra time producing and selling locally sourced goods such as foodstuffs, snacks and handicrafts. The problem is that until very recently the vast majority of these groups did not possess the necessary skills or knowledge to market their products outside their small communities. However, a thoughtful new initiative has helped to change that.

With support from Oxfam, the Wanita Social Enterprise was formed in 2016 by a number of women's groups in the Deep South. Meaning "women" in Malay, Wanita was established to empower women who are directly, or indirectly, impacted by the conflict by helping them to scale up their business endeavors and market their products to a wider consumer base.

The program takes a "bottom-up" approach to foster financial autonomy. As such, Oxfam provides support through training and by helping members to build leadership, marketing and management skills.

Another major focus is on improving market linkages by helping to select the best quality products and promoting those products to a wider audience. By establishing Wanita as a recognizable brand, the aim is to make its high quality, ethically-sourced products more appealing in domestic urban marketplaces.

An additional hope is that the story behind Wanita can also help raise awareness about what is transpiring in the Deep South. Many of the women who participate in the program have indeed lost husbands or loved ones to the conflict. But others also feel the impacts of the war in ways such as restrictions on movement, the jailing of a breadwinner from the family, or the emotional strain of simply being caught up in a violent situation one cannot escape.

Aliya Mudmarn, a project officer with Oxfam's Women's Economic Empowerment initiative in the Deep South, said it's important for these women "to gain better control over their economic livelihoods."

We believe that when they are empowered and have the right economic tools in their hands, it can also lead them to express their views, improve their political participation, and even give them a voice in the Deep South peace process, Aliya said.

Wanita now has 56 women's groups under its umbrella. Oxfam's work in the Deep South actually predates the founding of Wanita, and since 2012 more than 685 women have been given access to funding to help them set up their enterprises or get production under way. Each women's group typically receives around 20,000 to 30,000 baht in funding with the expectation that it will be paid back to Oxfam within two years (without interest). According to Oxfam, all of the women taking part in the funding project have increased their incomes, while some 32 percent report that their income has increased "significantly."

In what Aliya described as a "win-win situation," Wanita is also now partnering with private enterprises such as Rayawadee Resort, Punpuri Organic Spa and the Yuna clothing brand to co-brand and sell products such as bags and prayer embroideries. March 2017 also saw the establishment of the first official Wanita shop, which opened in Pattani town.

▶**PIONEERS**

ChangeFusion

History: Founded in 2001 by royal patronage as a nonprofit institute under the Thai Rural Reconstruction Movement Foundation

Location: Bangkok

Key features: A comprehensive, wide-reaching consultancy, advocacy and investment organization supporting social enterprise in Thailand, ChangeFusion develops seed fund programs, offers low interest loans, and organizes networking and resource-sharing opportunities for social entrepreneurs

ChangeFusion, a nonprofit organization under royal patronage, is providing the valuable support infrastructure social entrepreneurs need to make their ventures a reality. Operating largely as an investor and advocate for social ventures in a wide range of sectors including rural development, ChangeFusion offers funding, consultancy, resources and networks for social entrepreneurs tackling social and environmental problems in Thailand.

ChangeFusion focuses on creating funding opportunities for social ventures, an indispensible element to any venture's success. In addition to providing seed funds through various programs and competitions, ChangeFusion has developed ChangeVenture, an arm of the nonprofit that offers funding to scale up operations. ChangeVenture's investments offer social enterprises equity, as well as loans with low interest rates. ChangeFusion has also partnered with BBL Asset Management, an investment firm, and the Khon Thai Foundation, a nonprofit organization, to create Thailand's first ESGC (environment, society, governance and anti-corruption) mutual fund. The fund, dubbed the Khon Thai Jai Dee fund in Thai, or BKIND for short, invests money only in companies listed on the Thai Stock Exchange that meet ESGC criteria. The fund also allots 40 percent of its management fees to public projects.

As a well-connected and government-backed institution, ChangeFusion is able to partner with powerful and reliable organizations to develop programs that benefit social ventures in sectors key to Thailand's development. ChangeFusion currently works with Oxfam, for example, on social enterprises promoting sustainable agriculture. It is also working with the Thai Health Promotion Foundation

According to a 2012 Thailand Social Enterprise Landscape Report, the country is currently home to roughly 116,000 social enterprises.

to develop technology solutions to improve public health services. It has also garnered sponsorship from the Swiss Agency for Development, Microsoft, and UNESCO to launch the Youth Social Enterprise Initiative competition, which rewards winning social enterprises with funds, mentoring and networking opportunities.

According to ChangeFusion's 2012 Thailand Social Enterprise Landscape Report, the country is currently home to roughly 116,000 social enterprises, and younger generations of businesspeople are engaging more and more in social ventures. Although the highest numbers of social enterprises can be found in the agricultural and forestry sectors, many more sectors are ripe for social entrepreneurship in Thailand. ChangeFusion already lists ventures in sustainable tourism, information technology, and education in its portfolio, but it sees ever more potential for growth.

ChangeFusion staff observe the natural tea-growing process in Nan province.

The DoctorMe app, a tech solution for health.

GROUNDBREAKERS

EDWARD RUBESCH *is program director of the Innovation Driven Entrepreneurship Center at the University of the Thai Chamber of Commerce, as well as director of MIT Enterprise Forum Thailand. Over the last 15 years he has helped technology researchers and social entrepreneurs launch their ventures, and has developed the MetaMo Incubation Program for Southeast Asia. He is also director of the Southeast Asian Region of the Global Social Venture Competition, the world's largest social enterprise business plan competition.*

Is social enterprise growing in Thailand? If so, how, and since when?
This is an interesting but complex question. If you define social enterprise like I do, as a well-run business that also delivers a social or environmental mission, then there have been a small number of social enterprises around for a long time. However, when we started the Global Social Venture Competition for Southeast Asia in 2007, it was still a very new term for most people. Now, however, everybody is using the terms social enterprise and social entrepreneur so the concept seems to have stirred a lot more interest. What is less clear is whether this interest has actually encouraged more social entrepreneurs to work through the challenges to create successful ventures. We see a lot of interesting projects. We'll have to wait a little longer to see if any of those projects blossom into successful businesses that attract impact investors.

Why is social enterprise important to Thailand? What can it contribute?
Business is an incredi-bly powerful force, with the potential to do both good and bad. If we can harness that force, and the management, financial, and communication tools of business, to target social and environmental challenges, there is the potential to fix some pretty big problems around us.

What are some of the obstacles in Thailand to their success and what kind of moves could the Thai government or business community make to help?
There is still a lot of misunderstanding about what social enterprise is. It is often looked at as something that is different from real business. Traditional business people sometimes look at social enterprises as lightweight forms of business, which can't really stand on their own two legs. Meanwhile people who want to do good or fix social problems don't necessarily like to be bound by management systems or success metrics...therefore, I really believe mindset is the thing that needs to change the most.

What support networks or infrastructure would help further social enterprise in Thailand?
There are a lot of formal activities going on already, and I hope those continue. But in fact the single biggest thing that can really encourage social enterprise is having a couple of big successes that other would-be social entrepreneurs can see as role models.

Are there other significant factors that come into play?
Innovation is incredibly important when it comes to delivering meaningful impact. Over the past few years I have been putting a special emphasis on encouraging innovation-driven entrepreneurs. Innovation, at its heart, is about positive change for people. We need to put a much bigger emphasis on scaling innovations to move beyond the many projects in Thailand, which are well-meaning in their goals, but never grow. Moreover, in the developing world, innovation and impact overlap almost entirely: delivering a solution that makes an actual impact is likely to require substantial innovations to succeed.

What are you trying to achieve?
I believe all the necessary pieces are here: there are many capable people, there are lots of resources, and there are certainly many, many problems that need to be fixed. What we need are good social entrepreneurs to come in and align those pieces into a successful venture – which is exactly what entrepreneurs are supposed to do. And at the same time, we hope to encourage impact investors to come in and back ventures up with the resources that they need.

Mentoring sessions on social enterprise attract an increasing number of young graduates.

RESTAURANTS

Innovative restaurants are rethinking the way they source their food

In the 1980s, healthy living emerged as a trend that evolved into a lifestyle for those with a passion for exercise, clean living and fresh food. The "slow food" movement arose around the same time as a backlash against fast food and its harmful impacts on the environment, health, local communities and small businesses. Taking a firm position against such unsustainable practices as razing rainforests for cattle farms, genetically modified foods, and monoculture, the movement for sustainability in the food industry has now come to encompass everything from organic and free-range agricultural products to the **farm-to-table movement**; from the denunciation of chemical additives to the rise of macrobiotic and even socially responsible restaurants.

If an overarching name for this far-reaching food revolution is elusive, at least some of its tenets and practices are easier to pinpoint. Bo.lan, a well-known, high-end Bangkok restaurant, is striving to become carbon neutral by 2018 and favors an approach that runs the gamut from serving organic produce to cutting down on waste and emissions. The Oyster Bar and Eat Me refuse to serve endangered or farmed species of fish. Sustaina is all about creating an organic farm that feeds a restaurant, supermarket and supply chain that distributes their innovative products to more than ten countries.

Each of these enterprises faces similar difficulties. It's not easy sourcing sustainable products in Thailand where supply lines in an unreliable infrastructure are often disrupted by shortages or breakdowns. Restaurants also produce massive amounts of waste and consume a lot of power through lighting and temperature control, and many rely on industrial dishwashers that use a great deal of water.

In Bangkok, one of the world's foodie capitals, this new restaurant trend is being driven mostly by Western chefs and a few prominent Thais like Duangporn "Bo" Songvisava, the award-winning

chef and co-founder of Bo.lan, as well as Sho Oga, the Japanese entrepreneur behind Sustaina. Ever so slowly, consumer awareness has been growing, thanks to such popular documentaries as *Food, Inc.* and *Super Size Me*, numerous books about nutrition and food security, and the proliferation of Western-style farmers' markets in Thailand's bigger cities.

The markets are not only sources of organic produce, artisanal bread, natural soaps, shampoos and cleaners, but they also function as community centers for workshops, performances and gatherings of like-minded consumers. In spite of these new markets in Bangkok and elsewhere around the world, and despite the popularity of healthy restaurants and lifestyles, the farm-to-table movement has not altered the landscape of how food is grown and produced. Few people understand this better than Dan Barber, the award-winning American chef and restaurant owner named one of the 100 most influential people in the world by *Time* magazine in 2009.

"More than a decade into the movement, the promise has fallen short," Barber wrote in *The New York Times* in 2014. "For all its successes, farm-to-table has not, in any fundamental way, reworked the economic and political forces that dictate how our food is grown and raised. Big Food is getting bigger, not smaller. In the last five years, we've lost nearly 100,000 farms (mostly midsize ones). Today, 1.1 percent of farms in the United States account for nearly 45 percent of farm revenues."

It's a sobering editorial on the real limitations of any food revolution. For every restaurant that strives for sustainability, another opens with little awareness or financial margin to support its key concepts. For the privileged minority who can afford specialty markets and high-end restaurants, this trend will continue to grow. But the poorer and middle-class majority will continue to get most of their calories from sources that are heavily processed and ecologically unfriendly.

Farm-to-table movement:

An international movement that promotes local food consumption. Supporting local and seasonal food production and agriculture, the farm-to-table movement often works in tandem with organic principles.

A chef's selection of ingredients is crucial for sustainability in the food industry.

The markets are not only sources of organic produce, artisanal bread, natural soaps, shampoos and cleaners, but they also function as community centers for workshops, performances and gatherings.

FEAST FOR THOUGHT
FOOD, INC.
Director: *Robert Kenner*
Release Date: *2008*

The premise of *Food, Inc.* reads like an Orwellian take on farming and agriculture, where a multinational has patented seeds, employing dozens of private investigators and a hotline to track down farmers accused of stealing them, where just 13 slaughterhouses have monopolized the US meat market and incubated a slew of killer viruses, where abattoirs are run like assembly lines and the illegal immigrants who staff them are treated only slightly better than the animals.

It looks like a dystopian vision of the future, except this is a documentary from 2008. Many viewers find this grisly film hard to stomach, but a note of optimism resounds in the growth of the alternative agriculture movement. Consumers are driving this movement, said Gary Hirshberg, the millionaire organic yogurt entrepreneur, who mentioned Wal-Mart's decision to stop selling a brand of milk that contained a synthetic growth hormone because of consumer outrage. "Individual consumers changed the biggest company on earth."

The rise of consumer power is the brightest spot in this dark and brilliantly made documentary.

▶**PIONEERS**

Sustaina Organic Restaurant

History: Harmony Life International was formed by Sho Oga in 1999 to operate the Harmony Life Organic Farm; in 2009 a restaurant and shop were opened in the same building

Location: Sukhumvit Soi 39 in Bangkok

Key features: Restaurant serves organic produce from their own farm; all waste is sent back to the farm for composting; menu is both Thai and Japanese

Top: In a bid to provide healthy dishes, Sustaina only uses produce from its rural organic farm.
Bottom (three photos): Sustaina dishes are made of organic products.

Sustaina must be one of the developed world's few restaurants stocked with fruit and vegetables from its own rural farm. The restaurant is in the capital, the farm a few hours out of town near Khao Yai National Park. The two feed off each other.

Many of the noodles, teas and tonics served in the restaurant come from the factory at Harmony Life Organic Farm. The factory also makes natural detergent used at the restaurant. Unlike its chemical cousins, Harmony Life's detergent is filled with effective microorganisms (EM) that eat the bacteria in dirty water, making the detergent biodegradable.

Founder and president Sho Oga, a Japanese entrepreneur with a degree in oceanography, is a passionate and knowledgeable advocate of all things organic, who is deeply concerned about the plunging nutritional values of agricultural products farmed with chemicals. "In the last 10 years the level of enzymes, vitamins and minerals in fruit and vegetables has dropped by 50 percent," he said.

That is one reason why he has developed such innovative products as the Enzyme Drink, now the company's second-most-popular product abroad, and the bestselling Moroheiya Noodles, made from the Egyptian green known as the "Pharaoh's vegetable." Sho traveled to Egypt himself to bring the seeds back to plant in Thailand.

Since then, Harmony Life has proven itself an international success, distributing its products in more than 10 countries and in some 300 branches of the Whole Foods supermarkets in the US alone.

Sho set up the Bangkok restaurant in 2009, which also contains a shop and supermarket, and serves as a meeting place for workshops on organics. Every month, Sho and his colleagues organize two-day workshops on the company farm, teaching organic farming techniques to an average of 600 farmers and other visitors every year.

Could this be the future of healthy food? A farm, restaurant, supermarket, online shop, distribution network, product developer and training ground all under the umbrella of one dynamic, multi-purpose company?

"In the last 10 years the level of enzymes, vitamins and minerals in fruit and vegetables has dropped by 50 percent."

Sho Oga, founder of Sustaina

▶**PIONEERS**

Bo.lan

History: Founded in 2002 by Duangporn "Bo" Songvisava and Dylan Jones

Location: Sukhumvit Soi 53 in Bangkok

Key features: Organic menu featuring a wide range of Thai dishes; repurposed wood for furniture; recycling and composting systems; LED bulbs and wastewater management

Bo and Dylan are a Thai-Australian couple who met while working in the kitchen of London's nahm, the first Thai restaurant to ever receive a Michelin star. After relocating to Thailand, they became local purveyors of the "slow food" movement and started their restaurant, Bo.lan, the name based on an amalgamation of their names.

With so much competition in the Bangkok dining scene and numerous cheap options on the street, their joint venture required the building of a different brand.

Bo.lan's tagline is "essentially Thai," and while the restaurant's repertoire

includes different dishes from all over the kingdom, based on ancient recipes culled from yellowing cookbooks, sustainability is the main flavor of the cuisine. To achieve this, the menu changes with the seasons. "We have to be adaptable," Dylan said. "So if it's sustainable, it's also seasonal, which means you can't sell something all year round. To understand sustainability you have to look at the bigger picture, which is nature itself."

One of the essentials of the farm-to-table movement is sourcing foods locally. This ensures fresh ingredients and minimizes the restaurant's carbon footprint. When Bo.lan began some six years ago, the only chemical-free staple they could easily get was rice from Raitong Organics. "If you wanted organic lemongrass or Thai basil you had to buy a huge amount, like 50 kilos, export quantities. These days it's easier. Many smaller farmers have grouped together. So we can call the farmers directly to see what's available rather than scouring local markets," said Bo, who earned a master's degree in gastronomy in Australia and won the first award for Asia's Best Female Chef at the 50 Best Restaurants in Asia Awards in 2013.

As for their eco-friendly practices, the staff use LED bulbs to save energy, reuse water from the kitchen for the garden and bathrooms, compost leftover food, and sort glass and plastics into separate bins for recycling. Training the staff to do the separating and sorting has been the biggest battle in this area, Bo admitted.

Bo.lan's up-cycling efforts include turning 100 percent of its used oil into soap for use in the restaurant, or as souvenirs for guests. Leftover stalks of herbs like basil and lemongrass are also used in the process. By extracting the pectin and oil from used limes and adding some baking soda, Bo.lan has been able to start making their own multi-purpose cleaning agent and laundry liquid. Ash from charcoal is also collected and given to artisans who make Bo.lan's collection of tableware and glassware.

There are limits for a small business like theirs. The skyscrapers of Sukhumvit

Road are sun blocks, so solar panels are ineffective. Their ground-level location is not high enough for turbines to catch the wind and channel it into electricity. To reach their goal of becoming carbon neutral by 2018, the owners of Bo.lan began purchasing carbon credits in 2015.

Bo.lan's philosophy of sustainability revolves around a holistic approach that benefits everyone. "For us, sustainability means that it's sustainable for everyone involved. It's sustainable for the farmer, the one who's producing it. It's sustainable for the middleman, if there is one. And it's sustainable for the consumer and the restaurant owners like ourselves," said Dylan.

"To understand sustainability you have to look at the bigger picture, which is nature itself."

Chef Dylan Jones

Top: Bo and Dylan of Bo.lan. Above: Exterior of Bo.lan in Bangkok. Right: A dish from Bo.lan.

The "Catches" of Sustainable Seafood

Unbeknownst to many, our oceans and fish are in grave danger. If there is not a reevaluation and slowdown of commercial fishing, marine scientists have forecast that the planet's fish stocks will be mostly gone by 2048.

To raise awareness about this crisis, Billy Marinelli, owner of The Oyster Bar in Bangkok, organizes regular screenings of *The End of the Line*. The documentary film sweeps around the world illustrating how oceans everywhere are being overfished: in Canada's Newfoundland the collapse of the cod fisheries wrecked the economy of the maritime provinces; similar catastrophes rocked African countries. In Japan, bluefin tuna, a delicacy, is in danger of extinction – about 90 percent of the bigger species have already been fished.

Having studied marine biology and worked in the seafood business for much of his life, Billy is an evangelist for sustainability and practices what he preaches. At his restaurant, no endangered species nor farmed salmon are served, no MSG or other chemical additives are used. For Billy, sustainability in the seafood trade is also defined by the way fish are caught. "The fish that I use is all hook-and-line caught, or caught with a hand or gill net. This also weeds out the slower and stupider creatures which are the ones that tend to get caught. It's like a form of Darwinism, the survival of the fittest," he said. "When you fish in a sustainable way that keeps the gene pool strong."

Achieving such standards of sustainability is crucial in Asia, where around two-thirds of the world's seafood is consumed and where fishing fleets are notorious for destructive techniques like bottom trawling. These nets scoop up everything in their path, including coral reefs, eggs and juvenile fish, so that 70 percent of each haul is referred to as "bycatch" or "trash fish," and ends up as animal feed or sold under names like "sea bass."

Fortunately, efforts are underway in Thailand to produce higher quality fish in sustainable ways. The Earth Net Foundation's Small Scale Fishers and Organic Fisheries Products Project was started with funding from the EU to address this shortage. Overseeing the project is Supaporn Anuchinacheeva, who has worked on sustainability issues with Oxfam. The main goals of her current project are supporting small-scale fishermen who are doing conservation work, and getting them better prices for their products, such as 70 percent of the value instead of the 20 to 30 percent they are usually offered by middlemen. The project also ensures that higher quality seafood comes to Bangkok consumers.

Another organization, the Organic Agriculture Certification Thailand (ACT), is promoting safe, sustainable, and socially responsible catches. Seafood can only be certified organic if it meets four criteria. First, the fish must be caught using responsible practices, such as targeting the species and not scooping up bycatch. At no point during the process can fishermen use chemicals; the fish can only be packed in ice. The fishing area cannot be polluted, so it must be far from industrial or agricultural areas. The last criterion is traceability – they must be able to trace who caught each fish.

Marketing these products has proven to be one of the biggest challenges. Though these fishers have been selling some seafood to Bo.lan and the Plaza Athenee hotel, it's difficult to guarantee a certain amount of a fish species because of their catch-as-catch-can method. This makes supermarkets a hard sell too. In response, Earth Net has set up a stall at the monthly Bangkok Farmers' Market to hawk 12 different kinds of seafood from five different villages.

As our oceans empty in what is the biggest mass extinction of species since the age of the dinosaur, we should ask ourselves: Do we know where that seafood on our plates came from, who caught it, which middleman bought it and who packed it under what sort of conditions?

▶**PIONEERS**

Eat Me

History: Founded by Australian siblings Darren and Cherie Housler in 1998

Location: Convent Road, Bangkok

Key features: The menu details where many products are from and how fish was caught; no endangered species of fish are served

At Eat Me, one of Bangkok's top restaurants, executive chef Tim Butler has designed an ever-shifting menu where diners can often see at first glance where their food comes from.

Tim and his team source whatever they can from around Bangkok. But one of the biggest roadblocks to the progress of the farm-to-table movement in Thailand is the haphazard or unreliable nature of the local supply chain. Another challenge is that the high-end clientele who frequent such establishments often desire exotic foods and flavors, like lamb and beef from Australia and

---

> **"It's unthinkable for me to source the cheapest products out there. I wouldn't serve them in my home to my family or friends so I wouldn't serve them in the restaurant."**
>
> **Tim Butler, Eat Me chef**

---

black truffles from Italy. Eat Me's average customer does not know nor care that the tomatoes have been grown without chemicals. Very few have even asked the chef if a dish is organic, said Tim.

In Thailand, it's not the consumers who are driving this movement toward healthier fare. It's mostly been Western chefs like Tim, a native of Portland, Maine, who have taken it upon themselves to find the best quality meat and produce. "It's unthinkable for me to source the cheapest products out there," Tim said. "I wouldn't serve them in my home to my family or friends so I wouldn't serve them in the restaurant."

Eat Me is a case study in the problems any independent restaurant runs up against in pursuing a more sustainable agenda. First and foremost is food waste. The leftover food from diners could be composted (as Bo.lan does), but that's not possible for an operation with limited outdoor space. And because municipal authorities do not supply bins for separating waste items, the restaurant's staff separates all recyclables.

Since many restaurants use industrial dishwashers, water usage is another area where it is difficult to cut back. The same goes for electricity, which is also used in copious amounts for cooking, air-conditioning and lighting.

Small restaurants, however, police themselves in other ways that have more to do with fiscal prudence than ecological duty. "We try and make use of 110 percent of the products. We're an independent restaurant. Everything I order needs to be used, otherwise we won't survive," said Tim.

Top: Eat Me restaurant. Middle: Chef Tim Butler. Bottom: Dishes served at Eat Me.

GREEN FINANCE AND BANKING

The institutions and financial tools behind green growth

Economists argue that one driving force of non-sustainable production and consumption is that the negative impacts of such consumption do not carry a financial cost to those that create those adverse effects. Economists refer to this as "negative externality costs." The logical consequence is that financial decision-making would be "greener" if such external costs were "internalized" and made part of the decision-making process itself.

But what about financial institutions, whose products are largely immaterial? What role can they play in sustainable development? The answer is a very large one. By offering access to capital or incentives to catalyze environmentally friendly and socially responsible financial decisions, financial institutions can have a massive impact in encouraging change and spurring innovation in these areas. Internationally, demand for such financial products has been on the rise, especially in recent years, as both individuals and companies become more eco-conscious.

Today, green finance is beginning to reshape the corporate landscape as the private sector commits more investment to sustainability. In the energy sector, for example, capital is flowing away from the fossil fuel business into the low-carbon industry and alternative energy projects. A leading global hedge fund based in Norway decided it will no longer invest any money whatsoever in fossil fuels. Similarly, investors around the world are looking toward myriad opportunities presented by the sustainable development movement. Many of these investment strategies and instruments are unique, requiring a long-term view of the investment. Microfinance, for example, is predicated on the belief that small loans to those in need will, over the long-term, lift entire communities out of poverty and

Green finance is beginning to reshape the corporate landscape as the private sector commits more investment to sustainability.

SCB's classic Talad Noi Branch, located on the Chao Phraya River.

spur local economies. More and more investment firms are now factoring in the potential impacts of climate change in their assessments of risk and opportunity.

While green finance encompasses all the ways individuals and institutions can use capital to promote sustainable behavior – such as green loans, ethical investments, and environmentally friendly finance policies like carbon pricing – green banking, on the other hand, aims to transform the banking industry itself into an environmentally conscious one and refers to the myriad ways banks and financial institutions can cut down on the carbon footprints and energy use of their operations and customers. Green banking reaches consumers through initiatives such as mobile and online banking to minimize carbon footprints, solar-powered ATMs for energy efficiency, and green credit cards where the issuing bank buys carbon-offset credits or funds eco-friendly projects every time the credit card is used.

Together, green finance and green banking offer the products, services and incentives that create sustainability through every dollar spent, invested or transacted. This creates public benefits such as

World Bank Green Bonds

In 2008 the World Bank introduced its Green Bond initiative to raise funds for development projects that mitigate climate change or help affected people adapt to it. Designed in partnership with the European financial group Skandinaviska Enskilda Banken (SEB), the bonds provide access to green investments through a triple A-rated credit fixed-income product. Since the initiative's inception in 2008, the World Bank has issued nearly US$9.7 billion through more than 125 green bond transactions in 18 currencies.

Eligible projects fall into two categories: projects that target mitigation of climate change and projects that target adaptation to climate change, such as infrastructure that prevents climate-related flood damage.

Other issuers including development banks and corporations have joined the green bond market. This has expanded the investor base, leading to greater transparency and reporting on climate change projects. In 2016 the market reached some US$80 billion.

Green bonds are useful to developing countries like Thailand, which are in need of buildings, transport infrastructure, water and energy systems, and farms and food supplies that can withstand the impacts of climate change, such as rising sea levels and extreme weather patterns. The bonds currently support more than 60 projects across 20 countries, from solar and wind installations, reduced-emission renovation of power plants and transmission facilities, green transportation, eco-friendly farming, and clean water and irrigation management.

reducing global warming, raising standards of living and preserving natural capital. These practices also create private benefits for financial institutions in the form of reduced lending risks, stranded assets and reduced operational costs from minimized paper and energy use. This not only increases the efficiency and profitability of operations, but it also improves the bank's image and competitive edge, ultimately attracting a growing number of customers and investors demanding green products and services.

In recent years, banks across Asia have started taking steps toward green finance and banking. They are investing an increasing share of their financial portfolio in the low-carbon economy, launching new green financial products, including green bonds, and expanding green banking operations. Siam Commercial Bank (SCB), KASIKORNBANK and Bangkok Bank are among those in Thailand that have successfully incorporated green banking in their day-to-day financial activities. SCB says it was the first bank in Thailand to offer ATM customers the option of not receiving a transaction slip and has also launched a "slipless process" at the counter, eliminating handwritten forms for 80 percent of branch transactions, and saving a large number

of trees in the process. Similarly, Bangkok Bank's Bualuang iBanking service saves approximately 80 million sheets of paper per year. Bank policy also encourages entrepreneurs to develop or use environmentally friendly technologies; use alternative energy; and undertake sustainable agricultural activities. Bangkok Bank provides information and special low-interest-rate loans to help facilitate customers' sustainable business and network expansion. Meanwhile, KASIKORNBANK has extended its credit facilities to renewable energy projects like solar, wind and biomass power generation and to electric power-saving programs, among other endeavors.

►PIONEERS

KASIKORNBANK

History: Founded
June 8, 1945

Location: Headquarters
in Bangkok, operating
nationwide and
internationally

Key features: A leader
in sustainable lending
and green banking; allocates a percentage of net profits to sustainable
development projects annually;
offers green loans and green credit;
provides paperless banking

Of all Thai banks, KASIKORNBANK (or KBank for short) does the most lending to the green economy. In 2013 the bank began allocating 1–1.5 percent of its annual net profit for sustainable development projects, which has become a yearly commitment. Since the first year of implementation, KBank says it has seen improvement in its sustainability performance, evaluated by three dimensions: economic, environmental and social.

One example of putting these plans into action is shifting its lending practices to support environmentally and socially responsible projects. The bank has extended its credit facilities to renewable energy projects like solar, wind and biomass power generation and to electric power-saving programs, like the K-Top Up Loan for Energy Saving (Lighting Solutions), the K-Green Building Program, and the K-Energy Saving Guarantee Program. From 2013 to 2016, the bank put a total of 623.5 million baht toward financing renewable energy and environmental conservation projects. These initiatives are helping Thailand adapt to a greener future by financing the foundation and infrastructure for more sustainable development.

The bank's energy-saving policies and practices can also be seen through-out its business operations. A leading example is its paperless, energy-saving K-ATM service, which allows customers to withdraw and transfer funds, check balances, make purchases, complete bill payments, and change PIN codes through more than 7,000 terminals throughout the country. The K-ATM to Reduce Carbon Footprint program was successfully implemented to reduce the volume of paper used in ATM slips, and the bank estimates it could reduce the overall amount of ATM slips by up to 30 percent, decreasing its carbon emissions by 6,600 tons per year. Digital banking practices such as K-Cyber Banking and K-Mobile Banking have also helped make the bank more energy efficient.

Meanwhile, internal sustainability campaigns aim to improve the bank's operations and consumption habits. KBank's K-Cost Excellence program encourages departments within the bank to propose ways to conserve natural resources in KBank products and services. The bank's pilot Green at Heart energy-conservation campaign involved a competition at the bank's three main office buildings, Rat Burana, Phahon Yothin and Chaeng Watthana to achieve the highest-percentage reduction in average electricity usage.

During the three months of the campaign, KBank successfully reduced electricity costs, raised resource awareness among staff, and reduced consumption by almost 120,000 kilowatt hours, reducing greenhouse gas emissions by almost 110,000 $kgCO_2e$ (kilograms of carbon dioxide equivalent) – the equivalent of planting around 10,000 trees.

ENVIRONMENTAL BENEFITS OF E-BANKING

By switching to electronic bills, statements and payments, each year the average household can:

 Save **6.6 POUNDS** of paper   Save **0.079 TREES**   Avoid the release of **63 GALLONS** of wastewater into the environment

 Avoid the use of **4.5 GALLONS** of petrol to mail bills, statements and payments   Avoid producing **171 POUNDS** of greenhouse gas missions.

Saving this amount of greenhouse gas emissions is the equivalent of:

 Not driving **169 MILES**   Planting **2 TREE SEEDLINGS** and allowing them to grow for **10 YEARS**

 Not consuming **8.8 GALLONS** of petrol   Preserving **24 SQ FT** of forestland

Source: Bank Green Today

▶PIONEERS

Select banks that offer "green loans"

History: Green loans have become increasingly available at Thai banks over the past decade

Location: Green loans available nationwide and internationally

Key features: Green loans help finance sustainable entrepreneurship, projects and purchases by offering fee discounts and low rates to borrowers. Individuals, entrepreneurs, SMEs and businesses may apply

Banks offer "green loans," often with special rates, to individuals and small businesses that are undertaking or expanding environmentally friendly projects. These could be projects that improve energy consumption, use or develop alternative or renewable energy projects, deal with waste recycling and disposal, or facilitate green home improvements, such as installing solar panels and insulating buildings.

Bangkok Bank's Bualuang Green Loan

Objective: Aims to improve the environmental impact of SMEs.
Criteria: Investment must be for one of the following: energy-saving projects and activities; development of alternative/renewable energy; green label products; waste recycling; development and production of bio-products derived from renewable biological resources.
Loan period: As per bank's terms.
Qualification of borrower: SMEs certified as "Green Label" by the Thailand Environment Institute.
Interest rate and fee: Minimum lending rate (MLR) or lower per year; fee is 0.25 percent of the approved credit line. Other fees may apply.
Other eco-friendly loans offered: Bualuang Energy Saving Loan, which offers low interest rates to SMEs reducing

energy waste by investing in energy-saving equipment or using agricultural products, leftover materials or waste-water to produce alternative energy.

KASIKORNBANK's K-Energy Saving Guarantee Program

Objective: Supporting energy efficiency projects managed by energy consultancy firms.
Criteria: Commercial loans, equipment leasing and hire purchase (HP) financing for energy efficiency projects.
Loan period: As per bank's terms.
Qualification of borrower: Energy Service Company (ESCO) that offers integrated services for implementing energy efficiency projects, and a guarantee that energy savings generated by the project will be enough to repay the loan. Must meet ESCO Selection Criteria. Both the project and customer must meet the bank's qualifications.
Interest rate and fee: Subject to bank's rules and regulations.
Other eco-friendly loans offered: The Top-Up Loan for Energy Saving (Lighting Solution) program, which features a long-term loan aimed at energy efficiency projects managed by ESCOs or suppliers

Bangkok Bank head office.

Krungthai Bank's Green Loan

Objective: To improve the environmental impacts of SMEs.
Criteria: Projects must save energy, use alternative or renewable energy, or reduce pollution through improvements to the business's site, machinery, or equipment.
Loan period: Term loan of up to 10 years; working capital as deemed necessary.
Qualification of borrower: Thai national who has an acceptable financial record and can make loan repayments.
Interest rate and fee: For term loan, MLR minus 1 percent during the first two years, then MLR from third year onward. For working capital, not lower than Minimum Overdraft Rate (MOR). Fees according to bank's procedure.
Other eco-friendly loans offered: Energy Saving Loan for renovation of site, machinery or equipment to save energy; the use of alternative/renewable energy; or the implementation of new or alternative energy projects. Environment Loans for the Private Sector for companies investing in waste disposal or air or water pollution treatment systems, and for businesses that treat or dispose wastewater.

THE PRICE OF POLLUTING

Climate finance is a specific subset of green finance that targets climate change mitigation. Within these parameters, carbon pricing is a major trend that promotes low-carbon industries and climate-resilient development. Targeting the financial bottom line of large-scale emitters, national and international policymakers use carbon pricing largely as a disincentive to creating carbon emissions.

What is it? Putting a price on carbon and the effects of its emissions allows polluters to decide whether to reduce their carbon dioxide emissions or to pay for these emissions through taxes or carbon permits. Carbon prices are roughly equivalent to the cost of damages caused per unit of carbon. Carbon permits are tradable and the total number of permits allowed per entity is capped. This emissions trading scheme (ETS) is known as a cap-and-trade system.

How did it originate? Carbon trading was first adopted as part of the 1997 Kyoto Protocol, which required member countries to commit to reducing greenhouse gas emissions.

How does it work? An ETS allows industries with low emissions to sell their extra allowances to larger emitters. The system also increases less-developed countries' access to energy. By creating supply and demand for emissions allowances, the ETS establishes a market price for greenhouse gas emissions, while the cap helps ensure that the required emission reductions will take place and keep the emitters within their pre-allocated carbon budget. Carbon taxes also set a price on carbon by defining a tax rate on greenhouse gas emissions, or on the carbon content of fossil fuels.

What is it worth? The total value of the global carbon market in 2015 was around US$52.8 billion according to Thomson Reuters Point Carbon, a nine percent year on year rise due to higher per unit prices. However, that figure pales in comparison when measured against the US$5.3 trillion value of annual fossil fuel subsidies.

Where does it apply? As of the beginning of 2017, there were 19 carbon trading systems in operation worldwide covering nearly seven billion tons of emissions, according to the International

Australia is one of the countries active in the carbon market.

Carbon Action Partnership (ICAP). The European Union Emissions Trading Scheme (EU ETS) is the world's largest carbon market with 2,039 million tons of carbon dioxide equivalent (abbreviated as $MtCO_2e$). However, China is poised to overtake the EU as it establishes the world's largest carbon market, which is expected to be in the range of 3-5 billion tons of carbon allowances per year initially. Emissions trading schemes are also currently active in the US, Canada, Kazakhstan, Japan, Australia and New Zealand. Carbon taxes have been introduced or are about to be implemented in parts of Canada, Chile, Costa Rica, Denmark, Finland, France, Iceland, Ireland, Japan, Norway, South Africa, Mexico, Sweden, Switzerland and the UK. According to the World Bank, carbon taxes and carbon trading implemented across the world cover almost six gigatons of this poisonous element, or about 12 percent of annual global greenhouse gas emissions.

Who are the key players? National and international policymakers establish carbon markets and emission trading systems on local and regional levels. Development banks around the world both monitor and participate in carbon pricing initiatives, while private organizations such as the Carbon Disclosure Project (CDP) seek to address carbon issues that remain neglected or unaddressed by governmental bodies by working on a company-by-company level.

GROUNDBREAKERS

ALAN LAUBSCH *is a risk and ecosystems researcher and a founder of the Natural Capital Alliance. Alan currently runs the Natural Capital Markets group at Lykke, which provides instant liquidity for Natural Capital Coins without transaction fees. Alan started his career as a researcher in JPMorgan's corporate risk management group in New York, after receiving a degree in Industrial Engineering at Stanford University in 1993. He is one of 25 founding members of the RiskMetrics Group (now part of MSCI), where he authored "Risk Management: A Practical Guide (1999)."*

What are some innovative trends and new developments in "green finance" that could pave the way for sustainable development?

Blockchain is the most significant innovation of our age. Nation-based governance has failed to avert a global "tragedy of the commons." Self-organizing Blockchain communities can establish peer-to-peer networks to take effective global action. For example, Lykke Exchange partnered with Worldview International Foundation to launch TREE as a digital token to represent mangroves and carbon credit rights. Lykke guarantees 24/7 conversion into major currencies with zero percent fees. TREE democratizes natural capital investment as a currency with purpose.

What role can banks and traditional financial institutions play?

Financial institutions play a vital role in deploying large pools of capital sustainably. Solar will dominate the energy landscape in the next decade and is also a great financing opportunity. Demand for green investment products will continue to increase with a US$30 trillion generational wealth transfer in the next three decades. Banks who help the transition from carbon intensive "black capital" to sustainable "green capital" will thrive, while those who miss this opportunity will become irrelevant.

It is estimated that financing for sustainable development will require annual investment flows of between US$5-7 trillion. Is it realistic to think that the UN's Sustainable Development Goals can be financed?

It's feasible with the right governance and incentives. US$7 trillion annually is less than one percent of global financial assets, and just 3.5 percent of today's global debt. To succeed, policy makers must put a price on carbon and ecosystems services. About one percent of average consumption could pay for all externalities, including carbon emissions. But we must act quickly to avert a point of no return, with climate risk scenarios that could reach five percent of GDP per annum. To succeed, we need leadership at all levels. Anyone can be part of the solution with a Net Positive pledge and by investing in natural capital. Once people see that green investing can be profitable, a giant wave of capital will be unlocked.

What do you see as the future of "green finance" in Thailand? Are there any special opportunities here?

Thailand has abundant green finance opportunities. A good start is to focus on the two biggest global drivers of carbon emissions: deforestation and coal-fired power. Restoring lost mangrove forests alone has the potential to add $2.7 billion in economic value to Thailand. Mangroves may be nature's best defense against climate change, protecting against soil erosion and flooding, while filtering water to allow sea grasses and coral reefs to thrive. Mangrove based Blue Carbon and Biodiversity credits could pay for the restoration, with co-benefits of higher tourism and fisheries revenues. Renewable energy financing is a tremendous opportunity. Coal is a stranded asset that can't compete against solar's 41 percent annual compound growth rate. Building new coal plants is an environmental and financial disaster in the making, especially in ecologically sensitive areas with mangrove forests (e.g., Krabi). Blockchain financing could give renewables a financial edge, for example by integrating carbon and biodiversity credits.

What is your view on the sustainability performance of banks, which on one hand claim they are pursuing sustainability, yet on the other hand continue to lend to environmentally destructive projects?

It's a good to see more banks embrace green finance, but sadly they still finance too many stranded assets. To remain relevant, banks must rapidly decarbonize, and immediately divest of the most harmful fossil fuel projects such as coal and shale oil. In a world with ever greater transparency and ecological awareness, every investment will be scrutinized. Unsustainable banks will be abandoned by consumers and saddled with stranded assets.

INDICES

The complicated art of measuring corporate sustainability

By virtue of its wide-ranging nature, sustainable development can be difficult to quantify. Attempting to provide an objective and independent gauge, so-called sustainability indices have been developed to measure the economic, environmental and social performance of companies. These specialized indices offer socially responsible investors a look into a company's operations for their **sustainable investment approach**, rating everything from corporate governance and risk management to IT investments, climate change readiness, energy consumption, supply chain standards and labor practices.

Investors who are interested in sustainability indices are not only concerned with ethics. They also believe that companies with a high score in sustainability offer higher returns. Disclosure of a company's sustainability performance allows investors to analyze a company's operational and external "costs" – for example, vulnerability to future regulatory risks (such as those related to carbon pricing) – as well as the company's purchasing and sourcing practices and the environmental impact of its operations. Companies that focus on improving these areas are thought to have a long-term vision that will lead to lower costs and higher profits.

The ethical investment movement is said to have started with the launch of the FTSE KLD 400 Social Index in 1990, since renamed the MSCI KLD 400 Social Index. It ranks companies on the US equity market with positive ESG (environmental, social and governance) characteristics. In 1999, the trend went global when the Dow Jones Sustainability Index World was launched. Other notable sustainability indices that followed include the FTSE4Good Index, Calvert Social Index and NASDAQ OMX CRD Global Sustainability Index.

Each index assesses corporate sustainability differently. For example, the basis of DJSI World's assessment is a questionnaire developed by RobecoSAM, an international sustainability investment company.

Sustainable investment approach:

An approach to investment where environmental, social and governance (ESG) factors, in combination with financial considerations, guide the selection and management of investments.

Best in class:

Refers to investment in sectors, companies or projects selected from a defined universe for positive environmental, social and governance (ESG) performance relative to industry peers.

Sustainable investing is gaining ground worldwide.

Questionnaire responses are supported by company documentation, which are verified through examination of media coverage, public documents and stakeholder input. Each year, DJSI World invites 3,400 publicly traded companies across a wide range of industry groups to participate in the questionnaire. Of these, the top 10 percent with the highest ranking in sustainability factors are included on the index.

The FTSE4Good indices also use a questionnaire to assess sustainability, but it is developed by EIRIS, a global ESG research institution. Answers provided by companies are backed by company interviews and examination of annual reports. Companies are then given a score of 0 to 5 on their ESG performance, with a score of 5 indicating industry-leading practices and 0 indicating no disclosure. The MSCI ESG, on the other hand, does not use a questionnaire but instead interviews companies and uses SEC filings, media coverage, and third-party research. All indices annually update company rankings, allowing investors to monitor performance over the long term.

However, to date there is no fail-proof way to accurately assess corporate sustainability. Much data

Sustainable Investing

A growing field, sustainable investment, or SRI (socially responsible investing), involves the screening of investments in terms of environmental, social and governance factors. Sustainable investors may exclude or divest negatively performing firms from their portfolios, or choose those that excel within their respective industries in terms of sustainability. Alternatively, they may simply apply "norms-based screening" to make sure the companies are in minimal compliance with international norms. Here are some other investment terms and practices related to sustainable investment:

Corporate engagement and shareholder action: Shareholders may exercise their clout and influence corporate behavior through direct corporate engagement. They may file or co-file shareholder proposals, or proxy vote guided by comprehensive ESG guidelines.

ESG integration: This involves an explicit appraisal of the environmental, social and governance factors of a company in the investment decision-making process. Firms must demonstrate their adherence through a transparent and systematic process.

Exclusion criteria: Should a company violate one of the exclusion criteria (for example, the company uses forced labor), this firm is excluded from the investment universe for SRI funds and may no longer be purchased by SRI funds.

Impact investing: According to the Global Sustainable Investment Alliance, a collaboration of sustainable investment organizations around the world, "impact" investments are targeted investments, typically made in private markets, aimed at solving social or environmental problems. Examples include community investing, where capital is specifically directed to traditionally underserved individuals or communities, or financing that is provided to businesses with a clear social or environmental purpose.

Sustainability-themed investment: This strategy targets specific sustainability issues, such as climate change, food security, water use, agriculture, alternative energy or clean technology and more.

are based on company documents and self-evaluations. The "**best in class**" approach taken by most indices means that companies are ranked against industry peers, not necessarily providing investors the full picture of how sustainable a company is. Nevertheless, the influence of sustainability indices is growing: 15 to 20 percent of the world's assets currently under management are estimated to be managed according to sustainable guidelines. A 2013 paper by Harvard Business School found evidence that highly sustainable companies significantly outperformed their counterparts over the long term in terms of stock market and accounting performance. Sustainability indices also have a proven impact on company profitability: rankings affected stock prices and the cost of capital. Billions of dollars are also invested in mutual funds and other financial products that track sustainability indices.

In Thailand, a local sustainability index, the ESG100, now ranks the top 100 sustainable Thai companies. Thai companies are also making news abroad: 15 Thai companies were included in the 2016 DJSI. All 15 were selected for inclusion in the DJSI Emerging Markets list and five also made it onto DJSI's global index known as DJSI World.

▶ PIONEERS

ESG100

History: Established by the Thaipat Institute in 2014

Location: Thailand

Key features: Sources data from public documents and Form 56-1 declarations to compile an index of Thai companies with enviable scores in environmental, social and governance areas

The ESG100 is Thailand's first sustainability index. It lists the top 100 Thai companies with "outstanding" ESG performance. Launched in 2014 by the Thaipat Institute, a public interest group that promotes socially responsible business, today it evaluates the sustainability of 621 publicly traded companies listed on the Stock Exchange of Thailand (SET) and the Market for Alternative Investment (MAI).

The index eschews the traditional data compilation approach taken by the DJSI, which asks companies to fill out questionnaires. Instead, Thaipat Institute ranks companies by using more than 11,500 data points taken from publicly available sources. This data is drawn from annual reports and other information compiled by the Capital Market Supervisory Board, the DJSI, the SET, the Securities and Exchange Commission, the Thai Listed Companies Association, the Thai Institute of Directors and the Thaipat Institute itself.

In 2016, the services industry had the highest representation on the ESG100, with about 19 companies on the list. Next was the property and construction sector with 16 companies, followed by four companies from the MAI. The

Pipat Yodprudtikan, director of Thaipat Institute.

100 companies on the index have a combined market capitalization of around 6.7 trillion baht, accounting for 50.7 percent of the Thai stock market's 13.2 trillion baht capitalization.

GROUNDBREAKERS

*Thailand's largest oil and gas refiner, **THAIOIL GROUP**, is one of the country's industry leaders in sustainability. It was listed on the 2016 DJSI Emerging Markets list and named Energy Industry Group Leader (of 159 companies worldwide).*

Is it important to Thaioil's investors and shareholders for the company to receive a rating on the DJSI? It is said that there are two types of investors and shareholders: one puts the focus highly on the company's financial performance, while the other is also looking for intangible performance. Thaioil uses the DJSI Sustainability Assessment as a tool for corporate risk management. Being a member of the DJSI implies that we have enhanced systems to identify, manage and control our risks compared to our peers in the same industry. This can assure our investors and shareholders that our manage-

ment systems can effectively handle most scenarios or cases that might lead to a negative impact and can generate returns to them in the long term. It is also important that Thaioil is reliable in its operations with a long-term commitment to business continuity and stakeholders' satisfaction.

Does being rated highly for sustainability improve Thaioil's competitiveness? Yes, but not directly. Being rated highly may attract attention and raise awareness of the company's existence among those investors who are concerned with sustainability performance. With

a global standard guaranteed, this makes Thaioil more competitive.

What steps did Thaioil take to get itself listed on the DJSI? Thaioil created a sustainable development master plan and five-year roadmap according to each DJSI category in 2013. We have goals and initiatives in place with strategic partners.

Has Thaioil identified any areas for improvement for future ratings? Yes. The top three issues are supply chain management, environmental management and creating shared value between the company and society.

Thai Companies on the 2016 Dow Jones Sustainability Index (DJSI)

COMPANY	DESCRIPTION	SUSTAINABILITY INDICES	SUSTAINABILITY PRACTICES
Advanced Info Services	Mobile and digital service provider. Founded in 1986.	DJSI Emerging Markets Index	Focuses on community engagement, socially responsible business, and environmentally friendly products and services.
Airports of Thailand	Manages and operates Thailand's internatinal airports. Established in 2002.	DJSI Emerging Markets Index	Strives to reduce carbon emissions, improve human resources and develop "Green Aiports"
Banpu	Coal, mining and energy. Founded in 1983. Thailand's largest coal mining company, with operations across Asia Pacific.	DJSI Emerging Markets Index	Gets good sustainability ratings for projects like its wastewater treatment system.
Central Pattana	Retail property development and management arm of Central Group, Thailand's largest retail developer. Founded in 1980.	DJSI Emerging Markets Index	Lauded for energy conservation, water recycling and waste management.
Charoen Pokphand Foods	Agriculture, feed production, food production and retail. Founded in 1978.	DJSI Emerging Markets Index	Prioritizes food security and safety, consumer health, and strives for ethical sourcing in its value chain.
IRPC	Petroleum, petrochemicals and energy. Founded in 1978. Subsidiary of PTT.	DJSI Emerging Markets Index	Projects on energy and water conservation, as well as campaigns to preserve river basins and mangrove forests.
KBANK	Banking and finance. Founded in 1945.	DJSI World Index, DJSI Emerging Markets Index	Takes a holistic approach toward sustainability through its "Green DNA" initiative.
Minor International	Hospitality/food and beverage. Founded in 1978. Operates over 1,500 restaurants and 100 hotels worldwide.	DJSI Emerging Markets Index	Promotes the conservation of natural resources and runs waste management and water recycling programs.
PTT	Oil, gas and energy. Thailand's largest public company. Formerly the Petroleum Authority of Thailand, which was set up in 1978 and privatized in 2001.	DJSI World Index, DJSI Emerging Markets Index	Promotes alternative/renewable energy projects such as biodiesel, gasohol and solar power. Currently in the middle of a massive reforestation program.
PTT Exploration and Production	Oil, gas and energy. Exploration and production business subsidiary of state-owned PTT. Founded in 1985.	DJSI World Index, DJSI Emerging Markets Index	Aims to improve energy efficiency, lower greenhouse gas emissions, use green technology, improve water treatment and lower fuel consumption.
PTT Global Chemical Co	Petrochemicals and chemicals. Subsidiary of PTT, founded in 2011. Thailand's largest petrochemical and refining company.	DJSI World Index, DJSI Emerging Markets Index	Purchases carbon offsets to achieve a net zero carbon footprint. Has been hailed for energy conservation efforts and alternative energy use.
SCG	Construction materials. Founded in 1913. Thailand's largest cement company.	DJSI World Index, DJSI Emerging Markets Index	Projects include building check dams to protect watersheds and conserve forests, supporting research on growing crops on saline land, recycling waste-heat and encouraging sustainability throughout their supply chain and business.
Thai Union Group	Food and beverage. Founded in 1988.	DJSI Emerging Markets Index	Projects focus on natural resources, marine resources and biodiversity. Also strives for ethical sourcing and carbon footprint reduction.
ThaiBev	Food and beverage. Founded in 2000.	DJSI Emerging Markets Index	Strives to conserve and recycle water while also minimizing emissions, air pollution, and effluent waste. Also has social projects in 62 provinces.
Thaioil	Thailand's largest oil and gas refiner, and supplier of petroleum products. Founded in 1961.	DJSI's Gold Class Industry Leader in Energy, DJSI Emerging Markets Index	Achieving sustainability through energy efficiency projects, reducing energy consumption and greenhousegas emissions, and investing in alternative energy, such as ethanol.

COUNTERING CORRUPTION
Business groups aim to end graft

A gathering of the Anti-Corruption Organization of Thailand.

Thailand's private sector has long been aware of the malaise of corruption and the toll it takes on their economic fortunes. On many occasions, corruption has derailed key investments in areas instrumental for the country to reach its full potential. This has also negatively impacted Thailand's competitiveness, as shown in recent surveys that forecast opportunities will be lost to other ASEAN nations that are less prone to graft.

In 2016, Thailand ranked 101st out of 176 countries (falling 25 places versus 2015) in the Corruption Perceptions Index, an annual report published by Transparency International, which is based on the perceptions of foreign businessmen. The ranking put the kingdom on par with Gabon, Niger, Peru, the Philippines, Timor-Leste, and Trinidad and Tobago. In the case of foreign investment, image may not be everything, but it certainly is important.

Fortunately, the lessons learned from other economies – particularly those of Hong Kong and South Korea – and their proven successes in tackling corruption to boost economic growth have been widely shared among large and small enterprises in Thailand, leading to the birth of Thailand's Private Sector Collective Action Coalition Against Corruption (CAC) and the Anti-Corruption Organization of Thailand (ACT).

Since the CAC's inception in 2010, more than 230 companies have been certified for their implementation of effective anti-corruption policies; a few of the stauncher ones have even declared "zero tolerance" for such misconduct. By mid-2017, almost 840 private companies had declared to the CAC their intention to run "clean" businesses.

To aid this anti-graft drive, the Thai Institute of Directors (IOD), a founding member of the CAC, offers anti-corruption courses to disseminate best practices to a wide range of enterprises. The IOD holds ten such

Corporate governance:

The system of rules, practices and processes by which a company is directed and controlled. Good corporate governance sees the interests of the many stakeholders in a company taken into consideration, from its shareholders and employees down through its customers, suppliers and the community.

classes each year, with more than 270 management-level executives taking part in 2016 alone. The IOD also strives to enhance the level of professionalism among company directors and other top-level executives by organizing an Ethical Leadership Program four times per annum. In 2016, more than 110 leading executives participated in this course.

The IOD and some of its founding members, such as the Securities and Exchange Commission and the Stock Exchange of Thailand (SET), have lauded these positive moves by the private sector. In this gradual way, good **corporate governance** is becoming a priority for companies that see the rewards it brings, like being listed on the Dow Jones Sustainability Indices. In 2016, 15 Thai companies made the list, the highest number out of any ASEAN state. For years, the indices have produced key indicators for investors to show which companies are excelling both ethically and environmentally and, in turn, which are best adapted to thrive in an investment climate that increasingly considers a firm's environmental, social and governance (ESG) standards.

In 2015, a total of 23 Thai-listed companies received the ASEAN Corporate Governance (CG) award

Pramon Sutivong, ACT chairman, speaks at the event "Hand in Hand: Reform the Fight for Sustainable Victory" on Anti-Corruption Day.

The ASEAN Corporate Governance Scorecard showed an average corporate governance score of 75 for Thai listed companies: the highest percentage within ASEAN.

and the kingdom's CG score of 87 was the highest within ASEAN. "Since the program's inception, Thai-listed companies have been outstanding in consistently applying corporate governance principles in line with the ASEAN standard," said Bandid Nijathaworn, president and CEO of the Thai Institute of Directors. "This is particularly in the areas of the Rights of Shareholders and the Equitable Treatment of Shareholders," two of five areas of that are analyzed. Regarding Disclosure and Transparency, Responsibilities of the Board and the Role of Stakeholders, there is still room for improvement.

In terms of financial institutions, the SET's partnership with the UN Sustainable Stock Exchanges Initiative has proven its seriousness about taking its governance standards to the next level. In fact, the SET became the first stock market in ASEAN to join 12 other stock exchanges from around the world that are equally committed to promoting long-term sustainability by enhancing corporate transparency and ESG integration.

No question, the heightened awareness of corruption and the methods used to counter it are now spreading from large-sized companies to smaller

enterprises. But to win this war the private sector cannot go it alone. Collaborating with other agencies like the National Anti-Corruption Commission, the Anti-Money Laundering Office, and ACT is crucial to uprooting corruption, and keeping it to manageable levels, if not eliminating it entirely.

A growing number of firms are among the 51 member organizations of ACT, whose mission is to gather information about graft and disseminate it to relevant authorities to raise awareness and encourage the spread of best practices from the corporate world to the political sphere. Of these practices the "Integrity Pact" is worth singling out, as private companies vying for government procurement projects must swear not to take any bribes.

One of ACT's biggest successes so far was its drive to get approval for the Facilitation of Official Permission Granting Act of 2015. For years the private sector had complained about the inefficient process of granting approval for factory permits. The rigmarole could be dragged out endlessly, and the process presented too much leeway for bribery to cut through all the red tape. The ramifications of this new piece of legislation may have a ripple effect on other government services, forcing them to streamline their operations and cut back on opportunities for graft too.

Although corruption is a perpetual problem in Thailand, affecting the private and public sectors and people from all walks of life, substantial efforts (still small scale but with much potential for growth) are being made to keep the excesses in check.

BANDID NIJATHAWORN, *president and CEO of the Thai Institute of Directors, has long played an important role in Thailand's private sector. He was an economist at the International Monetary Fund, then served as deputy governor at the Bank of Thailand. He is also on the board of directors of the Thailand Development Research Institute.*

What's your view on the state of corruption in Thailand? It's a big problem. People, including those in the public and private sectors, tend to break laws. The behavior encourages facilitation payments and collusion in public procurement projects, which results in projects that are too costly. It also kills business motivation.

What is the cause of corruption?

The deterioration of the public's trust in the public sector as the patronage system replaced the merit system amid weak law enforcement and poor systems of checks and balances. The public sector's three main elements are administration, state enterprises and local governments, in which all of their operations must be transparent.

How do we tackle corruption? We must deal with the behaviors of both givers and takers, putting them under control through a transparent work and decision-making process in the public sector. We also need to motivate government officers, private companies' staff and the general public to join the cause.

What is the private sector's role in eradicating corruption? The role is immense because private enterprises figure directly into the equation since they offer the bribes. If they changed their behavior by stopping such practices, this would diminish corruption.

How does Thailand's Private Sector Collective Action Coalition against Corruption (CAC) fit in this picture? One company cannot do this alone. There should be a large number of companies that collectively stop such payments and steer their behaviors towards clean and corruption-free business. Through this collective force, the private sector can push for a change in the public sector.

Hong Kong and Denmark
Set Anti-Corruption Benchmarks

Lessons in fighting corruption can be learned from countries that have faced and eradicated similar ills. Chief among these standard-bearers is Hong Kong.

In the 1960s and 1970s, the government of Hong Kong found itself limited in its ability to meet the demands of a hungry and expanding populace. This provided a fertile environment for the unscrupulous to operate with impunity. Bribes known as "tea money," "black money" and "hell money" were accepted as a necessary part of life. Ambulance crews were demanding "tea money" before picking up a sick person.

Hong Kong established the Independent Commission Against Corruption (ICAC) in 1974 to fight corruption with a strategy of law enforcement, education and prevention. To this day, the ICAC has a statutory duty to examine the practices and procedures of governments and public bodies. Today, Hong Kong is Asia's biggest financial center. In 2014, Transparency International put it in 17th place among 175 countries on the Corruption Perceptions Index.

Topping the list, however, was Denmark. According to the government-sponsored Business Anti-Corruption Portal, Denmark has very little corruption in its bureaucracy. That's because the country is tough on the guilty and the agencies or companies who harbor them. The Danish Criminal Code forbids active and passive bribery and most other forms of graft outlined by international organizations. Furthermore, bribing foreign public officials is forbidden, and entire companies can be held criminally liable for acts of corruption committed by individuals working on their behalf. Besides all the rules and regulations, the chief safeguard against such abuses of power in Denmark is often said to be the integrity of the people.

▶PIONEERS

SCG and KASIKORNBANK

History: SCG was founded by royal decree in 1913; KASIKORNBANK was founded as Thai Farmers' Bank in 1945

Location: Both multinationals have headquarters in Bangkok but also operate internationally

Key features: Comprehensive anti-corruption policies, including codes of conduct and committees that impact both national and international operations and supply chains

At several leading Thai companies, anti-corruption policies have been incorporated into operations as part of the companies' corporate social responsibility programs and commitment to their staff and shareholders. Prominent among this elite group are SCG and KASIKORNBANK (KBank), which have been certified by the CAC for raising the bar to combat graft.

At SCG, its anti-corruption policy serves as a guideline to ensure that no such acts will defile its image or derail its commitment to the sustainability of its enterprise. The policy is backed up by guidelines in its code of conduct,

corporate governance guidelines, and various operating manuals.

Under the policy, all directors, executives and employees in every country in which SCG operates are prohibited from engaging in any type of bribery or illegal payments in any business transactions, including such high stakes bids as government procurements in Thailand. The Board of Directors determines the policy, aided by the Audit Committee that deals with the accuracy of financial reports. The management team is responsible for ensuring all staff and related parties are aware of the policy.

All directors, executives and staff are told to exercise caution when it comes to offering gifts or other perks like dinner and drinks. In these matters they are duty-bound to obey the code of conduct. Any such transactions must be transparent or made in accordance with the law.

The staff is also required to notify supervisors of any acts of misconduct they witness; the company guarantees immunity to whistleblowers. Any staff caught committing such misdeeds will face company discipline or a court of law. Striving to sustain the organization's anti-corruption culture, SCG also applies these same standards to all the businesses in its supply chain. "Corruption has undermined the Thai economy and society for such a long time. It's time

> *"Corruption has undermined the Thai economy and society for such a long time. It's time to end it."*
>
> **Kan Trakulhoon, former chief executive officer of SCG**

to end it," said Kan Trakulhoon, SCG's chief executive officer, at a CAC-hosted conference.

KBank takes a similarly strong approach to graft in its ranks. An independent supervisory board monitors transactions and reports any cases of fraud to the management team. Each month the Operational Risk Committee meets to review the threats of corruption in the organization and how best to contain them.

Policies that ensure transparency have also been put into place. The number of independent directors on the bank's board of directors, for instance, must exceed the number of executive directors. The independent directors also screen all the meeting agendas of the board. Senior staff are in charge of the internal control units while compliance officers are granted direct access to management.

At a training session for top executives arranged by the Thai Institute of Directors, the president of KBank, Teeranun Srihong, said that the most important prerequisite of an effective anti-corruption policy is the clarity of the guidelines, as well as the "tone from the top" that stresses zero tolerance for corruption.

Isara Vongkusolkit (left), chairman of the Thai Chamber of Commerce, and Banthoon Lamsam, KASIKORNBANK CEO, during an anti-corruption seminar held by the Thai Institute of Directors.

THE ROLE OF
GOVERNMENT

Education for Sustainable Development

Energy Conservation

Sustainable Transport

Sustainable Cities

International Partnerships

In Thailand, where budget planning tends to span three or four years at best, it's difficult to shift the mindset of politicians toward longer-term projects that are more durable and eco-savvy. When this strain of myopia and budget constraints are compounded by governments that come and go, the state's muscles are weakened even more.

One significant role of the public sector is to facilitate and enable the private sector to be the engine of growth, which benefits not only the economy on a macro level, but also individuals and communities every day. An effective way to do this is to offer businesses incentives or disincentives, as the state does with renewable energy, and which have proven quite successful.

While long-term public-private partnerships promoting sustainability may be challenging, these projects have tremendous potential. A case study in point is the ongoing extensions to the mass transit lines in and around Bangkok, as well as the construction of new lines that are part of a sweeping 10-year plan. The new mass transit lines could significantly cut down on traffic and clear the air in the chronically congested capital, as transportation is responsible for more than a quarter of all energy consumption related CO_2 emissions in Thailand. Almost all of this figure comes from the road sector, with the greater Bangkok area accounting for half.

In areas like these, the state holds the power, connections and instruments to make a huge difference. Through a subsidiary, the Provincial Electricity Authority (PEA) has invested substantially in its largest project to date, the LED Streetlight Upgrade Project, which requires replacing some four million high-pressure sodium street lamps with LEDs that could result in energy savings of up to 70 percent.

A constant refrain in Thai politics is decentralizing power and planning so that provincial authorities have the freedom to implement projects in their respective areas. Little by little, province by province, this transformation is taking place. Cities such as Phuket and Chiang Rai have taken charge of their own urban renewals. In the process they have won accolades for being "green cities." To truly develop Thailand sustainably, the public sector will require training a new generation of politicos and bureaucrats to think sustainably and be agents of change. What better place to start than with education reform? This is exactly what is happening at schools that train students in the Sufficiency Economy Philosophy, and through international partnerships with organizations like USAID, which sponsors programs in Thailand that help students to acquire 21st-century skills.

EDUCATION FOR SUSTAINABLE DEVELOPMENT

Training a self-reliant generation

Despite the overall shortcomings of Thailand's education system, one area in which the kingdom has proven to be forward-thinking is in the promotion of sustainability in schools. Nine years of applying the principles of King Bhumibol's Sufficiency Economy Philosophy (SEP) in the national education system have yielded some positive results. The goal of SEP within education reform, which is to promote sustainability practices and theories as a basis for national development, is also in line with Target 4.7 of the 2030 Agenda for Sustainable Development, which aims at ensuring that "all learners acquire the knowledge and skills needed to promote sustainable development."

Schools and educators that incorporate SEP into their curricula do their best to encourage students to cultivate mindsets and lifelong habits that ultimately support the building of a sustainable society. Students are encouraged to develop the kind of mentality and skill sets that foster a self-reliant, balanced lifestyle. Furthermore, SEP-based curricula is designed to foster a disciplined, moral and ethical outlook that reflects virtue, honesty and consideration for the greater good.

In SEP schools, learning through the process of doing (i.e., questioning, planning, acting and reflecting), utilizing collective local and global knowledge, and developing decision-making skills are of utmost importance. Students are taught to use reasoning, prudence and moderation when applying knowledge, and to do so with the aim of contributing to the betterment of their schools and communities. Because of the holistic nature of this approach, advocates of SEP say that it impacts the "head, heart and hands" of students – the intellectual, spiritual and practical aspects of education.

In the early stages of SEP implementation, schools used a curriculum intended to cultivate SEP-based behavioral principles in students. However, it was determined that classroom teaching alone was not enough to alter students' perspectives and actions, and so a "whole-school" approach was developed. This approach applies sufficiency thinking in all school activities including learning systems, extra-curricular student activities, school management and community relations. Furthermore teachers, school administrators, parents and community members are groomed to act as SEP role models, creating a school and community environment that helps to foster sufficiency mindsets in students.

Two levels of certification for sufficiency-based schools have been established by the Ministry of Education. The first level concerns an SEP curriculum and a whole-school approach. By December 2016, of the approximately 40,000 schools in Thailand, some 21,126 had been certified as having integrated SEP as a practical orientation in all aspects of school activities and daily lives.

The second type of school certification in SEP is the **Sufficiency Education Learning Center** (SELC) for schools that can offer advice, mentoring and supervision to other schools that aim to become sufficiency accredited. Today there are 121 accredited SELCs. In addition to the two formal certification levels, the Sufficiency School Center under the Foundation of Virtuous Youth (FVY) has created a "best-practice" status that serves to help sufficiency-based schools at the initial certification level to improve their quality of learning activities, innovation, quality performance and management. Those selected for the best-practice status receive opportunities to participate in various SELC programs and training. At the end of 2016, there were almost 400 best-practice SEP schools.

According to research conducted by FVY's Sufficiency School Center, students who attend

Sufficiency Education Learning Center (SELC): Special schools that are accredited to train and coach teachers and administrators of other schools that are striving for Sufficiency Education Philosophy (SEP) accreditation.

Thai students learn about agriculture at a Sufficiency Economy accredited school.

SEP schools demonstrate greater moderation, a better ability to efficiently utilize and share limited resources, and are more likely to volunteer in their schools and communities. They participate in and are proud of their local cultural activities, and show enhanced analytical and social skills. In addition, the research found that students from sufficiency schools demonstrated higher levels of proficiency than students in other schools in all five competencies, which include life skills, communication, logic, problem solving and IT literacy. In short, it can be said that they are acquiring 21st-century skills.

While academic grades are never the primary focus of sufficiency schools, the study determined that scores of students from SEP schools on the **Ordinary National Educational Test** (O-NET) tend to be higher than those of students from other schools.

According to the survey, parents have also noted the positive changes taking place in their children and in their communities. Where the SEP approach is taught, parents see schools and surrounding communities forging partnerships to identify and solve local problems. By working together, they can address issues that hit close to home and by incorporating SEP they can do so in a manner that is sustainable and equitable.

On a wider-ranging scale, there are some very similar initiatives that go back as far as the 1980s, such as the Foundation for Environmental Education (FEE). Recognized by UNESCO as a world leader in Environmental Education and Education for Sustainable Development, FEE has members in 73 countries around the globe. Much like SEP, the foundation's programs – including the Eco-Schools, Young Reporters for the Environment, and Learning About Forests programs – focus on educating youth to be more eco-conscious and take an active role in protecting the environment. For example, the Young Reporters for the Environment program aims to empower young people to take a stand on environmental issues they feel strongly about and also provides them with a platform to articulate their opinions via writing, photography or video.

Meanwhile, Thailand may want to consider following the lead of the European countries trail-blazing new ground in the drive to equip students with 21st-century skills like computer programing. From the United Kingdom to Estonia, education programs are increasingly incorporating coding as a fundamental component of primary and secondary curricula. The hope is that such forward-thinking initiatives will help to better prepare graduates for a globally connected, high-tech workplace environment.

Ordinary National Educational Test (O-NET):

A standardized test on eight major subjects administered by the National Institute of Educational Testing Service (NIETS) to assess the academic proficiency of 6th, 9th and 12th graders in Thailand.

Making Coding a Core Part of Curricula

A decade from now graduates will be applying for jobs that haven't even been invented yet. In most cases, these jobs will require IT skills. That's why major tech companies are urging that programing languages, or coding, be taught as a core part of primary and secondary school curricula. "We have to lay the groundwork for confidently navigating the digital world at a young age. Young people have to be capable of doing more than just using the apps on their smartphones. They should also know a programming language, because that's the only tool that will allow them to make their ideas reality," said Werner Struth, a member of the Bosch board of management.

According to a 2015 Microsoft survey, 75 percent of students in Asia Pacific wish coding were offered as a core subject in their schools, while 77 percent said they believe coding will be important to their future careers, and 63 percent said coding could help them better understand the digital world we live in. "As our world continues its evolution into one that is mobile-first and cloud-first, it is important for educators in the region to stop asking whether or not to offer coding as a subject — but how it can be integrated into the curriculum as soon as possible," said César Cernuda, President of Microsoft Asia Pacific.

Now, if the idea of Thailand's six-year-olds coding between recess and milk breaks sounds implausible, it shouldn't. Kids today are "Digital Natives" and grasp the functionality of personal devices far quicker than grown-ups in most cases. Plenty of toddlers can work out how to take a selfie and can navigate an interface before they even learn to read. The point being that it is no stretch of the imagination that they could begin learning the basics and establishing

> *"As our world continues its evolution into one that is mobile-first and cloud-first, it is important for educators in the region to stop asking whether or not to offer coding as a subject — but how it can be integrated into the curriculum as soon as possible."*
>
> **César Cernuda, President of Microsoft Asia Pacific**

a base from which to build upon at a young age. In fact, a growing number of countries are introducing coding as a core part of curricula. In 2012, Estonia launched a pilot program to teach coding to all primary and secondary students. Others soon began to follow suit. Today Bulgaria, Cyprus, the Czech Republic, Denmark, Finland, Greece, Ireland, Italy, Lithuania, Poland, Portugal and the United Kingdom have all incorporated computer programming into their curricula. Closer to home the likes of Australia, Malaysia and Singapore are also introducing coding in primary and secondary education.

The UK offers an interesting case study. Much like in Thailand, tech companies in the UK have long complained that the country is not producing enough qualified graduates. So, in 2014, the UK launched an ambitious new curriculum with mandatory computer science classes for all children between the ages of five and 16. The computing curriculum is divided up into three stages based on age. In Stage 1 (for 5- to 6-year-olds), children learn what algorithms are, create and debug simple programs, try to develop logical reasoning skills and begin using devices to create, organize, store, manipulate and retrieve data. By Stage 2 (for 7- to 11-year-olds), students are creating and debugging more complicated programs with specific goals in mind and coming to grips with more complex concepts like variables, and sequence, selection and repetition in computer programs. Stage 3 (for 11- to 14-year-olds) is where young learners are expected to be able to use two or more coding languages to create their own programs. They also learn simple Boolean logic, work with binary numbers, and study how computer hardware and software work together.

For such an ambitious initiative to have any chance of success, public-private partnerships are essential, as is providing teachers with the training and skills they need. For example, a great resource for educators unfamiliar with the UK's new computing curriculum is the Computing at School QuickStart Computing website, which offers Continued Professional Development materials and advice designed to help primary and secondary teachers plan lessons and execute the new curriculum. In the private sector, both Google and Microsoft played key roles in helping to launch teacher-training initiatives in the UK.

Thailand, too, could start by trying to tap a major tech company to help it develop curricula for a pilot program in primary and secondary schools. It could also seek to partner with a smaller organization, and the investment need not be prohibitively expensive. For example, with less than US$130,000 in funding from Google, Code Club set up its own training program to help UK teachers get up to speed on programing before the launch of the country's new computer science curriculum.

UNESCO's Education for Sustainable Development

Education for Sustainable Development (ESD) is a UNESCO global campaign that reflects its vision for a world where everyone can benefit from learning the values, behavior and lifestyles required for a sustainable future. ESD takes a holistic approach that identifies environmental sustainability with the sustainability of society. It is to be promoted both informally and through all educational levels. The aim is to build capacity for community-based decision-making, social tolerance, environmental stewardship, an adaptable workforce and improved quality of life for all, using techniques that promote participatory learning and informed thinking.

The key features of sustainable schools are listed here, which have been culled from evidence-based findings in the "Further Reading" sidebar.

1. A commitment to positive social, environmental and economic outcomes, focusing on socio-cultural dimensions of sustainability rather than a restricted focus on "green" agendas.

2. Visionary school leadership that encourages involvement and consensual decision-making.

3. Continuing professional development for teachers and all participants.

4. Extensive multi-stakeholder partnerships (i.e., community, government authorities, the private sector, school networks) that emphasize the active engagement of multiple actors in the joint redesign of basic operations, processes and relationships of school-related activities.

5. Curriculum committed to sustainability.

6. The school as a "learning organization," using participatory learning approaches for students and reflective practice for teachers. Sustainable education requires integrative, problem-based and exploratory forms of learning that invite participants to be critical, creative and change-oriented.

7. Whole-school approaches with sustainability practices in all aspects of school activities and everyone's lives. They coordinate sustainable learning activities between school and community.

8. Expertise in education for sustainability.

9. Appropriate political support (i.e., through the Ministry of Education and other governmental agencies).

Evidence-based learning for students is a part of ESD.

Primary students at Roongaroon School learn the traditional khon mask dance.

▶PIONEERS

Ban Don Kha School

Location: Si Saket province

Key features: Integrates SEP approach in curriculum and through learning stations and community practices

The farming village of Ban Don Kha in Si Saket province of Thailand's northeastern region has a small kindergarten and primary school. In 2009 a new school director arrived and found that the arid environment and buildings were not conducive to learning. To improve the situation, she managed a limited budget and resources efficiently with community support. She involved teachers in understanding change-making processes through implementing "After Action Review" and "Plan, Do, Check, Act" (PDCA) methodologies. Through self-study with documents and the Internet, visits to other schools and various training programs, the teachers and staff began to understand and practice SEP themselves and to communicate the SEP approach to students, parents, neighborhood communities and visitors. School activities and progress in making improvements

The Sufficiency Economy Philosophy is put into practice at Ban Don Kha School.

were communicated through local community and parental participation in meetings. Community help was enlisted to improve the local environment. Teachers made home visits with their students that helped prepare all involved in adopting appropriate SEP outlooks.

The Ban Don Kha School incorporated SEP in its curriculum mainly through the Society, Religion and Culture Project, and later integrated SEP into every subject. Participation in the project allowed students to practice systematic thinking and problem-solving skills while learning moral principles. Students learned how to balance four dimensions (material, social, spiritual

and cultural) in their lives. They applied resource management techniques using the "three R's" (reuse, reduce, and recycle). The school relates SEP learning with learning stations in the school such as the recycling bank, organic farming station, farm animal station and herb station. The whole community has become the school's learning space, including temples, shops, medical centers, rice mills and recycled-material shops. Parents and community members become student project advisors. Projects focus on building moral values and caring and sharing attitudes, volunteering for the community, and learning and preserving local culture.

Students in most sufficiency-based schools grow organic vegetables as a way to practically apply the SEP prinicples.

> **The whole community has become the school's learning space, including temples, shops, medical centers, rice mills and recycled-material shops. Parents and community members become student project advisors. Projects focus on building moral values and caring and sharing attitudes, volunteering for the community, and preserving culture.**

SEP schools seek to cultivate attitudes that will help students to form lifelong habits that support sustainability, such as recycling.

FURTHER READING

- *Toward a Sustainable Society: Cultivating Sufficiency-Mindset in Thai Schools*, by Priyanut Dharmapiya and Molraudee Saratun, 2016.

- *Sufficiency Economy: Philosophy and Development*, by Wibulswasdi, C., Piboolsravut, P. and Pootrakool, K, Crown Property Bureau, Bangkok, 2010.

- "Sufficiency School Centre: The Characteristics of the Sufficiency Economy Learning Centre," report prepared by Sufficiency School Centre in-house researchers, Bangkok, 2015.

- *Shaping the Education of Tomorrow: 2012 Full-length Report on the UN Decade of Education for Sustainable Development*, by Wals, A. E. J., UNESCO, Paris, 2012.

- *A Critical Study of Thailand's Higher Education Reforms: The Culture of Borrowing*, by Rattana Lao, 2015.

GROUNDBREAKERS

An interview with **PRIYANUT DHARMAPIYA**, *Director of the Sufficiency School Center, Foundation of Virtuous Youth, Bangkok.*

"World citizens are facing shared problems regarding environmental deterioration, conflict, and gradual decline in cultural heritage and spiritual values....In Thailand, we've applied SEP principles bestowed by King Bhumibol as a systemic process to restore balance in developing the country."

Why should Thailand instill a mindset in students based on King Bhumibol Adulyadej's Sufficiency Economy Philosophy (SEP)?

World citizens are facing shared problems regarding environmental deterioration, conflict and gradual decline in cultural heritage and spiritual values. There is unbalanced development among the social, environmental, cultural and economic or material dimensions of living. In Thailand, we have applied SEP principles bestowed by King Bhumibol as a systemic process to restore balance in developing the country. In order to be truly sustainable, we have to start with cultivating an SEP mindset in our younger generations so that they can have the knowledge and skills needed to build a sustainable society.

What are the successes so far?

The education sector has started to understand and apply SEP, with over 21,100 schools having been certified as "sufficiency-based schools". They are schools in a variety of contexts (e.g., in religion, size, geographical characteristics). We have gone through a learning process and have understood how to develop sufficiency-based schools. However, to ensure sustainability, we have to find ways to strengthen the SEP culture in these schools. Currently, 121 schools qualify as Sufficiency Education Learning Centers and have developed a very strong SEP culture, and are therefore able to coach and mentor other schools and educators.

What are the challenges?

We try to identify as many different models or building blocks as we need for developing sufficiency-based schools in different contexts. That will help in reducing the risks of trial and error when schools start implementing SEP. Furthermore, even though we cultivate a SEP mindset in students and their local communities, when students finish school and enter the wider society such as moving to other communities or provinces, their new surroundings might not be compatible with SEP thinking. Therefore, it might be difficult for them to sustain their SEP mindset and practices. The development of sustainability in Thai society has been slow partly due to a decade of countrywide political turmoil. We in the education sector can only start SEP with individual schools and surrounding communities. Success in this undertaking requires participation from other larger sectors of society. In fact, society at large, including the media, are the students' teacher, often influencing student behavior. We need more role models from all sectors.

Can SEP be applied in other countries?

Yes. SEP shares many similarities with the main concepts of international approaches to developing sustainability, such as focusing on balanced development between economic, environmental, cultural and social matters. SEP is also in line with international character-building methods such as learning by doing, learning through reflection or self-discovery, rather than memorization – together with what we call "21st-century skills (e.g., higher-order thinking, creativity and good citizenship)." In addition, SEP has its own special contribution to make a more complete and holistic framework of decision-making mechanisms (prudence, integrity, reasonableness and moderation, for example), instead of focusing on just one or a few decision-making elements. It's all about character-development focused education which many countries are implementing.

ENERGY CONSERVATION

Government sets targets to improve the nation's energy efficiency

As energy use has continued to increase year after year, Thailand has been compelled to develop proactive strategies to address energy efficiency. Since the passing of the Energy Conservation Promotion Act (1992), there have been concerted efforts toward the promotion of "energy conservation", meaning better energy efficiency or more economical use of energy (i.e., doing the same activities while using less energy). Among others, target areas for improvement include lighting, hot water production, cooling systems, transportation, and machine-intensive manufacturing processes.

The importance of such measures cannot be overlooked. As United States Senator Bernie Sanders wrote in 2012 of his own country, "Energy efficiency is the low-hanging fruit. Every day we are paying more for energy than we should due to poor insulation, inefficient lights, appliances, and heating and cooling equipment — money we could save by investing in energy efficiency."

Bangkok consumes about 30 percent of the whole country's electricity.

Such sage advice holds true for Thailand as well. Moreover, if handled smartly, energy conservation can play an important role in strengthening energy security, alleviating household expenditures, reducing production and service costs, reducing the trade deficit, increasing competitiveness, and reducing pollution and emissions of harmful greenhouse gases that contribute to climate change. "Improving energy efficiency and reducing energy demand are widely considered as the most promising, fastest, cheapest and safest means to mitigate climate change," wrote University of Sussex Professor Steve Sorrell, an energy and climate policy specialist, in his paper "Reducing Energy Demand: A Review of Issues, Challenges and Approaches".

Recently, Thailand adopted the Energy Efficiency Development Plan 2015-2036 (EEDP). It outlines several initiatives to help the government achieve significant energy savings, including short- and long-term policy targets, programs, activities and funding. The goal is to reduce Thailand's final **energy intensity** by 30 percent in 2036 (as compared to 2010).

Financed by the Energy Conservation Promotion Fund, Thailand's EEDP prescribes the following five strategic approaches to achieve its target:

Energy intensity: A measure of the energy efficiency of a nation's economy. It is calculated as units of energy per unit of GDP. Higher energy intensities indicate a higher price or cost of converting energy into GDP. Thailand has begun to reduce its energy intensity in recent years.

THE 2016 INTERNATIONAL ENERGY EFFICIENCY SCORECARD

(A ranking of 23 countries that represent 75 percent of all energy consumed worldwide)

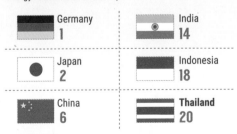

Germany **1**	India **14**
Japan **2**	Indonesia **18**
China **6**	**Thailand** **20**

Source: The American Council for an Energy-Efficient Economy (ACEEE)

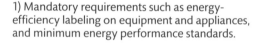

"Energy efficiency is the low-hanging fruit. Every day we are paying more for energy than we should due to poor insulation, inefficient lights, appliances, and heating and cooling equipment – money we could save by investing in energy efficiency."

US Senator Bernie Sanders

1) Mandatory requirements such as energy-efficiency labeling on equipment and appliances, and minimum energy performance standards.

2) Energy conservation promotion and support such as subsidizing energy savings activities.

3) Public awareness campaigns to change consumer behavior. For example, promoting efficient driving practices such as idling stops and gentle acceleration/slowing down instead of braking.

4) Promotion of research and technology development, including more energy-efficient and affordable LEDs and electric vehicles.

5) Human resource and institutional capacity development to create a local green job market and ensure long-term viability of energy savings programs.

In addition to meeting the target of reducing energy intensity by 30 percent, the EEDP has a long-term target of reducing overall energy consumption by 20 percent from projected business-as-usual levels in 2030. By achieving this target, the government anticipates reducing final energy consumption by 289,000 ktoe, avoiding 976 million tons of carbon dioxide emissions, and saving 5.4 trillion baht on energy expenditures. Funding for the program is to come from the Energy Conservation Promotion Fund.

The Bureau of Supporting Industries Development (BSID), the Department of Industrial Promotion, Thailand Board of Investment, Thai Subcontracting Promotion Association and Thai ESCO Association have collaborated to raise joint investment through "business matching" events for the energy efficiency market. For example, during Thailand Energy Efficiency Week 2017, qualified investors, subcontractors and leading manufacturers were brought together in an effort to foster partnerships. The idea is to help pair energy service companies with both government organizations and private companies that are interested in developing energy efficiency projects. The initiative provides incentive programs to encourage energy efficiency investment, as well as offering a broad range of energy solutions including design and implementation of energy saving projects; retrofitting; energy conservation; energy infrastructure outsourcing; power generation and energy supply; and risk management.

Ultimately, the government cannot go it alone on this issue. The concept of energy conservation must also be applied in business practices, and by individuals in their daily lives.

The ESCO Revolving Fund

Thailand's Department of Alternative Energy Development and Efficiency (DEDE), with financial support from Energy Conservation Promotion Fund, has established the ESCO Revolving Fund to encourage private investment in viable renewable energy and energy efficiency projects that lack financing.

The core objectives of the fund are to stimulate more than 1.25 billion baht in investments in renewable energy and energy efficiency; encourage more than 10 ktoe annual energy savings (or 250 million baht in savings) per year, promote and support private investment through Energy Service Companies (ESCO), assist entrepreneurs in minimizing their energy costs and achieving revenue from carbon credits; and provide financing to energy efficiency and renewable energy businesses.

Investment criteria is limited to 10 to 50 percent of total equity, and a maximum of 50 million baht per project. The investment period is capped at seven years. For the lease of equipment, the fund can pay 100 percent of equipment costs capped at 25 million baht per project for a five-year period.

EYE-OPENERS

THE ECONOMICS OF ENERGY EFFICIENCY: BARRIERS TO COST-EFFECTIVE INVESTMENT

Editor: *Steve Sorrell*
Date: *2004*

Why do organizations "leave money on the floor" by neglecting highly cost effective measures to improve energy efficiency? This question lies at the heart of policy debates over climate change and is a focus of continuing dispute within energy economics. This book explores the nature, operation and relative importance of different barriers to energy efficiency through a comprehensive examination of energy management practices within a wide range of public and private sector organizations. The editor and authors use concepts from new institutional economics to explain individual and organizational behavior in relation to energy efficiency, and identify the mechanisms through which such barriers may be overcome. In doing so, they are able to shed new light on the "barriers debate" and provide valuable input to the future development of climate policy. Notably, this document offers an impressive synthesis of contrasting disciplinary approaches and theoretical insights, including analysis of economics, behavior, organizational failures, rationality, adverse selection, moral hazards, risk, split incentives, and search and transaction costs.

▶PIONEERS

The Provincial Electricity Authority's LED program

History: The PEA is a state-owned enterprise established in 1960; the LED program was launched in 2014

Location: PEA's customer locations, which are all located outside of the Bangkok metropolitan area

Key features: A subsidiary of the PEA is promoting nationwide energy savings by replacing incandescent bulbs with LEDs

The Provincial Electricity Authority (PEA) operates as an electricity distributor to service Thailand in all geographic areas outside of the Bangkok Metropolitan area. Its subsidiary, PEA ENCOM, manages PEA's energy investments. PEA ENCOM is currently focusing on green energy investments, including LED lighting projects. Over the next 20 years, supported by the Energy Efficiency Development Plan, Thailand plans to promote research on more energy-efficient and affordable LEDs, as well as providing purchasing subsidies.

Current prices for LED lighting are higher than traditional lighting in terms of upfront costs, although money is saved on the electricity bill in the long-term. To bring the cost of LEDs down, government-run initiatives seek to increase economies of scale and create a local manufacturing market for LEDs. The hope is that LED lighting solutions will become cost-competitive enough to be more fully implemented in buildings, as well as becoming a competitive export commodity. The PEA estimates that LEDs will become cost-competitive in the residential market by 2019.

Under the PEA's LED program, PEA ENCOM markets LEDs to the PEA customer base of 15 million customers, with a focus on replacing highway lamps, streetlamps, high-mast lamps and railway lamps. This includes the LED Streetlight Upgrade Project, which seeks to replace four million high-pressure sodium streetlamps with LEDs by 2018. Full implementation will translate to energy savings of 70 percent.

Financing for this project is based on the Energy Service Company (ESCO) model, which engages a third party under a contract where payment is based on the cost savings derived from energy-saving measures.

Thanks to this and other government programs, Thai LEDs are becoming more cost-competitive against foreign products. The local LED market share has also grown, accounting for 12 percent of the lighting market in Thailand in 2016. The government estimates the country's LED market size to be close to US$2.44 billion, and has a projected growth rate of 30 percent by 2020.

▶PIONEERS

Thailand's Energy Efficiency No. 5 Labeling Program

History: Established in 1993 by the Electricity Generating Authority of Thailand (EGAT)

Location: Nationwide

Key features: A label that indicates the energy efficiency of a home appliance allows residents to participate in energy saving

The Electricity Generating Authority of Thailand developed the Energy Efficiency No. 5 Label to indicate on a scale of 1 to 5 how efficient a home appliance is. A rating of 1 indicates the lowest energy efficiency, while a 5 indicates the highest.

The No. 5 label applied to 28 household appliances as of the end

of 2016. These include refrigerators, air conditioners, electric rice cookers, electric kettles, fans and water heaters. To ensure viability of the No. 5 label, EGAT performs "random spot checking" of No. 5 appliances by purchasing and testing appliances. If those appliance models fail the test, EGAT has the right to recall the labels of that model and forbid its participation in the labeling scheme for one year. In 2012, EGAT found that only 81 percent of appliances tested passed the ratings test. By continuing random testing, EGAT hopes to ensure compliance, prevent counterfeit labels and boost consumer confidence in the label.

As a result of this program, EGAT estimates a total of 3.8 gigawatts of energy and over 12 million tons of carbon dioxide have been saved as of 2015. Using data derived from this program, the government has since pursued a national Minimum Energy Performance Standard program, which has instituted mandatory energy efficiency labeling for certain electrical appliances.

EGAT's Label No. 5 program forms a part of EGAT's broader Demand-Side Management Program, which includes additional components such as educational campaigns, demand response, air conditioner cleaning, efficient lighting, and thermal energy storage. Although EGAT earns revenues from electricity sales, it maintains that its avoided costs from implementing supply-side options are greater than implementing a demand-side management program. However, EGAT has conceded that because its revenues are tied to electricity sales and investment in infrastructure, there is little incentive to invest in lower-cost energy resources.

SUSTAINABLE TRANSPORT

Shifting away from private cars to mass transit networks is the key

The world is rapidly urbanizing, especially Asia. Over the next 20 years, India and China will welcome another 500 million urban dwellers into their cities. These burgeoning metropolises are the engines of their national economies, but they run the risk of stalling if clogged by congestion, pollution, greenhouse gas emissions and cars driven by the middle and upper classes hogging the roads. All told, the economic costs of transport externalities eat up an estimated 5 to 10 percent of GDP in developing countries.

24.7 percent of all CO_2 emissions in Thailand are from the transport sect

Eco car:

According to the Board of Investment in Thailand, "an eco car is a motor vehicle that emits either modest or no greenhouse gasses and as such is less harmful on the environment, in comparison to conventional internal combustion engine vehicles running on gasoline or diesel, or one that uses certain alternative fuels."

In 2014, the transport sector was responsible for 24.7 percent of all CO_2 emissions from energy consumption in Thailand, according to the World Bank. Almost all of these emissions were generated in the road sector, with the Bangkok Metropolitan Region accounting for about half. The percentages are not unusual; they reflect the general transport policy trend in emerging economies to focus on building roads for cars and trucks. However, in the next phase of the country's development, sustainable transport solutions have to feature more prominently, in order to maintain steady economic growth, safeguard the environment, and foster good health among urbanites. In addition, with Thailand's geographic location in the middle of the ASEAN member nations, the country is situated to become the logistical hub of Southeast Asia, making transport infrastructure all the more important for economic growth following the integration of the ASEAN Economic Community.

In 2014, the transport sector was responsible for 24.7 percent of all CO_2 emissions from energy consumption in Thailand.

Thailand is taking steps to integrate these approaches into its ongoing transport planning policies. The automobile tax structure, widely blamed for the proliferation of the heavy and inefficient pickup trucks on the country's roads, has been reformed and restructured to include taxes based on CO_2 emissions and energy

efficiency as of 2016. The automobile manufacturing industry is also being reshaped with the second phase of the **eco car** scheme, which prescribes stringent standards and promotes the production of energy-efficient vehicles that produce lower carbon emissions.

In Bangkok, massive extensions to the mass transit networks known as the BTS, or Skytrain, and the MRT, or subway, are being constructed, as well as new lines. Urban freight centers are being established to take the weight off the capital's roads from all the big trucks hauling goods.

In another massive development announced at the end of 2014, the government signed an agreement with Chinese state companies and financiers to develop an 867-kilometer rail corridor from Nong Khai to Bangkok and the industrial estate of Map Ta Phut, which will expedite the delivery of freight and reduce the wear and tear on the roads from truck traffic. It is the first step in the ambitious government scheme to build 3,000 kilometers of modern dual-track railway lines across the country.

In the capital, the Office of Transport and Traffic Policy and Planning is in charge of publishing traffic

management proposals focused on discouraging car use and promoting more public transport. One practical way to achieve that aim is by revamping Bangkok's bus network. At the moment, the Bangkok Metropolitan Transit Authority is upgrading its fleet with 3,183 buses that run on natural gas to replace most of the old diesel vehicles. GPS tracking and traffic management is being introduced on select bus routes, while the entire network is slated for route consolidation and optimization to improve efficiency.

At the start of 2017, an integrated ticketing system called the Mangmoom Card, or "spider" card in English, was unveiled. The universal card can be used as payment for the BTS, MRT, tollways, buses and Airport Rail Link. The Skytrain and subway stations are also being remodeled to make them more accessible for pedestrians and cyclists.

One can truly say that Thailand is at a transport crossroads. The plans and policies currently in place need to be further expanded and strengthened – and fully implemented in a timely manner – to provide a comprehensive framework for the sustainable development of this crucial sector in the decades to come.

Avoid – Shift – Improve

Sustainable transport solutions are the answer to the modern mobility conundrum: how to move even more people and goods while keeping overall costs and impacts low. In the sphere of policymaking, these solutions are usually categorized under the "Avoid-Shift-Improve" framework.

Electric cars and e-tuk tuks are driving cuts to fossil fuel consumption.

"AVOID" – WHAT IS IT? Reducing travel demand and bringing people, goods and services together with fewer and shorter trips.

HOW IS IT DONE? Mixed land-use planning means all amenities in business and residential districts are within walking distance; more efficient public transport networks and vehicles; improved route optimization and loading for trucks; holding video conferencing sessions instead of face-to face meetings.

"SHIFT" – WHAT IS IT? Changing behaviors so people are less likely to take so many car trips and more likely to use public transport

HOW IS IT DONE? Building extensive public transport networks in cities, promoting cycling and walking, discouraging car use through road charges and tax mechanisms, using rivers and railways instead of trucks to haul goods.

"IMPROVE" – WHAT IS IT? Developing advanced technology that makes vehicles more energy efficient and less prone to causing pollution.

HOW IS IT DONE? Using more electric and hybrid vehicles, cleaner fossil fuels and biofuels, promoting more efficient fuel injection systems, and trucks with more aerodynamic cabins and less energy-guzzling trailers.

►**PIONEERS**

Expanding Mass Transit Lines

History: Begun in 2011, the extensions should be completed by 2029

Location: Around Bangkok and neighboring provinces

Key features: When finished, a total of 13 Skytrain and subway lines will span some 500 kilometers in and around Bangkok

As a mega-city of around nine million inhabitants, Bangkok's future growth would be impeded without a safe and affordable mass transit network. The city is tackling this challenge by building a comprehensive public transport network with rapid rail transit at its core. The blueprints show an ambitious, interwoven mesh of 13 lines, spanning some 500 kilometers across the metropolitan region and into the surrounding provinces. The lines will connect with bus feeder systems, parking lots, cycling paths and pedestrian-friendly footbridges and sidewalks. All trips will be taken on a single ticket, and the entire journey can be pre-planned and updated via smartphones.

This may seem like an overly idealistic plan, but let us not forget that the world's largest metro network, which includes 364 stations and measures 588 kilometers in length as of 2016, was built in a little over two decades in Shanghai, starting from scratch.

Today, Bangkok boasts three mass transit rail networks that span some 111 kilometers. In 2016, an average of around 900,000 trips were taken on them each day. The BTS Skytrain notched up 600,000, the MRT subway 253,000 and the Airport Rail Link 47,000. The current expansion plans, some of which are already on track and in progress, are comprised of seven lines that will cover some 240 kilometers of new track by 2022 with a hefty price tag of around 600 billion baht.

The MRT Purple Line Bang Yai–Bang Sue route opened in August of 2016. The MRT Blue Line extension, which will pass beneath Chinatown and the Chao Phraya River, is nearing completion, although some of the 5 contractors involved are running behind schedule. Meanwhile, the Lat Phrao–Samrong section of the MRT Yellow Line is scheduled to begin operating in 2020, and the two BTS Green Line extensions along downtown's Sukhumvit Road should be up and running by 2018 or 2019.

Considering the usual hurdles of political turmoil, unpredictable macroeconomic conditions, the rivalries and parallel power bases of multiple operators, not to mention the cloud of corruption, it's impressive that such rapid progress has been made so far. Crossing the finishing line by the proposed deadlines is a different matter. Such a comprehensive mass transit system is the gold standard for modern cities. Indeed, it is difficult to imagine a sprawling steel and glass metropolis without such a system.

However, such initiatives have not come about without hiccups. There has been widespread talk of closing down Bangkok's Bus Rapid Transit (BRT), a system of dedicated roadways and buses. If it is indeed shut down, that means a new fleet of more than 3,000 NGV (natural gas vehicle) buses will end up stuck in the slow lane or a series of special bus lanes that are often not much faster. It also means that 25,000 daily riders will have to find alternative means of transport.

Along with the punitive price tag of expanding the Skytrain and subway will be other costs like revamping the Bang Sue Station as a connecting hub and improving the bus feeder system to name but two massive expenditures. In spite of such economic drawbacks, the new or expanded mass transit lines promise to give some important bypasses to the city's most traffic-clogged arteries, and that can only be seen as a good thing.

Expansion of Mass Rapid Transit's Green Line.

Bangkok's Bus Rapid Transit system was almost scrapped in 2017.

GROUNDBREAKERS

DENNIS HARTE, *the Dutch general manager of Tuk Tuk Factory (TTF) in Samut Prakan, has spent a decade designing and producing electric tuk tuks. TTF now exports its tuk tuks to Europe, Australia, New Zealand and the United States, and has recently begun selling its vehicles in Thailand. In 2017 TTF is set to manufacture around 1,000 electric tuk tuks.*

How did your work producing electric tuk tuks come about?

This actually started in 2007 with the modification of vehicles that had been imported into The Netherlands by another company named Tuk Tuk Company. These original Thai tuk tuks were hugely popular. But the vehicles were not very safe, not standardized (they required a lot of adaptations and testing) and, frankly, were causing too much pollution. At the time, I was a Masters of Engineering student conducting research for my thesis: "The Tuk Tuk of the Future." With a small group of engineers we started our own company: the Tuk Tuk Factory. Throughout 2008 and 2009 we experimented with converting tuk tuks to electric. Doing these conversions is a lot of (expensive) work, especially when the vehicles are not standardized. So at the end of 2009 I traveled to Thailand to build a new electric vehicle from scratch, along with a Thai partner. After nine months of development, we were able to begin production and sales. Out first ten vehicles were sold in The Netherlands and Germany.

How does the company try to make its supply chain more "green"?

At the end of 2017 we will be moving to a new factory, which will be powered by rooftop solar panels. We are striving to reduce waste by 90 percent by the end of 2018 through the use of reusable packaging, recyclable materials etc. We actively engage suppliers and try to set up our supply-chain in a green way. For the vehicle itself, we are also researching how to use recycled plastics for body parts and coconut husks as seat filling.

What are the benefits of owning an electric tuk tuk instead of the petrol-powered version?

There are several. First of all, electric vehicles require less maintenance. There are fewer moving parts and few lubricants/coolants are needed. Second, there is no exhaust or noise so a vehicle can be driven inside malls, buildings, airports, etc. The air quality and quality of life of the area will improve; the health of the driver and passengers is also improved. Finally, electric vehicles can be charged anywhere and allow a person to choose their preferred energy source. This could be renewable energy in the form of solar (either with panels on the tuk tuk's roof or on the roof of a home).

What are your key markets and how does Thailand fit into the picture?

Our production has grown from 30 vehicles per year to 350 per year in just five years' time. Our vehicles are mostly exported to Europe and the United States, and we have vehicles driving around in about 38 countries now. Unfortunately, only

a fraction of our sales have been in Thailand (less than 10 vehicles in total), but this will change. Our company is focusing more and more on the massive potential of the Asian market. Since the beginning of 2017, we have been working with Loxley Co on sales in the ASEAN region with Thailand as our main hub. We expect a lot to come from this. We are also working closely with the Electric Vehicle Association of Thailand, market experts – and have talked with Thai tuk tuk owners, drivers and customers – to ensure that we produce a vehicle of the right specifications for the Thai market, and one that meets the drivers' business needs. The government has also announced new licenses for electric tuk tuks, which will partially replace older polluting models.

What does the future hold for Tuk Tuk Factory?

These are very exciting times for electric vehicles and mobility solutions. Our team is being expanded both in Holland and Thailand and we have brought on a number of talented, creative minds. A separate business unit for food trucks and cargo/parcel build-ups has been set up. Our development team has completed a vehicle design update for the latest EU vehicle regulations and is now working on a totally new vehicle design for next year. Lithium batteries, solar cells and smarter vehicles will be more important to our customers and are part of this new design. Our suppliers are also being challenged to scale up with us. With the opening of our new factory, we plan to increase our capacity so that we are producing more than 60 tuks tuks a month by the end of the year.

Lightening the Ecological Load

High transport costs are lowering productivity and profits in Asia's export-oriented economies. Costing between 15 and 25 percent of GDP, they are two to three times higher than in the US, Europe and Japan. The environmental impacts are also staggering. Trucks account for less than 10 percent of the total number of vehicles in Asia but are responsible for about half of all the CO_2 and particulate matter emitted by the transport sector.

In response, major companies are greening their supply chains to reduce costs, mitigate delivery risks and improve brand image. Government and international organizations are also building platforms and developing policy to push the sector toward achieving greater energy efficiency. On the Green Freight Asia Network, members share methodologies for data collection and analysis, best practices and performance scorecards. It helps industry heavyweights like IKEA, DHL and UPS improve the CO_2 and fuel efficiency performances of their own fleets as well as their subcontractors'.

In the US, the SmartWay program helps truck operators compare technologies, like aerodynamic trailers or wire-based tires, and secure financing for improvements. Launched in 2004, it now covers 650,000 trucks (30 percent of US road freight), working with 3,000 shippers, carriers and logistics companies.

Thailand is taking steps to improve the efficiency of its logistics operations too. For example, the Logistics and Transport Management Application (LTMA) project, jointly developed by the Federation of Thai Industries and the Energy Policy and Planning Office, enhances logistic and transport management through technological advancements. In total, 104 companies with 5,396 trucks updated their operational practices by the completion of the first of three phases in May 2013.

EYE-OPENERS

ELECTRIC CARS RISE, FALL, AND RISE AGAIN
WHO KILLED THE ELECTRIC CAR?
Director: *Chris Paine*
Release Date: *2006*

In 1990, California introduced a Zero-Emissions Vehicle (ZEV) mandate to advance the research, development and marketing of "clean cars." Seven major car manufacturers, including GM, Toyota, Ford and Nissan, led the charge. During the 1990s they rolled out a few cars that ran on electricity.

However, the auto and oil industries lobbied successfully to soften the standards and include solutions that were closer to the traditional auto technology, such as hybrid vehicles. *Who Killed the Electric Car?* points to the root causes behind this resistance: low profit margins for the new technologies, lower maintenance costs for

electric vehicles, and the oil industry's fear that such a shift away from gas-burning cars would shrink its share of the fuel market. In the end, it all boiled down to money.

These power struggles are unraveled through the tragic yet captivating story of the EV1, General Motors' first plug-in electric car. When the program was cancelled in 2003, all of the 1,000-plus leased vehicles were recalled and destroyed.

The 2011 sequel, *Revenge of the Electric Car*, reveals the upbeat side of the story as Nissan, GM and the Silicon Valley start-up Tesla Motors ramped up their efforts in the 2000s.

SINGAPORE:
Gearing Down the Car

The city-state of Singapore has been at the forefront of sustainable transport in Southeast Asia for decades. Inhabiting such a small geographical space, it's a poignant example of how innovation and progressive policies emerge out of necessity. What are the lessons for the megacities of tomorrow?

- **Limiting personal car use.** Singapore introduced the world's first "congestion charging system" in 1975. Later adopted by London and Stockholm, the system was revamped in 1998 into a fully automated Electronic Road Pricing scheme, which charges variable rates (US$0.50–3.50) depending on the time of day and location. Cars are taxed at almost 100 percent and the aspiring motorist must also obtain an expensive certificate of entitlement. Unsurprisingly, only 15 percent of Singaporeans own a car.

A taxi in Singapore.

- **Build a comprehensive public transport network.** The current Mass Rapid Transit rail network stretches over 180 kilometers and will double in size by 2030. In addition, a comprehensive bus network with buses coming by every 10 minutes covers other areas and provides feeder lines for the rail network. Users can check connection times and other real-time information on the Mytransport.sg app.

- **Construct public spaces.** Numerous projects have enabled citizens to enjoy their leisure time and built a greater sense of community, from the Gardens by the Bay and enclosed centers for food stalls to the Jurong Lake Gardens and car-free historical district project. Today, around 80 percent of the city's residents live within a 10-minute walking distance from one of the city's 350 parks.

- **Promote walking and cycling.** Under the latest scheme to promote two-wheeled travel, the grid of bike lanes will be more than tripled from 230 kilometers to 700 kilometers by 2030. The goal is to make it safe and convenient for cyclists to travel around the entire island on a comprehensive network. Also in the works are more sidewalks and footbridges to encourage walking.

COPENHAGEN:
On the Road to Smart City Status

The world's foremost example of sustainable transport on a municipal level was not always such a frontrunner. Much like in the rest of post-World War II Europe, the car was the king of Copenhagen, driven by cheap oil and industrial growth.

The Danish capital's transformation started in 1962 under the leadership of the architect Jan Gehl and the city's administration. The philosophy was simple: to see how people use and interact with their landscape and then reshape the city around its residents. They soon found out that with fewer streets and parking spaces, there was more space for cyclists, pedestrians, cafes, shops and public squares. The change was gradual but determined: each year 1 to 2 percent of the parking spaces was reduced and more streets were closed off to vehicular traffic.

Today, Copenhagen has more than 1,000 kilometers of bike paths, some so wide and comfortable that they are called "cycling highways." You can take your bicycle for free on any metro and train.

The results of this transformation are striking. In the city center, where the transformation began, four times more people come through here every day than they did four decades ago. Almost 40 percent of the city's residents use a bicycle for inner-city trips, and another 38 percent get around by public transport or on foot.

By 2025 Copenhagen wants to be carbon neutral. To transform itself into a "smart city" will require harnessing the power of big data. Starting in 2015, a city-wide platform for gathering and analyzing big data in real time was established. Existing data sets, like demographics and statistics, will be merged with sensor-based information to measure traffic, air quality, bus routes and even the waste in garbage bins, to make Copenhagen a paragon of sustainability. By doing so, municipal authorities believe that they may be able to save some US$460 million per year.

SUSTAINABLE CITIES

Across Thailand, small cities have big ambitions

In a world where 54 percent of the population lives in urban areas, sustainable urbanization is a key consideration of sustainable development, especially in developing countries, such as Thailand, where the World Bank estimates that 98 percent of the world's urbanization is happening, and at a more rapid pace than in the past.

The importance of urban growth and its impact on the environment was officially recognized in 1987 by the Brundtland Commission, a UN-organized research group behind the seminal report on sustainable development known as "Our Common Future." But it was not until 2012 that the United Nations System Task Team included the concept of sustainable cities as a major issue in the pursuit of sustainable development. The UN identified four main pillars required to build a sustainable city: social development, economic development, environmental management and urban governance.

In Thailand, urbanization is dominated by Bangkok through what is known as the "primate city" phenomenon. By 2010, the capital had almost nine million inhabitants and boasted nearly 80 percent of the total urban area of Thailand. Bangkok's rapidly

Experts Weigh in on the Keys to Creating a Sustainble City

"In terms of sustainable urbanization policies, there are numerous policy options in different sectors that don't entail hefty spending by the municipality. Sometimes, achieving sustainable urbanization is just as much a question of changing consumption habits and lifestyle practices by engaging in education and community outreach – none of which are particularly expensive. But they all require political commitment; that is essential. Political commitment may come from the mayor, or other local politicians, or within the council more generally, but without it, sustainable urbanization policies will continue to be sidelined."
–Rowan Fraser, Architect and Urban Development Consultant

"Urban sustainability is possible not only because of policies and their implementation but also the citizens who reside in its vicinities. The citizens have to value their environmental resources; as they consume, they have to be able to sustain the resources. They have to understand the impacts of their activities on the environment. It is also important that citizens are satisfied with their surroundings and are willing to keep their own unique, authentic lifestyle including culture and traditions. Policies and their implementation are only one dimension, the quality of citizens such as honesty, integrity and ethics are crucial to bring the whole community toward sustainability."
— Dr. Nathsuda Pumijumnong, Faculty of Environment and Resource Studies, Mahidol University

"The greatest challenge...is the holistic and integrated approach required to tackle development issues. There are many instances of national government agencies, municipalities, regulatory bodies, NGOs, international agencies, and the private sector working in a fragmented way that is unable to effectively address socioeconomic and environmental concerns. This kind of compartmentalization and individualized way of working makes it difficult to implement policies because by nature, cities, communities, and all their socioeconomic and environmental problems are highly interconnected with one another, requiring multi-disciplinary approaches and cooperation from all stakeholders involved."
–Paht Tan-Attanawin, Oxfam, Asia Regional Centre, Bangkok

Phuket has been lauded for its efforts to cut down its negative environmental impacts.

growing population has nearly placed the city on par with the "megacities" of the world, or cities whose populations exceed 10 million. Although the world now has 37 mega-cities, the World Bank estimates that almost half of the world's urban population lives in settlements with less than 500,000 citizens. This model of what is often referred to as the **compact city** has become the focus of Thailand's sustainable city projects.

The concept of sustainable cities was introduced in Thailand in 2004 by the Ministry of Natural Resources and Environment and is based, in part, on the Sufficiency Economy Philosophy of King Bhumibol Adulyadej, which can be applied not only to rural development but also to urban development. Thailand's sustainable city project intends to strengthen local governments in environmental management by promoting public participation, developing capacity-building activities and transferring experiences from city to city. Project implementation started with 15 local governments as pilot sites and 485 local governments collaborating in the "Sustainable City Network." Some of the most successful Sustainable City Projects have been awarded by the ASEAN Working Group on Environmentally Sustainable Cities (ESC).

Building ESCs has been a top regional priority as well, since the East Asia Summit Environment Ministers Meeting (EAS EMM) of 2008. The ASEAN Socio-Cultural Community Blueprint 2025 was developed to improve environmental sustainability. One main objective is to meet the social and economic needs of the people without depleting natural resources. Specific actions to reduce pollution and improve air and water quality through regional or national initiatives have been proposed, including intensifying individual and collective efforts; encouraging transference of experiences, expertise and technology; promoting initiatives towards a "Low Carbon Society;" implementing "Compact City" and "Environmentally Sustainable Transport" concepts; developing internationally comparable measures and indicators for environmental sustainability; and introducing an ASEAN ESC Award as an incentive to promote these practices.

The ASEAN ESC Awards are bestowed by the ASEAN Working Group on Environmentally Sustainable Cities and endorsed by the ASEAN Environment Ministers. Three ASEAN ESC Awards (in the years 2008, 2011 and 2014) have been given to cities in Thailand thus far: Bangkok, Phuket and Chiang Rai city, respectively.

Compact city:
A popular concept among some urban planners and sustainable development thinkers. Essentially it advocates a high population density around mixed land uses, highlighted by outstanding public transport.

▶PIONEERS

Phuket City

History: Named an Environmentally Sustainable City in 2011

Location: Approximately 862 kilometers south of Bangkok

Key features: With a population of around 75,000 and millions of tourist arrivals every year, Phuket city's efforts to cut down its environmental impact have been challenging but rewarding. Phuket city oversees ongoing programs in air quality, waste management and public awareness.

Phuket province consists of 33 islands, including the island of Phuket, Thailand's largest island. Phuket is considered a world-class tourism destination. Phuket city is the capital of the province. The city has developed different strategies to encourage the development of a sustainable and healthy city. The ESC Award was given to Phuket for the city's efforts towards achieving:

Better air quality: Air quality was one of the biggest concerns in Phuket city. With the collaboration of the private sector, government agencies and communities, the city implemented an air quality monitoring system that measures key pollution factors to ensure ideal standards of air quality.

Additionally, in collaboration with the Traffic Police, Pollution Control Department and Land Transport Department, roadside inspections are carried out habitually to encourage the regular maintenance of automobiles for emissions reduction. This program proved to be very effective. In 2011, 79 percent of fuel-based vehicles and 87 percent of diesel vehicles were correctly maintained and achieved national emissions standards.

Waste management: To maintain good air quality and to simultaneously reduce solid waste accumulation, Phuket city designed the "Waste to Energy Program". This program produced 1.7 megawatts of electricity from solid waste incineration, accounting for 2.7 percent of total electricity consumed in the city. In order to achieve better long-term results, Phuket city also invested in better environmental education, not only through schools. Environmental activities were promoted through public information campaigns, which encouraged active participation in carbon emission reduction and better energy consumption in homes as well.

Monitoring of emissions has helped to clean up Phuket's air and make the city far more livable.

EYE-OPENERS

ENVISIONING A SUSTAINABLE URBAN FUTURE
SUSTAINABLE CITIES AND SOCIETY
Frequency: *Monthly Journal*
Co-editors-in-chief: *Professor F. Haghighat, Professor D. Niemeier*

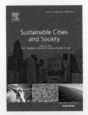

Considering that each country, municipality and city has different social, political, environmental, economic and cultural contexts and issues, it is important to review many and different case studies and research from around the world that focus on sustainable cities and societies.

Sustainable Cities and Society (SCS) is an international, monthly journal that concentrates on fundamental and applied research aimed at reducing the environmental and societal impact of cities. The journal includes a wide range of methodological and technical approaches from an impressive roster of academics, researchers, policy planners, and more.

Sustainable Cities and Society can be accessed online at http://www.journals.elsevier.com/sustainable-cities-and-society. Specific articles are fully searchable by topic, author or journal volume, presenting an excellent resource to read about your particular area of interest in building a sustainable city and society. It is published by Elsevier.

▶PIONEERS

Chiang Rai City

History: Named an Environmentally Sustainable City in 2014

Location: The northernmost city of Thailand, it is the gateway to the Mekong sub-region

Key features: Chiang Rai (registered population 70,201) has ongoing recycling, sanitary landfill, green space and ecosystem management projects

Chiang Rai province is considered the quieter neighbor of Chiang Mai. Chiang Rai city is the capital of the province. To become an Environmentally Sustainable City (ESC), Chiang Rai city initiated "clean and green land programs" emphasizing solid-waste management through the 3R's and increasing green areas.

The program outlines comprehensive strategies for efficient solid-waste management, including creating a cooperative network between private waste collectors, municipal waste collectors and private recyclers to encourage waste separation; creating recycling banks among schools and different neighborhoods; developing school programs to teach the 3Rs and to encourage activities where students transform solid waste into useful materials; setting up marketplaces to sell products made of recycled materials; developing waste collection with a fleet of trucks and volunteers who report any uncollected solid waste; and the improvement of a five-pit sanitary landfill by creating three wastewater treatment ponds to control leachates as well as seven groundwater-monitoring wells to prevent polluting leaks.

The city also developed a Green Area Management program to increase green space, conserve and protect biodiversity, maintain water balance and improve air quality. Chiang Rai has 41,462,100 square meters of green area and 17 public parks. To conserve this green space, Chiang Rai organized a community tree conservation activity where residents created a database of large plant species to increase people's awareness of their environment and to prevent deforestation. Chiang Rai's green efforts have been recognized internationally. The city was conferred the Zuangzhou International Awards for Urban Innovation in 2012, the UN-HABITAT's Certificate of Good Practice in Integrated Management for Biodiversity, and the World Health Organization's 1 in 1,000 World-Class Cities Prize.

Baan Dum, a northern-style house and a famous attraction in Chiang Rai.

ASEAN Certificates of Recognition

In 2011 the ASEAN Certificates of Recognition was created to distinguish the notable efforts made by small cities (20,000–750,000 people) and big cities (750,000–1.5 million people) in order to achieve "Clean Air," "Clean Land" and "Clean Water." The first Thai city to receive this recognition was Phitsanulok in 2011, in the category of Clean Land for Small Cities.

In 2014, two Thai cities received the certificate. Nakhon Sawan City was awarded in the Clean Water for Small Cities category. The city's major achievement was to monitor water-supply quality and effluent quality to protect raw water. The municipal water system has been randomly checked and certified by the Ministry of Public Health since the year 2000. After such monitoring, Nakhon Sawan's water has met with national and WHO standards.

Roi-et City was awarded in the Clean Land for Small Cities category. In order to be a "Zero Waste" city, the city implemented a solid-waste management program called the "No-Waste-Bin Streets" initiative. The city designated specific places and timing for collection using geographic information system (GIS) routing control and promoted the conversion of organic waste from markets into fertilizer and energy.

INTERNATIONAL PARTNERSHIPS

The role of NGOs in capacity building

Thais have a long history of openness to foreigners dating as far back as the Ayudhya Kingdom, which was founded in 1351. After Chinese and Indian traders settled in Ayudhya, explorers from Portugal arrived in 1512, representing the first Western engagement. By the 17th century, Ayudhya had become famous for its cosmopolitan nature, with many peoples from around the world represented.

The Europeans that followed the Portuguese also had a second agenda: to introduce Christianity to Thailand. The European and American missionaries of the 19th century tended to be unsuccessful in converting Buddhists, but they still had a large impact through the Western technology and medical knowledge they introduced. American missionaries founded the Bangkok Christian College as the first boys' school in 1852. The Thai government, seeing the importance of education, modeled public schools after those established by the missionaries. Thus, in a way, missionaries are considered to have created the first international NGOs in Thailand.

At the beginning of the 20th century, many Chinese immigrated to Thailand and founded philanthropic organizations still in operation today, such as the Poh Teck Tung Foundation, which works in accident and disaster relief. Thailand also launched its first homegrown NGO, Sapha Unalom Daeng (later the Thai Red Cross), to help the poor and sick. The Rockefeller Foundation arrived in 1915 and successfully collaborated with the Thai government in improving public health and medical education.

During Thailand's economic boom in the 1960s and 1970s, mass migration to Bangkok led to overcrowded slums, neglected rural communities and a widening income gap. Although the government focused on improving the national infrastructure and expanding rural development, its coverage extended to only 60 percent of the rural areas. NGOs responded with community development projects. They helped resource-poor communities become self-reliant, realizing that providing services without any capacity building would only increase the dependency on assistance.

The number of both local and international NGOs in Thailand grew during the 1970s as a result of conflicts within Southeast Asia, inspiring efforts by

A letter from Prince Mahidol to the Rockefeller Foundation.

Royal photograph of Rama VII and Queen Rambhai Barni being welcomed by the Rockefeller Foundation.

The laboratory of bacteriology at the Faculty of Medicine at Chulalongkorn University in the 1930s.

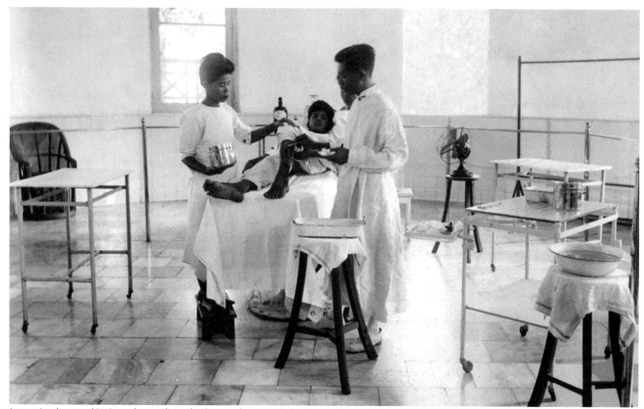
International partnerships in modern medicine development began in Thailand in the 1920s at Siriraj Hospital.

the Christian Children's Foundation (CCF), CARE Foundation (Rak Thai Foundation), Plan International and Save the Children. Oxfam launched its first project in Thailand to aid war refugees suffering from leprosy. Several German organizations such as the Friedrich Naumann Foundation engaged in political education. Amnesty International has been advocating human rights in Thailand since the political unrest of October 6, 1976, and was later joined by Human Rights Watch.

Environmental issues have also been addressed since the 1990s, when economic growth began to deplete natural resources. NGOs such as Worldwide Fund for Nature (WWF) and Greenpeace arrived in Thailand to support the conservation of wildlife and nature.

Before the 2000s, 27 international NGOs operated in Thailand, but many of them have transferred their projects to local NGOs due to Thailand's improved economy. In 2003, the Thai government announced that it had stopped accepting foreign aid so that the money could go to poorer neighboring countries.

This long history shows that international organi-

Currently, local NGOs face the challenge of becoming self-reliant, as many have received the majority of their financial aid from international organizations for almost 50 years.

zations have collaborated with local stakeholders in tackling an array of pertinent issues in sustainable development. These are just some examples of foreign support besides direct aid from the United Nations, the World Trade Organization, the Asian Development Bank and the governments of developed countries.

Currently, local NGOs face the challenge of becoming self-reliant, as many have received the majority of their financial aid from international organizations for almost 50 years. It is an interesting time to watch the movement of local NGOs, especially as Thailand faces new issues as the result of becoming an upper-middle-income country.

419

▶PIONEERS

USAID

Project: US Agency for International Development Connecting the Mekong through Education and Training (USAID COMET)

Purpose: This five-year (2014–2019) workforce development program supports universities and vocational institutions in teaching Thai students 21st-century skills

USAID COMET prepares students so that they can thrive in an increasingly tech savvy workforce.

In recent years, the Thai government and private sector have become increasingly focused on moving the country into its next stage of development through a program dubbed "Thailand 4.0," a platform focused on increasing the kingdom's competitiveness through more innovation, R&D and investment in emerging industries like robotics and the Internet-of-Things. While Thailand 4.0 makes sense on paper, without the human resources to make it a reality, it will most likely remain wishful thinking.

Enter USAID COMET, a project that recognizes the urgent need for better workforce development in Thailand and other countries of the Lower Mekong sub-region. With 70 percent of employers in the region looking to hire, but only 16 percent finding recent graduates who have the skills they need, USAID COMET prepares youth for employment by bridging the gap between education institutions and employers.

Observing a rise in innovative companies and startups in Thailand that are building up value-added products and services, for example, the project suggests ways to foster hybrid skills. As a recent labor market assessment by Bentley University noted, "the future of job skills is not one of hard science versus emotional intelligence. It's one where hard and soft skills come together to create the hybrid." The project also conducts an annual regional Labor Market Assessment to identify key growth industries and job market hiring trends. The businesses interviewed in

the 2017 Labor Market Assessment, for example, highlighted the importance of the following skills for their employees: innovation, critical thinking, creativity, English proficiency, teamwork and adaptability to new work situations and technologies.

USAID COMET supports universities and vocational schools to adapt their curricula and teaching approach to better meet these private-sector demands, and engages enterprises to strengthen work-based (hands-on) learning opportunities, such as internships. Through the MekongSkills2Work Sourcebook – USAID COMET's set of online experiential learning toolkits – USAID COMET promotes innovative education with technology-based solutions. Instructors are trained in dynamic, student-centered learning approaches, such as exploring real-world problems through project-based learning, and in modern strategies for complementing online learning with in-person classes. Out of the twelve USAID COMET leadership institutions, or "Mekong Learning Centers," in the region, two are in Thailand: Mahidol University in Bangkok and Maptaphut Technical College (MTC) in Rayong. MTC has a strong, symbiotic relationship with the Petroleum Institute of Thailand.

So what are the results so far? In 2016, over 34,000 students received training

based upon the MekongSkills2Work Sourcebook, and over 600 ASEAN youth competed in the Young Southeast Asian Leaders Initiative Innovation Challenge, pitching innovative solutions to some of the region's most pressing food security challenges. In Thailand, the Ministry of Education's Office of the Vocational Education Commission (OVEC) has expressed interest in taking the program model to other vocational schools and applying it in other private sectors such as agribusiness.

By 2019, USAID COMET will help 15 post-secondary education institutions equip 120,000 youth with the workplace skills demanded by businesses in the Lower Mekong countries of Myanmar, Cambodia, Laos, Thailand and Vietnam. These skills include adaptability to new technologies, working as a team, and communication and interpersonal skills – the 21st-century skills so often cited as lacking among Thai students and within the Thai workforce.

▶ **PIONEERS**

World Wide Fund for Nature

History: Founded in 1961 as the World Wildlife Fund, now known as the World Wide Fund for Nature

Location: Based in Switzerland, operating worldwide

Key features: The world's largest non-governmental environmental conservation organization, WWF strives to safeguard the natural world, take action against climate change and protect biodiversity. Operating in more than 100 countries, WWF is known for using the most current scientific research.

The Grand Palace before (top) and after (bottom) its lights were turned off to mark Earth Hour.

When Princess Juliana and Prince Bernhard of the Netherlands visited King Bhumibol Adulyadej and Queen Sirikit in Thailand in 1982, the king and queen expressed their concern about environmental issues and their willingness to support conservation work. As a result, WWF started working with the Thai government on several projects, including one on tiger conservation at Khao Yai National Park. In 1995, WWF set up an office in Thailand.

Protecting forests and keystone species plays an important role in sustainable development, especially in

Thailand, where only 28 to 30 percent of the country is still forest. Research indicates that forests provide ecological services that benefit local economies and communities – especially the poor.

In Thailand, WWF committed to protecting threatened key species like the tiger and the Asian elephant. It has implemented notable wildlife preservation projects in the Western Forest Complex and Kaeng Krachan-Kuiburi Forest Complex (covering a combined 22,546 square kilometers), as well as in Mae Wong-Klong Lan National Park in central Thailand.

WWF also worked with government agencies to launch its "Smart Patrol" project to provide new conservation technology in patrolling protected areas. The project has increased prey for tigers and decreased illegal poaching by 400 percent.

WWF has used community-based approaches to wetland management and restoration projects in Chiang Rai, Khon Kaen, and Nakhon Phanom provinces, where wetlands are important to local economies and employment. The lower Songkram River wetlands in Nakhon Phanom province, for example, have played a significant role in agriculture, tourism and transportation. WWF has co-hosted awareness-raising events in the community.

WWF also aims at increasing environmental awareness in its work with the general public. Bangkok has participated in WWF's 60+ Earth Hour since 2007, when lights are turned off for an hour. The 2015 campaign broke records, saving 1,768 megawatts of electricity.

Thai celebrities also participated in WWF's "Kill Ivory Trade" campaign in an effort to end Thailand's reign as the world's largest illegal ivory market. The campaign has been remarkably successful. Roughly 1.6 million people signed the petition against ivory trading, which ultimately led to the Thai government's prohibition on trading African ivory.

WWF works with Thai government agencies to support community-based wetland management and innovative wildlife conservation efforts.

▶PIONEERS

The Rockefeller Foundation

History: Founded in 1913 by the Rockefeller family, owners of Standard Oil

Location: Headquartered in New York, operating worldwide

Key features: Founded to "promote the well-being of humanity throughout the world," the foundation has given more than US$17 billion to support thousands of organizations and individuals

Kasetsart University's Suwan Farm in the 1970s shows an example of the Green Revolution in Asia.

Innovation and pioneering new development has always been at the heart of the Rockefeller Foundation's philanthropy. For decades, it has focused on eradicating diseases such as yellow fever and malaria. To promote food security and improve the incomes of vulnerable people, it has supported much research in sustainable agriculture. In the 21st century, the foundation has taken on new challenges, including climate-change resilience and impact investing.

Founder John D. Rockefeller was friends with Prince Mahidol, the father of both King Ananda Mahidol and King

Hookworm eradication was Rockefeller's first project in Thailand.

Bhumibol Adulyadej. Prince Mahidol and Rockefeller shared a similar interest in medicine. At the time, the foundation had launched a hookworm eradication initiative in the US and hoped to expand the project worldwide. It saw Thailand as a promising place to start. A partnership between the foundation and Thailand was officially established in 1915, and they launched the campaign in Chiang Mai, where 75 percent of the population was affected. By the time the project ended in 1928, it had treated some 300,000 people.

The campaign grew into an effort to improve medical, nursing and public health education. Both the prince and the foundation supported scholarships for medical and nursing students to study in the US. The foundation also sent professors and experts to reform the medical curriculum at Chulalongkorn University and Siriraj Hospital. Over the course of 20 years, the foundation invested more than US$15.6 million in these programs.

In the 1950s, the foundation initiated a project to increase the production of rice and other food crops. They also helped introduce a new breed of rice and farming techniques that yielded two

to three rice crops per year. Thailand's rice exports, which had remained stagnant at one million tons of rice per year from the 1920s to the 1970s, had increased to 10 million tons by 2010.

Now Thailand is confronting the consequences of its rapid economic growth and the effects of climate change. In response, the foundation has been expanding its focus for the past 20 years to connect people regionally to catalyze change with two major programs. The first aims to help the countries situated along the Mekong River Basin to work together to assess signs of possible disease outbreaks and share information. The other project is the Asian Cities Climate Change Resilience Network, which aims to cope with the effects of climate change and share resilience practices.

After a century, the collaboration between Thailand and the Rockefeller Foundation is still going strong, and the benefits are reciprocal. Despite a significant decline of foreign aid to Thailand, the foundation still continues its work here. And, as a global philanthropic organization, it can apply what it has learned from Thailand's developmental evolution to other countries.

▶PIONEERS

Thailand International Cooperation Agency

History: Established in 2004 by Royal Decree to serve as the focal agency for international cooperation under the Ministry of Foreign Affairs

Location: Headquartered in Bangkok, operating worldwide

Key features: Implements Thailand's international development cooperation programs by sharing technical assistance, expertise, and conducting training courses

Thailand has become more active in Africa in recent decades, helping Lesotho and other countries establish sustainable projects based on the Sufficiency Economy model.

Whereas Thailand was once primarily an aid recipient, it has evolved to become a significant donor, trade partner, technical advisor and provider of Foreign Direct Investment. With a commitment to assist its neighbors to achieve sustainability and prosperity, Thailand has been providing technical assistance and support to less well-off countries in the region for more than five decades.

Reaching far beyond Asia, Thailand is also pursuing a "Look West" policy to promote South-South cooperation through the sharing of experiences and best practices. In doing so, the kingdom plays an important role in bridging gaps among developing countries in the global south, and between developed and developing countries.

Thailand International Cooperation Agency (TICA) plays a crucial role in executing the kingdom's international development strategy, which is guided by Sufficiency Economy Philosophy (SEP) principles that emphasize the importance of human development, capacity building, fostering self-reliance and sharing best practices.

Often, Thailand's international development initiatives are carried out with the support and cooperation of bodies such as the Asian Development Bank, the United Nations Development Programme (UNDP), the United Nations

Population Fund (UNFPA), UNICEF, the Asia Foundation, Kenan Institute Asia, Konrad Adenauer Stiftung, Norwegian Church Aid, Stockholm Environment Institute and the Rockefeller Foundation.

With technical assistance from TICA, a number of countries have incorporated SEP concepts into their respective development initiatives, including Cambodia, Indonesia, Laos, Lesotho, Myanmar, Timor-Leste and Tonga. In most cases, these initiatives promote sustainable livelihood development and the use of integrated agriculture.

Through TICA, Thailand implemented a joint project with Lesotho during 2006–2010 to promote sustainable agriculture in the tiny African nation. The project focused on setting up a demonstration center for sustainable agriculture based on SEP practices. During the project, local farmers received on-the-job training from Thai experts in Lesotho and TICA organized a study visit program to Thailand for agricultural extension officers and farmers so they could learn alongside Thai farmers.

In Timor-Leste, bilateral cooperation with Thailand since 2003 has led to the establishment of an SEP Model Village and Technology Transfer Center, which is geared toward enhancing the capability of government officers through agricultural education and technol-

ogy training in crop production as well as promoting sustainable agricultural techniques that are in line with Timor-Leste's local needs and strengths. TICA also helped to develop a model village to demonstrate effective crop production and constructed a learning center.

More recently, in February 2016, technical cooperation between Thailand and Tonga led to the launch of the Agricultural Development Cooperation Project, which aims at cultivating sustainable agriculture in Tonga based on SEP. The three-year pilot project will include growing fruits and will be expanded to include initiatives such as fish farming.

TICA organizes Annual International Training Courses (AITC) for individuals from more than 40 countries, covering topics such as modern technology in sustainable agriculture, and the application of SEP in organic farming. Those hoping to attend these courses can apply for scholarships through Royal Thai Embassies and Consulates. Additionally, Thailand regularly hosts the annual Buakaew Roundtable International Study Visit, during which participants (mostly from foreign affairs or international development cooperation backgrounds in developing countries) can see firsthand how SEP can be applied to address sustainable development challenges relevant to their own countries.

EDITORIAL TEAM

Alex Mavro (contributing editor) is Chief of Operations at the Centre for Sustainability Management, Sasin Graduate Institute of Business Administration of Chulalongkorn University.

Amornrat Mahitthirook (writer: "Transportation") is a 20-year veteran of the *Bangkok Post* who covers the transportation sector.

Amy Wu (writer: "Trade") has over 20 years of professional journalism experience, including at *TIME*, *San Francisco Chronicle*, and *The Deal*.

Anchalee Kongrut (writer: "Soil", "Water", "Urbanization") has been a reporter at the *Bangkok Post* since 1997. She focuses on environmental issues and has a Masters in mass media studies from The New School for Social Research.

Anjira Assavanonda (writer: "Health") spent 16 years reporting local and international stories with a special focus on health and social issues for the *Bangkok Post*.

Apiradee Treerutkuarkul (editor and writer: "Gender Equality") spent the last 15 years as a print journalist and communications consultant for non-governmental organizations based in Bangkok prior to helping lead this project.

Apornrath Phoonphongphiphat (writer: "Trade") is a journalist with more than 20 years experience covering agricultural and industrial commodities.

Arianna Flores (writer: "Sustainable Cities") is a political scientist, who earned a Masters degree in Environmental Management and Technology at Mahidol University. Her recent research focuses on renewable energy, governance and the environment, and sustainable cities.

Ben Davies (writer: "Biodiversity", "Wildlife") is a Bangkok-based journalist and photographer. He regularly contributes articles on the international wildlife trade and is author of *Black Market - Inside the Endangered Species Trade in Asia*.

Benjapa Sodsathit (art director) received an MFA from Minneapolis College of Art & Design in Visual Studies and a BFA from Silpakorn University. She is co-founder of Palotai Design Co., Ltd., which serves clients locally and internationally.

Chanthipapha Sopanaphimon (designer) earned her BFA from Silpakorn University.

Christina Browning (writer: "Ethnic Lanna") is a researcher and writer interested in sustainable development, and health and wellness.

Deunden Nikomborirak (writer: "State-owned Enterprises") is a research director at the Thailand Development Research Institute. Her areas of expertise include competition policy, sectoral regulations, governance, anti-corruption strategy and services trade and investment.

Evan Gershkovich (writer: "Reforestation") has worked in Thailand as a freelance journalist covering forest conflict and as a communications specialist for a community forestry NGO.

Francis Wade (writer: "Tourism", "Conflict") is a freelance journalist based between Southeast Asia and London, with a focus on Myanmar. He worked as a journalist in Thailand for six years.

Greg Jorgensen (writer: "Commuting") writes about expat life in Thailand at GregToDiffer.com, and was co-creator and co-host of BangkokPodcast.com.

Ingo Puhl (contributing editor) is a co-founder of the South Pole Group, a World Economic Forum-recognized, clean-tech social enterprise, and an angel investor in companies that seek to create positive social and environmental impact at scale in the fields of food, consumer choice, consumer loyalty/reward and cause-related fundraising. He is also co-founder of Collaborative Designs, an investment company and accelerator that nourishes the execution of ideas that benefit communities and the planet.

Inhee Chung (writer: "Introduction") leads the Sustainability and Safeguards program at the Global Green Growth Institute. Previously, she managed projects on biodiversity, cleaner production and green buildings at ERM and UNEP.

Jim Algie (editor and writer: "Religion", "Forests", "Agriculture", "Sustainable Tourism", "Restaurants") is a Thailand veteran of Scottish-Canadian vintage who has authored a number of acclaimed books on the kingdom, and contributed a chapter to EDM's history book, *Americans in Thailand*.

Khan Ram-Indra (writer: "Introduction") is an environmental economist by training. He has worked on sustainable development, climate change, and clean energy for USAID, the British Embassy and private sector. He is currently serving as the Thailand Program Manager for the Global Green Growth Institute.

Kim Atkinson (editor: "Education for Sustainable Development") spent his career as an editor for UN agencies in Bangkok, Rome and West Africa, then returned to Bangkok to continue editing and teach writing.

Luxana Kiratibhongse (art director) graduated with a Fine Arts degree from Macalester College in the US. A graphic designer for more than 10 years, she has been commissioned by clients in Thailand and around the world for marketing, advertising and editorial projects.

Mark Fenn (writer: "Forest Conservation") was a British journalist based in Thailand.

Mick Elmore (writer: "Alternative Energy") is a journalist with over 30 years of experience. Based in Thailand since 1992, Mick earned his Southeast Asian Studies Masters at Chulalongkorn University and teaches there.

Molraudee Saratun (writer: "Education for Sustainable Development") is an Assistant Professor at Mahidol University. Her research focuses on human resource management, corporate sustainability and the Sufficiency Economy Philosophy.

Nelly Sangrujiveth (writer: "Energy Conservation") has a JD and LLM in Environmental Law from the University of Oregon. She has worked on various projects seeking to increase the implementation of clean energy in Asia through innovative financing and good governance.

Nicholas Grossman (editor-in-chief; writer "Climate Change") has produced over 10 books on Thailand, including *Thailand: 9 Days in the Kingdom*, *Chronicle of Thailand*, *Americans in Thailand*, *A History of the Thai-Chinese*, and *King Bhumibol Adulyadej: A Life's Work*.

Nikola Stalevski (writer: "Consumer Choices", "Sustainable Transport") develops and promotes sustainability interventions in Bangkok for international cooperation agencies (GIZ and others). His passions are climate change mitigation, cities and transport.

Nina Wegner (editor and writer: "Organic Revolution", "Integrated Farming", "Social Enterprise"), is a freelance journalist who writes about indigenous issues and corporate responsibility in developing countries. Her work has been published in *Al Jazeera*, *The Atlantic* and *The Huffington Post*, among others.

Noel Boivin (writer: "Heritage", "Personal Participation and Awareness"), a Canadian writer and communications specialist, worked in media in Bangkok for a decade before joining the United Nations Education, Scientific and Cultural Organisation (UNESCO) as a media and communications officer.

Patima Klinsong (writer: "Historical Preservation", "Family", "Poverty"), a graduate in Technical Communication from Illinois Institute of Technology, has been a journalist, writer and translator for over 15 years.

Patinya Rojnukkarin (art director) earned her MFA from Minneapolis College of Art & Design in Visual Studies and BFA from Silpakorn University. She has 17 years of graphic and motion graphic experience.

Pattraporn Yamla-or (writer: "Sustainable Business", "International Partnerships") is the co-founder of Sal Forest Co, Ltd., which furthers public discourse on sustainable business by conducting research on key sustainability issues as well as running workshops and events.

Phisanu Phromchanya (writer: "Labor", "Finance", "Corruption") has over 15 years of financial journalism experience with international wire services. He is now a consultant for an anti-corruption initiative in the private sector.

Purnama Pawa (assistant editor) holds a BA in Communications Management from Chulalongkorn University.

Raviprapa Srisartsanarat (writer: "Public Participation") is an international development specialist with almost 20 years of experience in designing, managing, monitoring and advising on development programs for international donors, NGOs, and Thai government agencies.

Siree Simaraks (designer) has over eight years of experience in graphic design. She holds a bachelor's degree in Graphic Design from the Faculty of Architecture, Urban Design and Creative Arts at Mahasarakham University.

Sofia Mitra-Thakur (writer: "Green Finance and Banking", "Indices") is a British journalist. She has worked for the Bangkok Post, South China Morning Post, The Telegraph, The Independent, and Engineering & Technology Magazine.

Surasak Glahan (writer: "Oceans and Seas", "Saving Marine Habitats", "Coastal Resource Management") has over 10 years of experience in journalism and communications. He has worked for the Bangkok Post, NGOs and inter-governmental organizations.

Surasak Tumcharoen (writer: "Area-based Rural Development") was a Bangkok Post political news reporter for over 25 years. He is currently a correspondent at Xinhua News Agency.

Sutawan Chanprasert (assistant editor/researcher) worked independently as a human rights researcher for three years prior to joining EDM. She holds an MA in International Development Studies from Chulalongkorn University.

Tibor Krausz (writer: "Urban Development", "Green Spaces", "Green Homes", "Green Buildings") is a journalist as well as a lecturer at Bangkok University's International College. He has also worked as a consultant for the UN.

Tom Metcalfe (writer: "Green Manufacturing", "Ethical Sourcing", "Waste Management") is a journalist and filmmaker with a focus on science, environment and Asia-Pacific region.

Warinthorn Kansupmits (designer) earned her BA in Communication Design from the Faculty of Fine Arts at Srinakharinwirot University and has over eight years experience in graphic design.

Wasant Techawongtham (writer: "Energy") was a former deputy news editor for Bangkok Post. Currently, he is a freelance writer and editor.

Will Baxter (lead editor; writer: "Inequality") is a journalist and photographer based in Southeast Asia since 2003. His work focuses on human rights, conflict, development and social issues.

Woranuj Watts (writer: "Competitiveness", "Manufacturing", "Education", "SMEs") graduated in history from Thammasat University and worked as a business journalist for 15 years. She is now a freelance writer and translator.

THANKS TO OUR ADVANCE READERS

Ariya Arunin
Lecturer, Department of Landscape Architecture, Chulalongkorn University (Topic: Urbanization)

M.R. Chakrarot Chitrabongs
Distinguished Scholar, Chulalongkorn University, Former Permanent Secretary of the Ministry of Culture (Topic: Heritage)

Supachet Chansarn
Lecturer, School of Economics, Bangkok University (Topic: Poverty/Income Inequality)

William Klausner
Senior Fellow, Institute of Security and International Studies, Chulalongkorn University (Topic: Buddhism)

Orathai Kokpol
Deputy Secretary-General, King Prajadhipok's Institute (Topic: Public Participation)

Sucharit Koontanakulvong
Head, Water Resources System Research Unit, Department of Water Resources Engineering, Chulalongkorn University (Topic: Water)

Usa Lertsrisantad
Program Director, Foundation for Women (Topic: Gender Equality)

Kiatanantha Lounkaew
Director, Dhurakit Pundit Research Service Center (Topics: SMEs, Poverty/Income Inequality)

Wimonthip Musikaphan
Deputy Director, National Institute for Child and Family Development, Mahidol University (Topic: Family)

Willem Niemeijer
Board Member, Khiri Group (Topic: Tourism)

Deunden Nikomborirak
Research Director, Economic Governance Thailand Development Research Institute (Topic: Corruption)

Sumeth Ongkittikul
Research Director, Transportation and Logistics Policy, Thailand Development Research Institute (Topic: Transportation)

Anand Panyarachun
Former Prime Minister of Thailand (Topics: Monarchy, Conflict)

Sompop Pattanariyankool
Head of Strategy Division, Office of the Permanent Secretary, Ministry of Energy (Topic: Energy)

Prempreeda Pramoj Na Ayutthaya
National Program Officer (HIV), UNESCO (Topic: Gender Equality)

Nipon Poapongsakorn
Distinguished Fellow, Thailand Development Research Institute (Topic: Agriculture)

Banyong Pongpanich
Director and Chief Executive Officer Kiatnakin Phatra Financial Group (Topic: State-owned Enterprises)

Pranee Srihaban
Expert on design & system on land development for Region 5 (Khon Kaen province), Land Development Department (Topic: Soil)

Jittima Srisuknam
Program Officer for Thailand and Lao PDR ILO Country Office for Thailand, Cambodia and Lao PDR (Topic: Labor)

Seree Supharatid
Director, Climate Change and Disaster Center Rangsit University (Topic: Natural Disasters)

Viroj Tancharoensathien
Senior Adviser, International Health Policy Program, Ministry of Public Health (Topic: Health)

Somkiat Tangkitvanich
Director, Thailand Development Research Institute (Topics: Competitiveness, Finance, Manufacturing, Trade)

Ronnakorn Triraganon
Manager, Capacity Development and Technical Services, RECOFTC, the Center for People and Forests (Topic: Forests)

Nalinee Thongtam
Biologist, Phuket Marine Biological Center (Topic: Oceans and Seas)

Supat Wangwongwatana
Coordinator, EANET Secretariat, Network Support, Regional Resource Center for Asia and the Pacific and Former Director-General, Pollution Control Department, Ministry of Natural Resources and Environment (Topic: Pollution and Waste)

Doris Wibunsin
Council Member, University of Thai-Chamber of Commerce, Vice-Chairman Council, Webster University (Thailand) (Topic: Education)

Ismail Wolff
Executive Director, ASEAN Parliamentarians for Human Rights (Topic: Conflict)

INDEX

Principal coverage of a subject is entered in **bold** and illustrations in *italics*.

A

Agenda for Sustainable Development 2030 30, **31**

agriculture **102-7**: Agrarian Revolution history 105; alternative agriculture movement 377; average age of farmers 106; Bank of Agriculture 115; challenges faced 106; ChangeFusion 374; chemicals, use of 62; community-supported agriculture 273; conversion of land into real estate 69; deforestation and 72; environment performance rank 17; Food and Agriculture Organization (UN) 64-5, 80-1, 306; forestry and 74, 75, 276, 286-7; Fukuoka's indispensable book 254; highland agriculture 291; importance of to sustainable development 104; insect farms 258; integrated farming **268-75**; land cultivated increases 65; large family farms 102-3; logging and 70; mangroves sacrificed for 73; "New Theory Agriculture" **270**; organic agriculture **254-67**; Organic Agriculture Certification Thailand (OACT) 380; percentage of GDP 18; percentage of total land area 16; percentage of workers involved in 101; rice farming 103; rice paddies 106; Rockefeller Foundation 422; rural development and 106; SMEs and 122; soil as lifeblood of **62-5**, 106; threats to 106; toxic agriculture 85; treated wastewater 305; water, importance of 66

Alternative Agriculture Network 97

Anand Panyarachun 35, 184, 186, 201, 211, 323

Annan, Kofi 211

architecture: BIG TREES 322; Crown Property Bureau 311; *Future of Architecture in 100 Buildings* 339; green architecture 240-5; 'Lanna' architecture 310; McDonough and Braungart 347; Olmsted and landscape architecture 325; religious architecture 221; vernacular architecture 241

Art of Designing Public Parks, The (Frederick Law Olmsted) 325

Article 112 **208** *see also lèse majesté* law

ASEAN: air pollution disputes 90; ASEAN Confidential 152; aviation industry 141, 145; Certificates of Recognition **417**; competition from 154; connectivity 153-4; corporate governance in 392-3; corruption in 392; Energy Awards 335; English proficiency in 168; free trade agreements 136, 139, 154; logistics costs 142; motor vehicles sales 228; Mutual Recognition Agreements (MRAs) 151; rail links 142, 144; Solar Power Company Group 342; statistics 16-19; Thailand primary education 164-165; Thailand's location in 408; Thailand's membership of 138; trade hub 138; visa cooperation 126; Wildlife Enforcement Network 299; women executives 195; Working Groups and Awards 414

ASEAN Economic Community (AEC) **113**: expected boost to Thai economy 136; integration of 106; labor sector and 151, 156; opportunities and challenges 145, 165; Thai SMEs and 125; Thailand and China 139, 142; Thailand's membership of 138

Asian Financial Crisis: currency flotation and 132, 134, 135; manufacturing and 108; medical tourism and 129; serious effect of 120; timelines 56, 135, 148

Ayudhya: Burmese sack 206; flood plain 69; openness to foreigners 418; renovation projects 310; timelines 75, 148, 169, 209, 215 221; tourist attractions 205

B

Baan Huay Hin **272**

Ban Don Kha School **402**

Ban Pred Nai Community Forestry Group **302**

Bandid Nijathaworn **394**

Bang Bua Canal Community **319**

Bangkok *see* urban development

Bangkok Bank 383, 385

Banjong Nasae **307**

Bank of Agriculture and Agricultural Cooperatives (BACS) 115

banking **382-87**

Bhumibol Adulyadej, King: core values of sustainable development 210-11; Coronation *209*; crop substitution programme 256, 268, 271; development of Deep South 201; Diamond Jubilee *207*; education reform 166, 168; extensive travels 254; forest conservation 284; fresh relevance given to monarchy 206; Golden Jubilee 277-8; historical sites quote 218; Huai Hong Khrai Center 276; New Theory farming 270; Phetchaburi River cleaning 305; physical and mental fitness quote 173; Rockefeller Foundation and 422; Sufficiency Economy Philosophy 36-7; timelines 209; Uthokawiphatprasit Watergate 309; water resources 68: wealth of monarchy 208; WIPO Global Leader Award 68; World Wide Fund for Nature 421

bicycling **250**; commuting by 142; Copen hagen 413; cycling park near airport 247; for commuting to work 337; KMUTT policy 247, 249; Lisu Lodge 364; mass transit lines and 410; Pun Pun Bike Share Program 251; Singapore promotes 413; Somsak Boonkam 367; Sukhothai encourages 312; YouBike 250

Big Trees **322**

Biodegradable Packaging for Environment Public Co Ltd **230**

biodiversity **82-5**; Amazon River 290; biomass energy and 61; Chiang Rai City 417; constant threats 82; decline of keystone species 71, 292, 293; forests 72, 276, 281, 282, 302; harmful cash crops 74; Kui Buri National Park 295; mangrove degradation 73; monoculture and 268; National Biodiversity Action Plan 83; National Marine Parks 76-7; organic revolution and 260; Pun Pun Center 265; rural development 254; Thailand's remarkable richness 82; tipping point 83; World Wide Fund for Nature 421; *Year of the Wolf* 299

biofuels **345**

Bo.Lan **379**

Buddhism 33, 204, **212-17**

C

Chai Lai Sisters Trek and Tours **365**

Chaiporn Phrompan **262**

ChangeFusion **374**

Chiang Rai **417**; Doi Tung Coffee 256-7; Doi Chaang 354

China **139**; baby milk scandal 350; Corruption Perception Index 156; demand affects rosewood forests 74; free trade agreement 136; Hydrology Institute 324; import percentage 137; pilot carbon markets 386; principal market for Thai farmers 63; railway networks 140, 142, 144, 156; rhinoceros horn smuggling 293; solar power 340; statistics 16-18, 20; tourism from 130; urbanization 408

Chiva-Som **330**

Chulalongkorn, King (Rama V): pacifies outer regions 203; reforms government bureaucracy 206; religious tolerance 207; timelines 105, 148, 169, 209, 215

cities *see* urban development

climate change **48-53**; agriculture and 101-2, 104, 106; biodiversity and 82-4, challenges 51; energy conservation and 404; finance and banking 382-3, 386; international cooperation 90; pollution 386, 404; rice production and 53; reforestation and 276-77; Rockefeller Foundation 422; Stephen Elliott 281; sustainability education and 167; Two-degree Tipping Point 50; water management and 67, 69; World Bank

Green Bonds 383
CO2 emissions: see also greenhouse emissions coal 61; tourism industry 363; transportation 396, 408
coal 49-50, 54-5, 59-61, 130, 387
coastal areas 306-309
commuting 246-51; see also transport balanced transport system 142; by bicycle 142; fuel quality law and 91; high speed trains 144; roads favoured for 140; subway and Skytrain 87, 140-1
competitiveness 119, 152-7; agriculture 106; corruption and 159, 161; 'creating shared value' (CSV) 328-9; crowdfunding 135; education and skills 150, 164; English language proficiency 168; policy and continuity 199; research and development 138; SMEs 123, 125; statistics 18; sustainability and 390; transport systems 141, 142, 145; Travel and Tourism Index 127; wages and productivity 146; World Bank quote 137
conflict 198-203; Deep South 203; pillars of peace 198; yellow shirts and red shirts 199, 202
consumers 228-33, 236; labelling 232
Copenhagen 413
Corporate Social Responsibility (CSR) 328-9
corruption 158-61, 392-5; bureaucracy and 113, 182; Corruption Perception Index 156, 392; healthcare 170; in depth journalism 237; lack of investment from 97; local authorities 92; mutual fund combating 374; National Anti-Corruption Commission (NACC) 119; political corruption 141; private sector 392; rail networks 87; scale of 153; SMEs 125; state-owned enterprises (SOEs) 114, 115, 117, 118; statistics 19; Thai Airways 119; transport 145; yellow shirts 199
Crown Property Bureau 64, 208, 311

D
Dairy Home 331
Defamation 183-4
Deep South 198, 200, 203, 215, 373
Dharmapiya, Priyanut 403
Doi Chaang Coffee 354
Doi Tung Development Project 256-7
Drought 48-9, 51, 64, 66-8

E
Eastern Seaboard: car manufacturing 140; industrial complex and port 94; industrialization 77; mangroves 70; Map Ta Phut industrial estate 86, 94, 144; refining and petrochemicals 108; soil type 62, 63; transplanting coral 300, 304

Eat Me 381
education 164-9, 398-403; basic rights 163; biodiversity and 85; Chiva-Som 330; competitiveness and 156; Doi Tung Development Project 256-7; for farmers 65; forest ecosystems 293; Human Development Foundation 316; innovation and 112, 125; Khiri Travel 366; King Bhumibol on 211; Klongdinsor 372; National Park Act 284; Panya Project 266; Phuket 416; political education 404; poverty and 153, 186, 187, 189; priorities 150; public health 422; public schools and the missionaries 418; Pun Pun Center 265; reforms required 197, 397; Rockefeller Foundation 422; rural areas 146, 207; sex education 177, 197; social enterprise supporting 364; social status and 185; southern Thailand 200; statistics 19; Sufficiency Education Learning Centers (SELCs) 398, 403; UNESCO for Sustainable Development 401; upgrade required 152; wildlife conservation 298; women and 192-3, 195
Elliott, Stephen 281
EnerGaia 324
energy 54-61, 240, 334, 339, 340-5, 385, 389-91, 404-7 see also renewable energy air conditioners 243, 337; bank credit for projects 383, 384, 385; biodegradable packaging 230; biomass in demand 103; capital flow in the sector 382; Chiva-Som 330; clean energy 61; CO₂ emissions 408; conservation campaigns 89; consumption of 58; ecosystems at risk 85; edible algae 324; electrical products 232; energy bars 258; Energy Complex 335; Energy Conservation Fund 341; Energy Efficiency No. 5 Label 407; energy-efficient labels 228; Energy Policy and Planning Office 412; energy savings 404-7; ESG 1000 representation 386; five major players 59; forests 276; Green Buildings 334; Green Campus 249; green homes 240; green manufacturing 346-7; landfills 92; LEED certification 339; level 5 certification 229; motor vehicles 337; older buildings 337; organic waste 417; paperless banking 384; protests 79; regulatory body 115; solar power 336, 338, 364, 382; statistics 17; streetlights 397; temperature and humidity issues 337; Thai Energy Efficiency labelling 232; timelines 117; Toshiba Semiconductor 348; transport issues 408; waste-to-energy technologies 357, 389-91, 416; water consumption and 67; Yak01 243
environment; value chains 152, 154
ESCO Revolving Fund 406
ESG100 390
ethical sourcing 350-5

Ethnic Lanna 353

F
families 176-9
Farmers' Friend Rice Community 264
farming see agriculture
finance 132-5, 382-7
FinTech 133
Flooding 67, 108, 241, 283, 309, 325, 387
Food, Inc (Robert Kenner) 377
Forest Man (William Douglas McMaster) 283
Forest Restoration Research Unit (Chiang Mai University) 280
forests 70-5, 276-83, 284-91, 300-5; carbon sinks, as 82; conservation 284-91; ecological research 421; felling 82, 90, 376; forest cover 70; Kui Buri National Park 295; mangroves 300-5, 306, 391; plantations 253; protection of 85, 204, 421; reforestation 276-83; shrimp farms 300; sustainable sources of produce 82, 84; timelines 75
Foundation for Consumers 236
Foundation for Environmental Education 399
Four Regions Slum Network 318
freedom of speech 180, 183, 185
Freeland Foundation 292, 299
Future of Architecture in 100 Buildings, The (Marc Kushner) 339

G
Galster, Steve 299
gender equality 192-7
Gibbon Rehabilitation Project 297
Great Transition, The (Lester R. Brown and others) 344
green buildings 334 - 9
Green Campus at KMUTT 249
green homes 240-5
green manufacturing 346-9
green spaces 320-5
Green World Foundation 246, 248
greenhouse gases 324 see also CO2 emissions

H
Harte, Dennis 411
health 170-5
heritage 218-23 see also preservation
hill tribes: corn planting 279; diseases affecting 172; forest rangers, as 291; government assistance 256; marginalization of 255; northern hill tribes 172, 212, 214; historic preservation see preservation
Hoi An 313
household debt 101, 134, 188, 262,

371, 431
housing: green homes **240-5**
Huai Hong Khrai Royal Development
    Study Center **276**, 282
Human Development Foundation **316**

I

indices **388-91**
industrialization 77, 94, 108–110
inequality 186, 191-2, 196-9
Innovation Coupon 111, 121, 138
internet 183, **241** *see also* social media
Intrachooto, Singh 338

J

Jim Thompson Silk **352**
Joint Management of Protected Areas,
    The (JoMPA) **286-7**

K

KASIKORNBANK **384-5, 395**; green
    banking 384; Learning Center 334;
    Research Center 129
key performance indicators 16–19
Khao Paeng Ma Reforestation Project 278
Khiri Travel **366**
Klongdinsor **372**
Korea: free trade agreement 136; Korean
    War 180; North Korea 199; SMEs 122;
    statistics 19; tackling corruption 392;
    timelines 157; Toshiba 348; tourists,
    as 130
Kui Buri National Park **295**

L

labelling **232, 407**
labor **146 - 51**
Laem Phak Bia Environmental Research
    and Development Project **305**
Laubsch, Alan 387
LEED (Leadership in Environmental and
    Energy Design) 334, **339**
lèse majesté law 183, 185, 208
LGBTIQ 163, 193-4, 197
Lisu Lodge **364**
Luckiest Nut in the World, The (Emily James)
    352

M

Mab Ueang Agri-nature Center **274**
Mae Klong Community Network **308**
Mae Wong Network **288**
Maejo Baan Din **242**
Maha Yu Sunthornchai **271**
mangroves **73**, 302-3 *see also* forests
manufacturing **108-13, 346-7**; competitive-
    ness challenges 137; difficult chal-
    lenges 96; green manufacturing **346-7**;
    increasing reliance on 90; innovation,

need for 152, 156; LED lighting 407;
    LEED certification 339; motor vehicles
    140, 408; multinationals 154; Plan Toys
    349; shift from 145; timelines 117;
    Toshiba Semiconductor 348; waste
    heat recovery 391
Map Ta Phut: air pollution 86; alternative
    to sought 79; controversy over 59;
    Eastern Seaboard industrialization 77;
    opens 94; rail links 142, 144, 408
mass transit lines **410** *see also* transport
media **183** *see also* press, the; social media
    Banjong Nasae 307; bias 201; Bud-
    dhism and 216; consumer preferences
    and 228; environmental concerns and
    344; financial pressures 237; LGBTI
    community 173; Local Alike 367;
    monarchy and 206, 208; pornography
    176; self-censorship 185; sufficiency
    economy participation needed from
    403; Thailand's active media 181;
    tourism, on 130
Metro Forest Project **325**
middle-income trap 152, 156
Millennium Development Goals (MDGs)
    170, 172, 194
Mitr Phol Bio-Power **360**
Monarchy **206-9**; absolute to constitu-
    tional 180, 198; Ayudhya rice fields
    67; Chinese and 137; health projects
    promoted by 172; rituals 218; school
    curricula and 167; steering Thailand's
    development 205; Thai flag and 212;
    timelines 117, 169, 209

N

Nan Model 259
natural gas **54-61**; discoveries of 77; global
    investment 61; Gulf of Thailand
    78; major driver of economy, a 76;
    monopolistic buyers and sellers of
    118; NGVs 410; percentage fuel mix in
    power generation 60, 61; reduction in
    dependency on 371, 408; shadow over
    its future, a 344, 348
New Theory **270**; brainchild of King
    Bhumibol 249, 268; Mab Ueang and
    274; Maha Yu and 271; principles of
    268; schools project 398; timelines 105
NGOs **418-9**
North Korea *see* Korea

O

Ocean of Life, The: The Fate of Man and the Sea
    (Callum Roberts) 301
oceans and seas **76-81, 301-3**
OECD 120-5, 156
oil **54-9**; Bangchak Petroleum 340; bio
    mass equivalent 390; declining pro-
    duction 85; emissions standards 412;
    Gulf of Thailand 77, 78; local protests
    79; motor oil 246; national grid and

391; oil spills 90; post World War II
    Copenhagen 415; price instability 345;
    reserves 76; shadow over its future
    347; Standard Oil 422; statistics 17;
    Thaioil **386**
*One-Straw Revolution: An Introduction to
    Natural Farming* (Masanobu Fukuoka)
    254

P

Pacific Asia Travel Association (PATA) **363**
Pak Phanang River Basin Royal Develop-
    ment Project **309**
Panya Project **264**
Phi Phi Model **368**
Phuket **416**
Pinkaew, Tul 238
Plan Toys **349**
Plant Banana Trees to Save the World **279**
pollution **90-5, 386** air pollution 62, 63,
    244, 344; Bangkok 86; biodiversity
    threatened by 82; coastal resource
    management and 306; dust pollution
    332; energy discoveries and 46; key
    factors 416; mangrove degradation and
    73; motor vehicles 409, 411; nocturnal
    light 320; Organic Revolution 260; Pol-
    lution Control Department 233, 416;
    pollution-free power 61; reduction
    of 357, 385, 391, 414; water pollution
    127, 130, 309; wind turbines 61
poverty **91-101**; alleviation 196, 292; com-
    munity forests 278, 279; conserva-
    tion and 293; Doi Tung 256; ethnic
    minorities 257; free market economics
    and 352; integrated farming and 377,
    379; New Theory farming 270; remote
    border villages 291; rice farmers 262;
    rural poverty 146; statistics 16; Thais
    lifted out of 162, 350, 382; workforce
    suffer from 150
Prasan Sangpaitoon **304**
Prayuth Chan-Ocha, General 123, 166
preservation **310-13** *see also* heritage
press, the: *see also* media first printing
    press 223; freedom of paramount im-
    portance 183; on polluted rivers 305;
    press freedom survey 201; restrictions
    on 185, 200; *Smart Buyer* 236
Princess Mother 172, 253, 256, 257
private sector **326-95**
Priyanut Dharmapiya **403**
provinces 14
Provincial Electricity Authority (PEA) 114,
    397, 407
PTT **59, 325**; assists with energy problem
    54-5; green buildings 334, **335**; Map Ta
    Phut project 94; Metro Forest 321, 325;
    monopolies 118; timelines 56, 117; top
    five SOE 114
public health: biomass energy and 61:
    ChangeFusion 374; Dr Carlos Dora 90;
    health insurance and 171; medical

tourism 129; monarchy supports 172, 207; obesity 170; pollution and 95; Rockefeller Foundation 418, 422; sexual abuse and domestic violence 192; timelines 173: Universal Coverage Scheme 174; water system 417
Public-Private Partnerships 111, 396
Pun Pun Bike Share 251
Pun Pun Center for Self-Reliance 265

R

Raitong Organics Farm 263
Rama V, King see Chulalongkorn, King
Rama IX, King see Bhumibol Adulyadej, King
religion 212-17
renewable energy 340-345; adder programmes 340, 341; algae, possible use of 324; Bangchak Petroleum 340; Bangkok Bank encourages 383; biofuels 345; biomass energy 61; biomass, use of 103, 324; business incentives 400; capital directed to 382; capital directed to 382; Development Plan 60; Energy Conservation Fund 341; energy saving loan company 385; feed-in tariffs (FIT) 55, 341; forecast for supply of by 2020 55; global investment 61; 'green loans' for 385; KASIKORNBANK 383, 384; leading resources for 390; Lisu Lodge 364; palm oil 405; percentage use 60; power plants 59; recent emphasis on 54; regional leader in 404; solar energy and 49, 241, 340; statistics 17; turning waste heat into 391
research & development 106, 109, 111-2, 138, 396
restaurants 376-81
Rockefeller Foundation 418, 418, 422
Royal Initiative Discovery Foundation 255, 259
Royal Project, The 105
royally initiated projects see also King Bhumibol Adulyadej; Doi Tung 256; farming 63; villages 254; water security 68; wide variety of 205, 206, 210
Rubesch, Edward 375
rural development 254-59

S

Sal Forest 333
Sampran Model 355
Sarus Crane Breeding Program 296
Schmidt, Sebastian 245
Scholars of Sustenance 359
seas and oceans 76-81, 300-5
Seub Nakhasathien 287
Siam Cement Group (SCG) 332-3, 336-7, 361, 391, 395; 100th Year Building 334, 336-7; anti-corruption policies 395; green manufacturing initiatives 334, 346; Map Ta Phut 94; paper packaging

346-7; photograph of plant 110; quality standards 232; soil treatment projects 64; waste heat recovery 391
Siam Commercial Bank (SCB) 132, 135, 379
Singapore 413; Bangkok and 344; decline in birth rate 177; Ease of Doing Business Index 137; electronic road pricing (ERP) 143; green space 320, 322; high speed train routes to 156; IT systems 138; Local Alike 367; logistic costs 142; medical tourism 129; organic product accreditation 262; Sakari Resources 59; Singapore Airlines 118; statistics 17-18; sustainable transport 413; Temasek Holding 119; Thailand's teenage pregnancies and 175; timelines 157; Toshiba 348
Singh Intrachooto 338
Sirikit Kitiyakara, Queen 291, 321, 421
Sirindhorn, Princess Maha Chakri 300
SMART Patrol 294
SMEs 120-5
social media see also media
Banjong Nasae 307; Big Trees 322-3; Blue Whale campaign 238; Facebook 239; Green World Foundation 248; Mark Kushner 339; new generation using 227; percentage using ; preservation of heritage through 213; raising public awareness through 243; restrictions on 185
Socialgiver.com 231
Social License 329
SOEs 114-9
soil 62-5, 267; biomass energy and 61; central region 254; compost 244; conservation of 291; diversity of crops 268; dry and alkaline 274; filtering waste 305; forests and 70-5; monarch's huge contribution 47; naturally fermented manure 274; organic revolution 260; pollution of 95; quality of 106, 206, 210, 264, 265, 325; rainwater 336; restoring fertility 271; soil erosion see below soil erosion; soil testing 262; timelines 105; types of 63; weed control 283
soil erosion: mangrove forests 300; monoculture and 268; plants and grass preventing 325; royal project 68; tree planting 276, 279, 283
Solar Power Company Group (SPCG) 342
Somsak "Pai" Boontam 367
Somsook Boonyabancha 315
Soontorn Boonyatikarn 344
Startups 121
State Owned Enterprises (SOEs) 114-119
statistics 16-19; commodities 105; competitiveness 152, 153; corruption 158; e-banking 388; education 164; families 176; forests 277; gender 192, 193; green spaces 322; health 170, 171; integrated farming 269; labor 146;

manufacturing 108, 109; organic farming 269; poverty 178; religion 213-4; rice farming 103; SMEs 120, 121; State Owned Enterprises (SOEs) 114, 116; tourism 126, 128; trade 136; transport 141; wildlife 292
Stock Exchange of Thailand: alternative for smaller firms 121; countering corruption 392; establishment of 132; Khon Thai Jai Dee fund 374; listings 120, 386; starts trading 135
South Korea see Korea
Srinagarindra see Princess Mother
Sufficiency Economy Philosophy (SEP) 36-7, 398-9, 402-3; educational system in 398-9, 402-3; formalized 211; Mab Ueang 274; Maha Yu 271; sustainable cities and 415; UN recognition 211
Sukhothai Historical Park 312
Sustaina Organic Restaurant 378
Sustainable Cities and Society (SCS) 416
Sustainable Development Goals (SDG) 30, 143

T

teenage pregnancy 175, 177, 179
Tetra Pak 357
Thaioil 56, 390, 391
Thailand 4.0 109, 111, 138, 157, 167, 420
ThaiPublica 237
The Oyster Bar 376
The Royal Project 105 see also royally initiated projects
TICA 423
timelines: agriculture 105; competitiveness 157; education 169; energy 56; finance 135; forestry 75; health 173; heritage 271; labor 148; monarchy 206, 209; religion 215; State Owned Enterprises (SOEs) 117; tourism 131
Toshiba Semiconductor 348
tourism 126-31, 362-9; advertising campaign 108; Ban Mae Kampong 285; ChangeFusion listings 374; coastal communities 306, 307; coral reefs, effect on 304; deep sea ports threaten 77; development plan with China 139; Doi Tung 256, 257; forest ecosystems and 281; goldsmiths 312; heritage factor 219, 312; hotspots 78, 141; huge employer, a 47, 76; import and export and 138; mangrove destruction 73; mass tourism 310-2; medical tourism 129; Mutual Recognition Arrangements (MRA) 151; Phuket 416; political upheaval 200; power stations that might affect 60; resilience of 97; revenue from 78; ruins 312; Songkram River wetlands 421; Southern Seaboard Development Plan and 79; Sri Lanna Natural Park 242; timelines 157; transport requirements 142; Vietnam 313; water, importance of for 66;

wildlife tourism 295
trade **136-9**
transport **140-5, 246-51, 408-13**; Chiang
    Mai 86; Chiva-Som's initiative 330;
    commuting **246-51**, 337; Green Bonds
    383; high speed trains **144**; LEED re-
    quirements 339; mass transit lines 397;
    Ministry of Transport responsibilities
    116; Phuket 416; ribbon develop-
    ment 84; SOEs and private enterprise
    119; Songkram River wetlands 421;
    sourcing local foods 362; statistics 18;
    Thailand a transport hub 137; two-
    stroke motorcycles 91
Tree Bank **289**
Tuk Tuks **411**
Tul Pinkaew **238**

U

United Nations (UN: Agenda for
    Sustainable Development) 30, 31; Ban
    Ki-moon quoted 377; climate change
    summit 277; development practices
    206; Doi Tung hailed by 253; King Bhu-
    mibol's Lifetime Achievement Award
    211; Industrial Development arm
    108; mangrove report 75; Millennium
    Development Goals 165, 168, 170,
    194; Office on Drugs and Crime 257;
    statistics 16–19; sustainable cities 414;
    UNCTAD 137; UNESCO 401; UNICEF
    178; World Intellectual Property Office
    (WIPO) 68; World Soil Day 64
Upcycling 228-229
urban development **86-9, 314–9, 408,
    414-7**; air conditioning 243; Bangkok
    89, 321; biodiversity and 82; climate
    change and 422-3; consumer culture
    186; cost of transport 162; density 243;
    documentary review 319; encroach-
    ment on rural areas 146, 309; farmers'
    markets 229, 372; flooding 53, 67;
    gardening in 263; green spaces **320-5**;
    habitat loss and 83; healthcare system
    175; historic cities and districts 203,
    218, 223, 310, 312; megacities 250,
    413; migration from rural areas 146,
    176; rapidity of urbanization 408;
    renewal 397; timelines 105; transport
    links to rural area 137; transport proj-
    ects 248, 409, 410; urban-rural divide
    168, 187, 238
Urbanized (Gary Hustwit) **319**
USAID **420**

V

Verapattananirund, Prateep 267

W

Wanita Social Enterprise **373**
waste **90-5, 356-61, 416**; agricultural 340,
    405; bagasse 230; composites from
    347; documentary 347; domestic
    93; energy wasted 240; Environment
    Loans 385; filtering 78; garbage bins
    413; generating power from 345; green
    buildings 334; healthcare purchased
    through 370; heat 61, **391**; illegal
    dumping 83; industrial **94**; lean man-
    ufacturing 346; precision farming 267;
    recycling and composting 244, 266;
    restaurants 376-81; seed quality 283;
    sustainability practices 391; Toshiba
    semiconductor 348; waste collectors
    417; waste management **356-61**, 405,
    416; wastewater 79, 82, 86, 305–6, 309,
    330; 'Zero waste to landfill' 332
Waste = Food (Rob van Hattum) **347**
Wat Khanon Shadow Puppet Troupe **313**
water **66-69**; access to 170, 186; agencies
    86, 89; Bangkok's consumption 86,
    89; Chiang Rai 417; Chiva-Som 330;
    coastal areas **306-9**; elephants 295, 298;
    flooding 309, 325, 387; forests and 70-
    5, 276-283, 284, 291; green buildings
    334, 337; industrial dishwashers 376,
    381; insect farms 258; integrated
    farming 268, 271; micro-reservoirs 64;
    monarch's role 47, 64, 204, 210-1, 270;
    oceans and seas **76-81, 300-1**; pollution
    356, 385; Provincial Water Authority
    115; quality of 91, 415; roofing 241;
    saltwater penetration 65; soil types 63;
    wastewater see above waste; water
    treatment systems 249, 305, 348, 391
Who Killed the Electric Car? (Chris Paine) **412**
wildlife **292-99**
Wonderfruit Festival **369**
Wolf Totem (Jiang Rong) **299**
Wongpanit Recycling **358**
World Bank Green Bonds **383**
World Wide Fund for Nature **421**

Y

YAK01 243

PICTURE CREDITS

Every effort has been made to trace copyright holders of images in this book. In the event of error or omissions, appropriate credit will be made in future editions of Thailand's Sustainable Development Sourcebook.

Adams Organic Bangkok CSA: 273
Ajjana Wajidee/Wanita Social Enterprise: 373
Akkaraych Petchampai/Rice Department: 104, 362-363
Alan Laubsch: 387
Amanuensis – Bureau Bangkok / Dominic Faulder: 56
Angelo Cavalli/Photolibrary: 215 bottom left
Apiradee Treerutkuarkul: 284 bottom left, 307 right
Associated Press: 211
Athit Perawongmetha: 101 bottom right, 201 top
Ayutt and Associates Design: 243
Bangkok Post: 43, 60, 74 top, 88 top, 95, 107, 131, 157, 160 top, 161, 165, 210 top, 409 right
Big Trees Project: 284-285, 323
Bo.Lan: 379 all
Bohnchang Koo: 124
Brent Lewin: 195
Bruno Barbey: 130 bottom left
Catherine Karnow: 310-311
ChangeFusion: 374 all, 375 bottom
Charoon Thongnual: 171, 200, 203
Chawalit Poompo/Rice Department: 260-261, 262 bottom left
Chien-Chi Chang/Magnum Photos: 149
Chiva-Som: 330 all
Community Organization Development Institute: 88 bottom left, 314-315, 319
Corbis: 27 right, 135 bottom left
Crown Property Bureau: 37, 283 top, 309 bottom right, 311 right
D.Light Design (DivatUSAID): 371 right
Daoruek Communications Co., Ltd.: 300
Dario Pignatelli: 153
Darunsikkhalai School: 401 left
Doi Chaang Original Coffee Co: 354
Doi Tung Development Project: 256 left, 257 all
Dome Pratumthong/WWF-Thailand: 421 bottom right
Dow Wasiksiri: 172 bottom right, 216 top
Dr. Chalermpol Kirdmanee: 63 bottom
Earth Net Foundation/Udomsak Bhatiyasevi: 78, 380
Eat Me Restaurant: 381 all
Eco-community Vigor Foundation: 65 bottom right, 267 all
EnerGaia: 324 all
Ethnic Lanna: 353 all
Evan Gershkowich: 278 bottom right, 279 bottom, 288 top left
Farmers' Friends Rice Project: 264 all
FORRU-CMU: 280-281 all, 288 bottom right
Foundation for Consumers: 236 all
Foundation for Virtuous Youth: 399, 402 all, 403
Four Regions Slum Network: 318 all
Freeland Foundation: 74 bottom left, 83, 298 center left, 299 top
Freeland Foundation/Kayleigh Ghiot: 85 bottom
Friends of the Asian Elephant Foundation: 298 top left
Getty Images: 27 left, 33, 34, 36, 62-63 top, 72 top right, 82 top left, bottom left, top center & bottom center, 91, 99, 100, 106, 108, 112 all, 113, 118 center, 122, 130 top right, 132, 134, 135 top right, 139, 143 left, 147, 151 bottom left, 155, 156, 159, 160 bottom

right, 166, 168, 178, 179, 181, 182 bottom, 183, 187, 199, 201 top right and left, 207, 212, 217 bottom, 219, 222, 239, 298 bottom right, 345, 356-357 center
Global Social Venture Competition: 375 top
Green Peace: 79
Green World Foundation: 248 top & bottom
Greg Gorman: 197 left
Huai Hong Khrai Royal Development Study Center: 69, 271 bottom left, 282, 291 bottom right
Ittikun Kanokkantrakom: 401 right
Jinnawat Pumpoung: 75
KASIKORNBANK: 335 bottom center, 339 top, 384
Khao Kwan Foundation: 262 top & bottom center
Khiri Travel: 366 bottom left
King Mongkut's University of Technology Thonburi (KMUTT): 111, 249 right
Klongdinsor: 372 all
Lisu Lodge: 364
Local Alike: 127, 367 all, 370-371
Luca Tettoni/Robert Harding/World Imagery/Corbis: 208
Mab Ueang Agri-nature Center: 274 all
Mae Fah Luang Foundation: 256 bottom right
Maejo Baan Din (Earth Home Village): 242 all
Melisa Teo: 377 top
Michael Freeman: 80
Michael Yamashita: 213
Mick Elmore: 342 bottom, 344 top
Nat Sumanatemeya: 76, 301 top
National Archive of Thailand: 215 center, 419
Nina Wegner: 265 all
Office of His Majesty's Principal Private Secretary: 68, 172 top left, 291 top
Office of the Royal Development Projects Board: 64 both, 173, 268-269, 270
Panya Project: 266 all
PATA Gold Awards 2015: 363 right
Petchpanom Jitman/Nitinarth Charoenpokaraj/Suan Sunandha Rajabhat University: 303 bottom left & right, 308 top right & bottom
Oz Go/Narrative Lost: 365 all
Phi Suea House: 244-245 all
Plan Creations Company Limited: 349 all
Prasarn Sangpaitoon/Rambhai Barni Rajabhat University: 303 center bottom, 304 all
Prasit Rodklai: 335 bottom left
PTT Metro Forest: 325
PTT Public Company Limited: 59 top left, 61
Pun Pun Bicycle Share Program: 251
Raitong Organics: 263 all
RECOFTC The Center for People and Forests: 303 top
Robert McLeod/Lantern Photography Co., Ltd.: 133 top
Royal Chitralada Projects: 105
Royal Project Foundation: 210 bottom left
S.C. Shekar: 216 bottom, 220 top
Sal Forest: 331 all, 333 bottom
Sampran Model: 355 all
Scholars of Sustenance: 359 all
Shutterstock: 2, 5, 7, 39, 40, 49, 51 all, 53, 54-55 all, 67, 73 all, 81 all, 82 left, 84, 85 top left, 87, 89, 92, 94, 101 center, 102, 110 left, 115 all, 118 bottom right, 119, 120-121, 123, 125, 126-127 center, 128 bottom right, 137, 138, 141, 142, 143 right, 150, 151 top and bottom right, 182 top, 185, 189, 190, 197, 202 bottom left, 221, 223, 228-229, 230, 233 top right, 234-235, 238 bottom, 240-241 all, 246-247 all, 248 center, 250 all, 258 all, 259, 276-277, 278 top, 279 top, 290 all, 293 right, 312, 313, 317, 321 right, 335, 340-341, 342 top left, 343, 366 top, 368, 385 top right, 388-389, 404-405 all, 406, 408-409 all, 410 all, 415, 416

Siam Cement Group: 65 top, 110 right, 328-329, 332 all, 336, 337, 361, 395 top
Siam Cement Group/Athit Perawongmetha: 233 bottom left, 350 bottom left, 350-351, 346-347, 347 right, 361
Sidekick: 238 top
Singh Intrachooto: 338 top left
Socialgiver: 231 all
Sueb Nakasathien Foundation: 70, 286 all, 287 all
SUPPORT Foundation: 196
Surasak Glahan: 308 left
Surat Osathanugrah: 220 bottom
Sustaina: 260 bottom left, 378 all
Tawan Thintawornkul: 249 left
Tetra Pack (Thailand) Limited: 357 right
Thai Airways International Public Company Limited: 117
Thai Airways International Public Company Limited/ Athit Perawongmetha: 145
Thai Creative Design Center: 333 top
Thai Health Center: 338 bottom left & right
Thailand International Cooperation Agency: 423
Thaipat Institute: 390
ThaiPublica: 237
The Anti Corruption Organization of Thailand: 392-393 all
The Crown Property Bureau/Kritsada Prathonsuriyakul: 275, 309 top
The Human Development Foundation (Mercy Centre): 316 all
The Private Sector Collection Action Coalition Against Corruption: 394, 395 bottom
The Rockefeller Foundation: 418 all, 422 all
The Royal Discovery Initiative Foundation: 271 top
The Thai Silk Company: 351 top right, 352 top & center
The Village Social Development Center (Wanakaset): 272 all
The Wild Animal Rescue Foundation of Thailand: 297 all
The Zoological Park Organization: 296
Tuk Tuk Factory: 411
Timothy Auger: 382-383
Tom Metcalfe: 358, 360 all
Toshiba Semiconductor (Thailand) Company Limited: 348 all
Tree Bank: 289 all
United Nations: 30
Univentures Public Company Limited: 335 bottom right
USAID COMET: 420 all
Wayuphong Jitvijak/WWF-Thailand: 295
Weerakarn Satitniramai: 320-321 top, 404-405 center
Wildlife Conservation Society Thailand: 294 all
Will Baxter: 129, 176-177,191, 193, 246 bottom left
Winniwat Traitrongsat/Rice Department: 255
Wonderfruit Festival: 369
WWF-Thailand & Department of National Parks, Wildlife and Plant Conservation: 288 bottom left, 292-293
Yann Arthus-Bertrand: 12 (14°00'N, 100°36'E), 24 (19°36' N, 99°41' E), 52 (13°59' N, 100°25' E), 77 (8°00' N, 98°22' E), 128 top left (16°30' N, 99°31' E), 306-307 (8°20' N, 98°30' E)

"It is no longer good enough for economies simply to grow. We must also end extreme poverty, a goal within reach by 2030. We must manage the economy to protect rather than destroy the environment. And we must promote a fairer distribution of prosperity, rather than a society divided between the very rich and the very poor."

Earth Institute Director Jeffrey D. Sachs